The Climate-Conflict-Displacement Nexus from a Human Security Perspective

Mohamed Behnassi • Himangana Gupta
Fred Kruidbos • Anita Parlow
Editors

The Climate-Conflict-Displacement Nexus from a Human Security Perspective

Springer

Editors
Mohamed Behnassi
International Law and Politics
of Environment and Human Security
College of Law of Agadir,
Ibn Zohr University
Agadir, Morocco

Center for Environment, Human Security
& Governance (CERES)
Agadir, Morocco

Fred Kruidbos
KRUIDBOS Ecologisch Onderzoeks- en
Adviesbureau
Helmond, The Netherlands

Himangana Gupta
Sustainable Landscapes and Restoration
Program, World Resources Institute (WRI)
New Delhi, India

Anita Parlow
Washington, DC, USA

ISBN 978-3-030-94143-7 ISBN 978-3-030-94144-4 (eBook)
https://doi.org/10.1007/978-3-030-94144-4

This Springer imprint is published by the registered company Springer Nature Switzerland AG
The registered company address is: Gewerbestrasse 11, 6330 Cham, Switzerland

Foreword 1: Climate Change or Climate Disruption?

When reading this impressive volume about the key dynamics of the nexus between climate change, conflict and displacement, one might wonder whether the label 'climate change' is the most appropriate one. The volume clearly shows how mankind is being launched into a new era. An era with an increasing gap between the demand and supply of basic human needs. An era in which our current way of living is not sustainable anymore and where current concepts can't provide the required solutions. Moderate forecasts show how our world population will probably increase by another 50% to some 11 billion people, all needing food, water, sanitation and a place to live. Meanwhile we are running out of resources to meet these increasing demands. Clean drinking water is becoming scarcer, resulting in food shortages and sanitation problems. It is often stated that water is the future oil. Climate change speeds up the process and aggravates that growing gap between demand and supply. Parts of our planet are becoming uninhabitable because of the increasing drought, flooding, sea-level rise and severe weather incidents, affecting human security in many ways. Climate change is clearly much more than just an environmental problem. It is also a risk multiplier that has a disruptive effect on our ways of living and our human security. It is a whole-of-society and an existential problem leaving no place on this planet untouched as this volume clearly shows from different regional perspectives.

The current COVID-19 crisis probably provides just an indication of the effects we can expect from a changing climate. This crisis should teach us how vulnerable we are on a global scale. We live in different neighbourhoods of the same global village. We can't hide behind oceans, national boundaries or sea dikes for changes in other parts of our intertwined and globalized world. Global problems demand global answers. The COVID-19 crisis should also teach us the importance of not ignoring the signs but to be prepared. Hardly ever before in the human history did we possess so much advanced knowledge of the next crisis awaiting us. We have the luxury of foresight as this volume makes painfully clear, which gives us a brief window of opportunity to be prepared, to prevent and mitigate the worse from happening, and to adapt in order to be able to deal with the unavoidable.

It is becoming evidently clear that climate change is the biggest challenge of this century, calling for a paradigm shift in our approaches to find the right answers. That paradigm shift starts with the fundamental understanding of the underlying social-ecological mechanisms we act upon, acknowledging the need to mainstream mitigation and adaptation efforts in our policymaking, in our action and outreach programmes, in our activities, and in the technologies we develop. It needs to be more than just another item on our world agenda. Courageous leaders are needed to look beyond the next election and to invest in the future of the next generation. Leaders who possess the political courage and wisdom as shown after the floodings in The Netherlands right after the Second World War when the dikes broke during a spring flood, threatening large parts of the population living below the sea level. In that period, the country was in a deep recession, struggling to recover from the destructions of the war. Political leaders decided to do more than just fix the broken dikes. Despite all the economic problems, they decided to invest in a future-proof coastal defence system called the Delta Works, a system that has protected the country for almost 70 years now and for which we still owe them.

Another part of that paradigm shift is about breaching the functional stovepipes, recognizing the need for comprehensiveness, and maximizing the synergy between departments and stakeholders. Breaching those stovepipes is hard to do when the political and public debate is polarized. However, climate change is a shared and existential threat that should unite us. None of us can deal with its effects by itself, we can only deal with it if we reach out and treat it through a collective effort and find ways to reinforce each other.

As an extension of existing capabilities to prevent further escalation of the current and future security challenges, the security sector, and especially defence organizations, needs to be part of both the comprehensive team and the solution. All over the world, security experts and military leaders experience the impact of a changing climate. They experience how increasing droughts lead to friction over access to drinking water, how access to water is being weaponized by extremist organizations, how herders and farmers are driven away from their traditional grounds leading to frictions elsewhere, and how fishermen turn into pirates because of depleting fishing grounds and warming seas. They also experience an increasing call for humanitarian assistance and disaster relief, and how climate change affects the geopolitical landscape, like in the Arctic area, where the melting ice triggers a new run for resources and like the energy transition leading to a new run for rare minerals and affecting the position of countries whose economies largely depend upon the revenues of fossil fuels. They are also aware of the fact that military organizations are amongst the largest emitters of CO_2 in any country.

An increasing number of leaders feel the need to help solve the problem, realizing that climate change is probably the biggest game changer of this century. They are concerned and want to do something about it, so they linked up in a new and unique global security network called the International Military Council on Climate and Security (IMCCS). Within 2 years, leaders from more than 40 countries out of all regions in the world joined this non-political and voluntary network, a good example of the required global collectiveness. IMCCS rapidly developed into a platform for security experts to exchange experiences and best practices, to build awareness of the climate-security nexus and to develop the role of the security sector. Four research institutes in Europe and the USA form the nucleus of the network and 16 other research centres are affiliated, resulting in a powerful scientific base. In the first 2 years of its existence, the IMCCS put its efforts in raising the awareness of security institutions around the world that climate change is also a matter of national and international security.

And they help change the tide. An increasing number of security institutions are recognizing the security implications of a changing climate. The IMCCS produced its World Climate and Security Report, the NATO developed its Climate and Security Action Plan, and the EU published a Climate Change and Defence Roadmap. Several countries have started recognizing the nexus and started realizing that the security sector has a role to play. Recently, the USA released four distinct climate security reports from the Department of Defense, the Office of the Director of National Intelligence, the Department of Homeland Security and the National Security Council. A clear signal of President Biden's intention to integrate and mainstream climate change planning across the national security institutions. All these reports underline the findings in this volume on how climate change affects human and physical security. It makes climate change also a matter of national security, and the security sector needs to develop its role as part of the whole-of-society efforts.

There are many misconceptions about climate change, such as 'our climate is changing slowly', 'it is all happening far away in the Arctic and desert areas', and 'it is an ecological change not affecting our way of living'. The label 'climate change' is probably not alarming enough since it suggests a gradual ecological process over a longer time period. This volume clearly proofs they are wrong. Like political regime shifts, ecological regimes can also undergo sudden catastrophic changes resulting in massive species loss and significantly changed habitats, most clearly seen in deforested landscapes. It frighteningly shows how large parts of our planet are becoming uninhabitable, resulting in regional frictions, large migration flows, and breeding grounds for organized crime and extremist organizations. You don't need to be a climate activist to be concerned about the potential disruptive effects of a changing climate on our way of living. We are facing global climate disruption leading to large-scale human insecurity. If ever there was a threat requiring global cooperation, this is it. This is not the time to hide in nationalist corners and protectionism. It is our collective responsibility to be prepared and to act while we still can.

Tom Middendorp

Tom Middendorp was the Netherlands Chief of Defence for 5.5 years and spent 38 years serving his country. He commanded soldiers on all levels, led a large multinational taskforce in the south of Afghanistan and was involved in over 20 different military missions as the director of operations. Middendorp led the defence organization through an intense period of transition and has extensive operational and strategic experience of building unity of effort with different nations, governmental- and non-governmental organizations, international institutions, and civil stakeholders in order to deal with a wide range of security risks. He has joined Clingendael and The Hague Centre for Strategic Studies (HCSS) as a strategic expert and is chairman of the International Military Council on Climate and Security. He is also the Netherlands' Special Envoy on European Defence Cooperation and a senior advisor in the areas of security, defence and strategic leadership.

Foreword 2: Tackling Climate-Related Distress, Conflict and Displacement Is the Need of the Hour

Climate change, migration, and conflict are age-old phenomena that started even before the birth of human civilization in its present form. Fossil and archaeological evidence suggest that modern humans moved out of Africa into Eurasia, Australia, and the Americas between 50,000 and 150,000 years ago. Beginning over 120,000 years ago, wild climatic swings triggered waves of human migration every 20,000 years, corresponding to periodic orbital variations. At the end of last glacial, *Homo sapiens* arrived in their new settings and outcompeted their near relatives – *Homo neanderthalensis* (Neanderthals) and *Homo erectus* – who became extinct ultimately. In those times, migration was not limited by national boundaries.

A major difference between then and now is the national boundaries and population explosion. World population at the end of the last glacial period, some 12,700 years back, was between 1 million and 10 million. Now it is over seven billion or 7000 million. While the last major migration, which is also termed as exodus from Africa, enabled man to colonize the world, there is hardly any place left for the next migration to take place. The last one led to human evolution, which placed man at the top of the pyramid, the next one may lead to extinction just like the Neanderthals and Homo erectus if we do not pay heed to the straws in the wind.

Climatic changes and their impact are happening at such a pace that they are eroding the resilience of human adaptability. The increasing frequency and intensity of natural disasters such as storms, drought, flooding, heat and cold waves, and disease are triggering shifts in the human and ecological systems. Climate change impacts are stressing the ecosystems beyond recovery in many parts of the world. This has made the poor and vulnerable countries a hub of climate-related displacement and conflict, especially in Africa and Asia. This burdens the neighbouring countries with influx. However, studies on climate and displacement nexus are currently limited, which makes it difficult to predict the long-term consequences of such linkages.

The IPCC AR6 demonstrates the urgency of tackling climate change issues. There would be increase in displacement of people, especially due to lack of resources and higher exposure to extreme weather events, particularly in low-income developing countries. Migration, in fact, is increasingly being perceived as an adaptation strategy as the climate patterns change. However, this may also increase the conflict risks and lower the adaptive capacity of the receiving community due to competing resources, in addition to socio-political unrest.

This volume, which is a compilation of research articles from 20 authors from different parts of the world, is unlike others on climate change and migration as it considers the important aspect of human security from a socio-ecological perspective. It explores local, regional, and global response mechanisms and evaluates their effectiveness, examines the latest trends in literature, and addresses the legal, security, and economic implications of the climate-migration-conflict nexus. It explores the linkages between biodiversity, natural resources, food, and health, also in the context of social and political issues that could exacerbate climate impacts. It is a forewarning of things to come if we do not mend our ways.

In this context, this volume also proposes pathways to manage the complex implications and build socio-ecological systems resilience with the engagement of local communities in the context of their needs. This book has the potential to steer further debate and scientific research on such an important topic.

Rajib Shaw

Professor, Graduate School of Media and Governance, Keio University, Japan; Senior Fellow of Institute of Global Environmental Strategies (IGES) Japan; Co-founder, Resilience Innovation Knowledge Academy (RIKA); Chairperson of SEEDS Asia and CWS, Japan; Chair of the United Nations Asia Pacific Science Technology Advisory Group (AP-STAG); and Coordinating Lead Author of IPCC (AR6 WGII). rajib.shaw@gmail.com; shaw@sfc.keio.ac.jp

Foreword 3: Sea Ice, Marine Ecosystems and Climate in the Arctic's Bering Sea: A Subsistence Marine Mammal Hunters' Roadmap for Research and Policy

Here in Savoonga, on St Lawrence Island on the Bering Sea, our ability to eat and to survive depends upon the seasons that bring the migrating fish, seals, walrus and whales for us to hunt and fish in the rich marine ecosystem that surrounds us and of which we are a part. Our lives are intertwined with the rich ecosystems that we have protected for more than 2,500 years as the marine mammals begin their long summer migrations deep into the Arctic North following the open waters created by receding sea ice. Everybody is poised for the hunt and fishing, with spirits rising all across the village. We prepare in many ways. Our elders remind us to stay safe and be aware that the ice locations are changing, so we cannot depend upon the ice maps passed to us from earlier generations. We prepare our hunting gear, with particular attention to sharpening our harpoon heads and knives to butcher our catches and meat with efficiency and protecting ourselves from getting injured by a dull knife. We have to be able to butcher a walrus, maklak (bearded seal) or whale quickly or the meat will freeze for the whole winter and we will be without food. Also, time is important because the northerly Bering currents are unforgiving. If you are slow, you can be stranded far from the Island with insufficient gasoline and lubricants to get you out and back. We bring rain gear to prepare our crew for any weather.

Before we push off, we scan the horizon, take note of the currents, communicate by radio or sat phone with other hunters, and double and triple check everything as we glass the horizon to chart a course. We know the ice systems and formations from previous years, but the sea ice is no longer as predictable with climate change. We make certain to steer toward the far end of the ice (*sivulitangani*) because that is where the marine mammals locate. Walrus favour locations ahead of the ice or at the tail end of the ice systems. From the high points on the ice, we look around for game. We mark each stop on our compass to pinpoint where home is located. We stay in touch with the other hunters by communications radio for safety and also the best location for concentration of herds of walrus, or seals or migrating whales. We check our fuel level constantly to make sure we have more than enough to make it home safely. It is harder now, with the ice breaking up earlier in the season, and into small pieces along the coast, it is harder to get home. Sometimes we don't go out because we might not get back. So, that leaves us with less food for the year.

We have to be careful to avoid alerting the mammals of our presence with our human smell. So, we look for a route that is downwind of the herd, so they won't smell our approach, but one that brings us close enough for the hunt. We have to be very quiet when approaching the herds with our motor running idle. We seek the ideal walrus for its meat and hide along with ivory for carving. The whales also hear our boats, so we don't even talk. And, we don't harpoon a pregnant or nursing whale.

When the walrus herds are away from the ice edge, we don't hunt them, instead, to save time and energy, we usually steer clear of pack ice where we could easily become trapped as the ice shifts rapidly. The ice can also act as a wind barrier that allows us to hunt in windier conditions. We keep looking for an opportunity to take advantage of the best conditions, sticking close to the ice edge, but away from pack ice.

When traveling north or northeast, depending upon ice patterns, we like to go point to point of the ice to get to the front of the ice, an approach that saves gas, time and energy so we can concentrate on getting to the game at or near the front of the ice. As I told an interviewer, "it all depends on the weather, ice and sea conditions and it takes constant observation, being vigilant, communicating with your crew on best practices and following through".

We rely on the ice to provide us safe passage and food. This is where I am most comfortable and happy when any hunting or whaling success happens, just like anywhere in the world. I always like to take the most direct route when heading home. But, storms or weather might mean we need to stick close to the ice edge. But, even in the fog, we can see the ice mirrored from the water. Sometimes we face challenges that sap our energy. But, the bonds with our fellow hunters are life long. We help and respect each other, the mammals, fish and whales to whom we give ceremonial thanks for feeding us, and allowing us to be part of this marine ecosystem for which we were taught to be stewards by oral history, ceremony and experience.

The fog limits your ability to see ahead so you use your other senses to look for mammals: sense of smell and hearing become very important. Communicating with others and knowing where they are become very important. Manoeuvring through the ice, you see the reflection of open water mirrored above the water in the fog and you can also see the bright reflection of the ice. This guides us through the ice, saving time and energy heading home. If it does not work out, we head away from the ice and look for an opening to the west, remembering to work with the current.

To get to the marine mammals, you learn where to go, learn their environment, their behaviour patterns and the most ideal conditions. Our traditional science and knowledge, in addition to academic science, have been used for more than hunting. Traditional knowledge can help us identify approaches to marine ecosystem management, as well as in developing appropriate regulations. Hunters and whalers know intimately the numbers and behaviours of migrating species.

The Kukuklget Mountains serve as our compass when we come towards shore. We stick close to the western edge of the ice to navigate the prevailing currents. Near shore, the ice tends to pack so we look for an opening to the west of the village and make better time working with the current. Lookouts on high points direct traffic for the best routes. Crews tend to form a group in tightly packed sea ice to manoeuvre over the ice as we look for the best path from the high point. Then, working cooperatively, when it's flat, we can glide our boats over the ice.

With climate change, hunters and whalers should be involved in the assessments of likely impacts to assist with the design of appropriate management measures. We have much to contribute and will have greater confidence that our ways of life will be adequately maintained and protected. Bowhead whale hunters have long pointed out that the whales can be affected by smells associated with human activity. Whales downwind can smell a human that leads them to dive or move away from the source of smell. This behaviour is the same for other mammals, as part of their ability to survive in the Arctic. As expressed in an earlier article, "biologists recently confirmed our observations that whales do in fact have a sense of smell, which had previously not been accepted by non-Yupik scientists".

That greater recognition being given to our observations in recent years is a welcome advance. And more can be done both to incorporate the knowledge itself and to improve the participation of traditional fishers, hunters and whalers in research and management. As I told a think tank several years ago, we need to be aware of the demands that this level of interest places on those who hold the information. Too many studies mean too many demands on our time and can lead to "research fatigue" and decreased interest in taking part. Paying people for their consulting is appropriate and can help sustain a willingness to participate in scientific studies and related activities.

The following four points might be included in increasing interest in the development of climate change policies, relevant science and new regulations:

- *First*, "traditional knowledge" and traditional science in Arctic context should be placed on a website for a general accessibility.
- *Second*, regular monitoring should occur to keep the knowledge of climate change impacts up to date. By monitoring new information, the science community will understand changes as they occur. Hunters, fishers and whalers spend a great deal of time out on land and sea, and cover a large territory with nuanced observations. This will provide a broader and different picture of ecosystems that are available from other methods such as remote sensing and space-monitoring methods.
- *Third*, because our science, observations and knowledge are often placed at a lower level of hierarchy than Western science, we are not paid for our observations and time spent to share our information for publication or regulatory changes. Further, we are generally not invited a seat at the table where climate-related decisions are made, even though we are on its frontlines.
- *Finally*, as conveyed to an American think tank, traditional hunters and whalers and our observations of environment, especially marine ecosystems, have sustained the lives of Arctic peoples "since time immemorial". Our values remain high today, especially to help figure out how to use our 1000-year approach to protect ecosystem balance with greater authority as the climate changes and

development increases. New ecosystems are happening daily, new air, new water and sea ice that can bring something new anytime; you just have to be close.

From my conversations with Anita Parlow, the author of the chapters on the Bering Sea in this publication, I know that this volume tells a story of people like us – Yupiks who live on St. Lawrence Island in the Bering Sea – who depend upon understanding, navigating and protecting the ecosystem balance that sustains and feeds us as we hunt, fish and whale. Each of the stories in this volume is different, depending upon what part of the world is being discussed. But, this book shows that all living species are tied together as part of nature and depend upon each other in what some call this "Turtle Island". The changes that we see – walruses and seals getting skinnier, more fish coming north, whale migrations almost out of reach given less sea ice, birds falling from the cliffs and more flooding – is at a scale we haven't seen in our lifetimes. This volume both confirms the changes by people most directly affected and offers Western science a roadmap for climate research and policy.

George Noongwook

References

Huntington, H. P., & Noongwook, G. (2013). *Traditional knowledge and the Arctic environment – How the experience of indigenous cultures can complement scientific research*. The Pew Charitable Trusts. Retrieved from: https://www.pewtrusts.org/-/media/legacy/uploadedfiles/peg/publications/report/arcarctictraditionalknowledgepdf.pdf

Huntington, H. P., Gearheard, S., Holm, L.K., Noongwook, G., Opie, M., & Sanguya, J. (2016). Sea ice is our beautiful garden: Indigenous perspectives on sea ice in the Arctic, In D. N. Thomas (Ed.), *Sea ice* (3rd ed.). Wiley. https://doi.org/10.1002/9781118778371

Siberian Yupik from St. Lawrence Island in the Bering Sea. Whaling Captain, Walrus Hunter, Former Chair of the Alaska Eskimo Whaling Commission, American Attendee to the International Whaling Commission, and Author.

Acknowledgments

The initial idea to publish a contributed volume on the climate-conflict-displacement nexus from a human security perspective stemmed from the outcome of the International Conference on Social-Ecological Systems: From Risks and Insecurity to Viability and Resilience (SES2019) organized by the Center for Environment, Human Security and Governance (CERES) in October 2019 in Marrakech, Morocco. From the onset of the project so far, humanity has gone through multiple crises that have triggered many shifting dynamics. In addition to an unprecedented global pandemic, which has demonstrated the consequence of a deep anthropogenic-induced disruption of the ecosystem's balance, coupled with a context of resource scarcity, climate urgency, disasters, increasing social inequalities, violent conflicts, and mass displacement, the relevance and originality of this book project are increasing over time. During a 2-year process, the scope and objectives of the publication have been refined in order to remain very sensitive to the evolution of many dynamics worldwide, especially in some particular regions covered by the book such as the MENA region.

It was my deep honor to serve as the chair of SES2019 and to lead the editorship of this volume. I would like to express here my deep gratitude and thanks to the editorial team members: Dr. Himangana Gupta (JSPS-UNU Postdoctoral Fellow at the United Nations University Institute for the Advanced Study of Sustainability (UNU-IAS) and the University of Tokyo, Japan); Fred Kruidbos (KRUIDBOS Ecologisch onderzoeks- en adviesbureau, Helmond, the Netherlands); and Anita Parlow (Founding Team Lead, Woodrow Wilson International Center for Scholars' Polar Code Program and Research Associate and Advisor, Harvard–MIT Arctic Fisheries Project, USA). Their pro-active and fruitful collaboration during the publishing process has made the edition of this volume a real passion and an exciting experience.

In addition, the achievement of this landmark publication is the outcome of the active contribution and engagement of prominent authors from a myriad of disciplines, research institutions, and continents. On behalf of the editorial team, my thanks and recognition are extended to all authors whose scientific contributions have ensured the relevance and value of this timely publication. In this vein, I would like to thank Dr. Mohamed Mliless for the effort and time devoted to evaluate the book's manuscript and for his thoughtful comments and suggestions towards improving its quality level.

The chapters in this volume are also the result of the invaluable contribution made by peer reviewers, who generously gave their time and energy to provide insight and expertise to the selection and editing process. On behalf of my co-editors, who actively participated in the peer-review process, I would specifically like to acknowledge, with sincere and deepest thanks, the work done by all peer reviewers, including Dr. J K Patterson Edward, Suganthi Devadason Marine Research Institute (SDMRI), India; Dr. Lakhvinder Kaur, Panjab University Chandigarh, India; Alis Daniela Torres, ICLEI – Local Governments for Sustainability, European Secretariat, Ecuador/ Germany; Raj Kumar Gupta, senior journalist, India; and Navjeet Kaur, University Librarian (Retd.), Panjab University, Chandigarh, India.

Finally, I would like to express my gratitude to Dr. Mohamed Mliless (Researcher in Ecolinguistics, Morocco) for the proofreading the final manuscript.

Agadir, Morocco Mohamed Behnassi

About the Publishing Institution

The Center for Environment, Human Security and Governance (CERES), Morocco

CERES, previously the North-South Center for Social Sciences (NRCS), 2008–2015, is an independent and not-for-profit research institute founded by a group of Moroccan researchers and experts in 2015 and joined by many partners worldwide. It aspires to play the role of a leading think tank in the Global South, and to serve as a reference point for relevant change processes. Since its creation, CERES managed to build a robust network involving various stakeholders such as researchers, experts, PhD students, decision makers, practitioners, and journalists from different spheres and scientific areas. These achievements are being rewarded by the invitation of CERES members to contribute to global and regional assessments and studies (especially Ipbes, Medecc, EuroMeSco, etc.) and the invitation of the Center to become a member of the MedThink 5+5, which aims at shaping relevant research and decision agendas in the Mediterranean Basin. The Center has organized so far five international conferences and several training/building capacity workshops, provided expertise for many institutions, and published numerous books, scientific papers, and studies which are globally distributed and recognized. These events and publications cover many emerging research areas mainly related to the human-environment nexus from a multidimensional, multiscale, interdisciplinary, and policy-making perspectives. Through its initiatives, CERES attempts to provide expertise to advance science and its applications, and to contribute to effective science and policy interactions.

Contents

Abbreviations and Acronyms

AEWC	Alaska Eskimo Whaling Commission
AIA	Aleut International Association
AMBCC	Alaska Migratory Bird Co-Management Council
AMOC	Atlantic Meridional Overturning Circulation
AMSA	Arctic Marine Shipping Assessment
ANCSA	Alaska Native Claims Settlement Act
ASTER	Thermal Emission and Reflection Radiometer
ATBA	Areas To Be Avoided
CA	Content Analysis
CAO	Central Arctic Ocean
CAR	Central African Republic
CCA	Climate Change Adaptation
CEJ	Critical Environmental Justice Studies
CELAC	Community of Latin American and Caribbean States
CERES	Center for Environment, Human Security & Governance
CF	Conventional Forces
CHCs	Cosmopolitan Harm Conventions
CMTS	Committee on Marine Transportation Systems
CODESA	Collective of Saharawi Human Rights Defenders
COED	Cost of Environmental Degradation
COP	Conference of the Parties
DA	Direct Action
DDR	Disarmament, Demobilization and Reintegration
DEM	Digital Elevation Model
DRC	Democratic Republic of Congo
DRR	Disaster Risk Management
EDD	Environmentally Driven Displacement
EDM	Environmentally Driven Migration
EDPs	Environmentally Displaced Persons
EJOLT	Environmental Justice Organizations Liabilities and Trade's
ENVI	Environmental Visualization

ERDAS	Earth Resources Data Analysis System
ESPA	Eastern Sudan Peace Agreement
EU	European Union
EWC	Eskimo Walrus Commission
FESO	Foreign Ecological Security Operations
FTZ	Free-Trade Zones
GCI	Gwich'in Council International
GEOBIA	Geographic Object-Based Classification
GNP	Gross National Product
GRC	Geneva Refugee Convention
GWEC	Global Wind Energy Council
GWR	Geographically Weighted Regression
HAB	Harmful Algae Blooms
HCP	High Commission for Planning
ICC	Inuit Circumpolar Council
ICC	International Criminal Court
ICJ	International Court of Justice's decision
IDP	Internally Displaced Persons
IMCCS	International Military Council on Climate and Security
IMO	International Maritime Organization
IOM	International Organization for Migration
IPCC	Intergovernmental Panel on Climate Change
LDCs	Least Developed Countries
LNG	Liquefied Natural Gas
LPI	Living Planet Index
LRA	Lord's Resistance Army
MA	Military Assistance
MDGs	Millennium Development Goals
MENA	Middle-East and North Africa
MMPA	Marine Mammal Protection Act
NAPAs	National Adaptation Programmes of Action
NDRE	Normalized Difference Red Edge Index
NOAA	National Center for Environmental Information
NSEDC	Norton Sound Economic Development Corporation
NSIDC	National Snow and Ice Data Center
NSR	Northern Sea Route
NTFPs	Non Timber Forest Products
OAS	Organization of American States
OCHA	UN Office for the Coordination of Humanitarian Affairs
ODA	Official Development Assistance
OLS	Ordinary Least Squares
PCB	Polychlorinated Biphenyl
PSC	Peace and Security Council
RIAC	Russian International Affairs Council
ROLAC	Regional Office for Latin American and the Caribbean

SAP	Structural Adjustment Program
SCMC	Sahrawi Center for Media and Communication
SDGs	Sustainable Development Goals
SES	Socio-Ecological Systems
SR	Special Reconnaissance
UCC	United Church of Christ Commission for Racial Justice
UNCLOS	United Nations Convention on the Law of the Sea
UNDRIP	United Nations Declaration on the Rights of Indigenous Peoples
UNEP	United Nations Environmental Programme
UNESCAP	United Nations Economic and Social Commission for Asia and the Pacific
UNFCCC	United Nations Framework Convention on Climate Change
UNGA	United Nations General Assembly
UNHCR	United Nations High Commissioner for Refugees
USGS	U.S. Geological Survey
WHO	World Health Organization
WIM	Warsaw International Mechanism for Loss and Damage Associated with Climate Change
WSRW	Western Sahara Resource Watch

List of Figures

List of Tables

About the Editors

Mohamed Behnassi is a full professor in the College of Law, Economics, and Social Sciences of Agadir, Ibn Zohr University, Morocco. He is a senior researcher of international law and politics of environment and human security. He holds a PhD in international environmental law and governance from Hassan II University of Casablanca, 2003, and a diploma in international environmental law and diplomacy from the University of Eastern Finland and UNEP, 2015. He is also an alumnus of the International Visitor Leadership Program of the Department of State, USA. Dr. Behnassi is currently the founding director of the Center for Environment, Human Security and Governance (CERES) – formerly North-South Center for Social Sciences (NRCS). From 2015 to 2018, he was the director of the Research Laboratory for Territorial Governance, Human Security and Sustainability (LAGOS) and head of the Public Law Department, College of Law, Economics, and Social Sciences of Agadir. Recently, he was appointed as expert evaluator to the National Center for Scientific and Technical Research (CNRST/Morocco), and selected as expert by the Intergovernmental Science-Policy Platform on Biodiversity and Ecosystem Services (IPBES) and the Mediterranean Experts on Climate and Environmental Change (MEDECC) to take part in global and regional assessments, respectively. Accordingly, he was among the lead authors of the 1st Assessment Report (MAR1): *Climate and Environmental Change in the Mediterranean Basin – Current Situation and Risks for the Future* (MEDECC, 2021). Dr. Behnassi has published considerable number of scientific papers and book chapters in addition to 17 books, including recent ones: *Social-Ecological Systems in the Era of Risks and Insecurity: Pathways to Viability and Resilience* (Springer, 2021); *Building Resilience for Food and Water Security Face to Climate Change and Biodiversity Decline: Perspectives from Asia, Middle-East and Africa* (Springer, 2021); and *Human and Environmental Security in the Era of Global Risks* (Springer International Publishing, 2019). Dr. Behnassi has organized many international conferences covering the above research areas, managed many research and expertise projects, and is regularly requested to provide scientific expertise nationally and internationally. Other professional activities include social compliance auditing and consultancy by monitoring human rights at work and the sustainability of the global supply chain.

Himangana Gupta is a JSPS-UNU postdoctoral fellow at the United Nations University Institute for the Advanced Study of Sustainability (UNU-IAS) and the University of Tokyo, Japan. She received her doctorate in environment science from Panjab University, India, in 2015. Dr. Gupta has worked on climate change and biodiversity policy and diplomacy, and is currently working on linkages between biodiversity, climate, and communities in socio-ecological production landscapes. Before this, she was a part of the National Communication Cell (NATCOM) of the Indian Ministry of Environment, Forest and Climate Change. She contributed to India's Second Biennial Update Report and several other publications of the Ministry. She is a certified expert in climate adaptation finance. Dr. Gupta is a university gold medalist and recipient of academic excellence award. She has published three edited books with Springer and has also written research papers in reputed international and national journals on climate policy, forestry, biodiversity, and women in climate change mitigation and adaptation.

Fred Kruidbos is a (population) biologist specialized in animal ecology. He is an independent researcher and the director of K-SN Ecological Services B.V. Based in the Netherlands, this company operates worldwide and consists of three branches focused on: (i) ecological research and consultancy, (ii) ecological products, and (iii) ecological safety and security services. This setup makes it possible to study problems from a multidimensional approach facilitating more holistic solutions. Mr. Kruidbos studied at the prestigious Wageningen University (population biology) and Fontys University of Applied Science Tilburg (teacher training). During and after his studies, he conducted research on various prey-predator and host-vector related studies. His current research is pragmatic in nature and focuses mainly on the interface between spatial planning and nature conservation. Recently expanded with safety and security issues of which his contributions to this book are an example. In addition to being a biologist, Mr. Kruidbos is a senior operational advisor attached to the Dutch army. During his military service (compulsory in the Netherlands at the time) he trained with the Dutch Special Forces and completed his operational period as a team member. In addition to his civil studies, he completed the training as reserve officer and senior officer at the Netherlands Defence Academy (NLDA). In this role, he has been involved in issues related to civil-military interactions as well as special operations for over 12 years. Recently, he has mainly focused on the climate-conflict nexus. Supported by the combination of his ecological and military knowledge, Mr. Kruidbos highlights issues from a different angle than usual within both disciplines. This broadens both scope and problem analysis, which leads to more inclusive solutions and new methodologies. An important part of his contribution is to direct the shift from kinetic reactive to non-kinetic pre-conflict interventions as well as post-conflict support and reconstruction. An example of this is his theoretical and operational contributions to the Dutch Provincial Reconstruction Team (focused on livestock recovery, agriculture, and ecology) of the province of Uruzgan, Afghanistan, which has made a significant contribution to shaping Dutch 3D policy.

Against the background of a theoretical introduction on ecology, his second chapter contribution to this book proposes to extend this approach with ecology and time as central factors.

Anita Parlow Esq, MSt, is a former Fulbright scholar from Iceland, researcher with the Harvard-MIT Arctic Fisheries Project, and founding team lead of the Woodrow Wilson International Center for Scholars' IMO Polar Program. She has taught international human rights law, arctic law, and federal Indian law, authored two books on Indigenous issues in the USA, and published op-eds and articles for the Anchorage Daily News, Arctic Today, The Polar Connection, the University of Maine Oceans Law Review, and the Economist. She recently advised and wrote environmental, social, and governance (ESG) and risk assessment reports for: the Port of Nome, Tanana Chiefs Conference – food security; and the Denali Commission – coastal communities and climate. Parlow has also advised companies such as Keystone, TransCanada, and Freeport McMoran on ESG matters. Parlow has spoken on Arctic issues in the USA, Northern Europe, Canada, China, and Russia. An Oxford post-law graduate in international human rights law and member of the Bar of the US Supreme Court, Parlow has conducted pro-bono mediations at D.C. Superior Court and the D.C. Office of Human Rights.

Chapter 1
The Climate-Conflict-Displacement Nexus from a Human Security Perspective – An Introduction

Mohamed Behnassi, Himangana Gupta, Fred Kruidbos, and Anita Parlow

Abstract It is now scientifically founded that climate change, induced and accelerated by anthropogenic activity, will happen regardless of any mitigation actions taken now. Related implications are disrupting the planet and its ecosystems and undermining the potential of the human species to continue to sustain itself as it has done for millennia. Related impacts and disasters are also triggering significant shifts in the inextricably interconnected social-ecological systems. Due to these shifting dynamics, there is a high probability for the emergence of new multi-scale conflicts or the transformation of the complexity and magnitude of existing ones, thus burdening governance systems, which are already exhausted by conventional challenges. In addition, these shifting dynamics not only threaten survival at species and community levels, but are also emerging drivers of human insecurity and displacement both within and across national borders. All these emerging dynamics are gradually capturing the attention of actors with the ability to trigger change processes. Against this background, this introductory chapter outlines the overall framework on which the whole volume is shaped while presenting the scope, objectives, and key findings of subsequent chapters. It unpacks the key dynamics of the

M. Behnassi (✉)
International Law and Politics of Environment and Human Security, College of Law of Agadir, Ibn Zohr University, Agadir, Morocco

Center for Environment, Human Security & Governance (CERES), Agadir, Morocco
e-mail: m.behnassi@uiz.ac.ma

H. Gupta
Sustainable Landscapes and Restoration Program, World Resources Institute (WRI), New Delhi, India
e-mail: gupta@unu.edu

F. Kruidbos
Kruidbos Ecological Research and Consultancy, K-SN Ecological Services B.V., LC, Helmond, The Netherlands
e-mail: info@kruidbos.com

A. Parlow
Fulbright Scholar Iceland, recent, Washington, DC, USA

M. Behnassi et al. (eds.), *The Climate-Conflict-Displacement Nexus from a Human Security Perspective*, https://doi.org/10.1007/978-3-030-94144-4_1

nexus between climate change, conflict, and displacement from human security and resilience perspectives and explores the various local and global response mechanisms to address the nexus, evaluate their effectiveness, and identify their implications for the nexus itself.

As the IPCC's Sixth Assessment Report makes clear, climate change, induced and accelerated by anthropogenic activity, will happen regardless of any mitigation actions taken now. Even under its modest conservative projections, global temperatures will rise by 1.5 °C. That may not sound like much, but it will double the frequency of droughts – from once every ten years to once every five. Worse still, a 2 °C-temperature rise – also a likely outcome without substantial emission reductions – will make droughts 2.5 times and other disasters more frequent and severe.

Currently, climatic changes are reshaping the planet, its ecosystems, and the ability of the human species to continue to sustain itself as it has done for millennia. Related impacts and disasters are triggering significant shifts in the inextricably interconnected human and ecological systems. Shifting dynamics are already underway, observed and experienced by a large part of humanity, increasingly investigated by scientists, and gradually capturing the attention of actors with the ability to trigger change processes. Such actors are currently developing diverse and multi-scale strategies, tackling both causes and consequences of climate change, including mitigation, ecosystem restoration, and resilience building.

Climate change-induced impacts and disasters such as sea-level rise, coastal erosion, drought, floods, increased heat and cold waves, and pandemics are already causing significant social and environmental disruptions. While medium and long-term implications remain mostly projections, both scientists and vulnerable communities predict a future of accelerated dynamics that are likely to bring the world 'dangerously close' to abrupt and irreversible changes. Inertia or incremental change is not anymore an option as a combination of natural resource scarcity, changes in species dynamics, shifts and decline in the resilience and carrying capacity of ecosystems, and biodiversity loss may generate potential complex and irreversible situations. In fact, such situations have already occurred since, for example, large parts of the tropical regions have turned into desert, which is known for its very different climate from that of tropical forests.

Due to these shifting dynamics, there is a high probability for the emergence of new multi-scale conflicts or the transformation of the complexity and magnitude of existing ones, thus burdening governance systems, which are already exhausted by conventional challenges. In addition, these shifting dynamics not only threaten survival at species and community levels, but are also emerging drivers of human insecurity and displacement both within and across national borders.

For a growing part of humanity, today's world is an insecure place, a landscape of various threats and crises on many levels. Such crises are complex, entailing multiple forms of human insecurity. When they overlap, they can grow exponentially, spilling into all aspects of people's lives, destroying entire communities and crossing national borders. This is exactly what distinguishes current conflicts and their consequences, such as displacements. Protracted crises, violent conflicts, natural disasters, persistent poverty, epidemics and economic downturns impose hardships and undercut prospects for peace, stability, and sustainable development.

In addition, despite that many crises and shifting dynamics are occurring on large scales, the most relevant and effective response mechanisms are proven to be community-based and location-specific. This calls out for processes that construct networks of local, national, and regional communities – if not global – to shape value systems that depart from the realities on the ground and build out. As current multilevel initiatives address changes in political, legal, and economic settings, the strength and power of bottom-up initiatives provide vital grist for the planet and its entire species, including humans. In this perspective, the human security approach is a proven analytical and planning framework that supports more comprehensive and preventive responses by governmental, intergovernmental, and non-governmental institutions. This approach helps cut across sectors, develop contextually relevant solutions, and adopt partnerships to make the world free from fear, want, and indignity. As noted in General Assembly resolution 66/290, human security is an approach that calls for "people-centered, comprehensive, context-specific and prevention-oriented responses that strengthen the protection and empowerment of all people".

Against this background, this contributed volume unpacks the key dynamics of the nexus between climate change, conflict, and displacement from human security and resilience perspectives. In addition, the volume explores the various local and global response mechanisms to address the nexus, evaluate their effectiveness, and identify their implications for the nexus itself. More precisely, the first set of chapters (1 and 2) examines the trends in the literature on the climate-conflict-migration nexus and explores the various implications of climate-related response mechanisms on the nexus areas. The second set of chapters addresses the legal, security, and economic implications of the nexus with a sharp focus on 'climate or environmental migration' from a regional perspective (3, 4, 5, and 6). The third set of chapters (7, 8, and 9), however, focuses on the Arctic region where climate change and geopolitical dynamics are crystalizing the vulnerability of both the marine ecosystem and indigenous communities depending on its resources for their existence, livelihood, and food security. The fourth set of chapters (10, 11, 12, and 13) investigates the linkages between biodiversity, natural resources, food and health security, conflict, and displacement and propose new pathways to manage the security implications of such linkages. The last set of chapters (14 and 15) focus on some particular types of resource-induced domestic and cross-border conflict, highlighting actors' strategies both in terms of discourse and practice.

Twenty authors from many disciplines and countries (Egypt, Germany, India, Morocco, Netherlands, Spain, Sudan, and the United States) have been provided with the opportunity to share relevant research, insights, and successful practices. They have explored various innovative approaches and provided valuable inputs on the climate-migration-displacement nexus in different parts of the world. In addition, the volume's content and approach are multidisciplinary (for fruitful interaction between numerous scientific fields) and relevant to policy-making processes (enabling interactions among experts and decision makers from different levels and spheres).

In Chap. 2, *Climate-Conflict-Migration Nexus: An Assessment of Research Trends Based on a Bibliometric Analysis*, Kumar Singh and Gupta claim that migration and conflict are issues of social and political concern, increasingly exacerbated by climate change impacts, started to be captured by researchers worldwide. Accordingly, the authors undertake a literature review using bibliometric indicators to examine the trends in research on the climate-conflict-migration nexus and, broadly, on climate migration and refugee concerns from 1990 to 2019. Using the Scopus database, the authors examined nexus publications during the selected timespan and could capture the geographic distribution, major sub-domains *viz*, consequence/impact, security/insecurity, policy, health, solution, migration and conflict, top authors engaged in this research area, and the top funders. Throughout the analysis, the authors developed a reflection on the major gaps in the research output and how this could shape the future debate on climate-related migration. The authors point out that there is less research on the solutions side. Thus, there is a need to reflect on how this fast-emerging concern can be tackled globally, given that each country is exposed to climate risks but with differentiated vulnerability and adaptive capacities.

In Chap. 3, *Climate Change Response Strategies, Conflict, and Human Displacement: Linkages, Evidence, and Potential Solutions*, Behnassi et al. claim that not only climate-induced impacts and disasters are increasingly affecting social-ecological systems worldwide by generating or amplifying risks and vulnerabilities, but also climate response mechanisms too may generate similar implications, which may increase the likelihood of conflicts and human displacements. Focusing on adaptation, as one of the mainstays of climate change governance, the authors based their analysis on the assumption that related policies and actions may, in some instances, generate unexpected outcomes and risks which erode resilience instead of building it. To substantiate this, they investigate the drivers, forms, and manifestations of maladaptation by reference to existing theoretical and empirical research. Accordingly, the maladaptation and migration nexus is analyzed from an inclusive governance and science-based policy making perspective. The authors have also identified and discussed the alternative approaches to address maladaptation to ensure resilience and sustainability through adaptation strategies. The needed guidelines to be considered while elaborating, implementing, and evaluating adaptation actions to ensure the achievement of expected outcomes and to avoid potential tradeoffs and negative implications, have been prescribed. Results correlate with the above assumption and show that some adaptation strategies and interventions in many contexts have undermined the balance of ecosystems and the resilience of local communities, creating additional conditions for increased human insecurity, including the compulsion to migrate.

In Chap. 4, *While Carbon Burns: The Debatable Journey of 'Environmental Refugee' as a Concept and Legal Dilemma*, Magdy and Yasser claim that anthropogenic climate change and its apocalyptic predictions opened for researchers a wide scene of conflicts, phenomena, and concepts, which would not have been available if climate change had not been initially considered as a problem causing social and environmental disruption, thus generating many victims. One way to protect the rights of such victims, especially displaced persons, was through conceptualizing their situation and giving them the status of 'environmental refugees'. However, the authors assert that such a concept did not pass smoothly through the academic, legal, and political circles due to its controversial nature. The legal gap regarding the protection of this category of refugees, in particular, allowed scholars and humanitarian organizations to look for alternative options and ideas. In this broad perspective, the authors attempt through their analysis to provide answers to: Who is the 'environmental refugee'? How did this concept emerge? In which context and through which phases of evolution? How was this concept criticized and why? Is there any legal instrument that recognizes such a concept or offers a protection for this vulnerable group? And what is the core mandate of UNHCR and IOM regarding climate-induced displacement? The authors also revisit the concept of 'environmental refugee', its emergence, evolutions, legal dilemmas, and criticism and investigate the extent to which such a concept can be employed to approach the ongoing massive waves of displacement around the globe.

In the same vein, Andrasko in Chap. 5, *Looking Ahead: A Human Security Perspective to Tackling the Potential for Widespread Environmental Migration in Latin America*, recalls that 25 million to one billion people globally could be displaced by 2050 due to climate change-induced extreme events and environmental degradation. The Latin American region could face similar impacts that remain relatively understudied, especially in the context of migration driven by climate change, environmental degradation, inequalities, and conflict. To fill such a gap, the author advocates for a regional approach to future pathways of environmental migration through a focus on sustainable management of natural resources and proactive climate change mitigation activities. According to the author, such pathways can make migration slightly easier within the region than outside and drive more cross-border migration than in other regions. Additionally, the author claims that addressing the reality of future environmental migration in an orderly manner requires long-term risk planning efforts capable of improving international and regional cooperation, capacity-building, and adaptive management of refugee resettlement programs. The author perceives migration as a valid adaptation approach in the face of rapid-onset events and for lowering exposure to slow-onset events, hence its importance as a part of climate change adaptation planning. Indeed, adaptive and forward-looking planning for environmental displacement is a way to avoid the anxiety, scrambling for funding sources, and sudden state of emergency in countries often associated with new refugee flows.

In Chap. 6, *Environmental Migration in the MENA Region: The Case of Morocco*, Ferreira Fernandes and Alves start by highlighting the importance of environmental factors in the process of human mobility as recognized by the IPCC, UNFCCC, and the Global Compact for Safe, Orderly and Regular Migration. However, the complexity of the drivers of migration, according to the authors, tends to limit its analysis from an environmental change perspective, due to the multifaceted decision of migration. Moreover, since migrants can contribute to climate adaptation strategies in their home communities, the authors assert that the analysis of the challenges and positive potential of migration should be relevant to the overall debate on adaptation. From this perspective, the authors undertake an overview of the studies that focus on the migration-environment-climate nexus in the Middle-East and North Africa (MENA). Their research highlights the gaps in the existing knowledge, which reflect the scarcity of studies in several MENA countries, despite being a hotspot of climate change and a home for excessive migratory movements compared to other regions. More particularly, the authors refer to the Moroccan case, a polymorphic country in terms of migration flows, which registers one of the highest frequencies of studies covering the climate-migration nexus.

In Chap. 7, *The Environment-Climate-Conflict-Displacement Nexus in the Arab Region: Implications and Recommended Actions*, Chougrani and Behnassi notice that environmental and climate-induced migration in the Arab region and its enormous impact on the national budget have received, to date, scant attention in the literature. To fill such a gap, the authors analyze the budgetary cost of environmental migration, taking into consideration the quality of local governance and the intersection of other significant drivers. Meanwhile, the feasibility of integrating and incorporating the cost of environmental migration into the Gross National Product (GNP) accounting has been demonstrated. Their analysis also shows a correlation between the constraints of internal and international displacement and the rapid environmental deterioration induced by natural resource depletion and scarcity, conflicts, climate change, disasters, desertification, and biodiversity loss. The extent to which environmental deterioration dynamics and internal conflicts are interlinked has been, therefore, examined. The analysis shows as well how and to what extent such conflicts affect the political boundaries of fragile states in the context of mass migration and the competition of global powers for influence, interests, and redistribution of roles.

Chapter 8, *The Changing Dynamics of Arctic Ecosystem and Indigenous Food Security: The Case of the Bering Sea Region*, provides a foundational context for subsequent Chaps. 8 and 9 of this volume. It focuses on the Arctic region where global warming is reshaping both the marine ecosystem and communities depending on its services. According to Parlow, the rapid melting of the Arctic sea ice is injecting fresh water into the world's oceans, thus causing a slowdown of the global ocean currents that circulate equator waters to the Arctic region, which in turn, are less able to cool the world's oceans. This circulatory slowdown generates erratic weather patterns and a 'feedback loop' thus melting more Arctic sea-ice and creating yet more global warming as atmospheric temperatures rise. In this particular context, the author tackles one of the complex existential crises through the lens of the Bering Sea, contextualizing the accelerating changes in one of the richest ecosystems worldwide. It also portrays the resilience actions by those who live on the front lines of the climate catastrophe. By further narrowing the focus on St. Lawrence Island, the author reveals the cascading changes to the Bering's complex food web and their negative impacts on Arctic coastal communities' food security. The author believes that since such communities contribute the least to the global carbon footprint, yet retain the highest level of knowledge on how to maintain the marine ecosystem's balance, the world might look more deeply to Arctic's Indigenous Peoples as a guiding force to address the existential threat currently facing the humanity. This chapter also describes how melting Arctic sea ice changes ocean current and the earth's temperatures through a self-perpetuating 'feedback loop' that accelerates climate.

In Chap. 9, *Arctic Geopolitics, Cross-Boundary Soft-Power, Ecosystem Protection, and Human Security in the Bering Sea and Strait*, Parlow addresses a myriad of complex and interrelated dynamics affecting the condition of the Arctic region, which serves as the Earth's thermostat, through a special focus on the Bering Sea. In this region, global warming is causing a cascading series of changes to the marine ecosystem, Indigenous communities, and the unfolding geopolitical and commercial interests amongst Russia, China, and the United States that, together, are reconfiguring the region. As the previously ice-bound globally accepted 'Zone of Peace' is now a paradox, the emerging dynamics raise significant questions regarding strategic cooperation and environmental strategies. Indeed, the United States and Russia have recently agreed that cooperation on marine ecosystem protections, regardless of contentious marine boundary issues, is vital. However, this geo-strategic stand also raises vital questions for coastal communities that vitally depend on marine resources. In addition to this, the author also addresses the global implications of a warming Arctic region, due to greenhouse gas emissions more than economic development, especially that the 'feedback loop' triggers more global warming and potentially catastrophic changes in both marine and terrestrial ecosystems.

In Chap. 10, Decolonization, *Food Sovereignty, and Climate Risks: The Case of St. Lawrence Island in the Bering Sea, Arctic*, Parlow recalls that human-induced warming is increasing the sea-ice melt and permafrost collapse and altering species migrations on land and sea. This impacts the local flora and fauna in the Arctic and also the Indigenous peoples who depend on hunting and fishing activities for survival. However, the author believes that past colonization – and its contemporary legacy – is the main impediment to Indigenous peoples' ability to address climate damage and protect both Arctic marine and terrestrial ecosystems in their areas of customary use and occupancy. This chapter discusses the era of the Russian and European contact to today's spate of international, American, and other Arctic national domestic laws that would appear to restore Native peoples into domestic and international decision-making. As this, sadly, is generally not the case, the author describes the context of current and ongoing developments by Native peoples to play a greater role in shaping Arctic law and policymaking. She presents the successful trajectory of multi-scale initiatives through which Native peoples seem to recover the right to self-determination, sovereignty, and a relationship with traditional lands and waters. For the author, the millennial story that is unfolding in St. Lawrence Island in the Bering Sea is the story of an island that is exposed to an existential threat of climate change. St. Lawrence Island, says the author, is an example of the courage and resolve of humanity that is currently engulfed in catastrophe, and a true story of a people whose self-determination and sovereignty must include meaningful participation in decision-making processes.

In Chap. 11, *Spatial Distribution and Geosimulation of Non-Timber Forest Products for Food Security in Conflict Area*, Deafalla et al. focus on conflict-affected areas where civilians often suffer from loss of livelihoods and increased food insecurity. Disruption of food systems and markets, in the case of the Nuba Mountains of Sudan, resulted in higher food prices and shortages of water, food, and fuel. Armed conflicts impacted the soil structure and destroyed farmland, mills, storage facilities, and machinery. Besides that, increasing insecurity and roadblocks prevented humanitarian convoys from reaching the most vulnerable. Therefore, the authors attempt to explore, analyze, and predict the spatial relationships that affect the collection of Non-Timber Forest Products (NTFPs) using geostatistical analyst tools. They use disaggregate statistics to test and model such a relationship, making it a viable methodology for studying NTFPs, associated with household food supply in Nuba Mountains. The models used in the study allowed the pattern of association to be visualized and all statistical values to be spatially represented on maps. The authors concluded with some recommendations to guide decision makers and development agencies in making interventions that will have sustainable and equitable implications.

In Chap. 12, *Ecology of Zoonotic Pathways Indicating Conflict and Mass Migration*, Kruidbos attempts to illustrate how ecosystem stability theory has been explained by early models, evolving towards ideas about how biodiversity and functional overlap of species create ecological system resilience. For the author, such a theory explains how damage to existing systems can lead to virtually irreparable ecological regime shifts that, in turn, lead to serious damage to safety and security. Through the so-called dilution effect, the author explains the outbreak of the new coronavirus (SARS-CoV-2), which now causes the zoonotic pandemic disease COVID-19, as a symptom of very serious damage to the stability of ecosystems worldwide, clearly illustrating the worldwide social-ecological conflict. The author also shows that biodiversity loss, habitat alteration, and climate change are all interrelated, influenced by, and crucial to the human existence. Therefore, a profound understanding of ecological fundamentals is critical to any strategy involved in maintaining long-term viability between humans and nature. The author claims that the loss of species and habitat over the last decades has increased to such an extent that it has now even affected the climate system. Climate change, in turn, affects resource availability, therefore influencing conflict behavior. From such a perspective, the author tries to provide both a theoretical background and an introduction to possible solutions to human-induced conflict and mass migration.

In Chap. 13, *Tree Species Classification of the Conflict Regions of Sudan Using RapidEye Satellite Imagery*, Deafalla and Csaplovics focus on conflict-affected contexts, where forests not only provide multiple benefits to local communities but also help in environment conservation and climate mitigation. However, if such resources are not sustainably and equitably managed, this will contribute to further environmental degradation and global warming. Focusing on the case of Sudan, the authors report that Non Timber Forest Products (NTFPs), in particular, started to gain considerable importance as inputs for industry, medicine, and food consumption, particularly during extreme events such as drought and famine. Therefore, information about the land cover, particularly the tree species, is crucial on technical, economic, and ecological grounds. Forested areas often cover large expanses making them difficult to analyze using traditional and costly methods of creating forest inventory. With the advent of remote sensing (RS), the scope of effective planning and management of natural resources has considerably widened. In this perspective, the authors claim that vegetation mapping may be a primary requirement for various management and planning activities at the landscape level. Their study focused, therefore, on developing methods of tree species identification in conflict areas using aerial hyperspectral data. In fact, through the use of RS images, researchers in forest resource use have inspired relevant decision and policy-making processes. However, the authors think that NTFPs are not adequately treated since there are no clear mechanisms and extension services to improve their quantity and quality. Furthermore, there is a huge lack of reliable data on the geo-location of trees species, their production and trade, and the number of people involved, which makes it hard to assess the effective contribution of NTFPs to rural livelihoods. Therefore, the authors attempt to map the NTFPs species and investigate the role of these products under specific circumstances like armed conflicts within the perspective of guiding the design of sound NTFPs management.

In Chap. 14, *Securitization of Human-Induced Environmental Conflict: Implications for the Military*, Kruidbos builds on the findings of Chap. 11 in this book and believes that the Anthropocene era, marked by global change and a precipitous slope of species extinction, requires unorthodox measures to counter some shifting dynamics. In addition to the enormous increase in the global human population, the author identifies several specific actors that can be held responsible for such shifts. The ecological effects generated by these actors differ in nature and manifest across different organizational levels. Since these processes are likely to have a significant impact on the future international safety and security landscape, the author assesses the potential of the securitization of human-induced environmental conflict. Such a framework offers the possibility to restructure the existing multidisciplinary and multidimensional deployment of military capabilities with a specific focus on conflict prevention. Within this perspective, the author identifies wildlife crime issues such as illegal wildlife trade and habitat destruction as an argument for the use of special military capacity. This is justified by the fact that such crimes are one of the root causes of future international destabilization that may result in a fluid mosaic landscape of diverse state and non-state actors. To support his reasoning, the author highlights the importance of new capabilities from a multidisciplinary approach, in which diplomacy-development-defense is intertwined with ecology and time (3D-ET) as a means to reduce human-environmental conflict.

On a local scale, Bourhim, in Chap. 15, *Resolution of User Rights Related Conflicts in Collective Rangelands Through Negotiation*, focuses on rangeland-related conflicts by reference to the case of Morocco. Such areas are mostly marginal lands for annual crops due to many factors including drought, low soil fertility, risk of erosion, etc. in addition to their legal status which hinders their development because of intensive farming. Collective rangelands, however, existed in a harmonious state, as there was a balance between the supply of fodder resources and livestock needs. However, the conditions of these areas have been increasingly deteriorating. The author identifies several changes in these areas, including the decline in the roles and importance of traditional institutions, cumbersome administrative procedures in terms of transhumance licenses, the non-compliance with the regulations, illiteracy, the negligence observed by local population and beneficiaries, and the contested legitimacy of the law. These changes contribute to the emergence of major conflicts whose solutions could be found in deal-making or formulation of charters through negotiation with the potential to organize the access to land and the exercise of user rights. The success of these measures largely depends on the approval of sedentary and transhumant parties. Based on empirical evidence, the author claims that customary norms could form the basis for a concerted regulation of collective user rights to ensure the sustainability and resilience of collective rangelands as vulnerable areas.

Focusing on the same context, Mliless and Larouz, in Chap. 16, *Reporting International Conflicts Through the Environmental Discourse,* analyze the Moroccan Sahara conflict as a case study. While environmental justice is commonly the framework guiding the actions and objectives of most international environmental NGOs, the case of the Environmental Justice Organizations Liabilities and Trade's (EJOLT) seems unconventional for the authors since its approach distorts the reality of the Moroccan Sahara. Against the expectations and principles of environmental justice, the EJOLT has targeted many achievements made in the region in economic and social development and preferred taking a pro-Polisario disinvestment stand, claiming that environmental justice should prevail, making a separatist movement benefit from the disputed resources. To demystify such a discourse, which considers the exploitation of the region's resources as a 'reinforcement of the occupation by Morocco', the authors assess the EJOLT's homepage and extract its narratives using the content analysis (CA) method. The findings revealed that the EJOLT uses a biased discourse – mostly distorted and defamatory – that failed to provide the real picture of the population in the region, therefore triggering violence, insecurity, and instability in the region, and by extension serving potentially interested agendas.

The collection of these diverse case studies shows that many developing country regions are already facing climate-related migration and displacement, which are exacerbated by several socio-economic and political drivers. Most of these risks emerge from socio-ecological imbalances and disruptions. Therefore, this volume has been designed in complementarity with another volume recently published under the title, *Social-Ecological Systems in the Era of Risks and Insecurity – Pathways to Viability and Resilience* (Springer 2021), co-edited by Behnassi M., Gupta H., El Haiba M., and Gopichandran R. This volume further analyzes the various risk dynamics and forms of insecurity currently threatening social-ecological systems and how it could lead to conflicts and displacement, reflecting on potential pathways to achieve resilience. This volume particularly focuses on countries from the Global South – especially Africa and Asia – given their specific situations in terms of environmental and climate vulnerability, resource decline, and security implications. But, spillover effects to the developed country regions, which might face large-scale influx of migrants, could be expected and should be accepted in order to deploy appropriate methods for the benefit of all concerned parties.

This volume, therefore, provides a timely contribution to the scientific debate covering the key dynamics of the nexus between climate change, conflict, and displacement from human security and resilience perspectives. The volume's chapters address the theoretical background of the nexus while expressing the need to build more migration-sensitive infrastructure and policies. Empirical cases directly or indirectly relevant to such a nexus are also presented and discussed. In addition, the volume explores various response mechanisms to address the nexus-related challenges, evaluate their effectiveness, identify their implications for the nexus itself, and develop pathways susceptible to guiding future change processes. We do hope that these two volumes contribute to the advancement of science and knowledge while positively impacting the wellbeing of human societies and the balance of ecosystems on the ground through informed and insightful policy-making processes.

Chapter 2
Climate-Conflict-Migration Nexus: An Assessment of Research Trends Based on a Bibliometric Analysis

Neeraj Kumar Singh and Himangana Gupta

Abstract The research on climate-conflict-migration issues started picking up around 2004, 12 years after the United Nations Framework Convention on Climate Change (UNFCCC) was established in 1992. The impacts of climate change started to become more visible and related migration and conflict became issues of social and political concern, which has been captured by researchers worldwide. This chapter examines the trends in research on the climate-conflict-migration nexus (326 publications), and broadly on climate migration and refugee concerns (10,138 publications) from 1990 to 2019, using bibliometric indicators. The study materials were sourced from the Scopus database, a bibliographical and citation database which provides access to quality content. For the 326 nexus publications, we capture the geographic distribution, major subdomains *viz,* consequence/impact, security/insecurity, policy, health, solution, migration and conflict, top authors engaged in this research, and the top funders. This helped us understand and reflect on the major gaps in the research output and how this could shape future debate on climate-related migration. The findings reveal that research publications have been rising exponentially in the last few years, with a noticeable peak in 2018. The cumulative world output on climate migration and conflict issues increased from 724 (7.14%) publications in the 15 years between 1990 and 2004 to 9413 (92.84%) in the succeeding 15 years from 2005 to 2019. The United States of America record the highest number of publications in this research area. So far, there is less research on the solutions side, thus there is a need to reflect on how this fast-emerging concern can be tackled globally, given that each country is exposed to climate risks but with differentiated vulnerability and adaptive capacities. Large-scale cross-border migration can create resource deficits, thus worsening existing conflicts or shaping new ones.

N. K. Singh
Panjab University, Chandigarh, India
e-mail: neeraj.singh@pu.ac.in

H. Gupta (✉)
Sustainable Landscapes and Restoration Program, World Resources Institute (WRI), New Delhi, India
e-mail: gupta@unu.edu

M. Behnassi et al. (eds.), *The Climate-Conflict-Displacement Nexus from a Human Security Perspective*, https://doi.org/10.1007/978-3-030-94144-4_2

Keywords Bibliometrics · Climate change · Conflict · Human security · Migration

1 Introduction

Migration was largely the result of a change in demand and supply of location-specific amenities (Graves, 1980). Urban settlements were themselves driven by climatic conditions (McLeman & Smit, 2006), and a further change in these conditions is driving settlements from one place to another. Climatic conditions are being viewed as one of the important drivers of migration and also conflict (Abel et al., 2019; Reuveny, 2007). As per the Intergovernmental Panel on Climate Change (IPCC), the dynamics of the interaction of mobility with climate change, just like other elements of human security, are complex and so a direct causation cannot be established. But it recognizes that extreme weather events are the most direct pathway from climate change to migration, displacing populations at least in the short term (IPCC, 2014). The IPCC's first assessment report in 1990 pointed out that the gravest effects of climate change are likely to be human migration as millions could be displaced by shoreline erosion, coastal flooding, and severe drought but new areas to which they flee may not have support services enough to accommodate them (IPCC, 1992).

Beine and Parsons (2015) find that environmental factors may indirectly impact migration and that natural disasters beget greater flows of migrants to urban areas. There is a similar pattern of shift to the cities as a result of weather shocks in India (Sedova & Kalkuhl, 2020) and many middle-income countries (Cattaneo & Peri, 2016). But it may still be very difficult to tag a refugee or migrant as a climate migrant (IPCC, 2014). The United Nations High Commissioner for Refugees (UNHCR) does not recognise the word 'climate refugee', mentioning that climate change would first cause internal displacement before it reaches a level where it displaces people across borders (UNHCR, n.d.). There have been discussions on securitisation of climate refugees; in the event of mass movements due to global warming, securitisation may act as a potential strategy to protect climate refugees (Boas, 2015). It is thus increasingly being recognised as an issue of human security (Bettini, 2014) caused by environmental insecurity and the uncertainty of response to the ongoing changes. The impacts of climate change are deemed to directly affect basic human rights like life and livelihood (Koubi et al., 2020).

The causal relationship between climate, conflict, and asylum-seeking has been established just recently (Abel et al., 2019). The debate on climate-induced migration often drew on a security lexicon and was more prevalent in media and policy circuits, but was fiercely contested by critical scholars (Bettini, 2014). However now, researchers around the world are studying these topics and publications on the climate-conflict-migration nexus have exponentially risen since 2008. Although the nexus, especially in the context of disasters, is still not well understood (Brzoska, 2019), the research done so far covers the nexus issues in the context of environmental damage due to forced displacements, bound to also increase conflicts in the future due to climate change and food insecurity (Koubi, 2018; Koubi et al., 2020; Tafere, 2018), socio-economic transformations and conflicts resulting from climatic changes, and food and water insecurity (Endfield et al., 2018; Levy, 2019). The intensity and gravity of such conflicts are yet to be understood.

In this chapter, we focus on the temporal distribution of research output on the climate-conflict-migration nexus which amounts to 326 publications. Such publications had been sourced from the Scopus database, a bibliographical and citation database which provides access to quality content. For the 326 publications covering the nexus areas, we show the geographic distribution, major sub-domains *viz*, consequence/impact, security/insecurity, policy, health, solution, migration and conflict, top authors engaged in this research, and the top funders. In the latter part of this chapter, we reflect on the major gaps in the research output and how this could shape the future debate on climate related migration. The findings show that although the nexus issues and dynamics are being recognized now, there is lesser research on how the related challenges could be tackled in the short run. Each country is exposed to climate risks but with differentiated vulnerability and adaptive capacities. Large-scale cross-border migration can create resource deficits, thus worsening existing conflicts or leading to new ones.

2 Materials and Methods

This research is based on bibliometric indicators. Bibliometrics is a method for measuring, monitoring, and studying scientific output (Campbell et al., 2010; Gumpenberger et al., 2012). It enables the mapping and expansion of knowledge on a particular area of research, showing connections between the main publications, authors, institutions, themes, and other characteristics of the field under study (Gumpenberger et al., 2012; Vogel, 2014). When carrying out the bibliometric analysis of a research field, the first step is to evaluate the available databases, their suitability, and the consequences of using one or another (Sánchez et al., 2017). Scopus and Web of Science enable the extraction of essential data for conducting a bibliometric analysis. The data for the present study was sourced from the Scopus database. Scopus is among the largest curated abstract and citation databases, with a wide global and regional coverage of scientific journals, conference proceedings, and books, while ensuring only the highest quality data are indexed through rigorous content selection and re-evaluation by an independent Content Selection and Advisory Board (Baas et al., 2020).

The data on publications were collected from 1990 to 2019, considering that at the international level, the issue of migration due to climate change was first raised by IPCC in 1990 in its first assessment report (IPCC, 1992). It was, thus, more worthwhile to capture scholars' interest in the subject after 1990. The main search strategy for the global output was formulated, where the keywords 'Climate change' AND 'migration' OR 'refugee' AND 'Climate change' OR 'conflict' AND 'climate change' were placed in the 'Article title, Abstract, Keywords', and further limited the search output so retrieved to period '1990–2019' within 'date range tag'. The search string (shown below) for sourcing research output in the subject yielded 10,155 global publications on climate change and migration from the Scopus database:

(TITLE-ABS-KEY ('Climate change' AND 'migration') OR TITLE-ABS-KEY ('refugee' AND 'climate change') OR TITLE-ABS-KEY ('conflict' AND 'climate change')

However, this search string did not show the nexus between climate change, migration, and conflict. It also included articles that mention plant and animal species migration. Therefore, an additional string was used to specifically reflect on the migration and conflict nexus in the context of climate change:

(TITLE-ABS-KEY ('climate change' AND 'migration' AND 'conflict') OR TITLE-ABS-KEY ('climate change' AND 'refugee' AND 'conflict')

This string yielded just 326 publications, but covered the nexus issues showing how climate change could put human security at risk because of rising conflicts. Therefore, to make the analysis more streamlined from the nexus perspective, it was focused on these 326 publications. This main search strategy was later refined by 'Country/Territory Tag' 'Author Name Tag', 'Source Title tag', 'Affiliation Tag', and 'Funding Sponsor' to get data distribution by countries, author-wise, source title wise, affiliation wise, and funding/sponsor wise.

To further study the specific issues related to the climate-conflict-migration nexus, ten sub-domains were selected, which include consequence/impact, security/insecurity, policy, health, displacement, force, vulnerability, risk, exposure, and extreme events/disaster. These keywords were entered in the search field and 'Article title, Abstract, Keywords' fields were selected, and combined with the main search string using AND operator. For example, to retrieve the total number of publications on the consequence/impact subdomain, the following search string was used:

(TITLE-ABS-KEY ('climate change' AND 'migration' AND 'conflict') OR TITLE-ABS-KEY ('climate change' AND 'refugee' AND 'conflict') AND TITLE-ABS-KEY ('Consequence' OR 'impact') AND PUBYEAR > 1989 AND PUBYEAR < 2020.

3 Results and Analysis

3.1 Overall Publication Output and Growth

Global research in the domain of climate change and migration returned 10,155 publications in 30 years from 1990 to 2019 (Fig. 2.1). The annual output increased from seven publications in 1990 to 1057 in 2019. The cumulative world output in climate change and migration research in 15 years (1990–2004) increased from 724

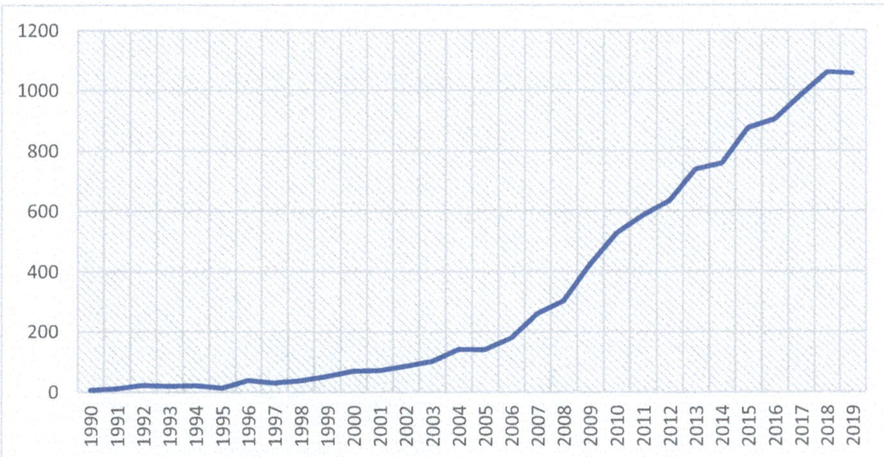

Fig. 2.1 Overall publication output and growth. (Source: Developed by the authors)

(7.12%) to 9431 (92.87%) during the succeeding 15 years from 2005 to 2019. This clearly shows that the interest of the scholars in this particular research area has increased in the last 15 years. This may also mean that the problems and conflicts induced by climatic changes have risen over time. This is alarming in the sense that though we do not yet have proper definitions of 'climate refugee' or we cannot give a particular refugee the 'climate refugee' tag, migration flows triggered by climate risks may continue to rise. The lack of international protection and preparedness of many countries would further have a negative impact. However, the next section shows particularly which issues are more alarming and have been more broadly captured before and after 2005.

3.1.1 Subdomains and Major Topics Covered by the Researchers

From the selected sub-domains (Fig. 2.2), the consequence/impact was the highest consideration by the scholars, especially in recent years. The publications raised rapidly after 2015, citing concerns of climate change impacting health, food availability, and increased disasters. The next three major sub-domains are policy, risk, and security/insecurity. Although there is no defined trend for these three sub-domains, publications continue to rise over time. For 'security', there was a sudden peak in 2010, the reasons for which are difficult to ascertain. The publications on 'security/insecurity' discuss national security implications, protection of climate refugees, enhanced international cooperation, and food security (Beniston, 2010; Janes, 2010; Nicholson, 2010; Schiermeier, 2010). Conversely, Hartmann (2010) raised the question of how portraying climate change as a security threat could militarise the provision of development assistance and distort climate policy. 'Disaster/ extreme events', 'health' and 'displacement' were other important issues captured

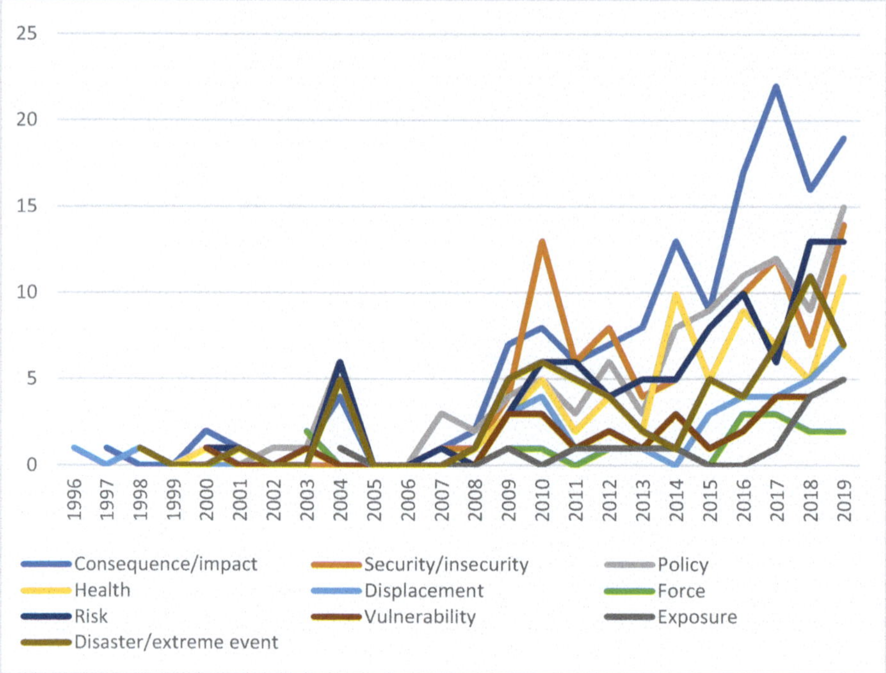

Fig. 2.2 Various sub-domains covered by researchers. (Source: Developed by the authors)

by scholars. Displacement is seen as a manifestation of vulnerability (Beniston, 2010), which may be caused by climate change, land degradation, freshwater availability, and extreme weather events (Naude, 2010). A study on migration in Africa finds that civil and interstate conflicts, lack of democracy, and poverty are the most important drivers of mass population displacements, whereas climate change has just an indirect effect (Bayar & Aral, 2019). Another study finds that climate change will give impetus to already existing factors like political instability (Naude, 2010), including conflicts. Koubi et al. (2020) find that most research on climate-conflict nexus highly focuses on the effects of climate variability and also on natural disasters to a lesser extent. This might lead to interstate as well as low intensity conflict. To solve future crises, integration of climate and conflict policy areas is needed, but there are obstacles when it comes to translating policy into practice (Gyberg & Mobjörk, 2021). While policy is currently the second highest sub-domain being stressed on in climate-conflict-migration publications, 'exposure and vulnerability' was not so much highlighted, which gives an understanding that there is a lack of direct scientific evidence and projections as to how, when and where climate change could cause the maximum impact on migrations, based on differentiated vulnerabilities. This might also be the reason for gaps in policy and practice. Unless policy is backed by science, chance of implementing good practices is a far-fetched dream. According to Mach and Kraan (2021), the climate-conflict dynamics include the inevitability of science–society interactions in climate-related research, the relevance of which will keep increasing for climate–conflict research as global warming intensifies. There is, therefore, a need for more concrete research on risk, vulnerability, exposure and health impacts which would form a stronger base for policies that can avoid conflict and forced displacements.

3.1.2 Prominent Keywords in the Researches

All research papers had three main keywords – climate, conflict and migration as that were the primary criteria for selecting the papers. The keywords 'human', 'conflict', and 'refugee' were the most cited in publications (Fig. 2.3). The publications discuss the increased risk of conflict as a result of climate change, which may also cause forced displacement or migration (Levy, 2019; Webersik, 2012), and could be violent in nature (Koubi, 2018). Changing agricultural production patterns can shape conflict risk, while added impact of climate change can cause social unrest, establishing a strong climate-migration-unrest link (von Uexkull & Buhaug, 2021). For example, migration from Mexico to the United States is reduced with higher levels of rainfall (Puente et al., 2018). Importantly, health figures as 'Health Hazard' and 'Public Health'. It is also the 5th highest studied sub-domain (Fig. 2.2).

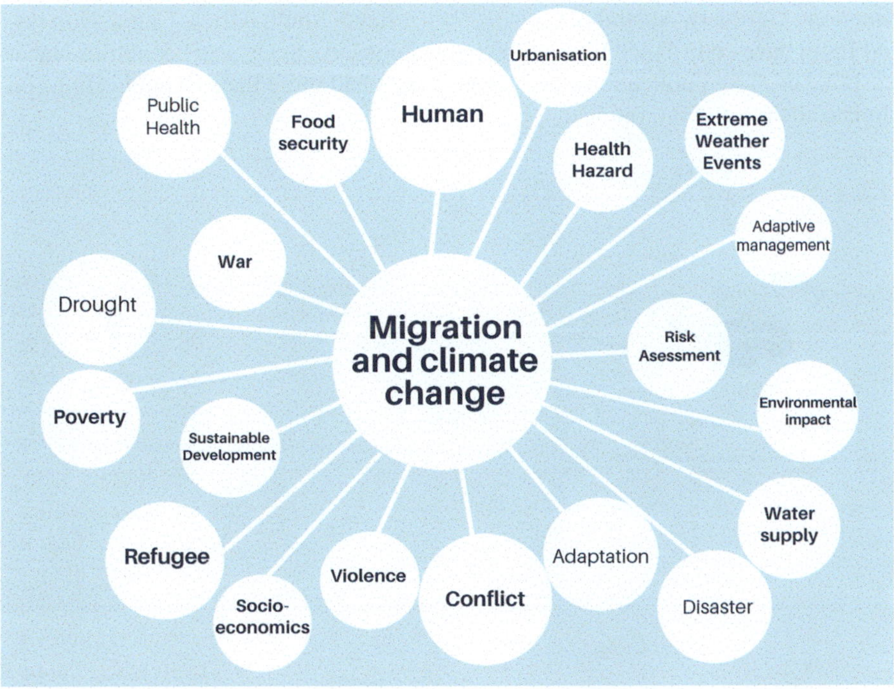

Fig. 2.3 Top keywords figuring in the research outputs. (Source: Developed by the authors)

Displacements could create health hazards for refugees (Bayar & Aral, 2019; Janes, 2010; Levy, 2019), as there might also be a lack of enough health facilities in receiving regions. It may also create security risks for receiving countries (Cattaneo & Peri, 2016). Disasters, extreme weather events, and drought that are similar in nature have been discussed in more than 40 research publications.

All other keywords have almost equal weightage quantitatively based on their reference in the publications. But it is important to note that climate related migration is linked to various subjects ranging from socio-economics to sustainable development. This seems to be extending to and overlapping with a larger chunk of other activities, and that we are able to see only the tip of the iceberg.

3.2 Geographic Distribution of Research Outputs

Researchers from more than 74 countries are engaged in climate-conflict-migration research, which may relate to their national or local circumstances under study or may provide a global perspective (Fig. 2.4). Of these, the top five countries contributed a 71% share to the global output in 30 years from 1990 to 2019. The United States contributed 25% of global share, followed by the United Kingdom, Germany, Australia, and Canada (with 18%, 14%, 9%, and 6%, respectively) from 1990 to 2019. During this 30-year period, the top ten countries (United States, United Kingdom, Germany, Australia, Canada, Switzerland, South Africa, China, Norway, and Italy) have contributed 92% of the publications. As many as 64 countries had at least one or more publications, and ten countries had more than 10 publications on climate change and migration.

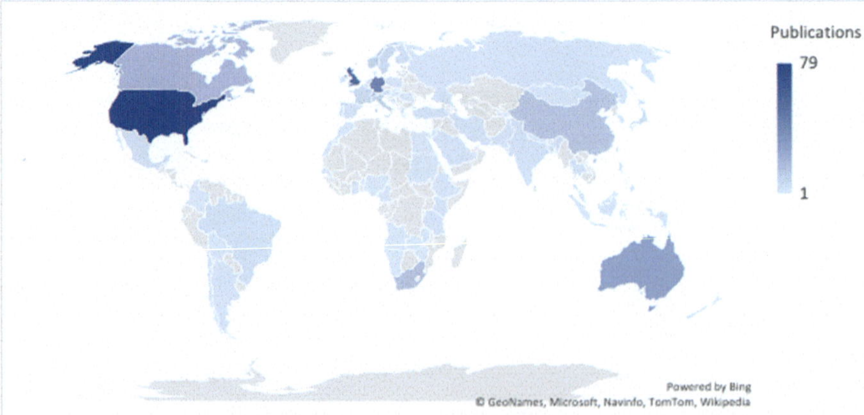

Fig. 2.4 Geographic distribution of publications on climate change and migration. (Source: Developed by the authors)

There were no publications from the top two countries on this topic until 2000; that is why instead of taking 30 years window, only 20 years windows were taken. The cumulative world output in climate change, conflict and migration research for the United States in the first ten years (2000–2010) increased from 14 (17.72%) to 65 (82.27%) and during the succeeding ten years from 2011 to 2019. Similarly, for the United Kingdom, the cumulative world output in the first ten years (2000–2010) increased from 11 (18.33%) to 49 (81.66%) and during the succeeding 10 years from 2011 to 2019. The United States collaborated with 34 countries on this topic (Fig. 2.5a). The top five countries with which the United States collaborated were United Kingdom (8 publications) Germany (6 publications), Canada (4 publications), Belgium (3 publications), and Brazil (3 publications). These five countries contributed a 30% share to the publications. Similarly, the United Kingdom collaborated with 35 countries on this topic (Fig. 2.5b). The top five counties with which the United Kingdom collaborated were Australia (8 publications), USA (8 publications), Germany (7 publications), South Africa (7 publications), and Norway (6 publications). These five countries contributed a 69% share to the publications.

As evident, most developed countries are more active in climate-conflict-migration research. The reason is probably their anticipation of immigration flows from neighbouring countries. There have also been collaborative researches with countries in Africa and Asia, which are most vulnerable. Although there is a pressing need to study the conflict potential as a result of increasing resource-related migrations in the developing countries, they have not been so forthcoming on discussing this topic. It is clear that they are still struggling for more basic needs of survival, but since they may face the maximum impacts of climate-induced migration and conflict, preparedness for this may soon become as essential as basic needs.

3.2.1 Funding Organisations

The bibliometric profile of the top ten funding organisations is given in Fig. 2.6. In all, a total of 74 funding organizations worldwide funded nearly 326 climate change-conflict-migration research publications during 1990–2019. Out of 326 research publications on this topic, 57 were funded by 74 funding agencies, and the remaining 269 were put under the 'undefined' category by the Scopus database.

The top funding organisation is Deutsche Forschungsgemeinschaft (Bonn, Germany), which has funded five publications, followed by the European Commission which has funded four. The top ten funding organisations contributed an 8% share to the global output in 30 years from 1990 to 2019. In fact, there was no funded publication from these organisations until 1999. The total funding from these top ten agencies in the first 15 years (1992–2005) is only two publications, which increased to 21 in the next 15 years (2006–2019). The Deutsche Forschungsgemeinschaft, which is the first top funding agency for this topic, started funding research from the year 2011 and funded only one research publication each on 2011, 2012, 2016, 2017, and 2018. The European Commission, which is the second top funding agency in this area, started funding research in 2009. The

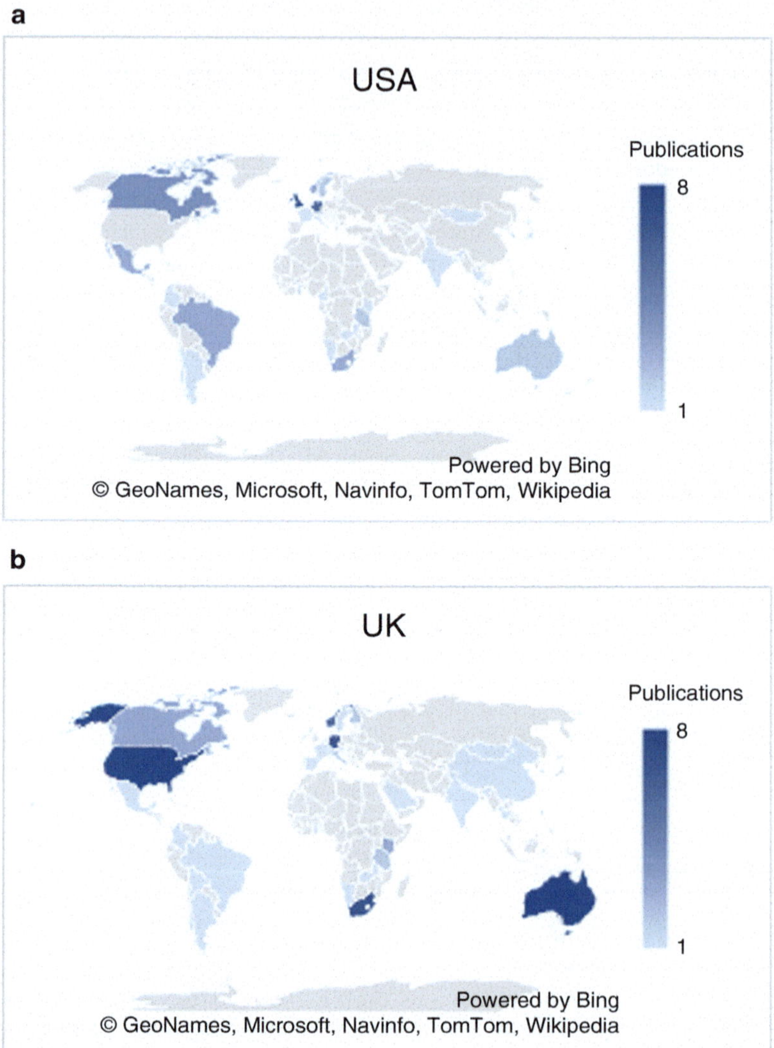

Fig. 2.5 (**a** and **b**) USA and UK Collaboration with other countries on climate change and migration. (Source: Developed by the authors)

highest number of publications were funded in 2018 (14), followed by 12 in 2019. This recent funding, especially from Germany and the European Union, points to the fact that there had been large-scale migration of Palestinians due to the political instability and war in their own country, leading to widespread conflict (BBC, 2017; Rommel, 2017). If a similar situation arises in the face of climate change, or if

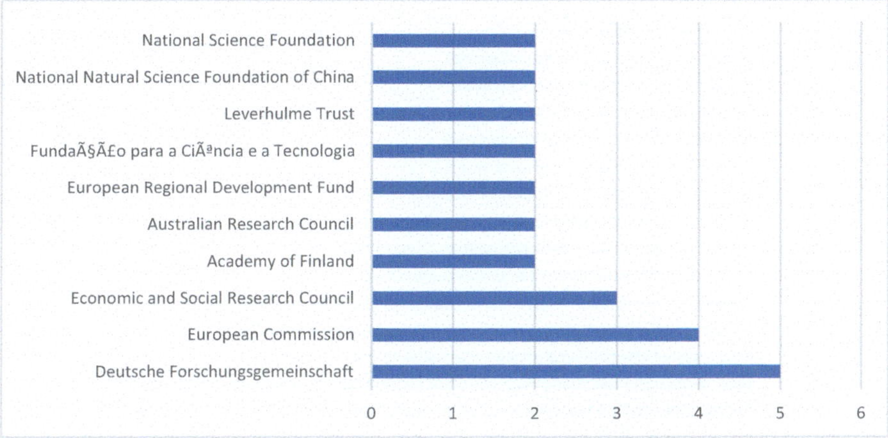

Fig. 2.6 Top ten funding organisations. (Source: Developed by the authors)

climate impacts intensify the existing problems, preparedness would be a must for countries providing asylum, especially to avoid getting into a situation of violent conflict. In addition to financial resources needed to support refugees, additional resources are needed to curb violence, creating further economic instability on part of the supporting country.

3.3 Authors and Journals Engaged in Research on Climate Change and Migration

3.3.1 Authors and Impact of Their Research

As many as 160 authors participated in global research on climate-conflict-migration nexus during the period 1990 to 2019. The productivity in this research area of the top 15 most productive authors varied from 02 to 06 publications. Together, they contributed 60 (18%) global publications during this period. The Scientometric profile of these 15 authors is presented in Fig. 2.7. Germany is home to the highest number of highly cited researchers. Out of a total of 60 research publications by the top 15 authors, 19 works, representing 6% of total publications, were published from Germany. However, the highest cited author is Butler, Colin D. from the Australian National University, Canberra, Australia with 55 citations for his 06 research papers on this topic. For 60 research publications out of the total of 326 published by these 15 authors, *Global Environmental Change*[1] was the preferred journal.

[1] https://www.journals.elsevier.com/global-environmental-change

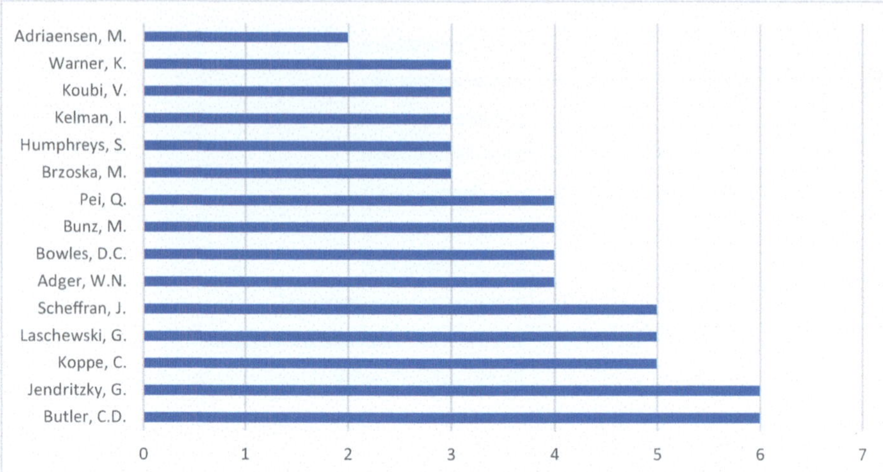

Fig. 2.7 Profile of the top 15 most productive authors on climate change and migration. (Source: Developed by the authors)

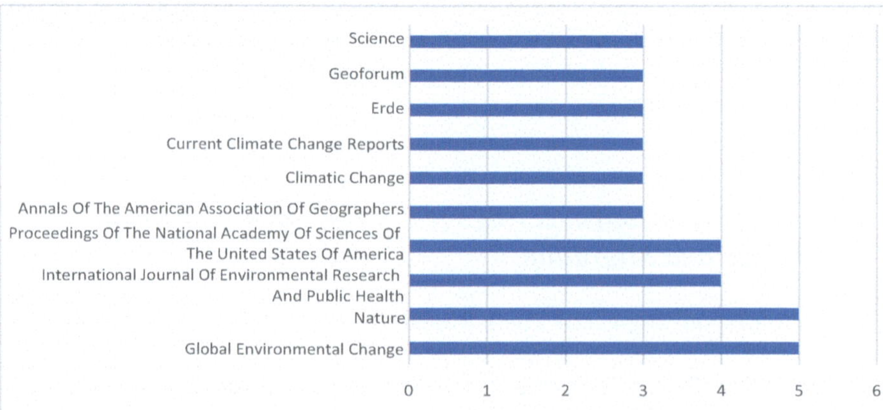

Fig. 2.8 Top ten most productive journals. (Source: Developed by the authors)

3.3.2 Most Productive Journals

The bulk of publications on climate-conflict-migration research (54%, 177) appeared in journals, and the remaining appeared as book chapters (20%, 64), reviews (11%, 36), or books (6.44%, 21). The top ten most productive journals (Fig. 2.8) published 47 papers, and together they account for a 14.41% share of total output during the period 1990 to 2019. The most productive journals in the subject are *Global Environmental Change* (5 papers) followed by *Nature* (5 papers). It is heartening to see that most such research is being covered by top-ranked journals that have a wider reach. Such reach will help bring awareness among the academic community regarding specific challenges their countries might face and further spearhead research.

4 Future Pathway

There are no robust global estimates of future displacements, but there is significant evidence that planning and increased mobility can reduce the human security costs of displacements induced by extreme weather events (IPCC, 2014). The lack of global estimates and research on differentiated vulnerabilities, however, may make it difficult to consider this a part of a national plan or a security issue. Implications of gradual climate change and the potential of conflict are also important research gaps (von Uexkull & Buhaug, 2021). The literature trends point to four major findings: (i) Displacements and conflicts are bound to rise not only because of climate change but as a compounded effect of political instability or poverty, in addition to weather uncertainties; (ii) there is already evidence of migration from rural to urban areas and to other countries, specifically because of climate change, and arising conflicts due to non-acceptance by the local population; (iii) there are minimal or no adaptive mechanisms that are already developed or are in the process of developing to manage the influx or stop the outflow of the population as a result of climate change, and avoid conflict situation; and (iv) there is a lack of research, particularly on how this nexus could pose more complicated questions in the future by increasing vulnerabilities, making it difficult to predict climate-related migration and conflict for the most vulnerable countries.

This shows that in addition to working towards mitigating climate change, there is a need to strengthen adaptive capacities in rural areas of low- and middle-income countries. *Second*, there is a need to assess the resource availability in the context of a country's own rising population and the possibility of influx from neighbouring countries. *Third*, strategies must be developed at the national level to manage displacement and conflict issues arising as a result of climatic changes. *Fourth*, more scientific research to predict future changes is needed in order to identify the regions that could be worst hit from outflux and influx, and whether those regions are prepared enough to tackle the resulting conflicts.

5 Conclusion

This quantitative assessment helped in understanding the trends and inclination of research pertaining to the climate-migration nexus, with its implications for conflict. The study reveals that publications in this research area have been rising exponentially during the last few years, and were the highest in 2018. The cumulative world output on climate migration and conflict issues increased from 724 (7.14%) in 15 years (1990–2004) to 9413 (92.84%) publications during the succeeding 15 years from 2005 to 2019. There is a higher inclination of scholars towards impact, health, security, and disasters. However, there have been lesser studies on the exposure and vulnerability side, thus making it difficult to predict future movements. Also, climate crises are likely to cause larger migration movements towards cities, and also influence the receiving countries which are already suffering from conflicts due to political instability. This will increase the chances of major and minor conflicts in resource deficient or disaster-stricken countries, exacerbating existing conflict situations (e.g. political instability, religious conflicts). Still, there are fewer researches originating from more climate vulnerable countries compared to abundant researches from countries that may actually bear the influx of refugees. The United States and the United Kingdom had the highest number of publications in this area, while the European Union has been the largest funder of such kind of research projects. This shows that the recognition of climate- migration-conflict nexus is still in its infancy in countries that are going to undergo the most challenges. There needs to be a push for higher academic research in this field to help get prepared for an uncertain future. The research needs to reflect on how this fast-emerging concern can be tackled globally given that each country faces serious threats induced by climate change, despite their differentiated adaptive capacity.

Acknowledgements The authors are thankful to Mr. Raj Kumar Gupta for going through this manuscript and suggesting appropriate changes.

References

Abel, G. J., Brottrager, M., Crespo Cuaresma, J., & Muttarak, R. (2019). Climate, conflict and forced migration. *Global Environmental Change, 54*, 239–249. https://doi.org/10.1016/j.gloenvcha.2018.12.003

Baas, J., Schotten, M., Plume, A., Côté, G., & Karimi, R. (2020). Scopus as a curated, high-quality bibliometric data source for academic research in quantitative science studies. *Quantitative Science Studies, 1*(1), 377–386. https://doi.org/10.1162/qss_a_00019

Bayar, M., & Aral, M. M. (2019). An analysis of large-scale forced migration in Africa. *International Journal of Environmental Research and Public Health, 16*(21). https://doi.org/10.3390/ijerph16214210

BBC. (2017, February 26). Germany hate crime: Nearly 10 attacks a day on migrants in 2016. *BBC News*. https://www.bbc.com/news/world-europe-39096833. Accessed 26 Apr 2021

Beine, M., & Parsons, C. (2015). Climatic factors as determinants of international migration. *The Scandinavian Journal of Economics, 117*(2), 723–767. https://doi.org/10.1111/sjoe.12098

Beniston, M. (2010). Climate change and its impacts: Growing stress factors for human societies. *International Review of the Red Cross, 92*(879), 557–568. https://doi.org/10.1017/S1816383110000342

Bettini, G. (2014). Climate migration as an adaption strategy: De-securitizing climate-induced migration or making the unruly governable? *Critical Studies on Security, 2*(2), 180–195. https://doi.org/10.1080/21624887.2014.909225

Boas, I. (2015). *Climate migration and security: Securitisation as a strategy in climate change politics*. Routledge.

Brzoska, M. (2019). Understanding the disaster–migration–violent conflict Nexus in a warming world: The importance of international policy interventions. *Social Sciences, 8*(6), 167. https://doi.org/10.3390/socsci8060167

Campbell, D., Picard-Aitken, M., Côté, G., Caruso, J., Valentim, R., Edmonds, S., et al. (2010). Bibliometrics as a performance measurement tool for research evaluation: The case of research funded by the National Cancer Institute of Canada. *American Journal of Evaluation, 31*(1), 66–83. https://doi.org/10.1177/1098214009354774

Cattaneo, C., & Peri, G. (2016). The migration response to increasing temperatures. *Journal of Development Economics, 122*, 127–146. https://doi.org/10.1016/j.jdeveco.2016.05.004

Endfield, G., Klein, J., Nash, D. J., Pribyl, K., & Hannaford, M. (2018). Climate, conflict and society: Changing responses to weather extremes in nineteenth century Zululand. *Environment and History, 24*(3), 377–401.

Graves, P. E. (1980). Migration and climate. *Journal of Regional Science, 20*(2), 227–237. https://doi.org/10.1111/j.1467-9787.1980.tb00641.x

Gumpenberger, C., Wieland, M., & Gorraiz, J. (2012). Bibliometric practices and activities at the University of Vienna. *Library Management, 33*(3), 174–183. https://doi.org/10.1108/01435121211217199

Gyberg, V. B., & Mobjörk, M. (2021). Integration conundrums: Framing and responding to climate security challenges in development cooperation. *Sustainability, 13*(5), 2582. https://doi.org/10.3390/su13052582

Hartmann, B. (2010). Rethinking climate refugees and climate conflict: Rhetoric, reality and the politics of policy discourse. *Journal of International Development, 22*(2), 233–246. https://doi.org/10.1002/jid.1676

IPCC. (1992). *Climate change: The IPCC 1990 and 1992 Assessments*. World Meteorological Organization and United Nations Environment Programme.

IPCC. (2014). *Climate change 2014: Impacts, adaptation, and vulnerability. Part A: Global and sectoral aspects. Contribution of Working Group II to the fifth assessment report of the Intergovernmental Panel on Climate Change* (C. B. Field, V. R. Barros, D. J. Dokken, K. J. Mach, M. D. Mastrandrea, T. E. Bilir, et al., Eds.). Cambridge University Press.

Janes, C. R. (2010). Failed development and vulnerability to climate change in central Asia: Implications for food security and health. *Asia-Pacific Journal of Public Health, 22*(3 Suppl), 236S–245S. https://doi.org/10.1177/1010539510373008

Koubi, V. (2018). Exploring the relationship between climate change and violent conflict. *Chinese Journal of Population Resources and Environment, 16*(3), 197–202. https://doi.org/10.1080/10042857.2018.1460957

Koubi, V., Behnassi, M., Elia, A., Grillakis, M., & Turhan, E. (2020). Human security. In W. Cramer, J. Guiot, & K. Marini (Eds.), *Climate and environmental change in the Mediterranean basin – Current situation and risks for the future. first mediterranean assessment report*. Union for the Mediterranean, Plan Bleu, UNEP/MAP.

Levy, B. S. (2019). Increasing risks for armed conflict: Climate change, food and water insecurity, and forced displacement. *International Journal of Health Services: Planning, Administration, Evaluation, 49*(4), 682–691. https://doi.org/10.1177/0020731419845249

Mach, K. J., & Kraan, C. M. (2021). Science–policy dimensions of research on climate change and conflict. *Journal of Peace Research, 58*(1), 168–176. https://doi.org/10.1177/0022343320966774

McLeman, R., & Smit, B. (2006). Migration as an adaptation to climate change. *Climatic Change, 76*(1–2), 31–53. https://doi.org/10.1007/s10584-005-9000-7

Naude, W. (2010). Forced migration from Sub-Saharan Africa: The conflict–environment link. In T. Afifi & J. Jäger (Eds.), *Environment, forced migration and social vulnerability* (pp. 43–55). Springer. https://doi.org/10.1007/978-3-642-12416-7_4

Nicholson, F. (2010). Protection and empowerment: Strategies to strengthen refugees' human security. In A. Edwards & C. Ferstman (Eds.), *Human Security and Non-Citizens* (pp. 82–124). Cambridge University Press. https://doi.org/10.1017/CBO9780511808371.004

Puente, G. B., Perez, F., & Gitter, R. J. (2018). The effect of rainfall on migration from Mexico to the United States. *International Migration Review*. https://doi.org/10.1111/imre.12116

Reuveny, R. (2007). Climate change-induced migration and violent conflict. *Political Geography, 26*(6), 656–673. https://doi.org/10.1016/j.polgeo.2007.05.001

Rommel, I. (2017). "We are the people". Refugee-'Crisis,' and the drag-effects of social habitus in German society. *Historical Social Research/Historische Sozialforschung, 42*(4, Volumes per year), 1. https://doi.org/10.12759/HSR.42.2017.4.133-154

Sánchez, A. D., de la Cruz Del Río Rama, M., & García, J. Á. (2017). Bibliometric analysis of publications on wine tourism in the databases Scopus and WoS. *European Research on Management and Business Economics, 23*(1), 8–15. https://doi.org/10.1016/j.iedeen.2016.02.001

Schiermeier, Q. (2010). "Climate wars" claims disputed. *Nature, 467*(7312), 145–145. https://doi.org/10.1038/467145a

Sedova, B., & Kalkuhl, M. (2020). Who are the climate migrants and where do they go? Evidence from rural India. *World Development, 129*, 104848. https://doi.org/10.1016/j.worlddev.2019.104848

Tafere, M. (2018). Forced displacements and the environment: Its place in national and international climate agenda. *Journal of Environmental Management, 224*, 191–201. https://doi.org/10.1016/j.jenvman.2018.07.063

UNHCR. (n.d.). *Climate change and disaster displacement*. UNHCR – The UN Refugee Agency. https://www.unhcr.org/asia/climate-change-and-disasters.html. Accessed 23 Nov 2020.

Vogel, R. (2014). What happened to the public organization? A bibliometric analysis of public administration and organization studies. *The American Review of Public Administration, 44*(4), 383–408. https://doi.org/10.1177/0275074012470867

von Uexkull, N., & Buhaug, H. (2021). Security implications of climate change: A decade of scientific progress. *Journal of Peace Research, 58*(1), 3–17. https://doi.org/10.1177/0022343320984210

Webersik, C. (2012). Climate-induced migration and conflict: What are the links? In K. Fog Olwig & K. Hastrup (Eds.), *Climate change and human mobility: Challenges to the Social Sciences* (pp. 147–166). Cambridge University Press. https://doi.org/10.1017/CBO9781139235815.010

Chapter 3
Climate Change Response Mechanisms and the Risk of Increasing Vulnerability: Conceptual Background and Pathways of Change

Mohamed Behnassi, Gopichandran Ramachandran, and Gitanjali Nain Gill

Abstract Climate-induced impacts and disasters are increasingly affecting social-ecological systems worldwide by generating or amplifying risks and vulnerabilities. Yet, climate response mechanisms too may generate similar implications, which may increase the likelihood of conflicts and human displacements. Focusing on adaptation, as one of the mainstays of climate change governance, this chapter aims to highlight that related policies and actions may, in some instances, generate unexpected outcomes and risks which erode resilience instead of building it. To substantiate this, the analysis investigates the drivers, forms, and manifestations of maladaptation by reference to existing theoretical and empirical research. Accordingly, the maladaptation and migration nexus is analyzed from an inclusive governance and science-based policy making perspective. Alternative approaches to address maladaptation to ensure resilience and sustainability through adaptation strategies are identified and discussed as well. Throughout the chapter, the analysis prescribes the needed guidelines to be considered while elaborating, implementing, and evaluating adaptation actions to ensure the achievement of expected outcomes and to avoid potential tradeoffs and negative implications. Results show that some adaptation strategies and interventions in many contexts have undermined the

---

M. Behnassi (✉)
International Law and Politics of Environment and Human Security, College of Law of Agadir, Ibn Zohr University, Agadir, Morocco

Center for Environment, Human Security & Governance (CERES), Agadir, Morocco
e-mail: m.behnassi@uiz.ac.ma

G. Ramachandran
Environment & Sustainability, NTPC School of Business, Noida, Uttar Pradesh, India
e-mail: gopichandran@nsb.ac.in

G. N. Gill
Professor of Environmental Law, Faculty of Business and Law, Northumbria Law School, Northumbria University, Upon Tyne, UK
e-mail: gita.gill@northumbria.ac.uk

M. Behnassi et al. (eds.), *The Climate-Conflict-Displacement Nexus from a Human Security Perspective*, https://doi.org/10.1007/978-3-030-94144-4_3

balance of ecosystems and the resilience of local communities, creating additional conditions for increased human insecurity, including the compulsion to migrate.

Keywords Adaptation · Climate change · Maladaptation · Mitigation · Social-ecological vulnerability

1 Introduction

The Intergovernmental Panel on Climate Change (IPCC) states that a continued global warming of 1.5–2 °C will be exceeded during the twenty-first century unless strong reductions in carbon dioxide and other greenhouse gas emissions take place in the coming decades (IPCC, 2021). Empirical evidences indicate that climate change related impacts and disasters have the potential to compromise development achievements and severely affect social-ecological systems by triggering or amplifying risks and vulnerabilities. These, in turn, increase the likelihood of tension, conflicts, and human displacement in vulnerable regions.

The global crisis caused by the COVID-19 pandemic and the various climate-induced disasters and hazards have severely impacted all aspects of life. The pandemic has progressed *pari passu* with global scale ecosystems perturbations: oceans continuing to warm, glaciers melting, and many ecosystems continuing to collapse. These ongoing risk dynamics are challenging and urge societies at all scales to rethink development and security approaches. It is equally important to look at recovery through the lens of social, environmental, and climate justice, sustainability, resilience, and human security. This calls for a paradigm change inspired by evidence-based governance for common good.

Countries are increasingly faced with the challenge to make immediate and significant policy choices and take unprecedented action to protect lives and avoid catastrophic societal consequences. A coherent response to climate change through structural changes in production, consumption, and mobility will strengthen countries' abilities to prepare, respond, and recover. Interconnected adaptation strategies that leave no one behind while restoring ecosystems, would further address and inculcate resilient and sustainable recovery. These include efforts to enhance innovative climate financing, food and water security, disaster mitigation and preparedness, climate-resilient livelihood development, nature-based solutions to development, ecosystem conservation, and climate-conscious governance. Decoupling development pathways from carbon emissions, overuse of natural resources, and waste generation are extremely important to ensure a resilient and sustainable recovery.

Nevertheless, many climate response mechanisms are not sufficiently shaped to reduce emerging risk dynamics and their implications. A growing trend in literature supports the assumption, based on empirical evidence, that inappropriately designed adaptation, mitigation, climate finance, science and technology transfer, may generate more social and environmental vulnerability and human insecurity with the potential to trigger more conflicts and displacements. Some interventions are undermining the balance of ecosystems and the resilience of local and indigenous communities – 'resilience grabbing' – while depriving such communities of their rights to land, water, food, and biodiversity on which their livelihoods are based. For example, voluntary or forced migration or displacement of affected communities are amongst the most disruptive of such externalities. This is mainly due to interconnected routes and the scale of impact felt across geographies and ecosystems, which makes the world ever smaller. The United Nations High Commissioner for Refugees (UNHCR) in its 2021 report identifies climate change as one of the drivers of refugee movements (UNHCR, 2021). The UNHCR's perspective to define "persons displaced in the context of disasters and climate change" reinforces the need to address issues especially faced by disproportionately vulnerable communities. Migration may be viewed either as a failure of adaptation or as an adaptive strategy to tackle vulnerability. Such elements, as exposure to the challenge and system sensitivity to perceive, respond and develop adaptive capacities to sustain transitions, determine outcomes. This is also true of displaced people at the site of climate emergency and related hotspots that experience extreme stresses.

This chapter assesses climate change strategies – with a focus on adaptation – by identifying the mechanisms through which negative effects can unfold in climate actions, especially maladaptation. The objective is to highlight that these strategies, if not appropriately elaborated, implemented, and evaluated, may generate new vulnerability risks that can trigger both conflicts and displacements. The chapter also presents guidelines that should be considered while elaborating, implementing, and evaluating climate response mechanisms to ensure expected outcomes and to avoid potential trade-offs or negative outcomes.

2 When Adaptation Increases Vulnerability: The Risk of Maladaptation

Adaptation is one of the mainstays of climate change governance. Its importance is increasingly recognized not only by high-level stakeholders including governmental bodies and the scientific community, but also by those citizens who perceive climate change as one of the most significant development and security challenges facing the world. Adaptation may take multiple forms in most countries impacted by climate change. It is designed to empower and increase the climate resilience of vulnerable groups and ensure food, water, livelihood and health security, biodiversity conservation, and increased disaster risk reduction and management. Recently, however, there are debates about the effectiveness of adaptation and the climate adaptation community's approach. Several large-scale systematic literature reviews have analyzed studies of climate change adaptation and recognized that there is a glaring lack of engagement regarding the potential for adaptation measures to exacerbate and/or redistribute risk and vulnerability (Lang, 2019).

According to Magnan et al. (2016), there is a consensus that concerted efforts are needed on priority to assist social-ecological systems threatened by ongoing and partly irreversible climate change across local to global scales. The problem is, however, that not enough is currently understood about the forms and functions of initiatives that enhance the ability of such systems to adapt on a sustainable basis. In other terms, there is still an insufficient understanding of the appropriate initiatives that enhance the ability of such systems to adapt in the long run, and those that have little or no positive impacts or negative impacts. As a result, although these efforts can potentially foster adaptation in the short term, there is a risk that they affect territories', sectors', and people's long-term capacities and opportunities to cope with and manage the impacts of climate change. Such kind of adaptation interventions tend to backfire resulting in increased vulnerability, thereby making maladaptation an area of concern.

Lang (2019) highlights the fact that the concept of 'maladaptation' can be traced to Scheraga and Grambsch (1998) who focused on adverse outcomes due to adaptation measures. It was also mentioned in the IPCC Third Assessment (2001), although at the time, it was specifically deemed as 'inadvertent' negative outcomes. However, as Magnan et al. (2016) claim, despite growing adaptation action worldwide, there appears to be little concern about the risk of maladaptation. There is a possibility that initiatives taken in the name of adaptation might not only waste financial resources, but could also aggravate the consequences of one-off and gradual climate-related changes.

To counter such trends, there is a growing literature that provides frameworks for understanding, undertaking, and assessing mal/adaptation actions. For Barnett and O'Neill (2013:88) maladaptation is perceived as the negative externality of an "action taken ostensibly to avoid or reduce vulnerability to climate change that impacts adversely on, or increases the vulnerability of other systems, sectors or social groups". In the same vein, Juhola et al. (2016:139) define maladaptation as the "result of an intentional adaptation policy or measure directly increasing vulnerability for the targeted and/or external actor(s), and/or eroding preconditions for sustainable development by indirectly increasing society's vulnerability". Even the IPCC in its Fifth Assessment Report (2014:1769) changed its perception to maladaptation as "actions that may lead to increased risk of adverse climate-related outcomes, increased vulnerability to climate change, or diminished welfare, now or in the future". For Magnan et al. (2016), maladaptation describes an action that results in an undesirable and unintended outcome(s). This leads to an increased vulnerability that the action was meant to reduce. Schipper (2020) categorizes maladaptation into three types – infrastructural, institutional, and behavioral – and offers relevant examples. Infrastructural maladaptation includes cases where for example the building of a seawall in coastal areas in Fiji, has made people vulnerable to climate hazards due to changes in sediment deposits and impacting the marine ecosystem. Institutional maladaptation includes situations of reduction in knowledge base, social capital, and risk awareness to mitigate uncertainty. This is a consequence, for instance, of the non-engagement of farmers within their networks due to reliance on agricultural climate insurance. Behavioral maladaptation encompasses changes in attitudes or behaviors. For instance, in northern Ghana, farmers migrated from rural areas due to the lack of rainfall, resulting in labor shortage in a good farming season.

According to Lang (2019), traditional conceptualizations of adaptation typically address adaptation within the status quo. Accordingly, adaptation is framed as complicated, short-term, and apolitical, with a greater control and/or understanding of the outcome. Moreover, the dominance of incremental, linear adaptation and technocratic solutions has been a growing concern for many researchers and practitioners. Critics argue that these adaptation approaches fail to address the root causes of differential vulnerability. These are largely social and political in nature. Schipper (2020) argues that these root causes perpetrate systemic inequalities often driven by historical power struggles and imbalances. Citing Bradshaw and Fordham's study (2013), Schipper states that vulnerability among Bangladeshi women and girls is greater than men's during disasters such as floods. Escaping to higher grounds without a male partner is not permitted due to unequal, gendered power relations, thereby increasing vulnerability to the negative effects of the hazard.

Lang (2019) asserts that maladaptation as a concept has existed in the climate change sphere for two decades. Yet the challenge is the lack of consensus on the exact meaning of maladaptation; notwithstanding suggestions to tackle it (IPCC, 2014). Magnan et al. (2016) remarked that sometimes there is, interestingly, only a fine line between success and failure vis-à-vis adaptation. Lang argues that this ambiguity is partly due to the difficulty to assess what is meant by adaptation, e.g. what does effective adaptation look like? What does the risk of maladaptation mean in reality? And, by that same metric, what does unsuccessful adaptation look like and is unsuccessful adaptation equivalent to maladaptation?

These questions are being addressed by those researchers who attempt to define the maladaptation concept. Lang (2019) believes there is a general agreement on five major aspects of maladaptation; that, it results from intentional adaptation policy and decisions; there are explicitly negative consequences; it consists of a spatial element, known as spatial spillovers – maladaptation does not necessarily occur in the geographic space or within the targeted group since it can extend social and geographic boundaries; and it consists of a temporal element – adaptation actions taken today can be maladaptive in the future.

Barnett and O'Neill (2013) developed a framework of analysis of maladaptation. First, they identified five manifestations of maladaptation, namely: increasing emissions of greenhouse gas; disproportionately burdening the most vulnerable; high opportunity costs; reduced incentives to adapt; and path dependency. They further argue that different adaptation actions carry their own unique risks of maladaptive outcomes. Those that seek to increase adaptive capacity are the least likely to be maladaptive. The ones who seek to decrease sensitivity carry medium risk, and those that seek to reduce exposure carry the greatest risk to be maladaptive.

Eriksen et al. (2021) have critically reviewed the outcomes of internationally-funded interventions aimed at climate change adaptation and vulnerability reduction. The authors demonstrated that some interventions inadvertently drive maladaptive outcomes. This is because they reinforce, redistribute or create new sources of vulnerability through four mechanisms. They relate to a shallow understanding of the vulnerability context, inequitable stakeholder participation in both design and implementation, a retrofitting of adaptation into existing development agendas, and a lack of critical engagement with how 'adaptation success' is defined. In addition, an important lesson from past adaptation interventions is that within current adaptation development paradigms, inequitable terms of engagement with 'vulnerable' populations are reproduced and the multi-scalar processes driving vulnerability remain largely ignored.

Climate change poses a significant threat to the livelihood and security of poor and natural-resource-dependent communities. Vulnerable people worldwide, especially in the Global South, are exposed to greater climate change risks mainly due to their higher level of poverty and lack of human, institutional, economic, technical, and financial capacity to respond to its impacts (IPCC, 2014; Mahapatra, 2019; Robinson, 2020). Therefore, although the impacts of climate change encompass a set of global phenomena, they must be analyzed and managed through local realities and manifestations. In other words, though the hazard seems global in scale when risk is considered, it is local when resilience is taken into consideration. In such a perspective, Rahman and Hickey (2019) indicate that climate vulnerability represents a highly complex public policy challenge for governments due to its interaction with diverse social, political, economic, and ecological factors across scales that result in context-specific interpretations of climate vulnerability. The policy challenge is further exacerbated when rural livelihoods depend on multiple land use practices within shared social-ecological systems as adaptation actions related to one practice affects the others. In such cases, it becomes likely that national and regional-level adaptation plans will result in maladaptive trajectories if local context and properties are not carefully considered.

To demonstrate the importance of the stated issue to public policy, Rahman and Hickey (2019) used the case of climate change adaptation planning in Bangladesh. In this context, the authors claim that beyond policy-driven adaptation actions, it is also important to recognize that the climate-affected communities of Bangladesh are developing their own local innovation-based adaptation actions. Although these innovations are often challenged by local resource availability and politics of access, learning from and promoting these 'grassroots' innovations have the potential to avoid government policy-driven maladaptation. From an empirical perspective, the authors focused on the northeastern floodplain region, an area dominated by wetland ecosystems, high climate vulnerability, and diverse and complex land use practices. Community-level adaptation actions were undertaken in contrast to national-level adaptation planning strategies and actions. They showed how the planned adaptation actions taken by the government may actually end up being maladaptive, either by shifting or rebounding vulnerability. They concluded that government adaptation planning would benefit from a greater focus on learning and scrutinizing the autonomous adaptation of communities to climate stress before making significant resource allocation decisions.

3 Managing the Risk of Maladaptation: Alternative Approaches

Several conceptualizations and pathways have been developed to address maladaptation. For Eriksen et al. (2021), unless adaptation is rethought, transformational interventions may even worsen vulnerability instead of tackling it. Therefore, instead of designing projects to change the practices of marginalized populations, learning processes within organizations and with marginalized populations must be placed at the centre of adaptation objectives. The authors ask if scholarship and practice should take a post-adaptation turn akin to post-development, by seeking a pluralism of ideas about adaptation. This could be along with critically interrogating how these ideas form part of the politics of adaptation and potentially the processes (re)producing vulnerability. The authors caution that unless the politics of framing and of scale are explicitly addressed, transformational interventions risk having even more adverse effects on marginalized populations than current adaptation. In order for adaptation interventions to contribute to equitable and sustainable vulnerability reduction, and to measure progress in this endeavor, the authors claim that structures around financing, planning, implementation, monitoring and evaluation of interventions that frame climate intervention processes may need to be redesigned.

Magnan et al. (2016) argue, in addition, that starting with the intention to avoid mistakes and not lock-in detrimental effects of adaptation-labeled initiatives is a first key step to the wider process of adapting to climate variability and change. Therefore, the authors advocate an ex-ante anticipation of risk of maladaptation as a priority for decision makers and stakeholders at all scales. They affirm that a challenge for future research consists in developing context-specific guidelines that will allow funding bodies to decide on support for adaptation with a low risk of maladaptation.

In the same vein, to address the risk of maladaptation and advance the field by offering an improved conceptual understanding and more practice-oriented insights, Magnan (2014) and Magnan et al. (2016) highlight four main dimensions to assess the risk of maladaptation – that is, process, multiple drivers, temporal scales, and spatial scales. They also present three examples of frameworks – the *Pathways*, the *Precautionary*, and the *Assessment* frameworks – to capture the risk of maladaptation on the ground as part of a holistic planning agenda (Fig. 3.1).

The framework of transformational adaptation is equally important. Barrott (2015) indicates there is a growing tendency within the climate change adaptation research community to discuss adaptation using the language of transformation. This reflects a sense that the current status quo will not secure a sustainable future, especially in light of the insufficient progress to mitigate the causes of anthropogenic climate change. Transformational adaptation is viewed as a paradigmatic shift that addresses complex problems, power imbalances and social justice, and takes a longer view of adaptation. Accordingly, Lang (2019) claims that thinking in terms of transformative adaptation can help shift vulnerability from being situated in the context of climate to being understood in the context of societies and political

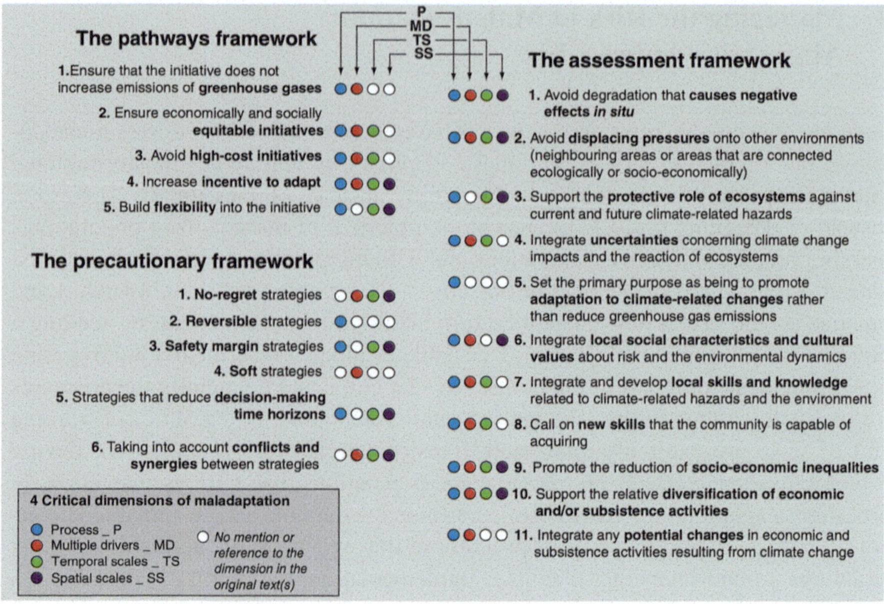

Fig. 3.1 The four critical dimensions of maladaptation in the *Pathways, Precautionary*, and *Assessment* frameworks. (Source: Magnan et al. 2016)

economies. Moreover, it would raise questions of whose knowledge counts in decision-making and what mechanisms define the subjects of adaptation.

For O'Brien (2012), transformational adaptation lacks an established definition because it means different things to different people or groups, and it is not clear what needs to be transformed and why, whose interests these transformations service, and what will be the consequences. However, Lang (2019) thinks that some existing definitions of transformative adaptation may provide some underpinnings. In such a perspective, Revi et al. (2014), quoted in Lonsdale et al. (2015), consider transformative adaptation actions which are recognized for their potential to address root causes of poverty and failures in sustainable development, including the need for rapid progress on mitigation. For the World Resources Institute (2018), transformative adaptation approaches are fundamental, systemic changes that help protect development gains, maximize resilience investments, and reduce the escalating risk of conflict from climate change.

From the same angle, Lang (2019) asked how transformative adaptation can occur – can it be planned, or is it a strictly organic process? Pelling et al. (2015:3) argue that transformation can result from "adaptive actions that have the capacity to shift existing systems (and their component structures, institutions, and actor positions) onto alternative development pathways, even before the limits of existing adaptation choices are met". That is, the system doesn't have to fail in order for transformational adaptation to occur. Others emphasize 'windows of opportunity' which trigger transformation at particular moments.

In this perspective, Lonsdale et al. (2015) highlight three areas that should be strengthened to increase the potential and capacity for transformational adaptation (Fig. 3.2). The *first* area pertains to the *capacity for systemic inquiry*. Indeed, the interconnections between players in any given system are complex, and poorly designed attempts to make changes can have negative unintended consequences or introduce new failures or inequalities. Supporting transformational adaptation requires, therefore, the capacity to inquire systematically. This means to inquire into a system of interest, to understand the history of that system (e.g. around sources of control, legitimacy, and knowledge), and challenge the assumptions that underpin existing structures and ways of doing things. Certainly, reproducing 'solutions' without assessing what holds the current system in place may result in simply reinforcing existing failures and inequality dynamics. By developing a more detailed sense of the system as it currently exists, we can design interventions and feedback mechanisms that enable us to learn as ideas for system improvements are put into practice

The *second* area relates to *leadership for transformation*. People in positions of leadership make choices between investing time and resources in day-to-day maintenance activities and those focused on coping in the future. This requires the capacity to shift between the details of current activities while maintaining an awareness of the bigger picture, and being conscious of long-term goals when making short-term plans. In this setting, the capacity to 'cultivate uncertainty' about the situation

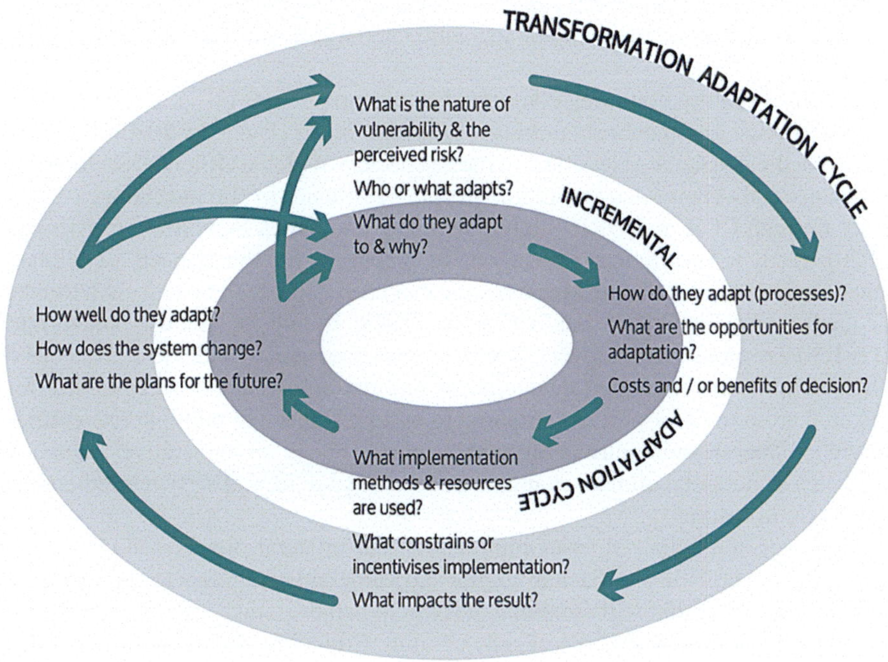

Fig. 3.2 Transformative adaptation cycle. (Source: Lonsdale et al. 2015)

of interest may also be essential to avoid abrupt assumptions about a situation and to limit the exploration of its complexity and the range of possible improvements that could be tried. The leader also plays an important role in articulating the issue and encouraging system-wide (or even inter-system) participation.

The *third* area is about *learning from practice* which implies: being present and noticing things, cultivating uncertainty, and learning from experience; creating opportunities to reframe understanding based on practice; and informing the development of new approaches. To achieve such objectives, the learning process should be facilitated to ensure it provides sufficient challenge – through the incorporation of dissonant information or opposing views – and support – to encourage wide participation by including seldom heard and disparate voices.

Lang (2019) claims that the scale and complexity of adaptation decisions mean that maladaptation is likely to occur and will occur in the future, despite efforts to avoid it. Therefore, including assessments that prioritize *learning* in monitoring and evaluation of decision-making is needed. Certainly, the only way for practitioners to enable better and more effective adaptation decision-making is by learning and sharing the knowledge of what does not work.

Adaptation actions from a strong multi-sectoral disaster risk management (DRR) perspective are equally important in the landscape of resilience management. These relate to land use, building codes, agriculture, environmental protection, energy, water, food, health, and livelihood security, education, poverty reduction, and climate change adaptation over the medium term. This also calls for a simultaneous focus on legal and institutional capacities for integrating climate change adaptation (CCA), and innovative financing of adaptation at the sub-national and local government levels.

In 2009, wealthy countries committed to mobilize $100 bn in annual climate finance to assist low-income countries to address climate change by 2020. However, a report dedicated to the assessment of the last decade of global Official Development Assistance (ODA) invested in building people's resilience to climate change – especially through CCA and DRR- highlights many anomalies. According to Alcayna (2020), there is insufficient investment in preparing for the impacts of climate change, especially for countries and people that need it most. To reverse such trends, the author recommended that: CCA and DRR should be mainstreamed into COVID-19 response and recovery. Recovery packages should endeavour to advance climate-smart, risk-informed development and donors should screen funding for potential areas to 'dual-purpose' funding to build resilience to multiple risk entities through bilateral and multilateral sources too. This is an imperative when the dynamics of potential climate change-induced risks are increasingly unpredictable, and severe in nature.

Moreover, the author stresses multilateral and bilateral donors need to take a long-term and holistic approach to fragile and vulnerable countries to support them with adaptation as this will simultaneously help achieve other development goals. Additionally, donors should specifically commit to doubling the assistance provided to the most climate vulnerable, Least Developed Countries (LDCs) by 2025. Financial mechanisms need to be reformed as well as strengthening the decision-making power of affected people, particularly marginalized groups, and reporting should include improvements in how donors track 'mainstreaming' of climate finance and the quality of such interventions.

4 The Maladaptation-Migration Nexus: Dynamics and Governance

One of the important consequences of maladaptation is climate-induced migration. Questions about opportunities to infuse the precautionary approach and assessments of potential risks of maladaptation are raised with the call for a much-needed body of empirical evidences to guide mitigation and adaptation plans. In this perspective, the Hugo Observatory Habitable project[1] is designed to interpret location-specific migration related drivers, outcomes, and mitigation strategies. This will also collate much-needed empirical evidences on the individual, synergistic, and antagonistic impacts of drivers and the scope for preventive management. The Council of Europe (2021) in its report by the Committee on Migration, Refugees and Displaced Persons acknowledges that the impacts of climate change tend to spark conflicts and, therefore, the felt need to combat disruptions to ameliorate hardships. Such important public policy aspects as human rights and rights-based protection should enable the return to normalcy. The Migration Information Source presents a good compilation of events and outcomes in this context[2].

Sector- and location-specific adaptation strategies are gaining significant attention in India, for instance, through the State Action Plans on Climate Change (Dinshaw et al., 2018). Singh et al. (2021) recently deliberated on adaptation effectiveness based on 11 guiding principles. One of them is to avoid maladaptation. Importantly, they acknowledge difficulties in defining adaptive capacities and resultant adaptation and, therefore, deeper theorization of proposed correlates. Vinke et al. (2020) present an interesting typology of the reactive and proactive dynamics of migration; the latter being governed and facilitated. Decisions to migrate are also guided by an assessment of resources that may be available at the alternative destination. In most cases, affected peoples are likely short of resources to adapt to a relatively more hostile environment. The effects cascade to reduce time at hand to recover.

Limited natural resources, such as drinking water, become scarcer in many places that host refugees. Crops and livestock struggle to survive under environmental and climate related stresses that, in turn, affect livelihoods; with humongous mitigation implications. The latter is, therefore, a threat multiplier that tends to exacerbate prevailing tensions, which may trigger conflicts. The 2018 UNHCR

---

[1] https://www.hugo.uliege.be/cms/c_5744502/en/hugo-habitable

[2] https://www.migrationpolicy.org/programs/migration-information-source/special-issue-climate-change-and-migration

Report, Part II Global Compact on Refugees[3], highlights the need to enhance refugee self-reliance that enable better access to solutions, safety and dignity aligned with the Comprehensive Refugee Response Framework (UNHCR, 2018). The Global Risks Report 2020 (World Economic Forum, 2020) presented the likelihood-impacts matrix. Involuntary migration is an inevitable outcome.

The recent Climate-fragility risk brief on Sudan defines the complex underpinnings of conflicts attributable to climate change and related perturbations (Foong et al., 2020). Competition for scarce resources, displacement, and livelihood challenges that accentuate maladaptation and inadequate governance, reinforce challenges. Community stabilization and peace building are as important as ecosystem-level protection strategies. Schaer (2015) presents a case of maladaptation in Senegal. This is with reference to challenges that diffuse coping strategies of flood-affected and, therefore, call for enhancement of basic services and infrastructure as a forerunner to longer-term resilience.

UNHCR et al. (2020) highlighted that a significant majority of forcibly displaced, especially lodged in rural settlements, do not have adequate access to energy. This is reportedly true because energy is not prioritized in humanitarian assistance. The UNHCR Global Strategy for Sustainable Energy 2019–2024 is probably the best framework to address this issue. It cites the experiences of some countries on offsetting impacts. This integrates access to power, water, and enhanced health services.

Podesta (2019) argued that the migration-climate nexus appears to be growing. This is evident through challenges faced by internally displaced people, and others in comparable stress circumstances across the globe. The framework of the Warsaw International Mechanism for Loss and Damage Associated with Climate Change (WIM) and The Global Compact for Safe, Orderly, and Regular Migration are useful templates for preventive strategies. They are however yet to be mainstreamed at scales for any significant outcome. Issues of legal protection to climate refugees are yet to be resolved.

One of the earliest discussions on the role of technology in adaptation was presented by the United Nations Framework Convention on Climate Change (UNFCCC) in 2006. The cited reference set the context about the numbers of people expected to be displaced and defined iterative processes to infuse adaptation planning with respect to coastal areas, water resources, infrastructure, public health, socio-economic dynamics, and physical integrity of systems. The International Organization for Migration (IOM) (2008) defined the interplay of government policies and community-level resilience preparedness as important non-climate drivers that influence the degree of vulnerability, centred on region-specific population, poverty, and governance dynamics. These should be addressed with commensurate speed and scale further determined by the ability to manage future flows of climate migrants; including the opportunity to develop and implement a national adaptation programme of action. The latter is dependent on the quality and quantity of baseline

[3] https://www.unhcr.org/climate-change-and-disasters.html

data at hand and limited by the ability to assess perturbations including emissions and their impacts in the future. These are needed to tackle temporary and long-term migration with implications for pressure on urban infrastructure and services, economic growth, conflicts that worsen health challenges, education, and related social indicators among the affected themselves.

FAO (2011) highlighted six important facets of adaptation planning concerning food produced and not consumed as an externality that compounds food security issues. This includes implications for embedded energy, especially in harvest, storage, and distribution, in addition to other elements of the supply chain. Financial mechanisms to support the deployment of energy efficiency and renewable energy are essential for this purpose due to its multi/cross-sectoral landscape. These, in turn, could strengthen conservation agriculture, soil health to sustain productivity of locally adapted crops, and cropping systems bordering in precision farming and improved water management.

McLeman and Hunter (2010) presented a set of cases to differentiate the phenomenon and outcomes of migration decisions and processes across continents. The temporal and spatial characteristics of migration stimuli could influence responses with respect to the age-profile of the affected. Conversely, the role of favourable conditions to attract populations is equally important in determining the persistence of migration. Prolonged stresses appear to completely erode the adaptive capacity of communities and, therefore, the preparedness to adopt temporary relocation. Socioeconomic groups, as important as demography, are influenced by ownership considerations. Expected future changes in exposure, especially of food and water access, disaster, health and economic considerations, influence priority-setting within policy-making domains.

Werz and Conley (2012) stressed the fact that it is quite difficult to resolve social, political, and related strategic questions across scales of time and space. They present a multi-continent perspective driven by environmental crises, political conflict, and competition for more scarce resources; aligned with newer policy challenges posed by movements. This involves the role of non-state actors and non-traditional sources of conflict and instability, including climate change. This is true also of larger geopolitical security concerns wherein climate change is seen as a threat multiplier.

Cases from across the globe – including Africa, South and Southeast Asia, the UK, South America, and the Middle East – substantiate this perspective. Such thrust areas, as water security, agricultural production, and power generation, appear to be at risk. A truly sustainable approach to security should address traditional security threats and the counter opportunity to promote the individual well-being of people. This sets the context for deliberations on maladaptation against the backdrop of the 5th Assessment Report of the IPCC (2014) that called for urgent efforts to support socio-ecological systems threatened by climate change. Importantly, the processes to secure such adaptation outcomes appear to be unclear. This has implications for maladaptation that could enhance vulnerability over the long term. As mentioned earlier, Singh et al. (2021) presented eleven practice-oriented guidelines through an assessment framework that can be applied at the local level. This refers to other frameworks from a historical perspective. The call is, however, to develop quantifiable indicators (Magnan, 2014; Magnan et al., 2016).

In a study by the United Nations Economic and Social Commission for Asia and the Pacific (UNESCAP) et al. (2014), covering climate change and related migration issues in the Pacific, it argued that environmental change can contribute to individual's decision to migrate; though economic and social reasons may dominate. While climate change can cause a reduction in land, livelihood or habitat security, reduced precipitation or increased disease vectors could reduce the habitability of alternative sites. On the other hand, voluntary migration of individuals and households could reduce environmental pressure when coupled with improved *in situ* adaptation strategies, population management and climate-resilient development.

One of the most recent debates on the characteristics and drivers of migration calls for a deeper understanding of the phenomenon as an adaptive response. The case in point is climate and food security. This interfaces with varied parameters such as ecosystem functions, socio-economic considerations, and technical capabilities to adopt alternatives. It is, therefore, not easy to design strategies that impact food, economic, nutrition, and livelihood security without trade-offs. The variety and intensity of changing community demographics, vulnerability, inclement weather, and seasonality of migration enhance such challenges (Jacobson et al., 2019). This is further complicated when migration is transient because of the interplay of impacts due to climate change and political conflicts. While climatic conditions affect drought severity and influence the occurrence of armed conflict, the resultant asylum-seeking process is probably transient and limited in time and context (Abel et al., 2019).

Perera et al. (2015) focus on the link between energy access and adaptation to climate change with special reference to resilience; wherein mal-adaptation is also a possibility. They further present a menu of nine dimensions of resilience to demonstrate whether access to energy builds resilience. They appear to dominate across three scenarios that saw complete, partial, and lesser impacts. Most importantly, on the other hand, they demonstrate that maladaptation has not been directly attributable to established energy access. Yet the uncontrolled use of energy leads to overexploitation of water or disrupting landscapes to extract natural resources and other societal mechanisms. The authors call for deeper insights about linking energy interventions with climate change adaptation. The moot question, however, remains if adequate scales of such interventions have been achieved and sustained across geographies. An interesting insight pertaining to the initiatives of the power sector is derived from Ouranos – Climate Scenarios and Services Group (Braun & Fournier, 2016). This relates to eleven important cases of adaptation strategies evolved by the power sector to tackle challenges posed by climate change. Floods, storms, sustainability of climate-resilient infrastructure and its performance are some of the adaptation targets. These are intricately linked with the scope to enhance the resilience of affected communities and economies. The authors, however, present an interesting set of barriers: companies lose their zeal due to the absence of a political will to acknowledge and foster transitions; the economic motivation to adapt decreases accordingly; and lacunae in climate impact projections distort preparations to adapt over the long term.

The case of sustainability of renewable energy projects in sub-Saharan Africa highlights some systemic barriers by Ikejemba et al. (2017). The aspect of sustainability by itself is reportedly not adequately defined in projects. This called for a special focus on four governance parameters including community involvement. Nazrul and Winkel (2017) further elaborate on the dynamics of social inequality that have to be addressed to enable adaptation- related transitions. These are relevant within and across countries to smother challenges faced by affected communities. The authors present three main routes that exacerbate inequalities: increased exposure to challenges; proportionate susceptibility; and resultant reduced coping abilities. Burke and Stephens (2018) present an interesting public policy overview of the links between renewable energy systems, energy democracy, and political economy. IRENA's 2019 report to the G20 highlights these and related aspects of energy access and governance. They include a robust policy milieu, supporting regulation, financial mechanisms, commensurate resources, capacity building, appropriate knowledge and skill products and services innovation.

Additional barriers to adaptation had been identified by Ampaire et al. (2017). These pertain to limited technical capacity and finances and political interference continually, with special reference to adaptation by smallholders. The National Adaptation Programmes of Action (NAPAs), however, provide the setting for an integrated approach across ecosystems as an attempt to enhance adaptation outcomes. These include the need to tackle land degradation and drought, improve community water access and use practices, and concomitantly manage other bioresources. The NAPAs have also generated other policies in some cases for engagement-centered outcomes. The need for such integrated approaches has been highlighted also by Gemenne and Blocher (2017), especially when the migration continues abated due to the failure to adapt to changes. This creates opportunities to reduce migration pressures, aligned with the assessments of adaptive capacities, vulnerabilities, and outcomes of maladaptive processes. The implications of the forms and functions of the destination appear to be equally pervasive. This is especially when we differentiate migration by victims of climate and environmental stresses from migration as a strategy to meet basic needs. Adaptive actions are known to be non-linear, that they are changing over time, and that they are not always necessarily positive. Short-term coping strategies can help mitigate harm and enhance adaptation but may not help sustain such positive impacts over a longer period of time.

McLeod et al. (2019) argue that migration is the last resort for some communities in Pacific islands. This emerges after detailed assessments of inherent trade-offs and limitations in adaptation. These could be based on traditional practices and recent technological advances to enhance resilience. Governance measures to manage community-based communication networks that define priorities appear critical for success. Work et al. (2019) implicate governance-related systemic issues in maladaptation outcomes in Cambodia from development projects. In a similar governance perspective, Feltmate and Fluder (2018), in a recent case study, highlight the benefits of preventive action in Canada through concerted action by several stakeholders. They indicate the need to intervene at all stages of the project lifecycle and sustain focused partnerships, community engagement, and scalability options.

The concept of planned relocation was presented by McAdam and Ferris (2015). This was against the backdrop of the Cancun Adaptation Framework and the Nansen Initiative on Disaster-Induced Cross Border Displacement; rightly highlighting the paucity of insights about scale, coping capacities, timing, and thrust areas that need immediate attention. They raise important questions about the policy perspective on voluntary or forced relocation, evacuation, or resettlement. Can be it seen as a preventive measure within countries to reduce the risk of larger scale displacement in the future? Can it also enable significant rebuilding at the re-located places if the return is not possible? An interesting preemptive measure is proposed wherein free and informed consent of the communities concerned guides relocation. This relates to information about proposed options, procedures, benefits, and risks in such transitions.

Incorporating mitigation and adaptation measures in more strategic ways, aligning the paths to get the most sustainable outcomes, will be probably one of the most policy challenges in the future. Accordingly, the IPCC in its 2012 Special Report on managing the risks of extreme events and disasters to advance adaptation highlighted the fact that adaptation and mitigation can mutually reinforce for optimal gains. The most important policy opportunities pertain to exposure and vulnerability determine risk and emergent impacts; through preventive strategies. These are based on insights from a large number of cases from across the globe; emphasizing potential low-regrets measures including focused risk communication, improved natural resources, and land-use management that complement ecosystem planning and climate-proofing infrastructure.

The cases cited above demonstrate the need for a deep dive into the cause-effect relationships to design integrated adaptation and mitigation strategies. Importantly, such strategies by themselves should be immune to extant and emerging stresses through mutually reinforcing policies and plans that do not work at cross-purposes. While the groundswell of evidence appears to be growing about maladaptation, the lack of holistic assessment frameworks and systems-scale qualitative and quantitative correlates has to be addressed on priority. The emerging IPCC 6th Assessment Report (2021) will probably set the impetus for such a concerted effort.

5 Conclusion

The above analysis has been built on the assumption, increasingly supported by growing empirical evidence, that climate response mechanisms may generate social-ecological vulnerability. These in turn may trigger both conflicts and human displacements. Focusing on adaptation as a core element of climate change governance, we could demonstrate how related policies and actions may, in some instances, generate unexpected outcomes and risks that can erode resilience instead of building it. The analysis has, therefore, thoroughly investigated the drivers, forms, and manifestations of maladaptation both theoretically and by reference to existing empirical research. Alternative approaches to address maladaptation and ensure the achievement of resilience and sustainability through adaptation strategies have been identified and discussed. In addition, the maladaptation and migration nexus has been analyzed from an inclusive governance and science-based policy making.

To conclude, the achievements and lessons learned from adaptation policies and interventions worldwide should be the blueprint to successful and transformative adaptation action. However, the dynamics of potential climate change-induced risks are becoming more and more unpredictable and severe. Therefore, inclusive and sustainable adaptation strategies, that centre the wellbeing and security of communities and the balance of ecosystems at the heart of all interventions, are crucial to all nations. The current climate-induced disasters that we are currently facing are signals that we should consider to examine how our adaptation policies can drive sustainability, resilience, and justice for both people and the planet.

References

Abel, G. J., Brottrager, M., Cuaresma, C. J., & Muttarak, R. (2019). Climate, conflict and forced migration. *Global Environmental Change, 54*, 239–249. https://doi.org/10.1016/j.gloenvcha.2018.12.003

Alcayna, T. (2020). *At what cost: How chronic gaps in adaptation finance expose the world's poorest people to climate chaos, Zurich Flood Resilience Alliance.* Retrieved from: https://europe.mercycorps.org/sites/default/files/2020-07/1087-PA-ZFRA-AtWhatCost-V15c-WEB.pdf

Ampaire, E. L., Jassogne, L., Providence, H., Acosta, M., Twyman, J., Winowiecki, L., & van Asten, P. (2017). Institutional challenges to climate change adaptation: A case study on policy action gaps in Uganda. *Environmental Science & Policy, 75*, 81–90. https://doi.org/10.1016/j.envsci.2017.05.013

Barnett, J., & O'Neill, S. J. (2013). Minimising the risk of maladaptation: A framework for analysis. In J. P. Palutikof et al. (Eds.), *Climate adaptation futures* (pp. 87–94). Wiley-Blackwell.

Barrott, J. (2015). Transformational adaptation: What it is, why it matters and what is needed, *We Adapt.* Retrieved from: https://www.weadapt.org/knowledge-base/governance-institutions-and-policy/transformational-adaptation

Braun, M., & Fournier, E. (2016). *Adaptation case studies in the energy sector – Overcoming barriers to adaptation.* Report presented to Climate Change Impacts and Adaptation Division, Natural Resources Canada. Retrieved from: https://www.weadapt.org/sites/weadapt.org/files/2017/september/ouranos_adaptation_case_studies_in_the_energy_sector_overcoming_barriers_to_adaptation.pdf

Burke, M. J., & Stephens, J. C. (2018). Political power and renewable energy futures: A critical review. *Energy Research & Social Science, 35*, 78–93. Retrieved from:. https://doi.org/10.1016/j.erss.2017.10.018

Council of Europe. (2021). *Climate and migration, report of the committee on migration, refugees and displaced persons*. Doc. 15113, Ref. 4521. Retrieved from: https://assembly.coe.int/LifeRay/MIG/Pdf/TextesProvisoires/2021/20210421-ClimateAndMigration-EN.pdf

Dinshaw, A., Ginoya, N., Preethan, P., Nambi Appadurai, A., & Gutierrez, M. (2018). *Mainstreaming adaptation in action: Case studies from two States in India* (Working Paper). World Resources Institute. Retrieved from: http://www.wri.org/publication/mainstreaming-adaptation-action

Eriksen, S., Schipper, E. L. F., Scoville-Simonds, M., Vincent, K., Adam, H. N., Brooks, N., Harding, B., Khatri, D., Lenaerts, L., Liverman, D., Mills-Novoa, M., Mosberg, M., Movik, S., Muok, B., Nightingale, A., Ojha, H., Sygna, L., Taylor, M., Vogel, C., & West, J. J. (2021). Adaptation interventions and their effect on vulnerability in developing countries: Help, hindrance or irrelevance? *World Development, 141*, 105383. https://doi.org/10.1016/j.worlddev.2020.105383

FAO. (2011). Energy-smart food for people and climate, *Issue Paper*. Retrieved from: https://www.fao.org/3/i2454e/i2454e.pdf

Feltmate, B., & Fluder, A. (2018). *Too small to fail: Protecting Canadian communities from floods*. Retrieved from https://www.intactcentreclimateadaptation.ca/wp-content/uploads/2018/10/Climate-Change-Adaptation-Projects-FINAL.pdf

Foong, A., Pohl, B., & Rüttinger, L. (2020). *Climate-Fragility Risk Brief – Sudan, Adelphi*. Retrieved from: https://climate-security-expert-network.org/sites/climate-security-expert-network.org/files/documents/csen_climate_fragility_risk_brief_sudan.pdf

Gemenne, F., & Blocher, J. (2017). How can migration serve adaptation to climate change? Challenges to fleshing out a policy ideal. *The Geographical Journal*, https://doi.org/10.1111/geoj.12205., retrieved from: https://collections.unu.edu/eserv/UNU:5960/Geoj_Journal.pdf

Ikejemba, E. C. X., Schuur, P. C., Hillegersberg, J. V., & Mpuan, P. B. (2017). Failures and generic recommendations towards the sustainable management of renewable energy projects in Sub-Saharan Africa (Part 2 of 2). *Renewable Energy, 113*, 639–647. Retrieved from. https://doi.org/10.1016/j.renene.2017.06.002

Intergovernmental Panel on Climate Change (IPCC). (2014). *Mitigation of climate change*. Contribution of working group III to the Fifth Assessment Report of the Intergovernmental Panel on Climate Change, 1454.

International Organization for Migration (IOM). (2008). Migration and Climate Change. *IOM Migration Research Series,* 31. Retrieved from: https://olibrown.org/wp-content/uploads/2019/01/2008-Migration-and-Climate-Change-IOM.pdf

IPCC. (2001) *Climate change 2001: Adaptation contribution of working group II to the third assessment report of the INTERGOVERNMENTAL Panel on Climate Change* [J. McCarthy, O. Canziani, N. Leary, D. Dokken, & K. White (Eds.)]. Cambridge University Press, http://www.ipcc.ch/ipccreports/tar/wg2/pdf/WGII_TAR_full_report.pdf

IPCC. (2012). *Managing the risks of extreme events and disasters to advance climate change adaptation* (A Special Report of Working Groups I and II of the Intergovernmental Panel on Climate Change) [C. B. Field, V. Barros, T. F. Stocker, D. Qin, D. J. Dokken, K. L. Ebi, M. D. Mastrandrea, K. J. Mach, G.-K. Plattner, S. K. Allen, M. Tignor, & P. M. Midgley (Eds.)]. Cambridge University Press. Retrieved from: https://www.ipcc.ch/site/assets/uploads/2018/03/SREX_Full_Report-1.pdf

IPCC. (2014) *Climate change 2014: Impacts, adaptation, and vulnerability. Part A: Global and sectoral aspects. Contribution of working group II to the Fifth Assessment Report of the Intergovernmental Panel on Climate Change* [C. B. Field, V. R. Barros, D. J. Dokken, K. J. Mach, M. D. Mastrandrea, T. E. Bilir, M. Chatterjee, K. L. Ebi, Y. O. Estrada, R. C. Genova, B. Girma, E. S. Kissel, A. N. Levy, S. MacCracken, P. R. Mastrandrea, & L. L.White (Eds.)]. Cambridge University Press, 1132.

IPCC. (2021). Summary for policymakers. In V. Masson-Delmotte, P. Zhai, A. Pirani, S. L. Connors, C. Péan, S. Berger, N. Caud, Y. Chen, L. Goldfarb, M. I. Gomis, M. Huang, K. Leitzell, E. Lonnoy, J. B. R. Matthews, T. K. Maycock, T. Waterfield, O. YelekçI, R. Yu, & B. Zhou (Eds.), *Climate change 2021: The physical science basis. Contribution of Working Group I to the sixth assessment report of the Intergovernmental Panel on Climate Change*. Cambridge University Press. In Press.

IRENA. (2019). *Climate Change and Renewable Energy: National policies and the role of communities, cities and regions* (Report to the G20 Climate Sustainability Working Group (CSWG)). International Renewable Energy Agency. Retrieved from: https://www.irena.org/-/media/Files/IRENA/Agency/Publication/2019/Jun/IRENA_G20_climate_sustainability_2019.pdf

Jacobson, C., Crevello, S., Chea, C., & Jarihani, B. (2019). When is migration a maladaptive response to climate change? *Global Environmental Change, 19*, 101–112. Retrieved from. https://doi.org/10.1007/s10113-018-1387-6

Juhola, S., Glaas, E., Linnér, B. O., & Neset, T. S. (2016). Redefining maladaptation, Environmental Science & Policy, *55*(1), 135–140, ISSN 1462-9011. https://doi.org/10.1016/j.envsci.2015.09.014.

Lang, A. (2019). Maladaptation: An Introduction, *WeAdapt*, https://www.weadapt.org/knowledge-base/vulnerability/maladaptation-an-introduction

Lonsdale, K., Pringle, P., & Turner, B. (2015). *Transformative adaptation: What it is, why it matters & what is needed. UK Climate Impacts Programme*. University of Oxford.

Magnan, A. (2014) Avoiding maladaptation to climate change: Towards guiding principles. *Sapiens,* Retrieved from http://sapiens.revues.org/1680

Magnan, A. K., Schipper, E. L. F., Burkett, M., Bharwani, S., Burton, I., Eriksen, S., Gemenne, F., Schaar, J., & Ziervogel, G. (2016). Addressing the risk of maladaptation to climate change. *WIREs Climate Change*. https://doi.org/10.1002/wcc.409

Mahapatra, R. (2019). *February 4*. Hindu Kush Himalayan region warming faster than global average. Retrieved from https://www.downtoearth.org.in/news/climate-change/himalayan-meltdown-hindu-kush-himalayan-region-warming-faster-than-globalaverage-63078

McAdam, J., & Ferris, E. (2015). Planned Relocations in the Context of Climate Change: Unpacking the Legal and Conceptual Issues. *Cambridge Journal of International and Comparative Law, 4*(1), 137–166. https://doi.org/10.7574/cjicl.04.01.137

McLeman, R. A., & Hunter, L. M. (2010). Migration in the context of vulnerability and adaptation to climate change: Insights from analogues. *Wiley Interdisciplinary Reviews: Climate Change, 1*(3), 450–461. https://doi.org/10.1002/wcc.51

Mcleod, E., Bruton-Adams, M., Förster, J., Franco, C., Gaines, G., Gorong, B., James, R., Posing-Kulwaum, G., Tara, M., & Terk, E. (2019). Lessons from the Pacific Islands – Adapting to climate change by supporting social and ecological resilience. *Frontiers in Marine Science, 6*, 289. https://doi.org/10.3389/fmars.2019.00289

Nazrul, I. S., & Winkel, J. (2017). *Climate change and social inequality* (DESA Working Paper No. 152 ST/ESA/2017/DWP/152). Retrieved from: https://www.un.org/esa/desa/papers/2017/wp152_2017.pdf

O'Brien, K. (2012). Global environmental change II: From adaptation to deliberate transformation. *Progress in Human Geography, 36*(5), 667–676. https://doi.org/10.1177/0309132511425767

Pelling, M. O'Brien, K. & Matyas, D. (2015). Adaptation and transformation. *Climate Change, 133*(1), 113–127. https://doi.org/10.1007/s10584-014-1303-0

Perera, N., Boyd, E., Wilkins, G., & Itty, R. P. (2015). *Literature Review on Energy Access and Adaptation to Climate Change, UK Department for International Development (DFID*. Retrieved from: https://assets.publishing.service.gov.uk/media/57a0896b40f0b652dd0001fe/LitRev-EnergyAccessandAdaptation-Final-2.pdf

Podesta, J. (2019). The climate crisis, migration, and refugees, *Policy Brief*, prepared for the 2019 Brookings Blum Roundtable. Retrieved from: https://www.brookings.edu/wp-content/uploads/2019/07/Brookings_Blum_2019_climate.pdf

Rahman, H. M. T., & Hickey, G. M. (2019). What does autonomous adaptation to climate change have to teach public policy and planning about avoiding the risks of Maladaptation in Bangladesh? *Frontiers of Environmental Science, 7*(2). https://doi.org/10.3389/fenvs.2019.00002

Robinson, S. A. (2020). Climate change adaptation in SIDS: A systematic review of the literature pre and post the IPCC Fifth assessment report. *WIREs Climate Change, 11*(4), 1–21.

Schaer, C. (2015). Condemned to live with one's feet in water? A case study of community based strategies and urban maladaptation in flood prone Pikine/Dakar, Senegal. *International Journal of Climate Change Strategies and Management, 7*(4), 534–551. Retrieved from. https://doi.org/10.1108/IJCCSM-03-2014-0038

Scheraga, J. D., & Grambsch, A. E. (1998). Risks, opportunities and adaptation to climate change. *Climate Research, 10*, 85–95.

Schipper, E. L. F. (2020). Maladaptation: When adaptation to climate change goes very wrong. *One Earth, 3*(4), 409–414. https://www.cell.com/action/showPdf?pii=S2590-3322%2820%2930483-8

Singh, C., Iyer, S., New, M. G., Few, R., Kuchimanchi, B., Segnon, A. C., & Morchain, D. (2021). Interrogating 'effectiveness' in climate change adaptation: 11 guiding principles for adaptation research and practice. *Climate and Development*. https://doi.org/10.1080/17565529.2021.1964937

UNESCAP, EU, ILO and UNDP. (2014). *Climate change and migration issues in the Pacific, United Nations*. Retrieved from: https://www.ilo.org/dyn/migpractice/docs/261/Pacific.pdf

UNFCCC. (2006). *Technologies for adaptation to climate change*. Retrieved from: https://unfccc.int/resource/docs/publications/tech_for_adaptation_06.pdf

UNHCR. (2018). *Report of the United Nations High Commissioner for Refugees – Part II: Global compact on refugees*, A/73/12. Retrieved from: https://www.unhcr.org/gcr/GCR_English.pdf

UNHCR. (2021). *Report of the United Nations High Commissioner for Refugees*, A/76/12. Retrieved from: https://www.unhcr.org/afr/615eb32c4.pdf

UNHCR, UNITAR and GPA. (2020). *Sustainable energy, displacement and climate resilience.* Retrieved from: https://www.unhcr.org/en-in/protection/environment/5f60a3c44/sustainable-energy-displacement-climate-resilience.html

Vinke, K., Bergmann, J., Blocher, J., Upadhyay, H., & Hoffmann, R. (2020). Migration as Adaptation? *Migration Studies, 8*(4), 626–634. Retrieved from. https://doi.org/10.1093/migration/mnaa029

Werz, M., & Conley, L. (2012). *Climate change, migration, and conflict – Addressing complex crisis scenarios in the 21st century*. Center for American Progress. Retrieved from: https://cdn.americanprogress.org/wp-content/uploads/issues/2012/01/pdf/climate_migration.pdf

Work, C., Rong, V., Song, D., & Scheidel, A. (2019). Maladaptation and development as usual? Investigating climate change mitigation and adaptation projects in Cambodia. *Climate Policy, 19*(1). https://doi.org/10.1080/14693062.2018.1527677

World Economic Forum. (2020). *The global risks report 2020, 15th Edition.* Retrieved from: https://www3.weforum.org/docs/WEF_Global_Risk_Report_2020.pdf

World Resources Institute. (2018). *Climate resilient and transformative adaptation for agriculture*. Retrieved from https://www.wri.org/initiatives/climate-resilient-and-transformative-adaptation-agriculture

Chapter 4
While Carbon Burns: The Debatable Journey of 'Environmental Refugee' as a Concept and Legal Dilemma

Ramy Magdy and Mai Yasser

Abstract Anthropogenic climate change and its apocalyptic predictions opened for academia a wide scene of conflicts, phenomena, and concepts, which would not have been available if climate change had not been discussed as a problem in the first place. Such a change, disastrous as it is, is not conducting itself on the environment as an abstract entity, but rather it disrupts livelihoods, wealth, economies, and governments. One important group of victims of this change is the individuals whose habitats are negatively impacted by climate change. One way to protect them and their rights was through conceptualizing their situation and giving them the status of 'refugees' who left their homes due to climate change and its disastrous effects. Yet such a concept of 'environmental refugee' did not pass smoothly through the academic, legal, and political circles, as from its early beginnings it was surrounded by debates and controversies. Moreover, the inevitable neediness for legal instruments to protect this category of refugees allowed scholars and humanitarian organizations to look for options and ideas to legalize the protection of environmental refugees. Who is the 'environmental refugee', how did this concept emerge, in which context and through which phases of evolution? How was this concept criticized and why? Is there any legal instrument that recognizes such a concept or offers protection for this vulnerable group? And what is the core mandate of UNHCR and IOM regarding climate-induced displacement? Answering these questions is the main objective of this chapter that revisits the concept of 'environmental refugee', its emergence, evolutions, legal dilemmas, and criticism. In addition to that, the chapter explores the question of environmental refugees in many cases, and

R. Magdy (✉)
Political Science Department, Faculty of Economics and Political Science, Cairo University, Giza, Egypt
e-mail: ramy.ahmed@feps.edu.eg

M. Yasser
Political Science Department, Faculty of Economics and Political Science, Cairo University, Giza, Egypt

Durable Solution Unit, UN Refugee Agency, Cairo, Egypt
e-mail: mai.ahmed2013@feps.edu.eg

M. Behnassi et al. (eds.), *The Climate-Conflict-Displacement Nexus from a Human Security Perspective*, https://doi.org/10.1007/978-3-030-94144-4_4

how such a concept can be employed to judge the ongoing massive waves of displacement around the globe.

Keywords Climate change · Conflict · Displacement · Environmental refugee · Refugee

1 Introduction

The concept of 'environmental refugee' was born 'debated'. The very relation of the concept with migration and refuge, and how these two phenomena are associated with fear and despair, made the concept from its early beginnings emotionally charged and, consequently, introduced the concept into a climate of debates and controversies (Boano et al., 2008:7). Yet what is paradoxical, as the chapter will also reveal, is that the crux of the debates and controversies on the concept do not focus on whether environmental refugees exist or not, but rather they focus on the very process through which such a category of people gets conceptualized. That is to say, the debate is not on the existence of the concept but on the nature of the relationship between environmental processes in general, and how might they would lead to migration (and asylum) (Morrissey, 2012:40). Generally, one can consider Ioane Teitiota as the first 'environmental' and (or) 'climate refugee'. He was first to apply in order to be assigned as a refugee for climate reasons since his Pacific island, 'Kribiate', is under the continuous threat of being flooded by sea water (Mohamed 2013:143). He is proceeding into judicial battles with the government of New Zealand for neglecting his claim (Mcdonald, 2015). However, the situation of Teitota is one in a series (and chain) of events and debates that formed environmental migration and refugees; what is climate change and the debates over it, what are the series of events leading to some formula of an 'environmental refugee'? This is the concern of the next two sections.

2 The Climate Debate: Arguing over a Destiny

One cannot develop an argument concerning environmental asylum without exploring the main debate beneath it regarding climate change. Climate change is one of the most pressing global concerns since the 1970s at least. Such change is mainly a negative change affecting the balance of the climate system due to rising levels of carbon emissions, which gradually increase the earth temperature, thus causing a disturbance of many natural and social processes, including the destruction of habitats and the worsening of the magnitude and severity of extreme events and disasters. Yet no matter how apocalyptic this might sound, the bottom line of the debate over the 'fate' of climate is concerned with the accuracy of scenarios predicted to come from negative climate change. Lucidly, B. Sudhakara Reddy and Gaudenz

B. Assenza, in some pro-trinity manner, delineate the contending stances in the climate debates to be of three groups and the theories over its truth to be also three (Reddy & Assenza, 2009).

As for the theories, the first one argues that a range of 5–10 °C of increased temperature will be witnessed by Earth over the next century or two and such a rise is largely induced by human activities. A counter theory claims that such a rise is just a variability, not a future trend, and even if some efforts of prevention are required, their cost is much higher than their benefit, which defies any impetus for action. A third middle way theory stresses that over millions of years the Earth climate is unstable and tends to go cooler and the data we have currently on climate change are some annual data, not ones over many decades or centuries, and this justifies the equal possibilities of ending up in an ice age or ending up warmer. So actually, everything could happen (Reddy & Assenza, 2009:2997). However, such a trio of perspectives might be in accord on one major point, which is the need to have more accurate measurements and longer-term data on climate change.

Against such clash of theories, there are three teams in the debate over climate change. They are: supporters, sceptics, and realists. *Supporters* underscore the urgency of climate change negative impacts and press for an immediate action, assuming that any procrastination might make the situation graver. On the other hand, *sceptics* doubt the accuracy and the sufficiency of the data that claim to have a complete picture of climate change or encourage for a quick action. Not to mention that they think that such an action, in light of the costs, might not be favorable economically. A third group includes climate *realists* who acknowledge the negative impacts of climate change and the need to take action, yet such an action should be thought in line with the development requirements of developing nations (Reddy & Assenza, 2009:2998). For them, the quests over climate change and development should not be thought and fought separately, and a proper understanding of the sustainable development needs should guide climate action (Reddy & Assenza, 2009:3005).

Overall, it might be argued that the debate centers on certain themes: *First*, the need to strengthen and enrich our scientific knowledge of climate change. *Second*, the acceptability of contemplating alternative explanations for climate change, as supporters of immediate action believe one cannot afford suggesting alternative explanations still in need of verification, while sceptics argue otherwise. *Third*, the urgency of the precautionary principle promotes the climate actors' need to take preventive actions even before the negative outcomes get fully unfolded. Precaution usually requires allocating resources for a certain need during times when that need is not so pressing relative to its counterparts. Sceptics in light of their doubt of climate data accuracy prefer not to take early measures (no rush) unlike the supporters who predict climate catastrophes to be severe, thus justifying pre-emptive immediate measures. *Fourth*, the study of the tradeoffs\benefits between climate action measures on the one hand and other political, social, and economic indicators on the other, which is seen as a complex academic area within this debate attracting accounts and analyses from all concerned researchers (Reddy & Assenza, 2009:3001–3003).

Against such uncertainties within the climate debate, conceptualizing the status of the people negatively impacted by such a change remains a difficult task. This requires understanding how, on the ground, climate change influences the status and fates of millions of displaced people called 'environmental refugees'.

3 History and Evolution of the Concept

First of all, one cannot approach the phenomenon of environmental asylum or refuge separately from the evolution of political migration and asylum granting in the modern state context. World War I was the opening episode for migration and giving political asylum on rights-based justifications; the post-war period and the global concern about the rights of people to defend themselves against persecution allowed for individuals to seek asylum under rights-based concern. However, linking migration with demographic negative events can be also traced back to the 1930s with Etienne Dennery's neo-Malthusian concern that the over-growing population of the Indian subcontinent will lead to uncontrollable migration waves (Dennery, 1931). In the 1940s, Vogt described the same phenomenon but in the North American prairies (Saunders, 2000).

With the outbreak of the cold war in the late 1940s and 1950s, migration and asylum rights were welcomed by the Western block countries to assert their privilege of being flourishing liberal capitalism(s) accepting refugees from an economically and politically oppressed Eastern block (Morrissey, 2012:42). Yet with the height of the cold war in the 1970s and the continuous flow of refugees, and during some episodes of entente, refugees' image changed from being a sign of privilege to being a suspect of subversion, as refugees began to be seen as a threat of socialist subversion. Into this climate of threats, environmentalism emerged calling for more awareness of the negative effect of pollution and its disastrous effect on Earth, crowning its attempt with the 'Earth Day' annual event starting from 1970. Nonetheless, such care for environmentalism, one must not forget, was born in the concomitant period of the cold war within which refugees were seen as a threat. These together made the very notion of 'refugees' a 'critical problem' since it was not received with a welcoming mood to alleviate the effects of pollution, but as a sign of concern worth of fear.

By the late 1970s and through the 1980s, with the Chernobyl disaster and increasing feeling of environmental threat, the United Nations Environmental Program (UNEP), coined for the first time the concept 'environmental refugee' in a paper written by El-Hinnawi (1985), in which he defined and tackled different aspects of the concept. El-Hinnawi was not the first to create the concept, ex-nihilo, as he was preceded in 1970 by Lester Brown, founder of the World Watch environmental agency. Yet the merit of El-Hinnawi's conceptualization is that he managed to bring the concept into academic and political debate(s), not to mention the publicity he had at the time since his paper was issued by an official international agency.

After the Soviet Union collapse, and the absence of any need of pride in accepting refugees, refugees in general were seen as a flood threatening society, a welfare junk, and as manipulators of public money. At the same time, in the early 1990s, Neo-Multhusian thought reasserted itself with a shift from superpowers world politics, to more concern with the problems of 'the globe'. This re-emergence during the period of massive political and economic shifts in the world order asserted the relation between population growth and economic crisis (Morrissey, 2012:36), leading to increased interest in social – including demographic – and environmental changes, and their effects. This led to flourishing the already expanding literature on environmental refugees. However, by the late 1990s, two changes happened: first, more research focus was directed towards environmental migration developing the literature from merely expounding the concept and giving general predictions to a more empirical search into the complexity of the environmental phenomenon and its relation with migration, in addition to developing 'hotspots' of migration and models for its study (Morrissey, 2012:37). Second, this interest in environmentalism and its complexity was reflected in literature with a wave of critics spanning all the 1990s. This critical wave was investigating the validity of concepts like 'environmental migration' and 'environmental refugee', and stressing their simplistic formation and their negligence of human ability to adapt to climatic changes (Suhrke, 1991; Lonergan, 1998).

The new millennium and its MDGs shaped the concern with environmental refugees in a 'governmental' discourse, which led to conceptualizing the problem of environmental migration as either a responsibility on governments to take or a responsibility that governments fear to take. The Western governments, specifically, found recognition of environmental refugees as a potential prelude for demanding compensation against their industrial pollution, and that compensation might be given to the South (Morrissey, 2012:36).

A mention should be given also to Norman Myers, the most prolific writer about the concept whose works span the period from 1989–2005, contributing to illuminating the concept and intensifying the debate around it. In general, the first two decades of the millennium, full enough with environmental and political problems widely introduced refuge as a problem and demanded care for environmental issues.

With the drafting and signing of Paris Agreement in 2015, climate change is deemed to be one of the global forces interacting with human mobility, most notably displacement. In the extraordinary climatic changes that the world witnessed many years ago, the Agreement stipulates on minimizing climate risks and risks emerging from not taking appropriate actions to manage climate impacts on society. Moreover, the Agreement's climate policy that was adopted since then stipulates on raising financial aid for around 100 billion dollars per year to foster adaptation options that keep people safe and ensure livelihood and food security for displaced persons. The Agreement emphasizes as well the seriousness of displacement induced by negative climate events like hurricanes, droughts, desertification, floods, and conflict upon food and water resources.

It argued that the displacement caused by climate change is multidimensional and complex. It is illogical to tease out environmental factors from the economic, social, political, and demographic aspects that shape an individual's decision to migrate. This requires holistically observing the phenomenon to develop a schema that includes many aspects of climate complications to minimize their risks.

On that score, the Agreement calls for a task force to develop schemas for integrated strategies to minimize and address climate-induced displacement. Moreover, human mobility has been framed as an issue of climate risk management in international decision-making documents under the UN Framework Convention on Climate Change (UNFCCC) and Paris Agreement included. Such a system for risk management has been developed by social science studies between the 13th Conference of the Parties (COP13) in 2007 and the COP24 in 2018 (Warner, 2018).

However, against its main calls, the Paris Agreement still does not cover the issue of displacement holistically. Instead, it emphasizes the right of displaced persons to integrate, work, and live in the host community without referring to any legal coverage and inclusion. It does not address the legal status of displaced people, mandate their protection or discuss the establishment of any new ground to recognize displaced persons due to climate change as 'refugees'.

4 Definition, Structure, and Criticism of the Concept of 'Environmental Refugee'

'Environmental refugee' as a concept, though severely debated, has diverse definitions, each trying to shape the core and limits of such a term. The most famous definition and the first publicized was that of El-Hinnawi in which he defined environmental refugees as "those people who have been forced to leave their traditional habitat, temporarily or permanently, because of a marked environmental disruption (natural and/or triggered by people) that jeopardized their existence and/or seriously affected the quality of their life" (El-Hinnawi, 1985:4).

Norman Myers defines it as: "People who can no longer gain a secure livelihood in their homelands because of drought, soil erosion, desertification, deforestation and other environmental problems, together with the associated problems of population pressures and profound poverty. In their desperation, these people feel they have no alternative but to seek sanctuary elsewhere, however hazardous the attempt. Not all of them have fled their countries, many being internally displaced. But all have abandoned their homelands on a semi-permanent if not permanent basis, with little hope of a foreseeable return" (Myers, 2005:1).

The International Organization for Migration (IOM) on its behalf, like most international organizations, prefers to define 'environmental migrant' instead of refugee, to avoid the legal repercussions of giving environmental victims the status of a 'refugee', and for this, it defines environmental migrants as: "persons or groups of persons who, for compelling reasons of sudden or progressive changes in the environment that adversely affect their lives or living conditions, are obliged to leave their habitual homes or choose to do so, either temporarily or permanently, and who move either within their country or abroad" (IOM, 2007:1–2).

In the same vein, which clearly indicates the controversial nature of the term, the United Nations High Commissioner for Refugees (UNHCR) prefers to use the name 'environmentally displaced persons' and defining them as those "who are displaced from or who feel obliged to leave their usual place of residence, because their lives, livelihoods, and welfare have been placed at serious risk as a result of adverse environmental, ecological or climatic processes and events" (Gorlick, 2007 in Boano et al., 2008:8). However, the UN Statistics Division, in its glossary of terms, defines the 'environmental refugee' as a "person displaced owing to environmental causes, notably land loss and degradation, and natural disaster" (UNSD, 2020).

Moreover, the UNHCR recognizes someone as a 'refugee' based on one of the five grounds that are mentioned in the refugee definition in 1951 Convention and, unfortunately, the environmental hazard is not among these grounds. A refugee's, according to this Convention, is someone who owing to well-founded fear of being persecuted for reasons of race, religion, nationality, membership of a particular social group or political opinion, is outside the country of his nationality and is unable or, owing to such fear, is unwilling to avail himself of the protection of that country; or who, not having a nationality and being outside the country of his former habitual residence as a result of such events, is unable or, owing to such fear, is unwilling to return to it (UNHCR, 1951).

Finally, Hamdawy (2015:116) defines environmental refugees as "those persons compelled by the environmental conditions to leave their normal habitats, or feel they are compelled to leave, for several reasons related to environmental deterioration, such as drought, desertification, water shortage, water pollution, and any other effects of environmental pollution that threaten their lives and livelihood directly, no matter whether their migration was for a long or a short period, inside or outside their state borders".

Based on this, one can distinguish between two main trends, identified by Suhrke (1991) in the writings on environmental refugees: maximalist trend and minimalist trend. The maximalist views the changes in the environment as the direct and sole reason for environmental migration, while the minimalist tends to problematize this relation between environment and migration and stress its complexity and multicausality (Rebert, 2006:4). Accordingly, Morrissey (2012:43) prefers to view the debate in the manner of discussion and discourse, and for this reason, he classifies the same trends as either proponents of the relation or critics. But in general, a clear understanding of the concept impels one to investigate its elements and (or) categories.

One can consider Bates (2002) in this regard. Working on elaborating El-Hinnawi's definition, Bates suggested a continuum through which one can do two things: first, it enables one to distinguish clearly between environmental refugees and environmental migrants; and second, it enables one to distinguish between the different categories of environmental refugees.

First of all, the margin of 'freedom' distinguishes an environmental refugee, from an 'environmental migrant' and 'a migrant'. Those who are driven by external factors, not of their own, and they were compelled against their voluntary will relocate are environmental migrant; on the other side of the continuum stands the 'migrant' who, having a variety of motives, such as economic improvement, chooses to relocate. Sandwiched between both is the anticipatory refugees (or the environmental migrants) who recognize the need to relocate before they become forced to do so; these are environmental migrants (Bates, 2002:467–468).

Within the category of environmental refugees, Bates distinguishes between refugees as a result of unintentional disasters that are unintentionally producing refugees like natural events and technological disasters (a Volcano or a reactor explosion), and refugees as a result of expropriation made by intentional acute environmental damages that are intentionally producing refugees (Dam building), and refugees as a result of gradual intentional deterioration of environment that is intentionally producing refugees (such as global warming, deforestation, and desertification).

Hamdawy, in turn, sets several conditions for classifying individuals as environmental refugees. They are: (1) the existence of a grave danger compelling them involuntarily to leave their homes; (2) leaving homes temporarily or permanently; (3) the existence of environmental deterioration that negatively impacts the lives and circumstances of these individuals; and (4) the existence of a destination that they moved/move (or) will move to inside or outside their national borders (Hamdawy, 2015:15).

However, these attempts to assert and conceptualize 'environmental refugees' did not pass unscathed. Many criticisms were directed against the term and the way it views the relationship between environment and migration. These criticisms are of three sorts: political, discursive, and academic. The political criticisms tackle the political repercussions that the recognition of such a concept may entail. Suhrke believes that this concept will distort the existent definition of traditional refugees, and thus will strain the already scarce resources available to them (Boano et al., 2008:10). Additionally, Castles' opinion is that the expansion of the definition of refugees will threaten the protection of traditional refugees (Castles, 2002 in Boano et al., 2008:10), while Lopez views the usage of such a term as a legal 'mistake (Boano et al., 2008:10).

The discursive critique includes these opinions that criticize the very discourse and vision underlying the work of those in favor of the concept of 'environmental refugee' or the 'maximalist theory'. The first discursive criticism was targeting the deep neo-Malthusian world view underlying this concept, it denotes how the maximalist view tends to simplistically perceive environment and its relation with society in terms of push-pull model, disregarding any complexity within each society that may play a greater role in shaping the result of this relation. The second criticism was concerned with how the maximalist theory tends to assume that all people are necessarily sedentary and under severe pressures they migrate, which is not the case since migration by itself was and (is) a normal practice with non-sedentary peoples who migrate in adaptation with environmental and seasonal changes. The third criticism, somehow, complements the second by stressing that the maximalist view ignores the dynamism in human life, that people migration is not necessarily – all the time – a sign of some problem. Individuals and social groups migrated through history and this is the dynamic part of human life, not to mention that not every adaptation to environmental change represents some exodus out of some environmental actual/or potential disaster. Finally, the fourth criticism scrutinize the historical apolitical nature of the maximalist view that tend to simplify the relationship between environment and migration ignoring the various political and historical factors that influence this relationship and drive its results otherwise (Morrissey, 2012:43–44).

In addition to the above critiques, it is important to note that a paper by UNHCR adopted by the IPCC on its website enumerated the different dilemmas within many dimensions of defining an environmental refugee (Black, 2001). For instance, the paper enumerates three instances for driving people into a situation of environmental displacement justifying their being 'environmental refugees': desertification, rising sea levels, and resources conflict. Each instance of them has its own complexities which make handling the concept of environmental refugee a more sophisticated task. For desertification, it is stressed that a displacement driven by desertification (in some cases) might not be really a displacement but a 'coping strategy' by groups accustomed to semi-arid surroundings. In addition, the concept of desertification itself is debatable due to the ability of some soils to recover or suffer variability not absolute desertification (Black, 2001:6). As for rising sea levels, it is argued that poor human management of the environmental situation might be the cause behind both the environmental deterioration and human displacement. This means that environmental migration might be, in essence, a human-induced displacement not an environmentally-induced one (Black, 2001:7). Furthermore, poor and arid environments igniting resources conflicts must be looked into with a more sophisticated vision; the paper mentioned famous examples in Rwanda and Somalia where resources conflicts were initially ignited by ethnic divisions or global powers proxy wars in which resources conflicts were only a side effect mobilized within larger battles for power and influence (Black, 2001:8–9).

As for the academic strand of the critique, it was concerned with the validity of the academic methods and research process through which the maximalist theory conducts its research. One can also enumerate three main points of criticism here. The first is that the maximalist view has a strong regional bias concerning their data which focus mainly on Africa and Asia, and thus their conclusions might not be well established or representative. The second is that the maximalist view usually bases its conclusions on brief and concise studies and no detailed studies into the relationship between environment and migration were available from this theory. The third, and last, point is that the term 'environmental refugee' over which the maximalist view construct their researches is by itself not acceptable in the international law due to its vagueness and due to the political repercussions, which though may not abort the maximalist enterprise, but lessen its legal and political acceptability and reflect a margin of conceptual irrelevance (Bates, 2002:466).

However, the author(s) of this research stand somehow with the maximalist view (and El-Hinnawi's definition), because, regardless of the complexities of social research, or the demands of political action and legal terminology, one is encountered with a flow of peoples and social groups who are threatened by the negative impacts of climate change, and only through stating clearly their case, loudly and outrightly, any action or publicity of their cause could be possible. In addition, defending the livelihoods of humans differs radically from just defending an academic, legal, or political argument. The momentum, enthusiasm, and resolution through which the defender of the rights of these people show his case justify his need to simplify the web of conceptual relationships in order to highlight the situation of these people and shorten the period needed to rescue them.

5 Conceptual Links and Legal Dilemmas

Certainly, one cannot tackle the concept of 'environmental refugee' without getting himself into a web of concepts that are linked and juxtaposed against the concept of 'environmental refugee'. It is easy to enumerate many of them; sufficient in this regard to mention Bates' (2002) work trying to distinguish environmental refugees from other similar concepts, in one way or another, ended up giving the reader of his work a wide array of concepts linked to environmental refugees: migration, displacement, environmental expelling, voluntary and forced migration refugee, environmental migrant, climate changes, and many others. So, for instance, the environmental refugee is the one forced or 'expelled' by an environmental threatening situation against his will. However, the environmental migrant finds himself compelled to leave his home, although by his will and through his decisions, but he moves within a severely limited set of choices. On the other hand, a migrant is the one who chooses to relocate in reaction to a variety of motives, one set among them is economic, by his free voluntary will. Therefore, the concept of freedom of choice greatly influences how we might consider a certain person an environmental refugee, an environmental migrant, and a migrant as such.

Nonetheless, understanding the abstract conceptual links of that term alone cannot give us a deeper understanding of its efficiency in reality. One way to explore the potentials of this efficiency is through explicating the different legal dilemmas involving that concept.

The seriousness of a threat to life and freedom is closely related to displacement due to the fact that human's basic needs are badly affected because of relocation. Legally, there is no legal status or coverage for such displacement. However, the international efforts tried to mitigate the waves of the displacement and prevent the 'environmental flights situation' or at least cope with it. Environmental flights situation is all situations of environmental change which cause a serious threat to life, livelihoods or freedom that possibly trigger involuntary displacement.

Prevention efforts mainly mitigate some the causes of the environmental displacement like the reduction of greenhouse gas emissions, providing sustainable energy use, reducing emissions through carbon budgeting which Paris Agreement and international efforts encourage upon recommendations by UNFCC and according to sustainable development goals (SDGs). Moreover, efforts are also invested into developing adaptation mechanisms like reducing the vulnerability of areas affected by the negative impacts of climate change by delineating risk-zone maps, improving land use, introducing and enforcing coping mechanisms with natural disasters, and improving early alerts system (Ammer & Stadlmayr, 2010).

As per the 1951 Convention, there is no ground that covers environmentally-induced displacement. Hence, none of these displaced persons are recognized as a refugee. However, some countries offer alternative forms of protection to persons who were not recognized as refugees under regional and international law, but who are still in need for protection due to their exposure to risks of serious harm in their countries of origin, and this harm is not related to persecution. These forms of protection are regulated by national legislations. They sometimes include people who are unable to return to their countries for practical reasons in the event of natural or ecological disasters, or where specific compassionate reasons prevail. Complementary and temporary forms of protection were developed to address this challenge. However, there is no coverage for these forms in the international law (Inter-Parliamentary-Union and UNHCR, 2017).

Some literatures also tackle the legal status of displaced persons and refugees upon climate changes in the national and international laws. Ammer and Stadlmayr (2010)'s compressive paper shows how different levels of law act upon the issue of displacement through the limited givens of the international law, humanitarian law, human rights law, regional laws, and national laws in order to mitigate causes of displacement by adopting regulations and obligations which are stemmed from the international environmental law. For instance, Francis's (2015) paper on 'Jordan refugee crisis' tackles refugee crisis in Jordan as a case study to shed light on the different levels of law regarding seeking asylum and refugee. On one hand, Jordan is hosting the second-greatest ratio of refugees to citizens compared to any country in the world and the fifth-largest refugee population in absolute terms since the outbreak of the Syrian conflict in 2011. On the other hand, the Jordanian government has surprisingly underarticulated asylum policy. Jordan did not sign any international convention or protocol that governs the treatment of refugees, including the UN's 1951 Convention and its additional 1967 Protocol. Jordan's lack of international obligations under these treaties affords the government a greater degree of agency in its policy responses to refugee influxes. Moreover, the country has legal obligation to respect the principle of non-refoulment which is widely considered as part of international customary law to which all nation-states must adhere.

Therefore, given Jordan's limited obligations under international law, refugees within the country remain legally uncovered. Jordan does not have a legal obligation to continue admitting refugees; thus, the physical protection of Syrians fleeing their war-zone country is at risk. The kingdom has increasingly turned Syrians away at its borders and reduced freedoms and services for refugees. The humanitarian community has also accused Jordan of repatriating refugees back to Syria, which considers as an explicit violation of the international law (Francis, 2015). In this score, as per Jordan's case study, the national law regulations override the international law level and are binding on the refugees in Jordan.

Such a dilemma of varying legal obligations introduces the question of the status of environmental *refugeeism* amid diverse legal levels. It is important to stress that lacking coherent priorities of legal obligations hampers the existence of the definition in the real life and hampers acting upon it. Without legally binding commitment, a concept becomes just a theoretical definition away from the real practice. It is, therefore, worth mentioning that no matter binding is the regional or national laws, it will never offer displaced people the very much missed international protection over and above narrow geographical scopes.

According to the international law, in the 1951 Convention, which is the main binding legal instrument in regard to seeking asylum and refuge, the refugee status is given to person who fled owing to the well-founded fear of being persecuted upon the five Convention grounds – sex, religion, nationality, membership of a particular social group, and political opinion – and that the person cannot avail him/herself of the protection of the country of origin and he/she is out of his/her country of origin. Hence, no refugee status is granted to a person who fled his/her country of origin upon climate change and natural disasters, and no international legally binding instrument in this score.

In light of the obligations under international human rights law, the state of origin has obligations towards people who are located in its territory or under its jurisdiction since it should preserve both their right to live and their dignity. Moreover, extra-territorial obligations of third states can be deduced from international assistance and cooperation obligations in the context of economic, social, and cultural rights. Those obligations are complementary and legally unbinding. The existence of these extra-territorial obligations with regard to the obligation to respect human rights is widely disputed because it is considered as unacceptable interference in the internal affairs (Inter-Parliamentary-Union and UNHCR, 2017:6).

Nevertheless, there are some obligations stemmed from international treaties on environmental (including climate) protection, like the UNFCCC, to prevent transboundary environmental change that negatively impacts humans and causes displacement and conflict upon resources.

On a European level, the European Convention on the Protection of Human Rights (ECHR) forms the strongest basis for human rights in the continent. In this Convention, the non-refoulement principle is interpreted very restrictively on the regional level, especially in the context of the regime of 'subsidiary protection' of the EC Qualification Directive. This mechanism is triggered by a decision of the Council of the European Community on proposal of the Commission; the mechanism has so far never been used.

On the regional and state level, few efforts giving rise to mitigation of environmental flight situations and coping mechanisms were included in national schemas and policies to reduce the negative impacts of such situations by receiving international displaced persons and providing protection for internally displaced persons. These obligations stem from international environmental law prevention of situations of environmental change by combating the harmful practices that lead to ruining the resources, adopting general environmental protection policies, and sustainable energy use.

Many states started to codify these policies through national regulations and obligations. These regulations and obligations are relevant for two reasons: firstly, they help avoid environmental change; and, as part of national regulations and obligations that citizens should respect, their violation is illegal. These national regulations and obligations are based on international norms on human rights protection (including international refugee law), international environmental law (including climate change law) as well as international norms on civil protection and disaster relief. Regarding coping and adaptation mechanism, many states and international organizations develop coping schemas and policies to reduce vulnerabilities in natural disaster-torn areas (Inter-Parliamentary-Union and UNHCR, 2017:11).

In view of the current lack of legal universal provision in the international law, efforts which are adopted by the States are derived from one of the human rights law's core principles, which is the right to a healthy environment. Since it is an indirect component of other human rights, they try to include it through the reduction of vulnerabilities and adaptation to natural disasters. These obligations are not stipulated within treaty law; they only exist when negative impacts of a natural disaster emerge. This limitation hampers immediate protection as the reaction can generally only be determined and taken after the damage has already occurred. Moreover, the inability or unwillingness of most states that are affected by internal displacement to assume their responsibilities towards internally displaced persons (IDP) remains a problem.

When it comes to internal displacement induced by climate change, IDP guiding principles play vital role with regard to the mitigation of internal environmental flight and stipulate on the rights of persons after their internal flight. But for displaced persons who cross the borders, the Geneva Refugee Convention (GRC), only under certain circumstances, is enabled in the case of environmental destruction used as a weapon against a certain group; this means it is only enabled when environmental destruction is a man-made incident based on persecution against certain group of people. However, since the current State's practice does not follow such an interpretation, legal bases other than the GRC need to be found in order to establish a binding rule by the virtue of the Convention (Inter-Parliamentary-Union and UNHCR, 2017:3–4).

Overall, codifying the environmental refugee status definition into binding laws cross-sectional issue, environmental flight cannot be regulated through measures taken regionally as it must be universal. Unification of norms and strategies in the areas of human rights, humanitarian (disaster) aid, and international environmental law need to be developed to offer optimal protection; all stages of prevention of environmental flight and mitigation of environmental flight need to be processed.

Many challenges will emerge, most notably, the unwillingness of some States to accept internationally binding norms and to agree only on the lowest common denominator or because States consider international binding laws and regulations as an interference into their domestic affairs and against the principles of the State absolute sovereignty. Besides, a legally binding instrument generally requires a lengthy drafting process, long period to reach consensus among parties and considerable long time to reach the ratification process and the entry into force of such an instrument. Nevertheless, only a binding instrument can develop concrete obligations of developing environmental refugee status and, in the case of violations, concrete sanctions are enabled.

In this perspective, UNHCR recognizes 'environmental refugee' in some contexts, which are nexus dynamics as there is potentially a situation where the refugee criteria of the 1951 Convention may apply. For instance, drought led to famine amid armed conflict on ground of race among different ethnic groups. Persons fled such a situation are recognized as refugees in the opinion of UNHCR on the ground of race not because of the drought or the famine. Overall, environmental refugee as definition is not endorsed by the UNHCR; yet it considered persons displaced in the context of disasters and climate change as a concern for UNHCR. However, no refugee status will be granted (UNHCR, 2018).

To sum up, UNHCR recognizes persons who fled climate change as refugees when nexus dynamics exist. Otherwise, it is not applicable. In this regard, efforts are needed to address such knowledge gap and come up with legal instrument to grant protection to whomever in need for it.

When it comes to the national State's contribution to the abovementioned issue, States who used regional and local refugee law to provide international protection in these complex situations have traditionally been limited. Below are some examples of different national State's contributions.

It has been done historically, Kenya recognized Somalis who arrived in 2011–2012 as refugee after the famine that was caused by the drought in southern Somali, secured them territorial entry and residency's permits. Ethiopia also maintained its historical stance in the same score of the Somali's famine in 2011, with access to its territory and grant refugee status. Somalis were recognized within the framework of Ethiopia's domestic refugee law, predominantly pursuant to broader refugee criteria. Most suggestion went to that Ethiopia may viewed the repercussions of climate disasters as potentially provoking claims that could satisfy the broader refugee criteria under the OAU Convention[1] (Weerasinghe, 2018).

In the aftermath of the 2010 earthquake in Haiti, Mexico as neighbor country implemented ad hoc measures within its migration framework to exceptionally permit specific categories of Haitians to enter and stay on a temporary and humanitarian basis. Moreover, the Mexican government secured them access to RSD process. Research indicates that in Mexico, some Haitians – who were victims of the 2010 earthquake – were recognized under broader refugee criteria on the ground of disruptions to public order. It appears that Mexico's refugee authority tried its best to assess Haitian claims under refugee law, including on how to impose broader refugee criteria on Haitian's case. Informants pointed out that assessing claims and complying with them with the boarder definition were tricky and difficult as Haitians suffered from five serious psychosocial harms and struggled to State consistent claims (Weerasinghe, 2018).

[1] Convention Governing the Specific Aspects of Refugee Problems in Africa.

It was believed that national governments while assessing the claims of displaced persons due to natural disaster, considered that the consequences of a disaster may do so, which is somehow a consideration of the dynamics nexus of the disaster. Both UNHCR and national governments consider the dynamics nexus directly or indirectly. However, national governments put it into practice and granted refugee status without waiting for the nexus to happen.

In conclusion, tackling the status of environmental asylum\refugeeism allows one to take a look at a complicated web of contending and diverse web of laws and obligations where international laws and organizations might not be in accord with the needs of a regional, nation-state, and the displaced groups situation. This burdens legal, humanitarian, and political circles with the duty of trying to find ways to consider the situation of environmentally displaced persons amid the currently entangled web of legal levels.

6 Environmentally Induced Displacement Case Studies

Not surprisingly that the world has been witnessing a massive influx of displaced people due to climate change amid holding UN Climate Action Summit 2019 that aimed at accelerating actions to implement the Paris Agreement which stipulates reducing carbon emissions within 12 years and holding the increase in the global average temperature to well below 2 °C.

"The climate crisis has arrived and is accelerating faster than most scientists expected. It is more severe than anticipated, threatening natural ecosystems and the fate of humanity" (Ripple, 2019). This statement was published in the latest edition of the bio science journal to point out the crisis which the world faces nowadays impacting the human's fates, displacement, and migration waves.

By tracing the global displacement waves, African countries will be on the top of the countries where climate change leads to massive displacement. Not to mention that the scarcity of water and food resources triggers conflicts aimed at securing basic needs. The Horn of Africa is currently experiencing many natural disasters like famine and drought, largely as a result of below average rains. Accordingly, dry conditions across the region spread out and have led to the deterioration of farmland, loss of livelihood means, and food and water scarcity in large areas in Somalia, Ethiopia, and Kenya where people have been unable to secure their basic food needs and were driven to relocate internally and across borders (Start-Network, 2019).

The numbers of displaced people that were driven by drought witnessed a massive increase from 739,000 in November 2016 to 46,000 in May 2017. On the contrary, the number of returnees is relatively lower since only 7,700 IDP were recorded as returnees in May 2017 (UNHCR, 2017). Given the aforementioned, the number of IDP returnees is still limited compared to the increase in the displacement witnessed, representing less than 1% of the drought-related displacements over 7 months (Start-Network 2019).

This makes climate-induced displacement a grave phenomenon that needs to be understood further if one wants to grasp climate change properly. For this reason, three case studies are to be explained in order to shed light on the complexities of climate, social\political pressures, and displacement. The three cases are the Somali case, the Nubian case, and the Syrian case. They are chosen mainly for their sophisticated mixture of integrating climate, socio-political pressures; and the need for displacement.

6.1 The Somali Case

It is important to indicate that the reason for choosing Somalia as a case study is because it is one of the Horn of Africa countries that are dramatically affected by drought and witnessed drought-driven displacement. Indeed, Somalia is one of the countries where people suffered from severe severe drought for many years. However, the severity was different till we reached the drought which occurred in 2015–2016 and was named as El Nino phenomenon. It had devastating impacts on the livelihood of local communities in Somaliland and Punt land, which result in out-migration, food insecurity, and death of livestock. This severe drought was a result of two consecutive seasons of poor rainfall. In the most affected areas, wide-scale crop failure and high levels of livestock deaths occurred. Moreover, malnutrition and drought-related diseases were raised. Somalia experienced its largest outbreak of cholera in five years, with 48,607 cases and 763 deaths reported in 48 districts since January 2017 (OCHA, 2017). Accordingly, competition and clashes over resources to secure the locals basic needs existed. It was pointed out in a report by the Office for the Coordination of Humanitarian Affairs (OCHA) 2017 that over 6.2 million people, roughly half of the population, needed humanitarian assistance (OCHA, 2017). Given the aforementioned factors, massive waves of displacement took place intensively to the neighboring country, Ethiopia. An estimated 738,600 people have been displaced by drought between November 2016 and May 2017 and relocated to Ethiopia (OCHA, 2017).

The ramification of this drought was severe especially for the population as 3.2 million people were severely food unsecured and three consecutive poor harvest seasons occurred because of the lack of rains. Malnutrition, one of the leading indicators of the emerging crisis, has reached emergency levels in many locations in southern and central Somalia because the prices raised sharply due to food scarcity. Repeated exposure to drought has sharpened the vulnerability of the population, which results in massive waves of displacement and increased conflicts and insecurities (OCHA, 2019).

Drought as a result of climate change drives displacement and raises protection concerns and challenges to secure basic human needs after losing socio-economic safety nets, particularly for the most vulnerable groups as women and children.

The Somali authorities not only failed to provide their population the protection they need but also hinder the ability of humanitarian workers to reach people in need by arrests, detentions, and suspension of programmes (Start-Network, 2019). Accordingly, affected people who could not benefit from their State's protection or the humanitarian assistance chose to relocate across borders. Even if displaced persons were fleeing the precarious situation to nearby countries like Ethiopia, they were subject to many risks and dangers due to the restrictions by the Ethiopian authorities which forced the displaced people to return to their place of origin (Yarnell, 2018).

Although Ethiopia maintained its historical stance in the same score of the Somali's famine in 2011, with authorizing access to its territory and granting refugee status to the displaced. Somalis were recognized within the framework of Ethiopia's domestic refugee law, predominantly pursuant to broader refugee criteria. However, that happened once and Ethiopia does not grant refugee status to those who fled after the drought occurred in 2016–2017.

No one of the Somali displaced persons was given the status of the refugee because they lacked the persecution aspect which is the core of the refugee definition according to the 1951 Convention. Although they fled the precarious situation of conflict and food and water scarcity, they are not considered refugees and, consequently, are in need of persistent protection according to international law.

6.2 The Syrian Case

Water scarcity is a critical issue facing the Middle East, more than any other region. Countries in this region have devoted a lot of money to resolve the diminishing supply of water. Water and food scarcity are migration and refuge nexus. Whenever displacement is found, food and water scarcities emerge, thus increasing the burden of the displaced people and deepening their vulnerabilities. Jordan witnessed the displacement influx of Syrians amid the outbreak of the conflict in Syria since 2011. However, Jordan is commonly regarded as one of the most water-poor countries in the Middle East with just 145 cubic meters of water per person per year, considerably less than the annual water shortage level of 500 cubic meters per person per year set by the United Nations (Gluck, 2013).

As the number of displaced persons became massive, camps and foreign assistance became inadequate to reduce the effects of the influx on Jordan's water supply. According to MercyCorps, in the Northern Municipalities, where the majority of Syrian refugees stay, the average daily water supply has dropped below 30 liters per person. 80 liters per day per person is needed to satisfy basic needs (MercyCorps, 2014). Moreover, many areas near refugee camps are forced to run out of water for days to come, resulting in an approximately 50% drop in the amount of water made available to Jordanian residents (Gluck, 2013). Hence, the citizens blamed it on the Syrians' displacement. Meanwhile, displaced persons were extremely affected given their multiple vulnerabilities. Discrimination acts spread out against the Syrian displaced persons as they considered them more privileged by the State to be provided by the water. However, theoretically and practically, accumulative acts of discrimination eventually shape persecution which is among the main core elements of the refugee definition. Water scarcity seems to be far away from the refuge's issues. However, it is a starting point for more vulnerabilities that have been added to the existent ones that increased the sufferings of displaced persons (Klingseis, 2016).

In the past year, the population of Azraq Camp has risen significantly from 15,000 to nearly 54,000 according to the UNHCR. This increase occurred in just a short period between March and May 2016, as people moved from the Jordan-Syria border area called 'The Berm', exerting considerable pressure to increase the scale of the water supply. The Jordanian government, from its side, tried to contain the problem. However, Syrian displaced persons continue to suffer from multiple vulnerabilities starting with their status as displaced people, poor conditions, and water scarcity which led to discrimination by the Jordanian population (ACTED, 2016).

Moreover, hospitals and schools don't have enough water to maintain sanitation standards. Mosques cannot perform the regular ablutions that are required. The pipelines run dry, particularly in hot, dry summers. Often, weeks go by until a drop comes out of the tap. The lack of water adds a lot of pressure on displaced people. So, it is not persecution which is the only reason to cause hardships for people or hamper their access to basic food and water needs, but also water scarcity and drought induced by climate change may become primary causes of hardships. Furthermore, persecution is not the only factor triggering conflicts, but also the water and food scarcity may cause disputes, hatred, and conflicts to secure resources. These combined presses a need to draft legal instruments and mobilize efforts to find ways to shed light on the impacts of climate change on displaced livelihoods and highlight this issue within the chaotic web of socio-political difficulties.

6.3 The Nubian Case

It is important to indicate that the reason for choosing the Nubian case, aside from the authors' personal preferences which cannot be denied in any social research, is the complexities regarding its evolution. The Nubians' forced migration and agreed relocation were but a series of reactions to four main events in the history of this regional group. The first was in 1902 when the British colonial authorities decided to build Aswan Low Dam, and then two consecutive processes of raising the dam in 1912 and 1933; all leading to submerging more of the Nubian villages and forced immigration of the Nubians accompanied by governmental decisions of compensation, which are seen by many Nubians as inadequate. The fourth and most crucial event is the relocation of the Nubian groups following the decision of building the

High Dam. Nubians in 1963–1964, after governmental promises of compensation and a discussion with President Nasser, decided to relocate to northern Aswan. However, Nubian accounts narrate that neither the governmental promises were totally fulfilled when the relocation was undertaken, nor the partial fulfillment of these promises was adequate in any decent manner, not to mention in comparison with how Sudanese Nubians were compensated (Shami, 1993; Eshhad, 2018; Noshokaty, 2013).

Back to our research, a question arises: how can we classify the Nubian case that shifted between forced migration to agreed relocation, and included broken promises, which had their disappointing results been clear from the outset, no relocation would have occurred? According to Bates' theory, the Nubian case would be a case of 'environmental refugees' in the three events before building the High Dam, and in the event of the High Dam relocation, it would be a case of environmental migrants; but the Nubian case would never be a mere case of 'migrants'. The crucial element that guides this classification is the margin of freedom which the migrants/refugees have when deciding to migrate or not. The less options they have the more they are 'refugees', the more these options increase, the more they become an environmental migrant. This process continues until we reach the status of potential 'migrants' who migrate to improve their future.

In the events of Aswan low dam which were initiated by official decrees, though with compensation, the Nubians had no room but to migrate because there were no chances of negotiations or rejection. While in the high dam construction process, which is quite complicated, some discussion occurred (Mohamed, 2013), and prior studies were undertaken (Shami, 1993), and a referendum was made (Noshokaty, 2013) which somehow indicates a relatively wider margin for freedom that makes Nubians more of 'environmental emigrants' not 'refugees'.

Additionally, during the events in which the Nubians were 'environmental refugees', in the light of Bates' theory, Nubians were refugees as a result of expropriation made by intentional acute environmental damages that are intentionally producing refugees. Not unintentional damages, nor a damage that unwillingly makes refugees, which underscores the responsibility for damage made against the refugee group. A damage of this type requires reparation of some sort. Even though the Egyptian government move of offering compensation was on time, the quality and quantity of this compensation was seen inadequate. This implies that the Egyptian government took its responsibility of helping Nubians, but in an inadequate manner.

Nevertheless, when it comes to everyday politics, what makes this slight change of status from 'refugee' to 'migrant' matter? Despite the legally disputed nature of the 'environmental refugee' concept, the status of a 'refugee' entails the responsibility of some sort, domestic and external. It somehow justifies a call for external partners/benchmarks to arbiter, intervene, or have some say. Being a refugee, although it might be a mere label, is a problem for sovereignty. Once one is considered a refugee, the State presupposed absolute sovereignty over him/her becomes questioned, and no state would like to face that.

7 Conclusion

The concept of 'environmental refugee' faces different criticism(s) and debates towards its political and legal validity/invalidity, the very discourse upon which it projects itself, and its academic validity. The very history of the concept shows how it was formed and re-formed in a climate of political and social tensions. This made many academics, jurists, and international organizations offer their views on the phenomenon of climate-induced displacement. In doing so, their contending aims, views, and conceptualizations not only helped illuminate the concept of environmental refugees but also showed the actual challenges that this concept and the people it represents are destined to face, especially amid the lack (or inadequacy) of any international legal instrument for protection. Moreover, the available protection methods are by themselves debatable and exposed to a dilemma of obligations over different legal levels. Understanding the dilemmas of the concept and its legal duels through the Somali, Syrian, and Nubian case studies showed how far such a concept remains in need of further efforts and how international actors have to find ways to fill the gap made by a climate situation never tackled legally before. Working on the concept of environmental refugee will ease dealing with emerging emergencies on account of climate change. Climate change is expected to worsen the threats that force people to displace and increase the severity of its resultant conflicts. Finally, the feasibility of labeling someone or some group 'refugee(s)' might look like a mere hassle over naming, yet at the core of this problem, the very sovereignty of the concerned State(s) is put into question.

References

ACTED. (2016, December 21). Providing clean and safe water in Azraq refugee camp in Jordan. *News and Press Release*. ACTED. Retrieved June 1, 2020, https://reliefweb.int/report/jordan/providing-clean-and-safe-water-azraq-refugee-camp-jordan

Ammer, M., & Stadlmayr, L. (2010). *Legal status and legal treatment of environmental refugees*. Environmental research, IPCC Website November 2010, federal ministry of the Environment, Nature Conservation and Nuclear Safety. Retrieved May 31, 2020, https://www.ipcc.ch/apps/njlite/srex/njlite_download.php?id=6671

Bates, D. (2002). Environmental Refugees? Classifying human migration caused by environmental change. *Population and Environment, 23*(5), 465–477.

Black, R. (2001). *Environmental refugees: Myth or reality? New issues in refugee research* (UNHCR Working Paper No.34). IPCC. Retrieved May 31, 2020, https://www.ipcc.ch/apps/njlite/ar5wg2/njlite_download2.php?id=11038

Boano, C., Roger, Z., & Morris, T. (2008). *Environmentally displaced people: Understanding the linkages between environmental change, livelihoods and forced migration* (Forced Migration Policy Briefing 1). University of Oxford.

Castles, S. (2002). Environmental change and forced migration: Making sense of the debate. *UNHCR: Working Papers, 70*, 1–14.

Dennery, E. (1931). *Asia teeming millions: And its problems for the West*. Jonathan Cape.

El-Hinnawi, E. (1985). *Environmental refugees*. United Nations Environmental Programme.

Eshhad. (2018, February). *Minority profile: Egypt's Nubians*. Eshhad, Center for The Protection of Minorities. Retrieved June 1, 2020, https://static1.squarespace.com/static/5947e4266a49635915ac0a31/t/5a79255f419202954ded4793/1517888880708/MP+

Francis, A. (2015). *Jordan's refugee crisis*. Carnegie Endowment for International Peace. Retrieved May 31, 2020, https://carnegieendowment.org/files/CP_247_Francis_Jordan_final.pdf

Gluck, C. (2013). *Syrian refugee influx adding to Jordan's water worries*. Oxfam Canada. Retrieved June 1, 2020, https://www.oxfam.ca/blog/syrian-refugee-influx-adding-to-jordans-water-worries/

Gorlick, B. (2007, May 16). *Environmentally-displaced persons: A UNHCR perspective*. www.ony.unu.edu/seminars/2007/16May2007/presentation_gorlick.ppt

Inter-Parliamentary-Union and UNHCR. (2017). *A guide to international refugee protection and building state asylum systems*. Handbook for Parliamentarians, UNHCR & IPU. Retrieved May 31, 2020, https://reliefweb.int/sites/reliefweb.int/files/resources/3d4aba564.pdf

IOM. (2007, November 1). *Discussion note: Migration and environment*. Ninety Fourth Session. Retrieved May 31, 2020, https://www.iom.int/jahia/webdav/shared/shared/mainsite/about_iom/en/council/94/MC_INF_288.pdf

Klingseis, S. J. (2016). *Syrian refugees: Are they a non-traditional threat to water supplies in Lebanon and Jordan*. Naval Postgraduate School. Retrieved June 1, 2020, https://calhoun.nps.edu/bitstream/handle/10945/50569/16Sep_Klingseis_Stephen.pdf?sequence=1&isAllowed=y

Lonergan, S. (1998). The role of environmental degradation in population displacement. *Environmental Change and Security Project Report, 4*, 5–15.

Mcdonald, T. (2015). The man who would be the first climate change refugee. in *BBC News: Asia*. Retrieved June 1, 2020, http://www.bbc.com/news/world-asia-34674374

MercyCorps. (2014). *Tapped out: Water scarcity and refugee pressures in Jordan*. Mercy Corps. Retrieved June 1, 2020, https://www.mercycorps.org/sites/default/files/2019-11/MercyCorps_TappedOut_JordanWaterReport_March204.pdf

Mohamed, T. (2013). *The Nubah. Forced migration misery continues (AlNubah.Ma'saat Al-Tahgeer Mostamerra (Arabic))*. The Revolutionary Socialists Movement. Retrieved June 1, 2020, http://revsoc.me/http://revsoc.me/politics/lnwb-ms-lthjyr-mustmr//lnwb-ms-lt

Hamdawy Mohamed. (2015). The environmental refugee from material existence to the hope of international legal protection, Al-Lage' Al Bee'ei Min Al-Wogoo Al-Maadeyy Ela Amal Al-Hemaya Al-Qanooneyya Al-Dawleyya (Arabic). *Revue Des Etudes De Droits, 2*(2), 109–148.

Morrissey, J. (2012). Rethinking the 'debate on environmental refugees': From 'Maximalist and Minimalists' to 'Proponents and Critics'. *Journal of Political Ecology, 19*, 36–49.

Myers, N. (2005). *Environmental refugees: An emergent security issue*. Prague: 13th Economic Forum, Session III: Environment and Migration, Organization for Security and Cooperation in Europe. Retrieved May 31, 2020, http://www.osce.org/eea/14851?download=true

Noshokaty, A. (2013). *Egypt Nubia : 50 years of Displacement*. Ahramonline. Retrieved May 31, 2020, http://english.ahram.org.eg/NewsContent/32/99/84158/Folk/Special-Files/Egypt-Nubia%2D%2Dyears-of-displacement-.aspx

OCHA. (2017, November 3). *Horn of Africa: Humanitarian impacts of drought*. issue 11. Retrieved June 1, 2020, https://reliefweb.int/sites/reliefweb.int/files/resources/HOA_drought_updates_snapshot_Nov032017.pdf

OCHA. (2019, December 1–31). *humantarian bulletin: Somalia* (OCHA Services). Retrieved June 1, 2020, https://www.humanitarianresponse.info/sites/www.humanitarianresponse.info/files/documents/files/december_2019_humanitarian_bulletin.pdf

Rebert, T. (2006, December 15). *The rising flood? Environmental refugees in a political ecology perspective*. Available at https://www.macalester.edu/geography/wp-content/uploads/sites/18/2012/03/rebert.pdf. Last accessed in 31 May 2020.

Reddy, B. S., & Assenza, G. B. (2009). The great climate debate. *Energy Policy, 37*, 2997–3008.

Ripple, W. J. (2019). World scientists' warning of a climate emergency. *BioScience, 70*(1), 8–12.

Saunders, P. (2000). Environmental refugees: The origins of a construct. In P. Stott (Ed.), *Political ecology: Science, myth and power* (pp. 218–246). Arnold.

Shami, S. (1993). The social implications of population displacement and resettlement: An overview with a focus on the Arab Middle East. *The International Migration Review, 27*(1), 4–33.
Start-Network. (2019). *Drought in the horn of Africa* (ACAPS Thematic report, OCHA Services). Retrieved June 1, 2020, https://reliefweb.int/sites/reliefweb.int/files/resources/20190722_acaps_start_thematic_report_horn_of_africa_drought.pdf
Suhrke, A. (1991). *Pressure points: Environmental degradation, migration And conflict*. Workshop on Environmental Change, Population Displacement, and Acute Conflict, Peace and Conflict Studies Program, University of Toronto and The American Academy of Arts and Sceinces, Ottawa.
UNHCR. (1951). *1951 Convention*. UNHCR.
UNHCR. (2017). *PRMN drought displacements to 31 May 2017* (Situation Report, OCHA Services). UNHCR. Retrieved June 1, 2020, https://reliefweb.int/sites/reliefweb.int/files/resources/57361.pdf
UNHCR. (2018). *climate change and disaster displacement*. UNHCR. Retrieved May 31, 2020, https://www.unhcr.org/climate-change-and-disasters.html
UNSD. (2020). *Environmental Refugee*. Environment Glossary. Retrieved May 31, 2020, https://unstats.un.org/unsd/environmentgl/gesform.asp?getitem=473
Warner, K. (2018). Coordinated approaches to large-scale movements of people: Contributions of the paris agreement and the global compacts for migration and on refugees. *Population and Environment, 39*, 384–401.
Weerasinghe, S. (2018). *International protection in the context of nexus dynamics between conflict or violence and disaster or climate change* (Legal and Protection Policy Research Series). UNHCR. Retrieved June 1, 2020, https://www.unhcr.org/5c1ba88d4.pdf
Yarnell, Mark. (2018, November). *The crisis below headlines: Conflict displacement in Ethopia*. Field Report, Refugees International. Retrieved June 1, 2020, https://reliefweb.int/sites/reliefweb.int/files/resources/FINAL%2BEthiopia%2BReport%2B-%2BNovember%2B2018%2B-%2BFinal.pdf

Chapter 5
Looking Ahead: A Human Security Perspective to Tackling the Potential for Widespread Environmental Migration in Latin America

Becca Andrasko

Abstract By 2050, between 25 million and 1 billion people globally could be displaced by extreme weather events, landscape degradation, and the environmental impacts of climate change. Environmentally displaced persons (EDPs) are people who are compelled to leave their homes in either the short- or the long-term because of an environmental hazard that affects their livelihoods, health, or physical safety. Latin America is a particularly important – and relatively understudied – region in which to explore migration driven by climate change, environmental degradation, inequalities, and conflict. This chapter advocates for a regional approach to future pathways of environmental migration in Latin America. Key to this approach is to first begin to sustainably manage natural resources and aggressively practice climate change mitigation activities within the region. These close ties could make migration slightly easier within the region than outside it and could drive more cross-border migration than in other regions. Long-term risk planning efforts must improve international and regional cooperation, capacity-building, and adaptive management of refugee resettlement programs, in order to address the reality of future environmental migration in an orderly manner. Migration should be considered as a valid adaptation approach in the face of rapid-onset events and for lowering exposure to slow-onset events and should be included in climate change adaptation action plans. Adaptive and forward-looking planning for environmental displacement is a way to avoid the anxiety, scrambling for funding sources, and sudden state of emergency in countries often associated with new refugee flows.

Keywords Climate change · Environmental migration · Latin America · Migration policy · Regionalism

B. Andrasko (✉)
Foreign Policy Magazine, Washington, DC, USA

M. Behnassi et al. (eds.), *The Climate-Conflict-Displacement Nexus from a Human Security Perspective*, https://doi.org/10.1007/978-3-030-94144-4_5

1 Introduction

No longer something for future generations to deal with, climate change and environmental degradation have already propelled migration in Latin America. Central American's northern triangle, made up of Honduras, El Salvador, and Guatemala, has long been a major source of migrants, both internally and to neighboring countries, notably to Mexico, Costa Rica, and the United States (Blitzer, 2019; Carte et al., 2019). There has been a spike since 2014 of Central American migration to the United States, primarily because of climate change impacts, particularly drought related to El Niño in the Dry Corridor that stretches across El Salvador, Honduras, and Guatemala (Steffens, 2018a, b). In 2018, 50,000 families were turned away at the U.S. border, thus doubling the number of the year before. By the first half of 2019, 66,000 families were apprehended at America's southern border (Blitzer, 2019). Some migrants that reach the United States' border surrender themselves to American law enforcement to claim asylum – which would formalize their status in the United States (Cheatham, 2019). About 13% of those that claimed asylum were granted status in 2018, which accounts for more than double the number of acceptances in 2015 (Ibid).

The three northern triangle countries, along with Nicaragua, are consistently listed among the twenty world nations that are most vulnerable to climate change (Bouroncle et al., 2017). Between 1992 and 2011, "Honduras was the world's most vulnerable country to extreme weather events, Nicaragua the 3rd most vulnerable, Guatemala the 11th, and El Salvador the 15th" (Bouroncle et al., 2017: 124). Aside from weather-related natural disasters, particularly hurricanes and cyclones, Central America has experienced increases in temperature, stronger dry seasons, and droughts but also more frequent floods and reduced soil fertility (Bouroncle et al., 2017; Steffens, 2018a, b). Agriculture is particularly vulnerable to climate change-related natural disasters and impacts (Bouroncle et al., 2017). Guatemala's economy has long been built on the back of an agro-export model that eschews smallholder production in favor of large-scale extractive agricultural industry, particularly crops like sugar cane and rubber (Carte et al., 2019). Two-thirds of Guatemala's land is held by only 2.5% of farms, leaving little land for almost half of the country to sustain themselves with rural, smallholder farming (USAID, 2010; Carte et al., 2019).

The seeds of these well-trodden migratory pathways away from Central America were planted decades ago, with neoliberal economic policies, resource, and land inequality. The changing climate, civilian-targeted gang violence, and the fallout of civil wars accelerated the dynamic that has left millions displaced (Carte et al., 2019). With climate change serving as a threat multiplier for existing issues in the region – and in countries across Latin American and the world – increased migration is a very real consequence of environmental degradation and climate change that the international community must consider when laying future plans and strategies.

This chapter will analyze the ongoing and likely future of environmental migration in Latin America. Section 2 begins with a discussion of the drivers, scope, and impacts of environmental migration, and then dive into original research on environmental conflict in Latin America that has the potential to catalyze migration. Section 3 lays out the history of the refugee regime, from the 1800s through the present day, along with international laws and cooperation mechanisms to address refugee flows. As for Sect. 4, it presents four potentially interlocking options for cooperation and compromise for future environmental migration pathways in Latin America and will advocate for a potential path forward based on a regional approach. Section 5 concludes with an assertion that compassion is the most important precondition for establishing successful plans for potential environmental migration pathways.

2 Environmental Migration, Now and in the Future

The International Office of Migration (IOM) estimates that by 2050, between 25 million and 1 billion people globally could be displaced by extreme weather events, landscape degradation, and the environmental impacts of climate change (Bassetti, 2019; IOM, 2009). A common prediction is that 200 million people are predicted to migrate due to environmental pressures by 2050 – or one out of every forty-five people (IOM, 2009; McMichael et al., 2012). This figure includes individuals who may be displaced across international borders or to another area within their country of origin (Bassetti, 2019). The majority of current environmental migration patterns are internal to states, but it is highly probable that migration to neighboring or far-flung states will likely become more of a norm in the coming decades (Cristiani et al., 2020). Latin America, sub-Saharan Africa, and Southeast Asia will likely produce 143 million environmentally displaced persons by 2050 (Kumari Rigaud et al., 2018). Approximately one-third (around 23 million people) of the 68.5 million people who were forcibly displaced in 2017 were displaced because of hazardous environmental conditions (Podesta, 2019). As the effects of climate change and environmental degradation become more intimately experienced by individuals, migration to climates or areas perceived to be safer is likely to follow, in most cases as the last adaptation strategy.

Environmentally displaced persons (EDPs), as these migrants will be referred to in this chapter, are people who are compelled to leave their homes in either the short- or the long-term because of an environmental hazard that affects their livelihoods, health, or physical safety. The term 'environmental refugees' or 'climate change refugees' has also been used since the 1970s to refer to, legitimize the study of, and raise awareness of environmentally displaced persons (Morrissey, 2012; Keyes, 2018). The concern about the potential increase of environmental migration has been recently raised because it has generally been framed as "a security issue for rich countries supposedly threatened with a flood of 'environmental refugees' from the South" (Boas, 2015 in Piguet et al., 2018: 3). For this reason, the term 'environmentally displaced persons' will be used instead of the more controversial 'environmental refugees'.

Critics, like Keyes (2018), have rightfully pointed out that it is difficult, and can be problematic, to untangle the reasons behind why certain populations choose to migrate. The term 'environmentally displaced persons' captures equally "the wealthy family selling frequently flooded property in South Carolina to move away from the coast, and the poor family who has to abandon land that is no longer arable in Niger and move to the capital in search of a livelihood" (Ibid: 3). Movements – which may be voluntary, compelled, for short-term gains, permanent, in the distant future, inside a country, and across international borders – are all included in the definition of environmental migration (Ibid). There have been very few concrete cases of long-term forced migration for purely environmental reasons, without taking into account economic or social reasons – the clearest case study is of several Pacific Islands that are slowly sinking into the ocean (Podesta, 2019). While many other case studies have been cited in which environmental factors primarily or potentially drive out-migration, there are almost always other reasons that migrants could identify as primary to their decision (Ibid). Like in the Central America example, insufficient land access, food insecurity, and loss of livelihoods for fishermen and farmers may be top-of-mind concerns for those emigrating, even though all of those reasons are affected or driven by degrading environmental conditions (Markham, 2019). When discussing migration from Central America to the United States, a Guatemalan forestry expert told a New Yorker reporter that "There are always a lot of reasons why people migrate…Maybe a family member is sick. Maybe they are trying to make up for losses from the previous year. But in every situation, it has something to do with climate change" (Blitzer, 2019: n.p.).

2.1 How Environmental Degradation Results in Displacement

Environmental vulnerability includes physical, economic, political, and social factors that affect people at the individual, local, and national levels. The environmental vulnerability affects rural communities more directly than city-dwellers, thus rural individuals make up the largest proportion of environmental migrants (Marshall, 2016). Behavior in the face of environmental vulnerability is correlated with the spatial area, as individuals from rural communities tend to move within a country or region, while city-dwellers move to farther, more desirable destinations (Mezdour et al., 2016). While environmental variability incentivizes moving, poor individuals without far-flung social connections may be unable to move or migrate (Black et al., 2001). Migration allows risk-takers the chance to have, but is no guarantee of, a better future.

Climate change and environmental degradation can compel migration along with two intertwined timelines – rapid-onset events and slow-onset events. Rapid-onset events include acute natural disasters such as hurricanes, tsunamis, or wildfires that tend to be discrete and occur over a short period (UNFCCC, 2012; Desai et al., 2018). The United Nations (UN) estimates that about 1.6 billion people "have lost their homes or livelihoods or have suffered other damage" since between 2000 and

2006 due to rapid-onset events, "which continues an upward trend over the past several decades and represents a four-fold annual increase, on average, from the decade of the 1970s" (Schwartz, 2006: n.p.). Slow-onset events, like sea level rise, land degradation, desertification, and stresses to the water supply, tend to erode the living conditions for people in an area over time to a degree that it is no longer possible to stay in such an area, thus driving out-migration (UNFCCC, n.d.; Warner, 2010). Unaddressed slow-onset events can create the underlying conditions for rapid-onset events: for instance, it could happen when gradual sea-level rise eventually floods houses situated along the shore, and leave communities vulnerable to other types of disasters as they recover from a natural disaster (Desai et al., 2018) (Box 5.1).

Box 5.1: Case Study – Natural Disasters in Latin America

According to the United Nations Office for the Coordination of Humanitarian Affairs (OCHA) (2020), Latin America is the second most disaster-prone global region after Asia. Between 2004 and 2019, 73% of the 124 short-term surge deployments by OCHA's Regional Office for Latin American and the Caribbean (ROLAC) were due to natural disasters. ROLAC has also deployed 33 times since 2015 in response to migration – notably to Venezuelan migration throughout the region – and/or political unrest. Between 2000 and 2019, 152 million people in Latin America were personally affected by the 1205 disasters that occurred in the region (OCHA, 2020). Natural disasters are highly correlated with migration displacement: for example, around 100,000 people moved from Haiti to South America following the 2010 Haitian earthquake, and flooding displaced people across international borders from Colombia to Ecuador and from Bolivia to Brazil (Cristiani et al., 2020). Brazil is also on the list of the top 20 countries to have the highest number of displaced people due to natural disasters in the period from 2008 to 2014 (Cristiani et al., 2020).

Floods are the most common natural disaster in Latin America, with 548 floods between 2000 and 2019, followed by storms and hurricanes (330), earthquakes (75), droughts (74), landslides (66), extreme temperatures (50), volcanic events (38), and wildfires (24) (OCHA, 2020). Climatologic and meteorological threats are globally considered the deadliest and most economically impactful natural disasters (Gu, 2019). Droughts affected 53 million people in Latin America between 2000 and 2019, making them the disaster that affected the most Latin Americans in that period (Ibid). The 2018 drought season was particularly dangerous in Central America's "Dry Corridor" in parts of Guatemala, El Salvador, Honduras, and Nicaragua, as crop yields were reduced 50–75% and 2.2 million people were left food insecure (Ibid). Droughts are related to the El Niño phenomenon, while cooler weather and flooding tend to be associated with La Niña, global climate patterns that are caused by cyclical changes in the Pacific Ocean's temperature and last between 9 to 12 months (Ibid; NOAA, 2021).

The number of natural disasters has dramatically increased over the last half-century, much of which is due to climate change (Caruso, 2017). According to government estimates, natural disasters cost governments across the world an annual average of $901 million, especially because of damage to costly infrastructure – and this is only likely to increase over the next century (Caruso, 2017). Twelve floods alone in the decade of the 2000s cost Latin American governments $1 billion in damages and all 548 floods in that decade affected 41 million people across Latin America (OCHA, 2020). It is estimated that flood damage, if flood risk remains constant, could reach upwards of $60 to 63 billion globally by 2050 (Gu, 2019). An economic analysis of risk to natural disasters by Gu (2019) found that Mexico City and Guadalajara, Mexico, Santiago, Chile, and Bogotá, Colombia were among the largest cities in the world with an outsized risk of mortality from natural disasters. Gu (2019) also found that the medium-sized Costa Rican cities of Alajuela and Heredia were two of the four world cities with the highest risk of large-scale economic losses from natural disasters. However, a study conducted in eight countries in Latin America in 2011 found that 60% of natural disasters between 1980 and 2010 happened in small cities with populations under 100,000 (UNISDR, 2011; Gu, 2019). As it is estimated that by 2030, 20–25% of urban population growth will be concentrated in smaller cities, it is essential to take into account risk and plan for resilience and vulnerability reduction in both megacities and smaller urban developments (United Nations, 2018; Gu, 2019).

Slow-onset events can also exacerbate and serve as a threat multiplier for violent conflict among communities (UNFCCC, 2012). For example, land degradation and desertification can result in a loss of livelihoods, which in turn can lead to conflict and displacement (UNFCCC, 2012; Desai et al., 2018). Percival and Homer-Dixon (2001: n.p.) identified three forms of environmental conflict:

- "Supply-induced scarcity is caused by the degradation and depletion of an environmental resource, for example, the erosion of cropland;
- "demand-induced scarcity results from population growth within a region or increased per capita consumption of a resource, either of which heightens the demand for the resource; and
- "structural scarcity arises from an unequal social distribution of a resource that concentrates it in the hands of relatively few people while the remaining population suffers from extreme shortage".

Commodity dependence is one facet of the 'resource curse' that tends to afflict natural resource-rich countries with low-income rates and worse development and democratic outcomes (Collier & Hoeffler, 2005; Raftopolous, 2017). The resource curse theory argues that "mineral and fuel abundance generates negative developmental outcomes in less developed countries" (Di John, 2011: 167). The link between commodity dependence and a higher propensity for conflict is strongly defined in the literature, although there is disagreement about the causal mechanism (Collier & Hoeffler, 2005). Economists argue that with poor economic indicators brought about by the 'resource curse', the opportunity costs of conflict are lowered and the potential spoils are more attractive (Ibid). Political scientists claim that the better explanation is that states that rely on commodities tend to be weak – they depend on patronage systems and do not bother to develop strong political and economic institutions and maintain the rule of law (Collier & Hoeffler, 2005; Humphries, 2005). Humphries (2005) notes that while 'resource curse' countries are more likely to experience conflict, the wars tend to be concluded more rapidly and with a higher likelihood for a decisive military victory than the average conflict because both sides (and often external actors) have economic interests tied to further exploitation of natural resources that cannot happen in the midst of conflict.

Environmental conflicts can also range in scale, from interpersonal conflict among individuals to international wars (Temper et al., 2015). For example, water scarcity and increasing natural resource scarcity were direct drivers of deteriorating economic conditions in Syria, which in turn led to the devastating Syrian civil war and migration flows to Europe (Gleick, 2014). The relevance of even small-scale environmental conflict to broader topics of security and international conflict is that a spark can ignite from a seemingly isolated incident that drives out-migration from the region and can destabilize a country or region. As Navas et al. (2018: 1) point out, "the common understanding of violence in environmental conflicts as a direct event in time and space is only the tip of the iceberg...violence can reach not only environmental defenders, but also communities, nature, and the sustainability of their relations".

2.2 Methodology

Original research by the author, conducted in 2020 for a master's dissertation from the London School of Economics, focused on the potential for environmental degradation in 26 Latin American countries to provoke or intensify existing environmental conflicts by 2075. The method applied to attempt to predict future environmental conflict was a machine learning approach called random forest analysis. At their core, machine learning algorithms employ computers to identify patterns in a data sample, and either describe the patterns or make predictions about how the existing patterns may influence future patterns. Random forest analysis is a method of predictive modeling that has seen a growing array of applications in recent years within the field of political science, and specifically in conflict modeling. Three separate climate change adaptation and mitigation scenarios were analyzed for all of the countries included in the maps, in order to show the effect of climate change mitigation policies on environmental conflict outcomes.

The final product of the machine learning algorithm was a series of heatmaps, made in a free geographical information software called QGIS, that depicted where and to what degree the algorithm predicted environmental conflicts would appear by 2075. Because of the density of points in the original environmental conflict dataset around the Andes mountain range, which stretches from Colombia down to Chile and Argentina, the majority of heatmaps displayed in this section predict the highest number of conflicts in this region. Specifically, southern Peru and northern Bolivia appear as major hotspots in every map, which means that the machine learning algorithm predicts the highest concentration and intensity of conflict in these regions. Across all of the heatmaps, the Andres region and particularly Peru and Bolivia, much of Mexico, and Patagonia in the Southern Cone emerge as the areas most ‘at risk’ for environmental conflict by 2075, based on the methodology used in this dissertation.

The principal source for the research on Latin America, and the source of the original dataset, discussed below was the Global Atlas of Environmental Justice, which is a collaborative map that tracks historical and current environmental justice conflicts and was created and is moderated by staff at the Institute of Environmental Science and Technology (ICTA) at the Universitat Autonoma de Barcelona (Temper et al., 2015). Points are included when they meet the following conditions: (1) there is an economic, social, or political activity that has impacted or has the potential to negatively impact the environment and people in a defined area; (2) one or more environmental justice organizations are involved in documenting and/or opposing the activity; and (3) there are one or more reports in the media about the activity and the organizational or popular backlash (Temper et al., 2015).

2.3 Environmental Degradation and Conflict in Latin America

Environmentally-driven conflicts tend to consist of “mobilizations by local communities [and/or] social movements…whereby environmental impacts are a key element of their grievances” and there is a potential for or impacts of negative environmental and social outcomes (Temper et al., 2015: n.p.). By this definition, environmental conflict mobilizations exist anywhere along a spectrum of intensity that ranges from complaints to government organizations about environmental impacts, to peaceful protests, to strikes, to blockades and land occupation, to outright massacres and wars (Ibid). Ten major forms of environmental conflict have historically affected or currently affect Latin America and are likely to drive environmental displacement in the future (Ibid):

- *Biodiversity Conservation:* The loss of biodiversity can affect the world in perpetuity, especially with the loss of keystone species. Biodiversity decline is of critical importance in Latin America as climates change – biodiversity losses exceed the limits of planetary sustainability more than any other human impact (Rockström et al., 2009; Young et al., 2010).
- *Biomass and Land:* Land (and land occupation) played a major role in Latin America’s longest-running conflict – the Colombian conflict – and changing landscape patterns can be seen as a counter-insurgency policy (Albertus & Kaplan, 2012; Suarez et al., 2018). Roughly thirty-five of the world’s fertile biomes have been already converted to agricultural land, and deforestation is Latin America’s most important land change trend – it can have immense impacts on water quantity and quality and biodiversity (Grau & Aide, 2008).
- *Fossil Fuel and Climate Justice/Energy:* Fossil fuel reliance is a massive catalyst of climate change, and fossil fuels are non-renewable and unsustainable in the long term and could contribute to conflict as the Earth gets more polluted and some countries rush to place blame on others (Rockström et al., 2009).

- *Infrastructure and the Built Environment:* As Latin America urbanizes in the next decades, up to a projected level of 90%, rapid, unsustainable infrastructure development will likely become more common (Paranagua, 2012). It is estimated that globally by 2050, there will be enough paved roads to loop around the world 600 times (Laurance et al., 2015).
- *Industry and Utilities:* Industrial processes can affect water quality and natural resource availability downstream (like dyes from textile manufacturing getting into the water supply). The heavy industry tends to be clustered in coastal areas, potentially affecting coastal biodiversity (Dixon et al., 2013).
- *Mineral Ores and Building Materials Extraction:* These materials are usually non-renewable resources, and their extraction is hard on the environment and those who live near mines. Extraction sites as extraction tend to generate vast quantities of toxic or harmful waste materials that can get into the water supply and food chains (Kossoff et al., 2014; Raftopolous, 2017).
- *Nuclear Energy:* Nuclear energy currently provides 13% of global energy, but challenges in expanding nuclear energy remain, which could lead to further minerals exploration and extraction (Zinkle & Was, 2013).
- *Tourism and Recreation:* Mass tourism can severely impact the economy and natural resource allocation of a region (Navas et al., 2018).
- *Waste Management:* Waste Management conflicts tend to affect a small area, although there can be widespread issues if toxic or harmful waste, often from even common materials like plastics, seeps into the water supply.
- *Water Resources:* As water resources dwindle in quantity and quality across Latin America (and the world), it is hard to overstate their likely importance to future conflicts. Agriculture, mining, industry, energy, etc. corporations continue to contaminate water resources, ignoring indigenous people and rural landholder protests, as climate change accelerates and formerly snow-capped landscapes in Patagonia and the Andes melt (Boelens et al., 2011). A worrying example is in Colombia, where the rare páramo ecosystem provides 70% of the water to the country and is under constant pressure from mining companies and ranching encroachment (Andrasko, 2019; Alonso et al., 2020) (Box 5.2).

Box 5.2: Case Study – The Potential for Water Depletion in Colombia's Páramos

Dubbed the "land of mist" by Spanish conquerors when they first encountered the unique high-altitude, tropical moors, páramos are located above the tree line but below the snow line and are considered one of the world's most important ecosystems (Altenhenne, 2015). In total, 35,770 square kilometers of páramos exist in only four Andean countries – Peru, Venezuela, Colombia, and Ecuador – and Colombia contains half of all páramos in the world (IUCN, 2010; Altenhenne, 2015). Colombia's páramos provide water that supports up to 70% of the country's population, including those who live hundreds of kilometers from the páramos (WWF, 2017). Due to their rich and highly endemic plant life, particularly a wide variety of mosses and shrubs, páramos act like sponges to store water, which makes them especially valuable for the provision of water-related ecosystem services

(Schmidt-Mumm & Vargas Ríos, 2012). They also capture an estimated 1000 tons of carbon dioxide per hectare in their rich soils and contribute to the mitigation of climate change – an ecosystem service that is disrupted when the soil is disturbed for productive activities like agriculture (WWF, 2018). Páramos are among the world's most vulnerable and fastest-evolving ecosystems, due to pressures from people and their high degree of sensitivity to climate change, particularly warming weather (IUCN, 2010; Diazgranados & Barber, 2017).

Some people who traditionally resided near but not in the páramos have, over the last decades, moved to higher altitudes to escape the heavy industry and armed conflict during Colombia's 50-year civil war and, in doing so, have inadvertently caused further damage to the delicate ecosystems (UK Research and Innovation, n.d.). Deforestation increasing throughout the country by 44% in 2017 alone, along with expanding land-grabs and illegal mining, are worrying trends (Steffens, 2018a, b). Around the páramos, mining, fracking, and the environmental degradation of their traditional agricultural land have challenged the traditional way of life for inhabitants (UK Research and Innovation, n.d.). Human encroachment, unsustainable land use, improper cleaning of wastewater, and other environmental pressures stemming from the increased use of the páramos are hindering the ability of these delicate ecosystems to provide essential ecosystem services (Siltar, 2015). Most of these negative environmental consequences occur in rural areas that lie far outside the direct control of the Colombian government (Steffens, 2018a, b).

Currently, in an effort to protect them, 36 of Colombia's total 37 páramos have been delimited and made into national protected areas (MinAmbiente, 2018). The controversial Ley 99 (1993) and Ley 233 (2018) delimit the páramos and ban agriculture and livestock-holding and other life-sustaining activities like hunting and chopping trees to produce charcoal (El Tiempo, 2018; International Conservation Caucus Foundation, 2018; MinAmbiente, 2018). This breeds resentment between páramo-dwellers and national and local governments, but also engenders conflict within communities, in which most members see both a need to protect the delicate landscape and to exploit it (to whatever degree) to survive. With the increased pressure on the páramos and on páramo-dwellers stemming from laws and regulations meant to protect the páramos, it is likely that Colombia will experience internal migration away from the páramos. This trend will likely lead to increased urbanization across Colombia – however, if Cresso et al.'s (2020) simulations are correct, Bogotá's current population of nearly eight million people and other urban centers in Colombia may see dramatically decreased water security. If Colombia's major cities no longer have a source of clean, accessible water, future migratory flows away from Colombia or to other areas within Colombia are more likely.

Figure 5.1 depicts the geographical extent of conflict points included in the Atlas, with the categories depicted as different colors – for example, orange points identify conflicts over Mineral Ores and Building Extractions, which is the category with the most conflict globally.

Conflict is most common in rural areas (66.7%) and least common in urban areas (13.73%), even though Latin America is the most urbanized of any continent (81% of the population live in urban areas) (World Bank, 2020). As shown in Fig. 5.2,

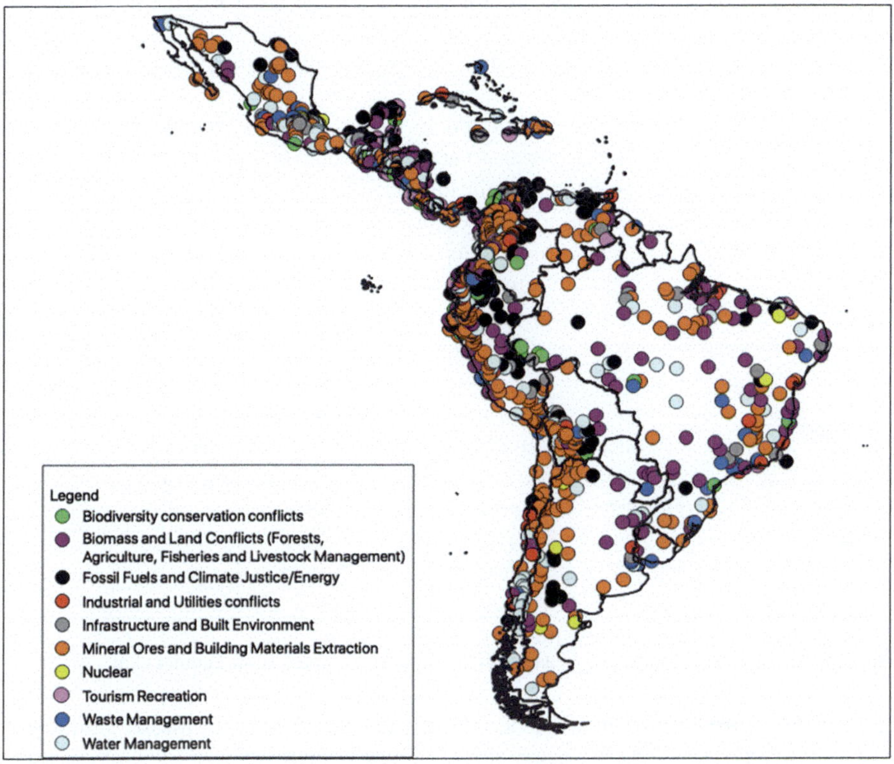

Fig. 5.1 Extent and distribution of conflict points in Latin America. (Source: Adapted from Temper et al., 2015)

2011 was the year in which the number of environmental conflicts peaked and conflicts have seen a precipitous decline since then, but this might be due to long-running but relatively recent conflicts getting recorded in the Atlas at higher rates.

Environmental conflicts situated in small countries like the Bahamas and Belize are on average 29.6% likely to have a positive outcome, compared to 17.5% in large countries. A positive outcome for environmental conflict included in the Atlas would be when the harmful activity is stopped or meaningfully mitigated and the protestors achieve at least the majority of their demands, without any major backlash to the protestors and social groups involved in the conflict. Caribbean countries have the highest likelihood of a positive outcome at 45.8%, countries in Central America and Mexico have a 26.2% chance of a positive outcome, and South American countries have a 16.1% chance. 42% of conflicts over land are medium-intensity and 38.3% are high-intensity, and a small majority of water conflicts are medium-intensity.

Across all 26 Latin American countries included in the analysis, land is the most contentious 'commodity' (27.9% of conflict points mention land), followed by water (22.1%), gold (20.1%), and electricity (15.5%). Rounding out the ten most

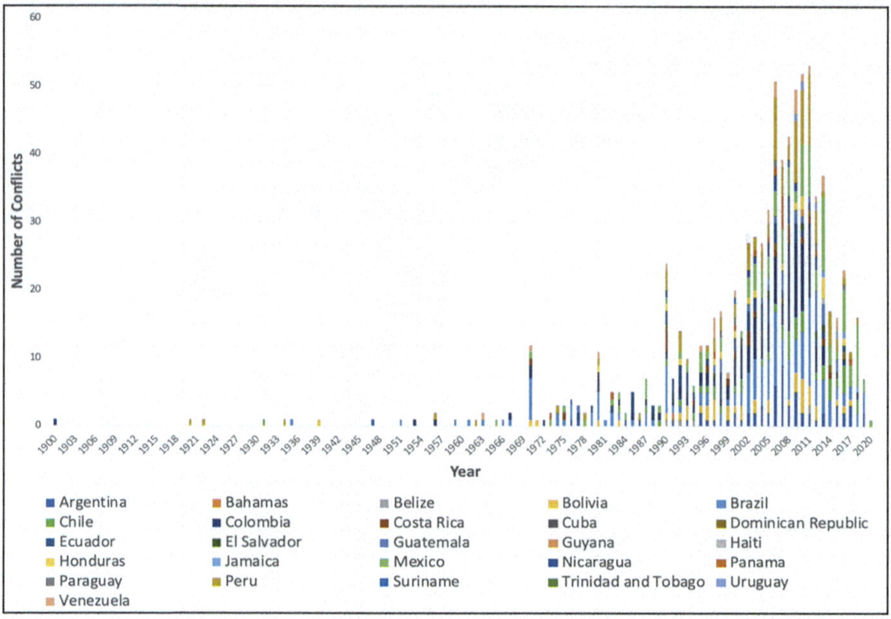

Fig. 5.2 Temporal distribution of Atlas conflicts. (Source: Adapted from Temper et al., 2015)

contentious commodities is silver, crude oil, copper, timber, domestic municipal waste, and natural gas. At least one other country is involved in 74.9% of environmental conflicts in the Atlas, especially in extractive conflicts. Canada is the country most commonly involved in conflicts and is particularly invested in high-conflict mining in Latin America; it is followed by the United States, Spain, China, Brazil, and the United Kingdom. While most of the environmental conflicts recorded in the Atlas involve extranational companies or states, the majority of extractive activities in some areas in Latin America are small scale or artisanal in nature, particularly for gold mining (Walter, 2016).

According to the Atlas, the social groups most commonly involved in environmental conflicts in Latin America are, in order: local environmental justice organizations; neighbors/communities; indigenous groups or traditional communities; social movements; and farmers. Of those five groups, farmers are marginally less likely to win positive outcomes from conflict in Latin America. Street protests/marches, official complaint letters and petitions, public campaigns, development of a network/collective action, and lawsuits/court cases/judicial activism are the most common forms of mobilization. While violent forms of mobilization like hunger strikes and self-immolation, threats to use arms, and kidnappings were each present in less than 1% of conflicts, conflict outcomes (which tend to refer to how those causing environmental degradation respond to mobilizations) were bloodier. Violent targeting of activists appeared in 6% of conflicts and murders or assassinations occurred in 0.05% of conflicts. Repression, corruption, and migration/displacement appeared frequently as well. More optimistically, the most common outcomes were the application of existing regulations and the strengthening of participation (among activists).

Judging by the historical trends for environmental conflict and projections for climate change in Latin America, it seems clear that environmental displacement will only grow in scale and importance by the end of this century. Ultimately, environmental conflict and displacement are about people – how people choose to prioritize and allocate ever-depleting resources and how people make decisions when faced with an unsatisfactory environmental outcome.

3 Historical and Theoretical Underpinnings of the Modern Refugee System

The current international law does not provide adequate protection to environmentally displaced persons (Betts, 2010). As is discussed throughout this chapter, the gap between the established definition of who can receive protection as a refugee, and the looming threat of environmentally driven displacement, is growing everwider (Ibid). This section sets the stage for further discussion of how environmental migration flows should be managed in the future in Latin America, by first discussing the theoretical foundations of the refugee regime, then by detailing the early history of refugee protection prior to World War II. The discussion then turns to how the current international refugee regime was established post-World War II.

3.1 Foundations of the Refugee Regime

The key building block and reference point of the current international system is the state. In a 1919 lecture, "Politics as a Vocation", German sociologist Max Weber famously declared that "a state is a human community that (successfully) claims the monopoly of the legitimate use of physical force within a given territory" (Weber, 1919: 1). States in the current Westphalian international system have 'prima facie' jurisdiction over their territory, which delineate physical space in which a certain set of laws apply (Goodwin-Gill, 2014). With borders comes a legally upheld definition of an 'other' – people who were born or live outside the state's territorial control (Motomura, 2020). As Motomura (2020: 471–472) astutely notes in reference to the United States, many liberal democracies have an inherent conflict between a general "belief in the equality and dignity of people…[and] the fundamental and intended effect of the national borders-to divide 'us' from 'them'".

National immigration laws serve as the legal architecture around which states can enforce their concept of borders (Goodwin-Gill, 2014; Motomura, 2020). States can choose which non-citizens they are willing to legally accept into their territory and which are forced to depart (Goodwin-Gill, 2014). States are also able, to a lesser extent, to control some migratory movement outside their borders, through interceptions and redirection of migrants clearly intending to cross into their country, especially in the sea (Goodwin-Gill, 2014; Benhabib, 2020). Additional methods to deter migration include administrative detention upon arrival, visa controls, and a constricting definition of who is a refugee and entitled to resources and who is not (Edwards, 2005; Echeverría, 2020). These sovereign rights are monitored and contained by international legal structures and treaties. States that are voluntarily party to Refugee Conventions (discussed later in the chapter) are pressured by soft international law to provide at least a minimal level of material assistance for refugees who end up inside their borders (Goodwin-Gill, 2014; Kaldor Centre, 2020) (Box 5.3).

Box 5.3: Case Study – Migration in Haiti and the Dominican Republic

Haiti and the Dominican Republic (DR) represent a 'natural experiment' in which to study EDM, as the countries share the island of Hispaniola in the Caribbean. From colonial times until 1960, Haiti and the DR seemed to be moving along the same trajectory, first as resource-rich colonies then as unstable authoritarian states. France created a highly extractive sugarcane colony in Haiti using African slaves, while the Spanish invested less in the colony that became the DR (Jaupart, 2018). Once Haitian slaves overthrew the French in 1804, they conquered their neighbors – during this period, Dominicans began to assert their identity as being distinct from Haitians due to their Hispanic roots, Catholicism, and Spanish language (Stoyan et al., 2016).

Unstable military dictators and American intervention plagued both sides of the island for more than a century until 1960, after which there has been a clear divergence between Haiti and the DR's relative trajectories. In 1960, the countries shared a real GDP per capita of $800, but by 2005, Haiti's real GDP per capita fell to $430 while the DR's jumped to $2500. Haiti's economy during the same period grew a paltry 1% per year as the country experienced decades of violent authoritarian rule, while the DR began a process of irregular democratization and outperformed the majority of Latin American nations, with a growth rate of 5% (Jaramillo and Sancak, 2009). This is not to suggest that democratization alone brought about economic stability – rather, the DR's structural and stabilization policies were market-friendly and reinforced stability, while Haiti experienced political volatility that did not encourage growth (Jaupart, 2018). The disparity in economic circumstances right across the porous border constitutes a major 'pull factor' for Haitian migrants.

While both countries are suffering from climate change impacts, Haiti is more vulnerable due to its poor infrastructure, weak government, history of environmental degradation, and poverty-stricken populace (Alscher, 2011). Natural disasters like hurricanes and earthquakes are worsened in severity by overcrowding, deforestation, wetland destruction, and climate change, all of which have human components (Singh & Cohen, 2014). Deforestation encourages short-term migration by reducing environmental resilience to natural disasters and forcing people to cross the Dominican border to search for fuel in Dominican forests (Amin & Goldstein,

2008). Haiti is particularly vulnerable to flooding, as hurricane-force rains run rapidly across the deforested landscape into fields, which contributes to soil erosion, driving crop failure, and increasing food prices (Singh & Cohen, 2014). This means that populations which previously relied on a particular crop, for example, may eventually not be able to grow it and find themselves with little food (Gioli et al., 2016). Lack of institutional capacity is another key decision driver for out-migration. Trust in the Haitian government plummeted after the 2010 earthquake and rebounded to levels far below the Latin American average (Stoyan et al., 2016). The Haitian government is unable to provide important social infrastructure like education for rural populations and secure employment opportunities. As of 2017, 14% of Haiti's total labor force was unemployed, with another 85% of the male and 90% of the female labor force engaged in vulnerable employment (World Bank, 2018). All of the variables mentioned are linked to poverty, which is a major driver of migration from Haiti.

Migration from Haiti to the DR already causes tensions and is likely to dramatically increase as Haiti is widely considered a 'hotspot' of EDM (Doran, 2011). A 2013 change of the Dominican Constitution rendered 100,000 Dominicans of Haitian descent stateless, which pushed thousands to migrate back to Haiti due to economic and social discrimination (Fendt, 2016). This cycle, of short-term migration to the Dominican Republic that becomes long-term migration, which is then penalized by the Dominican authorities, is likely to worsen over time as climate change impacts slam the island.

In the Weberian international system, states are obligated to protect and respond to the needs of people within their borders (Kozoll, 2004; Betts et al., 2012). This social contract breaks down when states are either incapable of protecting individuals or groups or are actively targeting them for certain immutable or protected characteristics like race and religion (Kozoll, 2004). Physical displacement outside of a country of origin is the primary indicator of vulnerability in a state-led international system, and humanitarian law is supposed to protect those individuals who reasonably decide to migrate in order to seek protection in other states (Kozoll, 2004; Betts et al., 2012). This is both because it is relatively easy to track and measure physical displacement and because people who have been forced to move are unlikely to be able to rely on their normal coping strategies to better their own circumstances (Betts et al., 2012). Refugees and forced migrants are generally likely to have left behind their communities, livelihoods, possessions, and, in many cases, language and culture (Ibid). If refugees, as noted Italian philosopher Giorgio Agamben (1998: 131) pointed out, "represent such a disquieting element in the order of the modern nation-state, this is above all because by breaking the continuity between man and citizen, nativity and nationality, they put the originary fiction of modern sovereignty in crisis".

3.2 Early History of Refugee Protection

International protection for refugees is not a new phenomenon – the precursor legal architecture for the current international protection system extends back to the 1800s. Charitable organizations focusing on supporting war efforts and aiding wounded troops began to be established in the nineteenth century, notably Florence Nightingale's work during the Crimean War (1853–1856) and the founding of the International Committee of the Red Cross in 1863 (Holborn, 1939; Hitchcock, 2014). The chaos brought by numerous European and Eurasian wars up to World War I "coincided with a growing sentimental view that civilized societies ought to respond to suffering with care and human kindness rather than indifference" (Hitchcock, 2014: 146). During this period, refugee camps became a viable option for population resettlement and control, notably during the South African War (1899–1902) (van Heyningen, 2010). The refugee camps exemplified a 'sentimental' view of the time that refugees and vulnerable populations should not be left to their own devices (Ibid).

World War I (1914–1918), the world's first example of an industrialized total war, produced four million refugees and resulted in a number of highly publicized humanitarian crises, like the 1914 Austrian and Hungarian invasion into East Prussia (De Vuyst, n.d.; Gatrell, 2008). Three key consequences of World War I shaped the modern refugee regime. First, the period after World War I was chaotic – several major empires fell, independence struggles raged, and national boundaries were reconfigured, often following ethnic lines, as a result of the war (Leonhard, 2018). Due to World War I and the conflicts that were initiated because of the end of such a war, the twentieth century became known as the 'century of the refugee', in that the scale of international forced migration surpassed all previous refugee flows (Myers, 2010). Secondly, since refugee flows came to be regarded as "an international problem, even a norm of the post-war order" during the twentieth century, people and governments warmed to the idea that large-scale social issues, like displacement from the war, required coordinated international responses (Burgess, 2016). The final consequence was the shift from a concept of charity, in providing primarily material support for war efforts and for the less fortunate within the boundaries of a state, towards a broader notion of human rights and humanitarianism that is inclusive, global, and applies in times of peace as well as times of war (De Vuyst, n.d.; Gatrell, 2008).

The League of Nations was established in 1919 as a component of the Treaty of Versailles, one of the peace treaties that formally ended World War I (Burgess, 2016). As the only supernational political authority capable of solving a problem which is beyond the power of exclusively humanitarian organizations, the League created a High Commission for Refugees to address population displacement and refugee flows, headed by the Norwegian Fridtjof Nansen (Holborn, 1939; Burgess, 2016). Nansen advocated for shifting international efforts away from short-term charity work towards promoting that refugees eventually become self-supporting and settled (Holborn, 1939; Heyward & Ödalen, 2013). Soon after Nansen's death in 1930, the modest but important Refugee Convention (1933) was signed by fourteen countries, from which originated the principle of non-refoulement, safe labor conditions, welfare and relief, and exclusion from reciprocity (Holborn, 1939; Jaeger, 2001). Non-refoulement means that host states generally do not have the right to return individuals to their home states if they have a credible fear of persecution – it is one of the main theoretical foundations of the modern refugee regime (Jaeger, 2001; UNHCR, 2010).

During this period, the underpinnings for a firm international legal framework of protection for refugees were established, through treaties, resolutions, and other legal instruments (Jaeger, 2001). Generally, each legal document served to draw a new refugee population into international protection or expand protection to known populations. An example of this is the progressively expanded protection for refugees from Germany between 1936 and 1939, as 150,000 German Jews had fled by 1938, along with hundreds of thousands of Jews from other European countries (Jaeger, 2001; Kalb, 2015). The Évian Conference notably provided international protection for the first time to would-be refugees who had not yet left their home countries "but who must emigrate on account of their political opinions, religious beliefs or racial origin" (Jackson, 1999: 21). Ultimately, the Évian Conference failed to generate substantial material support beyond international goodwill – the tiny Dominican Republic was the only country willing to take in refugees at scale, committing to accepting 100,000 refugees (Kalb, 2015).

3.3 The Modern Refugee System

World War II resulted in the largest refugee crisis the world has ever experienced. Contemplating after World War II, Holborn (1956: 210) wrote that "It is evident how difficult a task it was to select genuine refugees and displaced persons among the huge mass of uprooted humanity" (Ballinger, 2012). To rebuild the international system and cope with the one million refugees in Europe reeling from World War II, states founded the United Nations (UN) in 1945 (UNHCR, 2020). Former UN Secretary General Kofi Annan stated that the UN was "never intended to be a utopian exercise…it was meant to be a collective security system", building on American President and League of Nations founder Woodrow Wilson's hope that the League of Nations would bring "an organized common peace" (Schlichtmann, 2010).

Five years after the UN was founded, the UN High Commissioner for Refugees (UNHCR) was inaugurated in 1950 (UNHCR, n.d.). Like its precursor institutions such as the League of Nations, UNHCR was intended to fade once the European refugee crisis after World War II was resolved, optimistically assumed to be within three years (Betts et al., 2012; UNHCR, n.d.). The 1951 Refugee Convention Relating to the Status of Refugees (hereafter known as the '1951 Refugee Convention') is the key multilateral treaty that UNHCR is bound to uphold (Debergh Robinson, 2012). The 1951 Refugee Convention defines who is considered a legal refugee – only those who meet the legal definition of a refugee are required to be protected under international law (UNHCR, 2010). The 1951 Refugee Convention is built on the principle of non-refoulement, which was laid out for the first time in the 1933 Refugee Convention (Jaeger, 2001). People who fall under UNHCR's mandate, as defined by the 1951 Convention, must meet four conditions: (1) they must have fled their home state by crossing an international border; (2) be unwilling or unable to return to their home state due to (3) a 'well-founded fear of persecution'; and (4) the persecution levied against them must be based on their membership in a distinct and often immutable group, like race, religion, nationality, or political opinion (Docherty & Giannini, 2009).

The first condition, that refugees are those who have crossed an international border, excludes people who meet all other conditions but were unable to cross an international border for some reason, often because of poverty, and are displaced within their home countries (Betts et al., 2012). Monitoring and protecting 'people in refugee-like situations' is one of the most difficult and enduring challenges that UNHCR faces, and protection for internally displaced people was only extended in recent years (Betts et al., 2012). The second condition, of being unwilling or unable to return, particularly applies to people who are functionally stateless because of their inability to return, even in the short term (Alexander & Simon, 2014).

The 'well-founded fear' clause included in the Convention assumes that political authorities in the country of origin are aware of and are actively oppressing or threatening the claimant because of the political opinion (Storey, 2014). This puts the burden of proof onto the asylum-seeker to prove to immigration authorities that their fears are real and life-threatening in order to gain protection (Anderson et al., 2020). Perceived credibility of the asylum-seeker by immigration authorities is often a key determinate for whether they are granted asylum or not and requires a future-focused assessment of risk (Anderson et al., 2020). However, Alexander and Simon (2014) take issue with this interpretation of the 'well-founded fear' clause. They argue that fear of retribution is not a central reason to accept or deny an asylum-seeker and instead the 'unable to return' clause is a more important determinant that should serve for refugee protection for those asylum-seekers that do meet it (Ibid). Finally, the clause on 'group membership' can be difficult to prove – for example, political asylum on the basis of persecution of gender expression or sexuality (Greenberg, 2017).

Additionally, the definition of refugees at the time was restricted to people who were directly affected by "events occurring before 1 January 1951", which mainly referred to events that affected Europeans or colonial subjects (Jackson, 1999). This reflects both the erroneous belief at the time that the refugee crisis was primarily contained inside Europe, and the negotiations among state delegates when writing the Convention about what groups of people to exclude from protection because of their perceived status as 'political' migrants or in order not to stretch resources (Ballinger, 2012; Debergh Robinson, 2012). In fact, the total number of forced migrants after World War II may have surpassed 175 million (Gatrell, 2000; Ballinger, 2012). To address the needs of refugees not included in the Convention, a variety of *ad hoc* stop-gap measures were developed, including Allied-run refugee camps in Europe and India and Pakistan's bilateral treaties and legal instruments that regularized resettlement options for the 14 million forced migrants across the newly established border (Cohen, 2008; Debergh Robinson, 2012).

The 1951 Refugee Convention was extended to include protection for refugees outside of Europe with the Protocol relating to the Status of Refugees, signed in 1967 (Stone, 2018). The 1967 Protocol reaffirmed the conditions under which people can be considered refugees but removed its geographical and time-based limitations so the Convention could be universally applied (Hong, 2001). Any country that ratifies the 1967 Protocol is also explicitly agreeing to abide by the tenets of the 1951 Convention, as stated in Article 1 of the Protocol (Kaldor Centre, 2020). By the second half of the twentieth century, less-industrialized countries began to take up more space in UNHCR's mandate (Crisp, 2001). In the 1960s, UNHCR began to link the long-term goal of refugee resettlement with short-term development funding, "with its potential as a win-win situation for donors and asylum states and, in theory at least, for refugees as well" (Krause, 2016): 51. Refugees were viewed by UNHCR and other development agencies, at this time, as a bargaining chip that could be leveraged to plan for and promote sustainable development in the (often low-income) receiver countries, instead of as a burden to be solved (Glasman, 2017). Around the same time, UNHCR began to develop standard responses to address large-scale refugee flows in low-income regions, particularly in Africa (Crisp, 2001; Krause, 2016).

Figure 5.3 tracks the number of total refugees that were included until UNHCR's mandate and those outside UNHCR's mandate that were kept track of but did not meet the legal definition of refugees at that time. Note that the 'Internally Displaced

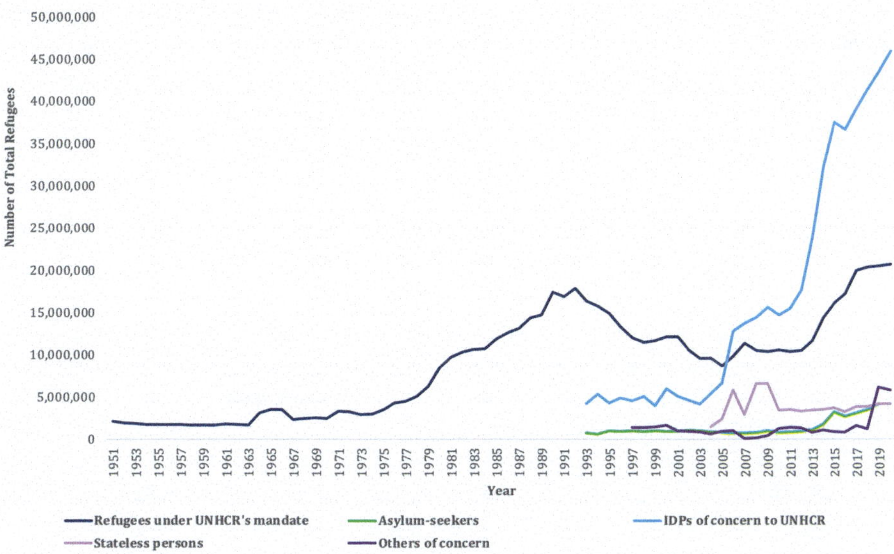

Fig. 5.3 Refugees and persons of concern from 1951 to 2020. (Source: Adapted from UNHCR, 2020)

People (IDPs) of concern to UNHCR', which makes up for the highest percentage of people of concern in recent years, meet many of the criteria to be considered a refugee, but have not crossed an international border.

The three durable solutions that UNHCR currently promotes are voluntary repatriation, local integration, and resettlement (UNHCR, 2021). Practically, voluntary repatriation can be highly risky to individuals returning to unstable and often unknown situations, and UNHCR's website acknowledges that less than 1% of total refugees of concern are submitted for resettlement in a country outside their country of origin (UNHCR, 2021a). Additionally, local integration is rarely a feasible option, especially for refugees in lower-income host countries – Egypt, for example, has passed a number of policies to limit refugee access to basic social services, including employment and education (Brun & Fábos, 2017). For the millions of refugees who are both under UNHCR's mandate and legal care and are not eligible for or have not accessed any of the three durable solutions of choice, most reside in camps (Turner, 2016). This includes short-term environmental migrants fleeing natural disasters, and people who have crossed international borders to avoid years-long civil wars in their home countries (Ibid). Refugees in camps are caught in a cruel contradiction: "first, they cannot settle where they are because they are supposedly 'on the move', on their way home or somewhere else in the future; second, they cannot remain 'on the move' as they possibly are not going anywhere, either now or in the near future" (Ibid: 4). Refugee camps are intended to be temporary solutions to temporary emergencies, but often cross over to long-term displacement (Ibid).

Long-term displacement, often in these camps, is the "new normal" and these durable solution options "are increasingly unsuitable for offering social, economic and cultural means for refugees to rebuild their lives and livelihoods" (Brun & Fábos, 2017: 1). Ironically, with a new buzzword and best practice defining each decade, and solutions that were designed to be temporary turning into decades-long fixtures, the search for truly durable, compassionate solutions remains unaddressed, such as Palestinians trapped in refugee camps in the Middle East.

3.4 Provisions for Environmentally Displaced People

UNHCR has also taken halting steps towards acknowledging EDPs but has yet to seek durable solutions. When applied to EDPs who flee across international borders, the 1951 Refugee Convention and its extension documents have major gaps (Williams, 2008). The Refugee Conventions were designed to address social and political persecution by governments or non-state actors, so most EDPs will only meet the first two conditions for being considered a refugee under the 1951 Convention (Ibid). With the burden of proof put on those applying for protection, it is difficult for EDPs to prove intent on the part of their governments or non-state actors (UNHCR, 1998).

Generally, environmental destruction affects people bound together by geography or living conditions, and thus would have an outsized effect on those living in unsafe locations due to poverty or close to natural features vulnerable to disasters or climate change, regardless of ethnic identity or political opinion (Höing & Razzaque, 2012). However, cases in which social groups were purposely induced to environmentally driven displacement (EDD) contradict the assumption that poverty and natural disasters are necessarily linked, like the Iraqi government in the 1990s methodically destroying marshes and possibly deliberately poisoning the water resources around which the Iraqi Marsh Arabs built their livelihoods and culture (King, 2006). Existing case studies like that of the Marsh Arabs constitute legal openings that could be exploited to provide for EDPs and fill existing gaps in protection (Kozoll, 2004) (Box 5.4).

Box 5.4: Case Study: Iraqi Marsh Arabs

The Ma'dan, also called the Marsh Arabs, are an ethnic group that primarily dwells in the Tigris-Euphrates marshlands in Iraq and Hawizeh marshes that overlap the border between Iraq and Iran in an area nearly the size of Wales (Adriansen, 2004; Priestley, 2020). Their unique culture is claimed to be descended from ancient Sumerian society and the majority of the Ma'dan are Shi'a Muslims (Adriansen, 2004). Much of the Ma'dan's livelihood was based on buffalo breeding, fishing, and rice cultivation, and swamp reeds formed the structures of their houses (Adriansen, 2004; Al Ahram, 2015a).

Until World War I, the Ma'dan lived in relative isolation from the rest of Iraqi society, as the marshes themselves provided an area outside of state control, despite weak ties to markets in towns adjacent to the marshes and some advancement into the swamps by Ottoman and British surveyors and militaries starting in the 1860s (Adriansen, 2004; Al Ahram, 2015a). By the 1920s and 1930s, the marshes and their dwellers had become targets of regular British air-bombing campaigns, following a 1920 tribal uprising (Al Ahram, 2015a). The water in some areas of the marshes had become undrinkable and many marsh-dwellers had left for nearby cities by the 1970s, following campaigns between the 1930s and 1970s to combat malaria by introducing pesticides, the introduction of invasive species for aquaculture, water salinization, and increased inequality resulting from a British policy of converting public land to shayks' private holdings in order to promote stability in these 'ungovernable' areas (Al Ahram, 2015a). In the cities, the Ma'dan were viewed with a mix of contempt and apprehension for their 'backward' nature (Adriansen, 2004; Al Ahram, 2015a).

All of these tensions came to a head with the Iranian Revolution in 1979, Iraq's war with Iran (1980–1988), and the Gulf War (1990–1991). The marshes proved to be forbidding places in which to wage war and served as key battlefields, particularly in the Iran-Iraq War, and also hosted groups of deserters (Adriansen, 2004; Al Ahram, 2015b). Within the context of these conflicts, the marshes became viewed as potentially dangerous, rather than a place that 'required' development (Ahram, 2015a). Iraq began to deport the Ma'dan population from the swamps in 1983, with the hope of developing oil resources in the marshes, promoting modern agriculture rather than the Ma'dan's traditional practices, and, primarily, wiping out the Ma'dan's culture (Ahram, 2015a). This policy resulted in more than 500 arrests and about 200 deaths within the first several days (Al Ahram, 2015a, b; Priestley, 2020). In March 1991, an uprising against President Saddam Hussein of Iraq was brutally crushed, and just weeks later, Hussein issued orders in late March 1991 to drain the swamps (Al Ahram, 2015b). These orders turned into a counterinsurgency campaign of genocidal actions against the Ma'dan and their marshes, including forced disappearances, torture, and executions for people and poisoning the marsh and burning reed-beds to kill off the ecosystem (Priestley, 2020). Two years later, between two-thirds and 90% of the original marsh area were completely dried out and 200,000 were displaced (Al Ahram, 2015b; Priestley, 2020).

Several decades later, there remains debate about how deliberate the policies against the Ma'dan were and whether the narrative of genocide and 'ecocide' was pushed for political reasons. Priestley (2020) and Al Ahram (2015a) both rightly point out that while the flashpoint for the more obvious genocidal actions was in 1991, the structural underpinnings of ecological degradation over centuries, under the guise of development, had been conducted by the Ottomans, British, and Iraqi governments. Adriansen (2004) reveals that few in the international community were paying attention to the plight of the Ma'dan until 2002–2003, when the narrative shifted from ecological destruction to a humanitarian justification made by the United States for invading Iraq.

As far back as 1993, UNHCR recognized four legitimate reasons for fleeing – political persecution, economic pressures, ethnic conflict, and environmental destruction (Höing & Razzaque, 2012). UNHCR dodged participating in policy or research work on EDD until 2007, as then-High Commissioner Antonio Guterres regarded EDPs as falling outside UNHCR's bounds (Ibid). In 2007, however, Guterres first gave a speech in front of donor states that linked climate change and displacement, which signaled a shift in the agency's position towards including EDD in its worldview, without taking exclusive responsibility for the issue (Hall, 2013; McAdam, 2014). By 2011, there was faint hope that "states would be favorable to an involvement of UNHCR in addressing protection gaps related to cross-border displacement as a result of natural disaster and climate change-induced displacement" (Hall, 2013: 100).

Policy instruments have been slightly ahead of hard international law in recognizing the plight of EDPs. For example, a clause in the Cancún Outcome Agreement, which was adopted in 2010, specifically characterized EDD and migration as a whole as an adaptation option available to individuals but avoids sticky questions of causality and accountability (McAdam, 2014). Previously, EDD was seen as a failure to adapt to changing circumstances, so the Cancún Outcome Agreement represented a policy shift (Sasser, 2010). The Nansen Protection Initiative and the Cancun Climate Change Adaptation Framework acknowledge this, stating that migration refers to:

> Human movements that are predominantly voluntary insofar as people, while not necessarily having the ability to decide in complete freedom, still possess the ability to choose between different realistic options. In the context of slow-onset natural hazards,

> environmental degradation and the long-term impacts of climate change, such migration is often used to cope with, 'avoid or adjust to' deteriorating environmental conditions that could otherwise result in a humanitarian crisis and displacement in the future. (Desai et al., 2018, p. 4)

Launched in 2012 and building on the Cancún Outcome Agreement, the Nansen Initiative identified appropriate humanitarian responses to environmental degradation, based on three pillars – international cooperation, consistent standards for how displaced people should be managed and treated, and operational concerns like funding mechanisms (Nansen Initiative, 2014). EDPs can potentially appeal to regional bodies for protection that UNHCR cannot provide, like the Cartagena Declaration and the Organization for African Unity Convention, but Keane (2004) argues that neither body recognizes migration as being solely caused by environmental reasons and, therefore, cannot offer adequate protection (Höing & Razzaque, 2012).

4 Policy Options to Address Existing Legal Gaps for Environmental Displacement

Because climate change defies simple solution-building, due to the immense uncertainties, conflicts of interest among stakeholders, along with interdependencies inherent in its study and response, it is frequently known as a 'super-wicked' global problem (Behrman & Kent, 2018). Environmentally driven migration (EDM) is a separate but related 'super-wicked' issue, as it touches on numerous other issues, including economics, law, and emotional factors, including climate change and environmental vulnerability.

4.1 What Policies Must Consider

Some policymakers and academics characterize EDM as a failure to adapt to changing circumstances, rather than as an adaptation option (Warner, 2010). Most national adaptation plans and policies currently fail to include migration and resettlement in their analyses and budgets, which is a major exclusion (Warner, 2010; Desai et al., 2018). A second complicating factor in durable solution planning for climate change risk is 'maladaptation'. Maladaptation refers to the perception that short-term solutions, like building a dike in a floodplain, encourage settlement patterns or behavior that actually increases long-term vulnerability (Oliver-Smith, 2009). This is relevant to climate change-driven migration because migrants may be trading one unstable living situation for another across a border (UNFCCC, 2012). There is little stopping governments from placing migrants in cheap, environmentally undesirable resettlement locations when selecting camp locations (Warner, 2010). Some resettlement locations lack sufficient access to basic services like schools, water treatment, and banks (Dun, 2009). In making the decision to move, migrants often abandon established jobs and livelihoods and can lose time-honored social networks that link migrants to new job opportunities (Warner, 2010).

Long-term risk planning efforts must improve international and regional cooperation, capacity-building, and adaptive management of refugee resettlement programs, in order to address the reality of future environmental migration in an orderly manner (UNFCCC, 2012). Migration should be considered as a valid adaptation approach in the face of rapid-onset events and for lowering exposure to slow-onset events and should be included in climate change adaptation action plans (IOM, 2009; Desai et al., 2018). As Desai et al. (2018: 4) point out, there is often a "'tipping point' at which communities shift from voluntary, adaptive migration into forced displacement. When their coping capacities are exhausted, they risk falling into a gradual process of impoverishment, eventually leading to their displacement". Adaptive and forward-looking planning for environmental displacement is a way to avoid the anxiety, scrambling for funding sources, and sudden state of emergency in countries often associated with new refugee flows.

The human cost of migration is particularly clear in how host countries and citizens react to migrants when they arrive. One major setback in the preemptive planning for potential future patterns for environmental migration is the "tightening of the distinction between the citizen and the non-citizen" that has occurred around the world in recent years (Benhabib, 2020). The United States, for example, has been violating the international principle of non-refoulement in mistreating Latin American illegal immigrants along the U.S.-Mexico border in a clear nod to rising nationalist sentiments stoked during the Trump era (Motomura, 2020; Nyabola, 2019; Benhabib, 2020). These anxieties tend to be deeply held and not easily dispelled with evidence, even when faced with economic research that suggests that migration improves economic outcomes across the host society (Motomura, 2020). This trend is clear across both rich and poor host countries throughout all world regions (Benhabib, 2020).

4.2 *Potential Policy Solutions*

Four potential – and possibly interlinking – solutions to long-term environmental migration challenges are feasible and foreseeable.

4.2.1 Extend Existing Documents

The first potential approach for how to manage future pathways of environmental migration advocates for UNHCR to extend the 1951 Convention to include EDPs or to liberally interpret existing documents (Höing & Razzaque, 2012). As António Guterres asserted, "the history of UNHCR is one of constant change and adaptation" in response to emerging crises and challenges (Betts et al., 2012: xx). UNHCR's policy mandate has expanded to include returnees, stateless people, IDPs, asylum-seekers, and people 'threatened with displacement' as the need arose, so there is historical and legal precedent for folding new groups of people into UNHCR mandate (Hall, 2012; UNHCR, 2020). While the current iteration of the 1951 Convention and more recent UNHCR documents do not explicitly include EDPs in their definitions of refugees, they do indirectly have sway over the way in which migration is conceptualized (Höing & Razzaque, 2012). By extending the 1951 Convention and potentially other landmark documents, UNHCR would legitimize EDPs as 'worthy' of protection, resources, and attention.

Proponents of this approach argue that extending or liberally interpreting the current legal framework would be pragmatic, because rather than drafting entirely new legislation that would likely lack political will and funding from states, UNHCR could use its influence and expertise (Betts, 2012). Hall (2012) argues that over time, driven by Guterres, UNHCR has been incrementally positioning itself to "replicate its moral legitimacy in new spheres", like managing EDPs. One camp of detractors portrays UNHCR as being unwilling to take responsibility for EDPs, due to fears that states would be averse to being legally bound to take on more refugees and wary of eroding sovereignty (Williams, 2008; Höing & Razzaque, 2012). Another group views the 1951 Convention as too-narrowly defining refugees as people who have been persecuted, as they perceive modern-day forced migration to be more diffuse and less persecutory in nature (Fitzpatrick, 2010). The third set of critics cites the legitimate worry that adding more categories of refugees to existing Conventions would thin protections for refugees included in the original definition, who are themselves vulnerable (Renaud et al., 2007).

4.2.2 UNHCR Ad Hoc Extension

Perhaps the simplest potential response is *ad hoc* extension of UNHCR's mandate, most likely in the form of expanded temporary protection for EDPs. UNHCR is currently pursuing this option for the EDPs that have already chosen to migrate internationally, by providing some *ad hoc* aid without also affording these environmental migrants a clear legal status or designation (Betts, 2010). Recognition for EDPs could exist along a sliding scale from acute to chronic displacement, which would allow for maximum flexibility in states' response options (Williams, 2008). This would include EDPs in UNHCR's protection sphere without needing to bind states into agreements and without institutional changes (McAdam, 2014). Time-honored seasonal labor or transnational settlement patterns, such as a farmer moving to a distant city during the dry season, is a strategy that individuals can use to remove themselves from an unstable situation and that could be integrated into an *ad hoc* extension of UNHCR's mandate (Lubkemann, 2005).

Supporters of this approach argue that IDPs were originally brought under UNHCR's mandate through *ad hoc* protections, but are now a central component of UNHCR's work and formally included in UNHCR's mandate (Loescher et al.,

2008; Gemenne & Brüker, 2015). Especially since some academics consider EDPs to be a subset of IDPs, there is some possibility that environmental migrants could experience a similar route to recognition as IDPs (Ibid). A downside of this *ad hoc* approach is that it does not create normative standards that UNHCR would be able to enforce, unlike most formally negotiated international agreements (Hall, 2012). Another downside to this approach comes from the complexities of the 'asylum-migration nexus', in which formalized migrants, informal or clandestine migrants, and refugees all use the same routes (Castles & Van Hear, 2005; Betts, 2010). This muddying of migration pathways could lead to hierarchies of need and politicization of what forms of migration, including migration due to environmental degradation, should be protected (Williams, 2008; Betts, 2010). Regardless, even simple, temporary solutions that would not be difficult to convince states to adopt must have sufficient political will, funding, capacities at the international level, and civilian and military support to function well and serve as a durable solution (Betts et al., 2013).

4.2.3 Soft Law Framework

The third policy option is to establish a soft law framework, drawing from the existing, highly effective Guiding Principles on Internal Displacement and national or regional action plans like the Cartagena Declaration (Ferris & Bergmann, 2017). The Guiding Principles extend protection to IDPs – a group of concerns that were previously functionally excluded from migration and humanitarian law (Ferris & Bergmann, 2017). The Guiding Principles are not legally binding and are instead more consistent with a set of norms (Ibid). They were drawn from disparate pieces of existing case law, so as to remain consistent with international human rights law and to deflect criticism (UNHCR, 2004; Betts, 2010). A similar framework to the Guiding Principles could bring EDPs into a sphere of protection and provide a set of norms for countries to follow (Mayer, 2011; Ferris & Bergmann, 2017).

This framework could be built on international human rights and humanitarian law bodies, not necessarily just on refugee law (Betts, 2010; Höing & Razzaque, 2012). Without such a framework, states and organizations do not have clear guidelines to follow for how to deal with EDPs, particularly about how to fill protection gaps and fulfil their international humanitarian obligations (Betts, 2010; Ferris & Bergmann, 2017). Both states and international organizations have illustrated, in various international forums, that there is a demand for clearer guidance on how to understand and apply their existing commitments – as long as they do not have to assume new obligations (Betts, 2010; Cohen, 2013). Betts (2010: 215), a vocal proponent of this soft law approach, argues that "What is required is simply: (a) an authoritative consensus on the application of these instruments to the situation of vulnerable migrants, and (b) a clear division of responsibility between international organisations *[sic]* for the operational implementation of such guidelines". This soft law framework could draw from two key elements: the strengthening of prevailing international human rights norms into clear principles for how to respond to the needs of different groups; and enhanced mechanisms for international collaboration, among states and organizations, to collectively and collaboratively implement these norms (Ibid).

Optimistically, soft law can provide a platform for states and international organizations to dialogue, resolve any uncertainties, and explore creative or innovative solutions without having to commit to formal hard law structures (McAdam, 2014). It is probable that, to successfully implement a soft law approach, at least one international organization and likely one global power state – or powerful regional body like the European Union – would have to lead the charge. This approach may also improve good will between states, international organizations, and UNHCR, through the negotiation process (Ibid). The soft law framework could lead to the implementation of hard law and ratification at the regional level, like in the African Union Convention, which would necessitate elevated political cooperation (Betts, 2010; Ferris & Bergmann, 2017).

Opponents contend that gaps in protection could persist or worsen from knitting together international law bodies into a soft law framework (Höing & Razzaque, 2012; Ferris & Bergmann, 2017). It seems unlikely that UNHCR would lead the formulation of a soft law framework, for fear of being bound to stretch their resources thin, so another influential actor would likely need to step up, even though many countries that have started to experience EDD are not powerful on the global stage. Another obvious pitfall is that without a hard law framework or another compulsory funding mechanism, it is unlikely that there will be sufficient resources to implement robust protection for EDPs (Mayer, 2011).

4.2.4 New Treaty

The fourth option is to start fresh with an entirely new treaty or protocol designed specifically to address EDD. This bold approach recognizes that EDPs are unlikely to be meaningfully included in existing refugee law, and a new treaty could be easily expanded or changed as the scale of EDD becomes clearer (Falstrom, 2002). This possibility has been suggested before, by academics and at the international level, including by the German Advisory Council on Global Change in 2007 (Biermann & Boas, 2008). Biermann and Boas (2008) suggest that a new treaty be established under the jurisdiction of the UN Framework Convention on Climate Change (UNFCCC). The international community could also create a consortium of funding and implementing agencies, including possibly the World Bank and the UNDP (Biermann & Boas, 2008). An alternative approach, recommended by Mayer (2011), is for the UN to adopt a framework that explicitly recognizes the rights of EDPs, and at the same time establish a new agency to be in charge of providing *ad hoc* protections to EDPs.

A treaty or protocol's executive committee could establish a running list, that any state could contribute to, of areas in imminent need of relocation assistance (Biermann & Boas, 2008). This would be consistent with sovereignty principles of the UN system, as it would rely on buy-in and acknowledgement from the affected country (Ibid). This approach could also improve burden-sharing among countries – the acknowledgement of 'common but differentiated responsibilities', as Biermann and Boas (2008) put it – and integrate developing countries into a "global mitigation regime of quantified reduction and limitation objectives" in a way that could increase developing countries' bargaining power on the international stage.

Advocates of establishing a new treaty or protocol to address environmental migration assert that doing so could build political and financial support, attract attention from states and international organizations, and link protection to the huge investment in climate change science (Biermann & Boas, 2008). Critics argue that constructing a new framework to address EDD would likely shift focus away from those fleeing natural disasters in the short term, who are already protected to some degree by UNHCR (Hall, 2012; McAdam, 2014). This may also stem from the conceptual difficulty of attributing long-term migratory patterns to specifically environmental causes, aside from short-term, rapid-onset reasons (McAdam, 2014). Other, perhaps more cynical, critics contend that establishing a new hard law framework seems only less likely over time, as states lack the political will and have pushed back against shouldering new legal commitments to EDPs (McAdam, 2014).

4.3 A Potential Way Forward: Towards a Regional Approach

Several of the policy options discussed above should be combined to create a maximally comprehensive response to EDD. In the short-term, the international humanitarian community could choose to temporarily extend protections and reevaluate the scale of the problem in the future. Soft law structures should be developed, once extended protections are in place, to guide states on how to address environmental migration and to develop a set of norms. EDPs may foreseeably be included in UNHCR's mandate in the next several decades, as the extent of EDD becomes felt more acutely by donor countries. By the middle of the century, EDD should be tackled through a new international treaty, once countries get accustomed to the scope and concept of EDPs through short-term increased protections. No matter what form of protection is extended to cover environmental refugees, both the causes and outcomes of environmental degradation and climate change must be addressed at the international level (Williams, 2008). This path towards acknowledgement and response for EDPs would doubtless be a significant achievement in humanitarian action at the global level.

A major hurdle for Latin American migrants and EDPs is that the international community generally frames Africa and Asia as the regions with the highest number of EDPs and, consequently, focuses on those regions, to the neglect of Latin America. There is some truth to this notion – India, Bangladesh, Philippines, China, and Somalia were the five countries with the highest number of people displaced due to natural disasters in the first six months of 2020 (Migration Data Portal, 2020). This pattern ignores the role that Latin American EDD may play on migration in the future, both in the region and across the world (Dingeman et al., 2017). This is why a collaborative, regional approach to address EDD, that can be tailored to the needs of Latin American states, cities, and people would be the best path forward.

4.3.1 The Benefit of Regionalism

While the prevailing international system places states as the building blocks and referent entities, this may shift by the end of the twenty-first century. Cities – particularly megacities – have taken center stage in the fight against climate change and serve as hubs of innovation (Esri StoryMaps Team, 2021). Bogotá, Buenos Aires, Lima, Mexico City, São Paulo, and Rio de Janeiro, which all have populations over ten million people, are set to urbanize even more through the century, and may become more important political and cultural entities than state governments (Esri StoryMaps Team, 2021). Additionally, supranational organizations may become more important over the course of the twenty-first century, especially when it comes to sharing, managing, distributing shared, and contentious natural resources and at the continental or regional level. Supranational regional organizations, like the European Union and the African Union, can unite states committed to shared goals and facilitate regionalism.

Regionalism occurs when governments band together based on regional, collectivist ties and eschew the pursuit of purely individualist interests, often with regards to free trade deals and open market schemes (Grugel, 2004). Regionalism was particularly prevalent during the period of heightened (neo)liberalism in the 1980s and 1990s (Ibid). Few organizations meaningfully unite most or all Latin American countries. Malamud and Gardini (2012) list nine Latin American bodies that each have some unsuitable characteristic, like the Ibero-American Community that incorporates Andorra, Spain, and Portugal, and "the processes of subregional integration (Mercosur, the Andean Community, the Central American Integration System) [that] are even less encompassing" than other regional organizations. Organizations like the Organization of America States (OAS) include the United States and Canada. While these two countries may eventually experience the effects of EDD in Latin America via increased migration, they likely should not be meaningfully included in process of a regional response to and planning for EDD, as they are the two countries that are most likely to be involved in environmental conflicts in Latin America. Malamud and Gardini (2012:117) do note, however, that:

> One potential exception stands out: the Rio Group, which [as of 2012] numbers 23 members, including all of Latin America, but also a few countries from the Caribbean. Yet, there is still a caveat: this organisation *[sic]* lacks a secretariat or permanent body, so if it did have a number it would have to be a cell phone.

As a spinoff of the Rio Group, the Community of Latin American and Caribbean States (CELAC) was founded in 2011 with the Declaration of Caracas (Soriano, 2019). Thirty-three states in Latin American and the Caribbean have joined CELAC by 2021, with the goal to unite Latin American countries through improved political relations and to resolve tensions through regional dialogue (CELAC, n.d.). CELAC provides a natural platform for Latin American countries to begin collaboratively planning for climate change and corresponding forced displacement. CELAC is a particularly useful organization because it includes a number of Caribbean countries, which will likely experience the impacts of climate change more rapidly (Watkins & Salinas, 2020). A collective platform like CELAC could also likely better attract international investment and generate research and interest into Latin American EDD than individual countries.

Particularly when it comes to environmental degradation, climate change, and environmentally driven migration in Latin America, it makes sense for Latin American states to develop a regional approach. Key to this approach is to first begin to sustainably manage natural resources and aggressively practice climate change mitigation activities within the region (Watkins & Salinas, 2020). Countries that share natural resources or directly benefit from ecosystem services from ecosystems that exist across international borders should develop joint action plans for how to mitigate and respond to climate change and its associated migration. This is true for countries that share the Amazon Rainforest, like Brazil, Peru, Ecuador, and Colombia, as deforestation, a higher incidence of forest fires, and extreme climate conditions are push factors for out-migration from the Amazon region (Maia & Shons, 2020). This is equally true for transnational, non-renewable natural resources like Lake Titicaca in Peru and Bolivia, which supports the urban population of La Paz, Bolivia and is dwindling in quality and quantity due to a high population density, intensive aquaculture, and heavy economic activity (Buytaert & Breuer, 2013; Archundia et al., 2017). Much international investment has been dedicated to specific environmental issues or ecosystems within Latin America, to the neglect of others, so it would make sense to centralize investment in CELAC or another regional body. Countries that are experiencing particularly acute climate change impacts or have had a recent natural disaster could perhaps apply to CELAC's centralized fund for short-term support or long-term financing options.

Once concrete, actionable, and comprehensive climate change plans have been settled upon, along with sufficient funding mechanisms from the international community, a regional approach to tackling environmental migration can then be negotiated. Several Latin American countries have already called for response and attention towards EDD in their national adaptation laws and plans on climate change (Watkins & Salinas, 2020). Peru's Framework Law on Climate Change "calls for addressing forced migration due to climate impacts, and… Honduras's National Strategy for Climate Change that proposes to establish a legal and institutional framework for migrations of climate origin as part of their adaptation strategies" (Ibid: n.p.). A unified regional approach would be a form of soft law that would codify norms and provide funding mechanisms for similar strategies to those of Peru and Honduras. The proposal of similar laws in Latin American countries shows that this approach would not be a major overstep.

Latin American governments should care about pre-emptively coming up with plans for how to fund and address EDD because, unlike most other world regions, Latin America has close linguistic and cultural ties. These close ties could make migration slightly easier within the region than outside it and could drive more cross-border migration than in other regions. This is already being tested with the five million displaced people from the Venezuelan diaspora, upwards of 80% of whom have chosen to stay in the region (IOM, 2020). Colombia, Peru, Chile, Ecuador, and Brazil all host large Venezuelan populations (Ibid). Colombia alone hosts more than 1.8 million Venezuelan migrants and, in a move praised by UNHCR and the international community, has recently announced an initiative to regularize migrants' status and provide them with protection for up to ten years (IOM, 2020; UNHCR, 2021b). The Temporary Protection Status will provide Venezuelans in Colombia "access to basic services including the national health system and COVID-19 vaccination plans…and access to the job market, which in turn serves to lessen the dependency of people on humanitarian assistance while also contributing to the country's post COVID-19 socio-economic recovery" (UNHCR, 2021b).

It would be essential for CELAC or another regional body to establish a scheme, like Colombia's Temporary Protection Status for EDPs, at the regional level so most or all states are party to burden-sharing and financing, instead of individual states bearing the cost of EDD. Colombia's Temporary Protection Status Initiative aims to reduce the human cost of migration in a way that could be applied to EDPs in the future. CELAC could also harness the shared sense of burden-sharing among Latin American states, particularly those in South America, from managing and responding to Venezuelan migration and direct it towards EDPs as the need arises. As Keyes (2019) points out that:

> [T]here is a solidarity among countries, as many are grappling with internal climate-change migration and adaptation…After decades of seeing the refugee framework as divided between countries 'creating' or sending the refugees, and countries receiving the refugees, this shared experience marks a profound change. (p. 26)

5 Conclusions

Latin America is a particularly important – and relatively understudied – region in which to explore migration driven by climate change, environmental degradation, inequalities, and conflict (Piguet et al., 2018). Latin America's economic history can be traced by way of natural resources and land use, stemming from European colonial rule in which native and Africa-born slaves worked on sugar, coffee, and rice plantations (Hall, 2008; Panizza, 2013; Navas et al., 2018). American corporate extractivism and political destabilization that fashioned 'Banana Republics' in the nineteenth century set the stage for a commodity boom in the early 2000s that propelled economies to grow at the highest rates in decades and then crash again (Hall, 2008; Panizza, 2013; Navas et al., 2018). The end of the commodity boom resulted in slowed job creation, higher levels of poverty in many Latin American countries, and slowed growth – trends that have gotten starker with the lockdowns and loss of economic activity caused by the novel coronavirus (Balakrishnan & Toscani, 2018). All of these commodity-driven paths to economic improvement have resulted in worsened environmental degradation over time, which will likely lead to EDD within the next decades of the twenty-first century.

In the aggregate, individual decisions to migrate add up to tens of millions of migrants moving every year, which inspires practitioners and scholars to pejoratively "speak of flows, streams, waves and trickles of migrants...the metaphors we use to talk about migration require us to think of migrants as an undifferentiated mass" (Turton, 2003: 10; Schon, 2019). This has implications for the way refugee populations are viewed and treated – by host countries, humanitarian organizations, and scholars – all of which tend towards myopic responses to migration because of a lack of clarity of the causes of flight and avoiding addressing additional, systemic push and pull factors (Lischer, 2007). Regardless of how migration is viewed from the top-down, the human cost of environmental migration is very often great. In the future, greater compassion among citizens should be hoped for when it comes to long-term migratory paths into their country from people fleeing unbearable environmental conditions or natural disasters.

References

Adriansen, H. K. (2004). What happened to the Iraqi Marsh Arabs and their land. *Danish Institute for International Studies*. http://www.jstor.com/stable/resrep13388

Agamben, G. (1998). *Homo Sacer: Sovereign power and bare life* (D. Heller-Roazen, Trans.). Stanford University Press.

Al Ahram, A. (2015a). Development, counterinsurgency, and the destruction of the Iraqi marshes. *International Journal of Middle East Studies, 47*(3), 447–466. https://www.jstor.org/stable/43997992

Al Ahram, A. (2015b, August 17). Lessons from the destruction of Iraq's marshes. *The Washington Post*. Available at: https://www.washingtonpost.com/news/monkey-cage/wp/2015/08/17/lessons-from-the-destruction-of-iraqs-marshes/

Albertus, M., & Kaplan, O. (2012). Land reform as a counterinsurgency policy: Evidence from Colombia. *Journal of Conflict Resolution, 57*(2), 198–231. https://doi.org/10.1177/0022002712446130

Alexander, H., & Simon, J. (2014). Unable to return in the 1951 refugee convention: Stateless refugees and climate change. *Florida Journal of International Law, 26*(3). https://ssrn.com/abstract=2643925

Alonso, D. L., et al. (2020). Assessment of mining activity on arsenic contamination in surface water and sediments in southwestern area of Santurbán paramo, Colombia. *Journal of Environmental Management, 264*. Available at: https://doi.org/10.1016/j.jenvman.2020.110478

Alscher, S. (2011). Environmental degradation and migration on Hispaniola Island. *International Migration, 49*(1), 164–188. https://doi.org/10.1111/j.1468-2435.2010.00664.x

Altenhenne, M. (2015). *Bogotá's unique water source at risk*. Deutsche Welle. Available at: https://www.dw.com/en/global-ideas-colombia-paramos-biodiversity-water-supply/a-18428938

Amin, S., & Goldstein, M. (Eds.). (2008). *Data against natural disasters: Establishing effective Systems for Relief, recovery, and reconstruction*. The World Bank.

Anderson, A., Foster, M., Lambert, H., & McAdam, J. (2020). A well-founded fear of being persecuted…But When? *Sydney Law Review, 42*, 155–180.

Andrasko, R. (2019). *"Here we love life and our planet": Applying a Bayesian belief network to conflicted Páramo landscapes in Boyacá*. University of Edinburgh.

Andriansen, H. K. (2004). What happened to the Iraqi marsh Arabs and their land? The myth about garden of eden and the noble savage. *Danish Institute for International Studies*. http://www.jstor.com/stable/resrep13388

Archundia, D., Duwig, C., Spadini, L., et al. (2017). How uncontrolled urban expansion increases the contamination of the Titicaca Lake Basin (El Alto, La Paz, Bolivia). *Water, Air and Soil Pollution*. https://doi.org/10.1007/s11270-016-3217-0

Balakrishnan, R., & Toscani, F. (2018). *How the commodity boom helped tackle poverty and inequality in Latin America*. International Monetary Fund. Available at: https://blogs.imf.org/2018/06/21/how-the-commodity-boom-helped-tackle-poverty-and-inequality-in-latin-america

Balcells, L. (2012). *Violence and displacement in civil war. Evidence from the Spanish civil war (1936–1939)* (Barcelona GSE Working Paper Series Working Paper, No. 603). Available at: https://digital.csic.es/bitstream/10261/59054/1/Violence%20and%20Displacement%20in%20Civil%20War.pdf

Ballinger, P. (2012). Entangled histories or 'extruded' histories? Displacement, refugees, and repatriation after world war II. *Journal of Refugee Studies, 25*(3), 366–386. https://doi.org/10.1093/jrs/fes022

Bassetti, F. (2019). *'Environmental migrants: Up to 1 billion by 2050', foresight: The CMCC observatory on climate policies and futures*. CMCC Foundation. Available at: https://www.climateforesight.eu/migrations-inequalities/environmental-migrants-up-to-1-billion-by-2050/

Behrman, S., & Kent, A. (2018). *'Climate refugees': Beyond the legal impasse?* Routledge.

Benhabib, S. (2020). Exile, statelessness, and migration: Response to my critics. *Philosophy and Social Criticism, 46*(1), 34–44. https://doi.org/10.1177/0191453719893556

Betts, A. (2010). Towards a 'soft law' framework for the protection of vulnerable irregular migrants. *International Journal of Refugee Law, 22*(2), 209–236.

Betts, A., Loescher, G., & Milner, J. (2013). *The United Nations high commissioner for refugees (UNHCR): The politics and practice of refugee protection*. Routledge.

Biermann, F., & Boas, I. (2008). Protecting climate refugees: The case for a global protocol. *Environment: Science and Policy for Sustainable Development, 50*(6), 8–17. https://doi.org/10.3200/ENVT.50.6.8-17

Black, R., et al. (2001). Migration as adaptation. *Nature, 478*, 447–449. https://doi-org.ezproxy.is.ed.ac.uk/10.1038/478477a

Blitzer, J. (2019, April 3). How climate change is fueling the U.S. border crisis. *The New Yorker*. Available at: www.newyorker.com/news/dispatch/how-climate-change-is-fuelling-the-us-border-crisis

Boas, I. (2015). *Climate migration and security: Securitization as a strategy in climate change politics*. Routledge.

Bouroncle, C., Imbach, P., Rodriguez-Sanchez, B., et al. (2017). Mapping climate change adaptive capacity and vulnerability of smallholder agricultural livelihoods in Central America: Ranking and descriptive approaches to support adaptation strategies. *Climatic Change, 141*, 123–127. https://doi.org/10.1007/s10584-016-1792-0

Brun, C., & Fabos, A. H. (2017). Mobilizing home for long-term displacement: A critical reflection on the durable solutions. *Journal of Human Rights Practice, 9*(2), 177–183. https://doi.org/10.1093/jhuman/hux021

Burgess, G. (2016). *The league of nations and the Refugees from Nazi Germany: James G. McDonald and Hitler's victims*. Bloomsbury.

Buytaert, W., & Breuer, L. (2013). Water resources in South America: Sources and supply, pollutants and perspectives. In *Proceedings of the IAHS-IAPSO-IASPEI Assembly* (pp. 106–113), Gothenburg, Sweden. Available at: https://iahs.info/uploads/dms/15575.16-106-113-361-05-Buytaert-Gothenburg-H04CORR.pdf

Carte, L., Schmook, B., Radel, C., & Johnson, R. (2019). The slow displacement of smallholder farming families: Land, hunger, and labor migration in Nicaragua and Guatemala. *Land, 8*(6). https://doi.org/10.3390/land8060089

Caruso, G. D. (2017). The legacy of natural disasters: The intergenerational impact of 100 years of disasters in Latin America. *Journal of Development Economics, 127*, 209–233. https://doi.org/10.1016/j.jdeveco.2017.03.007
Castles, S., & Van Hear, N. (2005, January 27). *The migration-asylum nexus: Definition and significance* Lecture presented at COMPAS, Oxford University.
CELAC. (n.d.). CELAC. Available at: http://celacinternational.org/celac-4/
Cheatham, A. (2019). *Central America's Turbulent Northern Triangle*. Available at: https://www.cfr.org/backgrounder/central-americas-turbulent-northern-triangle
Cohen, G. D. (2008). Between relief and politics: Refugee humanitarianism in occupied Germany 1945–1946. *Journal of Contemporary History, 43*, 437–449.
Cohen, R. (2013). *Lessons learned from the development of the guiding principles on internal displacement*. Georgetown University. Available at: https://www.alnap.org/system/files/content/resource/files/main/lessons-learned-gps-cohen-oct2013.pdf
Collier, P., & Hoeffler, A. (2005). Resource rents, governance, and conflict. *Journal of Conflict Resolution, 49*(4), 625–623. https://doi.org/10.1177/0022002705277551
Cresso, M., Clerici, N., Sanchez, A., & Jaramillo, F. (2020). Future climate change renders unsuitable conditions for paramo ecosystems in Colombia. *Sustainability, 12*(20), 1–13. https://doi.org/10.3390/su12208373
Crisp, J. (2001). Mind the gap! UNHCR, humanitarian assistance and the development process. *International Migration Review, 35*(1). https://doi.org/10.1111/j.1747-7379.2001.tb00010.x
Cristiani, F., Fornale, E., & Lavenex, S. (2020). Environmental migration governance at the regional level. In T. Krieger, D. Panke, & M. Pregernig (Eds.), *Environmental conflicts, migration and governance*. Bristol University Press.
De Vuyst, J. (n.d.). Refugee relief during the first world war. *Voices of War and Peace*. Available at: https://www.voicesofwarandpeace.org/portfolio/belgian-refugees/
Debergh Robinson, C. (2012). Too much nationality: Kashmiri refugees, the south Asian refugee regime, and a refugee state, 1947–1974. *Journal of Refugee Studies, 25*(3), 344–365. https://doi.org/10.1093/jrs/fes030
Desai, B., Ginnetti, J., & Sydney, C. (2018). *No matter of choice: Displacement in a changing climate*. Internal Displacement Monitoring Centre. at: https://www.internal-displacement.org/sites/default/files/publications/documents/20181213-slow-onset-intro.pdf
Di John, J. (2011). Is there really a resource curse? *A critical survey of theory and evidence', Global Governance, 17*, 167–184.
Diazgranados, M., & Barber, J. C. (2017). Geography shapes the phylogeny of frailejones (Espeletiinae Cuatrec., Asteraceae): A remarkable example of recent rapid radiation in sky islands. *PeerJ, 5*. http://dx.doi.org.ezproxy.is.ed.ac.uk/10.7717/ peerj.2968
Dingeman, K., Arzhayev, Y., Ayala, C., Bermudez, E., Padama, L., & Tena-Chávez, L. (2017). Neglected, protected, ejected: Latin American women caught by crimmigration. *Feminist Criminology, 12*(3), 293–314. https://doi.org/10.1177/1557085117691354
Dixon, J., et al. (2013). *Economic analysis of environmental impacts* (2nd ed.). Edited by The Asian Development Bank. London.
Docherty, B., & Giannini, T. (2009). Confronting a rising tide: A proposal for a convention on climate change refugees. *Harvard Environmental Law Review, 33*, 349.
Doran, A. (2011). Where should Haitians go – Why environmental refugees are up the creek without a paddle. *The Villanova Environmental Law Journal, 22*(1) Available at: https://digitalcommons.law.villanova.edu/elj/vol22/iss1/5
Dri, C. (2010). 'Limits of the institutional mimesis of the European Union: The case of the Mercosur parliament', Latin American. *Policy, 1*(1). https://doi.org/10.1111/j.2041-7373.2010.00004.x
Dun, O. (2009). Linkages between Flooding. In *Migration and resettlement in the Mekong Delta, Vietnam: Case study report for the EACH-FOR project*. United Nations University Institute for Environment and Human Security (UNU-EHS).

Echeverría, G. (2020). *Towards a systemic theory of irregular migration: Explaining ecuadorian irregular migration in Amsterdam and Madrid*. Available at: https://link.springer.com/chapter/10.1007/978-3-030-40903-6_2

Edwards, A. (2005). Human rights, refugees, and the right "to enjoy" asylum. *International Journal of Refugee Law, 17*(2), 293–330. https://doi.org/10.1093/ijrl/eei011

El Tiempo. (2018, June 29). *"Sin los pobladores es imposible proteger los páramos"*: Grisales. Available at: https://www.eltiempo.com/colombia/otras-ciudades/que-es-la-ley-de-proteccion-de-paramos-en-237360

El-Hinnawi, E. (1985). *Environmental refugees*. United Nations Environment Programme.

ESRI StoryMaps Team. (2021). *The age of megacities*. National Geographic. at: https://www.nationalgeographic.org/interactive/age-megacities/

Fendt, L. (2016, May 12). "I came here with nothing": Life in Limbo for Unwilling Migrants on Haiti's Border. *The Guardian*. Available at: https://www.theguardian.com/global-development/2016/may/12/haiti-dominican-republic-migrants-makeshift-camp

Ferris, E., & Bergmann, J. (2017). Soft law, migration and climate change governance. *Journal of Human Rights and The Environment*. https://doi.org/10.4337/JHRE.2017.01.01

Fitzpatrick, J. (2010). Revitalizing the 1951 refugee convention. In H. Lambert (Ed.), *International refugee law*. Routledge.

Gatrell, P. (2000). Forced migration during the Second World War: An introduction. In *Refugees, relief, and resettlement*. Available at: https://www.gale.com/intl/essays/peter-gatrell-forced-migration-second-world-war-introduction

Gatrell, P. (2008). Refugees and forced migrants during the first world war. *Immigrants & Minorities, 26*(1–2). https://doi.org/10.1080/02619280802442613

Gemmene, F., & Brucker, P. (2015). From the guiding principles on internal displacement to the Nansen initiative: What the governance of environmental migration can learn from the governance of internal displacement. *International Journal of Refugee Law, 27*(2), 245–263. https://doi.org/10.1093/ijrl/eev021

Gioli, G., Hugo, G., Costa, M. M., & Jürgen, S. (2016). Human mobility, climate adaptation, and development. *Migration and Development, 5*(2), 165–170. https://doi.org/10.1080/21632324.2015.1096590

Glasman, J. (2017). Seeing like a refugee agency: A short history of UNHCR classifications in Central Africa (1961–2015). *Journal of Refugee Studies, 30*(2), 337–362. https://doi.org/10.1093/jrs/few044

Gleick, P. (2014). Water, drought, climate change, and conflict in Syria. *Weather, Climate, and Society, 6*(3), 331–340. https://doi.org/10.1175/WCAS-D-13-00059.1

Goodwin-Gill, G. (2014). The international law of refugee protection. In E. Fiddian-Qasmiyeh, G. Loescher, K. Long, & N. Sigona (Eds.), *The Oxford handbook of refugee and forced migration studies*. Oxford University Press. https://doi.org/10.1093/oxfordhb/9780199652433.013.0021

Grau, R., & Aide, T. M. (2008). Globalization and land-use transitions in Latin America. *Ecology and Society, 13*(2). https://www.jstor.org/stable/26267952

Greenberg, R. (2017). The particular social group requirement: How the Asylum process is consistently failing LGB applicants and how an evidentiary standard of self-attestation can remedy these failures. *University of Maryland Law Journal of Race, Religion, Gender & Class* (1). Available at: https://digitalcommons.law.umaryland.edu/rrgc/vol17/iss1/9

Grugel, J. B. (2004). New regionalism and modes of governance — Comparing US and EU strategies in Latin America. *European Journal of International Relations, 10*(4). https://doi.org/10.1177/1354066104047850

Gu, D. (2019). *Exposure and vulnerability to natural disasters for world's cities*. United Nations Department of Economic and Social Affairs. at: https://www.un.org/development/desa/pd/sites/www.un.org.development.desa.pd/files/files/documents/2020/Apr/un_2019_techpaper4.pdf

Hall, N. (2013). Moving beyond its mandate? UNHCR and climate change displacement. *Journal of International Organizations Studies, 4*(1), 91–108.

Harrell-Bond, B. (1985). Imposing aid: Emergency assistance to refugees. *Contemporary Sociology, 18*(5). https://doi.org/10.2307/2073365
Heyward, C., & Odalen, J. (2013). *A new Nansen passport for the territorially dispossessed.* Advancing Political Science. Available at: https://ecpr.eu/Events/Event/PaperDetails/4601
Hitchcock, W. I. (2014). World war I and the humanitarian impulse. *The Tocqueville Review/La revue Tocqueville, 35*(2), 145–163. https://www.muse.jhu.edu/article/578168
Hoing, N., & Razzaque, J. (2012). Unacknowledged and unwanted? 'Environmental refugees' in search of legal status. *Journal of Global Ethics, 8*(1), 19–40. https://doi.org/10.1080/17449626.2011.635691
Holborn, L. (1939). The league of nations and the refugee problem. *The Annals of the American Academy of Political and Social Science, 203*(1), 124–135. https://doi.org/10.1177/000271623920300115
Holborn, L. (1956). *The international refugee organization: A specialized Agency of the United Nations—Its History and Work, 1946–1952.* Oxford University Press.
Hong, J. (2001). Refugees of the 21st century: Environmental injustice. *Cornell Journal of Law & Public Policy, 10*(2), 323–348.
Humphries, M. (2005). Natural resources, conflict, and conflict resolution: Uncovering the mechanisms. *Journal of Conflict Resolution, 49*(4), 508–537. https://doi.org/10.1177/0022002705277545
International Conservation Caucus Foundation. (2018). *Ley De Páramos Es Aprobada En El Congreso De La República De Colombia.* Available at: https://www.international conservation.org/es/america-latina/grupos/ley-de-paramos-es-aprobada-en-el-congreso-de-la-republica-de-colombia
International Union for Conservation of Nature. (2010). *Paramos.* Available at: https://www.iucn.org/content/paramos
IOM. (2009). *Migration, environment and climate change: Assessing the evidence.* F. Laczko & C. Aghazarm (Eds.). Available at: https://publications.iom.int/system/files/pdf/migration_and_environment.pdf
IOM. (2020). *Venezuelan refugee and migrant crisis.* Available at: https://www.iom.int/venezuela-refugee-and-migrant-crisis.
Jackson, I. (1999). *The refugee concept in group situations.* Martinus Nijhoff.
Jaeger, G. (2001). On the history of the protection of refugees. *International Review of the Red Cross, 843.* Available at: https://www.icrc.org/en/doc/assets/files/other/727_738_jaeger.pdf
Jaramillo, L., & Sancak, C. (2009). Why has the grass been greener on one side of Hispaniola? A comparative growth analysis of the Dominican Republic and Haiti. *IMF Staff Papers, 56*(2), 323–349. https://doi.org/10.1057/imfsp.2008.40
Jaupart, P. (2018). Divided Island: Haitian immigration and electoral outcomes in the Dominican Republic. *Journal of Economic Geography, 18*(4), 951–999. https://doi-org.ezproxy.is.ed.ac.uk/10.1093/jeg/lby030
Kalb, M. (2015). *Refugee crises and the sad legacy of the 1938 Evian conference.* Brookings Institute. Available at: https://www.brookings.edu/blog/order-from-chaos/2015/09/23/refugee-crises-and-the-sad-legacy-of-the-1938-evian-conference/
Kaldor Centre. (2020). *What is the 1967 Protocol relating to the Status of Refugees?* Available at: https://www.kaldorcentre.unsw.edu.au/publication/1967-protocol
Keane, D. (2004). The environment causes and consequences of migration: A search for the meaning of environmental refugees. *Georgetown International Environmental Law Review, 16*(2), 209–223.
Keyes, E. (2019). Environmental refugees? Rethinking what's in a name. *North Carolina Journal of International Law, 461.* Available at: https://scholarworks.law.ubalt.edu/all_fac/1082
King, T. (2006). Environmental displacement: Coordinating efforts to find solutions. *Georgetown International Environmental Law Review, 18*(3), 543–565.

Kossoff, D., et al. (2014). Mine tailings dams: Characteristics, failure, environmental impacts, and remediation. *Applied Geochemistry, 51*, 229–245. https://doi.org/10.1016/j.apgeochem.2014.09.010

Kozoll, C. M. (2004). Poisoning the well: Persecution, the environment, and refugee status. *Colorado Journal of International Environmental Law and Policy, 15*, 271.

Krause, U. (2016). Limitations of development-oriented assistance in Uganda. *Forced Migration Review, 52*. Available at: https://www.fmreview.org/solutions/krause

Kumari Rigaud, K., de Sherbinin, K., Jones, B., et al. (2018). *Groundswell: Preparing for internal climate migration*. The World Bank. at: https://openknowledge.worldbank.org/handle/10986/29461

Laurance, W., et al. (2015). Reducing the global environmental impacts of rapid infrastructure expansion. *Current Biology, 25*(7), R259–R262. https://doi.org/10.1016/j.cub.2015.02.050

Leonhard, J. (2018). *Pandora's box: A history of the first world war*. President and Fellows of Harvard University.

Lischer, S. K. (2007). Causes and consequences of conflict-induced displacement. *Civil Wars, 9*(2), 142–155. https://doi.org/10.1080/13698240701207302

Loescher, G., Betts, A., & Milner, J. (Eds.). (2008). *The United Nations higher commissioner for refugees (UNHCR): The politics and practice of refugee protection into the twenty-first century*. Routledge.

Lubkemann, S. C. (2005). 'Migratory coping in wartime Mozambique: An anthropology of violence and displacement in "fragmented wars"'. *Journal of Peace Research, 42*(4), 493–508. https://doi.org/10.1177/0022343305054093

Maia, A. G., & Shons, S. Z. (2020). The effect of environmental change on out-migration in the Brazilian Amazon rainforest. *Population and Environment, 42*, 183–218. https://doi.org/10.1007/s11111-020-00358-2

Malamud, A., & Gardini, G. L. (2012). Has regionalism peaked? The Latin American quagmire and its lessons. *The International Spectator: Italian Journal of International Affairs, 47*(1), 116–133. https://doi.org/10.1080/03932729.2012.655013

Markham, L. (2019, April 6). How climate change is pushing Central American migrants to the US. *The Guardian*. Available at: https://www.theguardian.com/commentisfree/2019/apr/06/us-mexico-immigration-climate-change-migration

Marshall, N. (2016). Forced environmental migration: Ethical considerations for emerging migration policy. *Ethics, Policy & Environment, 19*(1), 1–18. https://doi.org/10.1080/21550085.2016.1173284

Mayer, B. (2011). The international legal challenges of climate-induced migration: Proposal for an international legal framework. *Colorado Journal of International Environmental Law and Policy, 22*, 357–416. Available at: http://benoitmayer.com/files/The_International_Legal_Challenges_of_Climate-induced_Migration.pdf

McAdam, J. (2014). Creating new norms on climate change, natural disasters and displacement: International developments 2010–2013. *Refuge: Canada's Journal on Refugees, 29*(2), 11–26. https://doi.org/10.25071/1920-7336.38164

McMichael, C., et al. (2012). An ill wind? Climate change, migration, and health. *Environmental Health Perspectives, 120*(5), 646–654. https://doi.org/10.1289/ehp.1104375

Mezdour, M., Veronis, L., & McLeman, R. (2016). Environmental influences on Haitian migration to Canada and connections to social inequality: Evidence from Ottawa-Gatineau and Montreal. *Environmental Migration and Social Inequality, 61*, 103–115. https://doi.org/10.1007/978-3-319-25796-9_7

Migration Data Portal. (2020). *Environmental Migration*. Available at: https://migrationdataportal.org/themes/environmental_migration.

MinAmbiente. (2018). *Páramos, Gobierno de Colombia*. Available at: http://www.minambiente.gov.co/index.php/bosques-biodiversidad-y-servicios-ecosistematicos/paramos

Morrissey, J. (2012). Rethinking the "debate on environmental refugees": From "maximilists and minimalists" to "proponents and critics". *Journal of Political Ecology, 19*(1), 36–49. https://doi.org/10.2458/v19i1.21712

Motomura, H. (2020). The new migration law: Migrants, refugees, and citizens in an anxious age. *Cornell Law Review, 105*(2), 457–548.

Muscolino, M. S. (2015). *The ecology of war in China: Henan Province, the Yellow River, and beyond, 1838–1950*. Cambridge University Press.

Myers, K. (2010). The hidden history of refugee schooling in Britain: The case of the Belgians, 1914–18. *History of Education, 30*(2). https://doi.org/10.1080/00467600010012445

Navas, G., Mingorría, S., & Aguilar-González, B. (2018). Violence in environmental conflicts: The need for a multidimensional approach. *Sustainability Science, 13*(3), 649–660. https://doi.org/10.1007/s11625-018-0551-8

NOAA. (2021). *What are El Nino and La Nina?* Available at: https://oceanservice.noaa.gov/facts/ninonina.html

Nyabola, N. (2019). *The end of asylum: A pillar of the Liberal order is collapsing—But does anyone care?* Foreign Affairs. at: https://www.foreignaffairs.com/articles/2019-10-10/end-asylum

OCHA. (2020). *Natural disasters in Latin American and the Caribbean*. Available at: https://www.humanitarianresponse.info/sites/www.humanitarianresponse.info/files/documents/files/20191203-ocha-desastres_naturales.pdf

Oliver-Smith, A. (2009). *Sea level rise and the vulnerability of coastal peoples: Responding to the local challenges of global climate change in the 21st century* (InterSections No. 7). United Nations University Institute for Environment and Human Security.

Panizza, F. (2013). *Contemporary Latin America: Development and democracy beyond the Washington consensus*. Zed Books Ltd.

Percival, V., & Homer-Dixon, T. (2001). The case of South Africa. In P. F. Diehl & N. P. Gleditsch (Eds.), *Environmental conflict*. Routledge.

Piguet, E., Kaenzig, R., & Guelat, J. (2018). 'The uneven geography of research on "environmental migration"'. *Population and Environment, 39*, 357–383. https://doi.org/10.1007/s11111-018-0296-4

Podesta, J. (2019). *The climate crisis, migration, and refugees*. Brookings Institute. at: https://www.brookings.edu/research/the-climate-crisis-migration-and-refugees/#footref-5

Priestly, C. (2020). "We Won't survive in a City. The marshes are our life": An analysis of ecologically induced genocide in the Iraqi marshes. *Journal of Genocidal Research*. https://doi.org/10.1080/14623528.2020.1792615

Purseigle, P. (2007). A wave on to our shores': The exile and resettlement of refugees from the western front, 1914–1918. *Contemporary European History, 16*(4), 427–444.

Raftopolous, M. (2017). Contemporary debates on social-environmental conflicts, extractivism and human rights in Latin America. *The International Journal of Human Rights, 21*(4), 387–404. https://doi.org/10.1080/13642987.2017.1301035

Renaud, F. G., Bogardi, J. J., Dun, O., & Warner, K. (2007). *Control, adapt or flee: How to face environmental migration?* UNU-EHS.

Rockström, J., Steffen, W., & Foley, J. A. (2009). A safe operating space for humanity. *Nature, 461*, 472–475. https://doi.org/10.1038/461472a

Saunders, P. (2000). Environmental refugees: The origins of a construct. In P. Stott & S. Sullivan (Eds.), *Political ecology: Science, myth and power*. Arnold.

Schlichtmann, K. (2010). UN collective security and the transitional period: A myth over the founding and aims of the United Nations. *Journal of East Asia and International Law, 3*(1). https://doi.org/10.14330/jeail.2010.3.1.05

Schmidt-Mumm, U., & Vargas Ríos, O. (2012). Comunidades vegetales de las transiciones terrestre-acuaticas del paramo de Chingaza, Colombia. *Revista de Biología Tropical, 60*(1). Available at: https://www.scielo.sa.cr/scielo.php?script=sci_arttext&pid=S0034-77442012000100003&lng=en&tlng=en

Schon, J. (2019). Motivation and opportunity for conflict-induced migration: An analysis of Syrian migration timing. *Journal of Peace Research, 56*(1), 12–27. https://doi.org/10.1177/0022343318806044

Schwartz, E. (2006, March 23). A needless toll of natural disasters. *Boston Globe*. Available at: http://archive.boston.com/news/globe/editorial_opinion/oped/articles/2006/03/23/a_needless_toll_of_natural_disasters/

Siltar, H. E. (2015). *Water stewardship in Colombia*. Eschborn. at: https://www.ceowatermandate.org/files/Water_Stewardship_in_Colombia_Case_Study_GIZ.pdf

Singh, B., & Cohen, M. (2014). *Climate change resilience: The case of Haiti* (Oxfam Research Reports, pp. 1–35). Available at: https://www.oxfam.org/sites/www.oxfam.org/files/rr-climate-change-resilience-haiti-260314-en.pdf

Smith, A. (2009). Coerced or free? Considering post-colonial returns. In R. Bessel & C. B. Haake (Eds.), *Removing peoples: Forced removal in the modern world*. Oxford University Press.

Soriano, J. P. (2019). High expectations. Interregional agendas on global security challenges: East Asia, Europe and Latin America. *Revista Brasileira de Política Internacional, 62*(1). https://doi.org/10.1590/0034-7329201900106

Steffens, G. (2018a). *Changing climate forces desperate Guatemalans to migrate*. National Geographic. Available at: https://www.nationalgeographic.com/environment/article/drought-climate-change-force-guatemalans-migrate-to-us

Steffens, G. (2018b). In the Colombian Amazon, peace has environmental consequences. *Public Radio International*. Available at: https://www.pri.org/stories/2018-05-03/colombian-amazon-peace-has-environmental-consequences

Stone, D. (2018). Refugees then and now: Memory, history and politics in the long twentieth century: An introduction. *Patterns of Predjudice, 52*(2–3). https://doi.org/10.1080/0031322X.2018.1433004

Storey, H. (2014). What constitutes persecution? Towards a working definition. *International Journal of Refugee Law, 26*(2), 272–285. https://doi.org/10.1093/ijrl/eeu017

Stoyan, A., et al. (2016). Trust in government institutions: The effects of performance and participation in the Dominican Republic and Haiti. *International Political Science Review, 37*(1), 18–35. https://doi.org/10.1177/0192512114534703

Suarez, A., Arias-Arevalo, P. A., & Martinez-Mera, E. (2018). Environmental sustainability in post-conflict countries: Insights for rural Colombia. *Environment, Development and Sustainability, 20*, 997–1015. https://doi.org/10.1007/s10668-017-9925-9

Temper, L., Del Bene, D., & Martinez-Alier, J. (2015). Mapping the frontiers and front lines of global environmental justice: The EJAtlas. *Journal of Political Ecology, 22*, 255–278. Available at: http://jpe.library.arizona.edu/volume_22/Temper.pdf

Torres, J. R., & Castro, J. S. (2019). Venezuela's migration crisis: A growing health threat to the region requiring immediate attention. *Journal of Travel Medicine, 26*(2). https://doi.org/10.1093/jtm/tay141

Turner, S. (2016). What is a refugee camp? Explorations of the limits and effects of the camp. *Journal of Refugee Studies, 29*(2), 139–148. https://doi.org/10.1093/jrs/fev024

Turton, D. (2003). *Refugees, forced resettlers and other forced migrants: Towards a unitary study of forced migration*. UNHCR.

UK Research and Innovation. (n.d.). *Towards environmental reconciliation in Páramo land in Boyacá: resolving ecosystem trade-offs in post-conflict spaces*. Lead Research Organisation: King's College London. Available at: https://gtr.ukri.org/projects?ref=NE%2FR017999%2F1

UNFCCC. (2012). *Slow onset events*. Available at: https://unfccc.int/resource/docs/2012/tp/07.pdf

UNFCCC. (n.d.). *Slow onset events*. Available at: https://unfccc.int/wim-excom/areas-of-work/slow-onset-events

UNHCR. (1998). *Note on burden and standard of proof in refugee claims*. Available at: https://www.refworld.org/docid/3ae6b3338.html

UNHCR. (2004). *Guiding Principles on Internal Displacement*. Available at: https://www.unhcr.org/protection/idps/43ce1cff2/guiding-principles-internal-displacement.html

UNHCR. (2010). *Convention and protocol relating to the status of refugees*. Available at: https://www.unhcr.org/3b66c2aa10
UNHCR. (2020). *Refugee data finder*. Available at: https://www.unhcr.org/refugee-statistics/download/?url=7MSHnr
UNHCR. (2021a). *Solutions*. Available at: https://www.unhcr.org/solutions.html%20(UNHCR%20,%202021
UNHCR. (2021b). *UNHCR and IOM welcome Colombia's decision to regularize Venezuelan refugees and migrants*. UNHCR-IOM Joint Press Release. at: https://www.unhcr.org/news/press/2021/2/60214cf74/unhcr-iom-welcome-colombias-decision-regularize-venezuelan-refugees-migrants.html
UNHCR (n.d.). *History of UNHCR*. Available at: https://www.unhcr.org/en-us/history-of-unhcr.html.
United Nations. (2018). *The world urbanization prospects: The 2018 revision*. The United Nations. Available at: http://esa.un.org/UNPOP/wup/index.htm
United Nations International Office for Disaster Risk Reduction (UNISDR). (2011). *Revealing risk, redefining development. Global assessment report on disaster risk reduction*. UNISDR.
USAID. (2010). *Gautemala*. Available at: https://www.land-links.org/country-profile/guatemala/
Van Heyningen, E. (2010). A tool for modernisation? The Boer concentration camps of the south African war, 1900–1902. *South African Journal of Science, 106*(5–6). Available at: http://www.scielo.org.za/scielo.php?script=sci_arttext&pid=S0038-23532010000300014
Walter, M. (2016). *Extractives in Latin America and the Caribbean: The basics*. Inter-American Development Bank. Available at: https://publications.iadb.org/publications/english/document/Extractives-in-Latin-America-and-the-Caribbean-The-Basics.pdf
Warner, K. (2010). Global environmental change and migration: Governance challenges. *Global Enivornmental Change*. https://doi.org/10.1016/j.gloenvcha.2009.12.001
Watkins, G., & Salinas, A. G. (2020). *The climate crisis could drive massive human displacement in Latin America and the Caribbean*. Inter-American Development Bank. Available at: https://blogs.iadb.org/sostenibilidad/en/the-climate-crisis-could-drive-massive-human-displacement-in-latin-america-and-the-caribbean/
Weber, M. (1919). *Politics as a vocation*. Available at: https://www2.southeastern.edu/Academics/Faculty/jbell/weber.pdf
Williams, A. (2008). Turning the tide: Recognizing climate change refugees in international law. *Law & Policy, 30*(4), 502–529. https://doi.org/10.1111/j.1467-9930.2008.00290.x
Wisner, B. (2003). Sustainable suffering? Reflections on development and disaster vulnerability in the post-Johannesburg world. *Regional Development Dialogue, 24*(1), 135–148.
World Bank. (2018). *World Bank Open Data*. Available at: https://data.worldbank.org
World Bank. (2019). *Venezuelan migration: The 4,500-kilometer gap between desperation and opportunity*. Available at: https://www.worldbank.org/en/news/feature/2019/11/26/migracion-venezolana-4500-kilometros-entre-el-abandono-y-la-oportunidad
World Bank. (2020). *Population, total*. Available at: https://data.worldbank.org/indicator/SP.POP.TOTL. Accessed 22 July 2020.
WWF. (2017). A look at the natural world of Colombia. *World Wildlife Magazine*. Available at: https://www.worldwildlife.org/magazine/issues/winter-2017/articles/a-look-at-the-natural-world-of-colombia#popup
WWF. (2018). *Aprobada ley que protege los páramos en Colombia*. Available at: http://www.wwf.org.co/?uNewsID=330280
Young, J. C., et al. (2010). The emergence of biodiversity conflicts from biodiversity impacts: Characteristics and management strategies. *Biodiversity and Conservation, 19*, 3973–3990.
Zinkle, S. J., & Was, G. S. (2013). Materials challenges in nuclear energy. *Acta Materialia, 61*(3), 735–758. https://doi.org/10.1016/j.actamat.2012.11.004

Chapter 6
Environmental Migration in the MENA Region: The Case of Morocco

Carla Sofia Ferreira Fernandes and Fátima Alves

Abstract The importance of environmental factors in the process of human mobility is acknowledged by the IPCC, UNFCCC and the Global Compact for Safe, Orderly and Regular Migration. However, the complexity of the drivers of migration tends to limit its analysis from an environmental change perspective, due to the multifaceted decision of migration. On the other hand, migrants can also contribute to adaptation strategies to climate change in their home communities. In order to address the challenges and positive potential of migration, it is relevant to integrate it in the overall discussion of adaptation to climate change. This chapter aims to present an overview of the studies that focus on the migration-environment-climate nexus in the Middle-East and North Africa (MENA) region, highlighting the gaps in the existing knowledge, which reflect the lack of studies for several countries in the region, despite the foreseen impacts of climate change and the high prevalence of migratory movements in the region. A special emphasis is given to Morocco, a polymorphic country in terms of migration flows, which registers one of the highest frequencies of studies in the region for the climate change and migration nexus. The main source of information is the CliMig, which is a database that compiles the bibliographic resources addressing the migration-environment-climate nexus.

Keywords Climate change · Environment · MENA region · Migration · Morocco

1 Introduction

The Foresight Report (2011) highlights the contribution of environmental change to migration by framing its influence in parallel with the various drivers that affect migration: economic, political, social, and demographic. The interaction between

C. S. Ferreira Fernandes (✉) · F. Alves
Universidade Aberta, Lisbon, Portugal

Centre for Functional Ecology, Universidade de Coimbra, Coimbra, Portugal
e-mail: 1600737@estudante.uab.pt; fatimaa@uab.pt

M. Behnassi et al. (eds.), *The Climate-Conflict-Displacement Nexus from a Human Security Perspective*, https://doi.org/10.1007/978-3-030-94144-4_6

the various factors is complex and is further influenced by the existence of diaspora networks, recruitment agencies, legal frameworks that favour or counter migration, and economic conditions to finance the migration process. This highly complex process of decision-making at an individual and household level renders it difficult to isolate a sole factor of migration, thus limiting the extension of the designation of 'environmental migrants'. The Foresight Report considers that environmental change is equally likely to make migration less possible since migration requires forms of capital, which might be negatively impacted by the very same environmental change that is placing the populations at risk.

Economic-related circumstances are among the most common factors that trigger migration, while environmental migration has the potential to further deteriorate the economic situation of the communities that lack resilience when facing negative impacts. In particular for the MENA region, the populations will also see their food security affected by the decrease in agricultural productivity as a result of environmental changes, particularly due to the reduced availability of water as well as from the emergence of new pests, impossibility of traditional crops to grow under a warming climate and from a decline in soil fertility (Verner, 2012). A growing population will place further stress on the food sources, which will ultimately widen the reliance on food imports and increase the volatility of food prices, thus straining the economic situation of the population (Burke et al., 2011). Climate change can also contribute to an increase in human health risks associated with heat stress and vector-borne diseases (e.g. malaria and dengue fever) (IPCC, 2014; Stern, 2014; Leal Filho et al., 2016). Additionally, the economic resilience of the population might be negatively impacted by the decreasing revenues from a decline in the tourism sector due to rising temperatures and from the additional demand for cooling during the summer. The economic investment necessary to compensate for the potential destruction of infrastructure due to the sea-level rise might place additional burdens on the institutions. Finally, the adaptation to climate change itself might result in additional costs in the various communities which might increase the vulnerability and risk of impoverishment (Stern, 2014).

The environmental degradation of resources might also lead to localized conflicts and social tensions particularly when we verify an asymmetrical utilization of the spaces and ecosystems (including fresh water resources) by community members, public entities, and the private sector (Cândido Fleury et al., 2014). Therefore, it is not possible to simplify the causality between reduced resources and conflict, since it is important to account also for the role of institutions and their capacity to manage scarcity and resource distribution.

Migration is expected to expand in the future, independently of environmental change, however, since the extreme weather and slow-onset events are predicted to increase, the displacement associated with the response to those events will contribute further to migratory movements (Foresight, 2011). Already the region registers large numbers of international migrants, both as a destination area with five out of the top twenty countries that host the largest numbers of international migrants, and also as an origin of migrants with four out of twenty countries with the largest diaspora populations, according to the United Nations, Department of Economic and Social Affairs, Population Division (2016).

By presenting an overview of the studies for the MENA region that focus on the migration-environment-climate nexus, this study can contribute to identifying the gaps in the existing knowledge. In the first section of this chapter, the main concepts related to the topic of environmental migration are presented as well as an overview of the various institutional responses to climate change. In order to have an overview of the literature published for the entire region, the methodology focused on using CliMig (a database that compiles the bibliographic resources addressing the nexus) as a source of information and selecting the available studies per country. The results are presented subsequently, with a particular emphasis on Morocco, a polymorphic country in terms of migration flows, which registers one of the highest concentration of studies in the region for the climate-migration nexus. The analysis also confirms the concern with the lack of studies for several countries in the region despite the foreseen impacts of climate change and the high prevalence of migratory movements in the region.

1.1 General Concepts of Environmental Migration

The Foresight Report (2011) defines 'migration' as a broadly voluntary movement from one place to another for 3 months or more, while 'displacement' implies a less voluntary movement that might involve a need for protection and/or assistance. Migration also includes relocation, which usually involves entire communities changing their location. Migration can also take various dimensions, regarding its timescale: permanent, temporary, and cyclical depending on the willingness and possibility of returning to the place of origin. Additionally, a distinction is also made between migration that occurs within a country or across international borders (IDMC, 2018), and in the case of international migration, if there are transit countries in the process of migration.

Migration is a complex 'phenomenon' that is affected by different factors, which can be classified using multiple levels (Foresight Report, 2011). At a macro level, it is possible to identify economic, political, demographic, social, and environmental drivers for migration, so the environmental drivers (e.g. exposure to hazard, land productivity, and water security) are perceived as one among several factors that influence the decision of migrating. At a meso level, there can exist facilitators of and/or obstacles against migration, such as the legal framework that regulates migratory movements, the cost of moving, and the existence of diasporic links and social networks. At a micro level, the household and individual characteristics associated with ethnicity, marital status and level of education will provide the final layer of considerations in the decision to migrate or not. All the mentioned factors can be either push or pull or a combination of both, i.e. they can either be determined by the region of origin or by the region of destination, e.g. employment opportunities (economic driver) can either consist of lack of opportunities at origin (push factor) or abundance of opportunities at the destination (pull factor).

Furthermore, it is important to consider the specific situations faced by the population that is unable or unwilling to move. Climate change and land degradation can actually reduce migratory movements due to their negative impact on the resources and, therefore, on the ability of populations to seek other destinations. This population can be extremely vulnerable since they might remain in environmental risk zones (Zickgraf, 2018).

In environmental migration, the focus of the study is on the influence of environmental and climatic factors in the decision of migration. In that sense, it is relevant to distinguish between the migration that is a reaction to sudden-onset (extreme events that occur over a short period of time) and to slow-onset events (e.g. sea-level rise, increasing average temperatures, ocean acidification, glacial retreat and related impacts, salinization, land and forest degradation, loss of biodiversity, and desertification) (UNFCCC, 2011).

The challenges associated with the study of environmental migration relate directly to the absence of data on migratory movements, particularly the ones occurring at a national level, due to an unsystematic inclusion of migration related questions in national censuses. The same void in data can be identified for various environmental indicators, with the exception of rainfall variability (Icduygu & Sert, 2011). Both limitations converge when studying environmentally-induced migration which then amplifies the uncertainties that exist due to lack of data.

1.2 Institutional Responses to Climate Change and Migration

The link between climate change and migration was officially addressed for the first time at the 16th Conference of Parties (COP16) of the United Nations Framework Convention on Climate Change (UNFCCC) in December 2010, where the countries adopted the Cancún Adaptation Framework (Warner et al., 2013). The paragraph 14-f of the Framework – which specifies that "measures to enhance understanding, coordination and cooperation with regard to climate change induced displacement, migration and planned relocation, where appropriate, at the national, regional and international levels" – constitutes an acknowledgement of the link between climate change and migration, displacement and planned relocation, for the first time in an international agreement. In June 2011, the launch of the 'Nansen Initiative' marks the beginning of a new state-owned consultative process that aims to address the current and future implications of the climatic challenges for human displacement, in the context of sudden- and slow-onset disasters (The Nansen Initiative, 2015). By integrating different dimensions of human mobility in the climate change related policies, it is possible to contribute to a shift in the long-term approach to population shifts, governance of borders and mobility and local, regional and global planning (Warner et al., 2013).

The legal protection of the populations facing displacement due to natural or human-made disasters, including climate change, is specifically addressed by the Kampala Convention which was officially adopted in 2009 by the African Union member states (African Union, 2009), and constituted an innovative regional response to create a legal framework that addresses migration due to climate change.

Currently, several platforms include environment-induced migration in their discussions and policy recommendations. In 2015, the Paris Agreement mandated the creation of a Task Force on Displacement under the UNFCCC Warsaw International Mechanism for Loss and Damage associated with Climate Change Impacts (UNFCCC, 2016a) and in COP24 in Katowice, parties have agreed on the COP Decision 10/CP.24 that outlines the Recommendations from the report of the Executive Committee of the Warsaw International Mechanism for Loss and Damage associated with Climate Change Impacts on integrated approaches to averting, minimizing and addressing displacement related to the adverse impacts of climate change (UNFCCC, 2018). The United Nations Global Compact for Safe, Orderly and Regular Migration specifically addresses "Natural disasters, the adverse effects of climate change, and environmental degradation" as factors that influence migratory movements (United Nations, 2018). Other initiatives include the Platform on Disaster Displacement that follows up on the 'Nansen Initiative', the Sendai Framework for Disaster Risk Reduction which aims to reduce the risks and vulnerabilities associated with displacement in case of disasters (UNISDR, 2015) and the United Nations Convention to Combat Desertification which includes the objective to reduce the migration forced by desertification and land degradation (UNCCD, 2017). The International Organization for Migration (IOM) has decided to place the population's vulnerability to climatic events and environmental degradation in their operational efforts (IOM, 2014).

Additionally, the 2030 Agenda for Sustainable Development (2030 Agenda) includes Sustainable Development Goals (SDGs) and targets on both climate change and migration, though explicit links between the two are absent. UNHCR has also acknowledged the specific challenges posed by climate change in the 2014 Brazil Declaration and Plan of Action and underlined the need for adopting appropriate national and regional measures (UNHCR, 2017).

2 Methodology

In order to identify the current state of the knowledge on environmental migration in the MENA region, the source of the information was the CliMig Database[1] which is the most exhaustive bibliography on the topics of migration, climate change and environment and is hosted by the Institute of Geography of the University of Neuchâtel. The publications included in the database have various foci:

[1] The CliMig Database is available at https://www.unine.ch/geographie/climig_database

- Disaster risk reduction/early-warning system in environmental migration;
- Studies dealing with environmental migration with a special focus on gender issues;
- Migration as an adaptation response, as a strategy to cope with environmental degradations;
- Representations and perceptions of actors on climate change;
- Climate change induced statelessness;
- Studies dealing with immobility and trapped populations;
- Studies linking environmental migrations and conflicts.

The database also includes different types of environmental hazards, such as floods, rainfall variability, hurricanes and storms, sea-level rise, coastal erosion, desertification, heat waves, and land degradation.

The criteria applied in the search of the documents related to the region were exclusively geographical, meaning that all types of environmental hazards and foci of the study were included. The search was therefore done separately for each country of the MENA region.[2] The following countries were included in the search: Morocco, Algeria, Tunisia, Libya, Egypt, Lebanon, Syria, Jordan, Iraq, Israel, Turkey, and the Palestinian Territories. The Arabian Peninsula was also included: Saudi Arabia, Yemen, Oman, United Arab Emirates, Qatar, Kuwait and Bahrain.

The CliMig database allows to further enlarge the search for specific studies on environmental migration by selecting the geographical location, as specified in the previous section. The successive application of the search parameters provided the following results of studies per country (number of studies are included after the country's name): Egypt (16), Morocco (12), Syria (9), Yemen (7), Algeria (6), Turkey (6), Israel (4), Iraq (2), Jordan (2), Libya (2), Palestine (2), Lebanon (1) and Tunisia (1). The search using the following countries as parameters did not produce any results: Saudi Arabia, Oman, United Arab Emirates, Qatar, Kuwait and Bahrain. The studies that focus on multiple countries in the region are counted separately for each country, with a total of 48 studies. The analysis was done individually for each country, grouping the studies either by topic or region, depending on the relevance.

3 Results

The analysis of the literature available in CliMig for the MENA countries enables this study to have a holistic perspective on the region in terms of environmental change: human rights approach, recommendations to policy makers, securitization approach, response to and consequences of various climatic and environmental phenomena such as water scarcity, desertification and sea-level rise, different dimensions of human mobility such as rapid urbanization and forced displacement and

[2] Due to the various definitions of the geographical definition of MENA region, the decision regarding the countries to include in the region was done by the authors.

projections of future mobility patterns. In the following subsections, firstly, it is identified how the existing studies in the region fit within these wider categories and, secondly, the specific case of Morocco is analyzed in detail as an example of a country in the region with multiple environmental challenges and mobility profiles: origin, destination, and transit country.

3.1 Environmental Migration in the MENA Region

Several general studies on environmental migration have focused on the MENA region: for instance, the Foresight Report (2011) included workshops on the Mediterranean basin countries while the IOM publication *People on the Move in a Changing Climate* includes a chapter dedicated to the literature review of five countries in the MENA region (Algeria, Egypt, Morocco, Syria, and Yemen) (Wodon et al., 2013) and The World Bank published *Climate Change and Migration: Evidence from the Middle East and North Africa* focusing particularly on the previously mentioned five countries (Wodon et al., 2014).

Despite the integration of climate-induced migration in international agreements and acknowledged by the United Nations High Commissioner for Human Rights (2018), in general, the studies do not focus on the issue of the legal protection of the migrants, except for Kolmannskog (2015) that reflects on the legal protection of international migrants from the Horn of Africa in Yemen, thus demonstrating the lack of studies on the specific issues faced by internal migrants, which constitute the majority of the migratory movements. Kolmannskog (2015) focuses particularly on the challenges of granting asylum to migrants originally affected by climate disasters, and not on migration due to slow-onset events. Furthermore, regarding international migration, in general, a lack of legal channels for migration might lead to populations being unable to leave a specific location (Foresight, 2011), which might include one of the MENA countries that are used as a transit country, even if the migrant intends to seek asylum in a third country.

Several studies address general policy-related issues, either by providing recommendations for policy-makers (Westing, 1994; Warner, 2010; Foresight, 2011), including for international actors such as the United Nations and The World Bank (Werz & Conley, 2012) or by doing a policy analysis, particularly in regards to international migration (Sow et al., 2015) and adaptation to climate change (El-Raey et al., 1999). Westing (1994) and Tribak et al. (2019) include the environmental change and migration studies within a wider discussion on sustainability. In practice, a large number of migrants is moving into urban areas that face their own issues in regard to climate resilience and sustainability, which if not addressed, lead to a second cycle of vulnerability and environmental migration (Foresight, 2011).

There are situations when the population is forcibly displaced for the construction of infrastructure (e.g. dams) (Kadirbeyoglu, 2008) intended to improve conditions that are negatively affected by climate change (e.g. reduce carbon emissions as part of mitigation efforts or to capture more water to compensate for reduced water availability).

One of the most likely impacts of climate change in the North of Africa is the increase in the periodicity of droughts which impacts the migration from the southern Mediterranean countries (Magnan et al., 2009; Mulligan et al., 2013). However, and as stated by Magnan et al. (2009), it is essential to understand migration as a multi-causal phenomenon that depends not only on one factor (e.g. environment) but on various factors, namely socio-economic, that influence each other mutually. While the concerns with the reduced water availability are widespread across the MENA region and are most likely to be intensified with the prevalence of the climate change impacts, in Egypt the studies have mostly focused on the sea-level rise (Warner, 2010). In this country, the dependence for agricultural production on the coastal and deltaic areas of the Nile will be further constrained by the inundation and salinization due to sea-level rise (Agrawala et al., 2004; Warner, 2010) but other economic activities such as tourism will also be affected (El-Raey et al., 1999). But while Egypt is the country in the MENA region that is most studied in regards to the effects of sea-level rise, the region overall, together with East Asia, exhibits the greatest relative impacts (Dasgupta et al., 2007), which might be a factor for migration (Myers, 2002). Nevertheless, considering the particular case of Egypt and how environmental changes and land degradation influence the migratory movements, Afifi (2009a) claims that such factors alone are not sufficient to lead to transnational migration, and that the pull factors in the destination countries are a major part of the decision-making process.

The movements of populations within the borders, mainly the urbanization phenomenon, is the most common type of migratory movements, with the population leaving impoverished rural areas to seek opportunities in urban centers. Due to the multi-causal nature of migration, it is difficult to identify the environmental contribution to the decision to change countries or to move to larger urban areas. In Yemen, in the period that preceded the political crisis which started in 2015, Joseph and Wodon (2013) concluded that the climate variables affect migration, but not to the same extent as socio-economic considerations. De Haas (2011) focused mainly on international migration and concluded that the impact of environmental and demographic factors is limited and indirect.

The analysis of both climate change and movements of the population is often framed as a security concern with climate change at the origin of resource depletion, namely water and farming land, which are then presented as triggers for conflict and/or migration (the latter being perceived also as a potential cause for conflict). In parallel, armed conflict in general has systematically caused migration and displacement in the MENA region, particularly within the Mashriq countries (Egypt, Iraq, Jordan, Lebanon, Palestine, and Syria) (Icduygu & Sert, 2011). In general, there is a tendency to perceive climate change as a threat multiplier, with the possibility to exacerbate situations of insecurity (Werz & Conley, 2012) and be one of the factors that lead to conflict (Hsiang et al., 2013).

In general, demographic changes, including processes of urbanization and internal or international migration, might increase the likelihood of conflicts, but the link is dependent on the context (Goldstone, 2002). The North of Africa is particularly relevant when adopting a securitization approach to climate-induced migration due to its polymorphic role: origin, transit, and destination of migrants (Population and Development Review, 2010; Werz & Conley, 2012), while in the Middle East the Syrian conflict has been cited as an example of the impacts of climate change on migration and conflict (Kelley et al., 2015). However, this conclusion is not fully supported in the literature (Selby et al., 2017), despite several reports prior to the beginning of the Syrian crisis that point to worrisome levels of drought and consequent migration, and to the inadequate response from national and international actors (IRIN News, 2009; Ali, 2010). In Egypt, Afifi (2009b) establishes a link between water depletion and land degradation and migration, but excludes a direct impact on increased insecurity. The protracted conflict in Israel and Palestine occurs in a context of water scarcity, which will most likely be intensified by climate change. While the impact of water scarcity on conflict and migration might be difficult to assess, it is nevertheless a source of tensions at the local level (Selby, 2006), whose contribution cannot be overstated which would imply a downplay of the political and socio-economic drivers to the conflicts and migration in the region (Weinthal et al., 2015).

As with migration and climate change, water is often analyzed through a security lens, particularly in the countries of Eastern Mediterranean that face various concerns with conflict and migration as well as water scarcity which has been exacerbated by the extreme drought of the past two decades (Weinthal et al., 2015). In other situations of conflict in the region, namely in Yemen, Libya and Iraq, it would be relevant to have studies that analyze how resources, namely water, can contribute to migration and conflict or to cooperation.

Another perspective on the analysis of migration in relation to climate change is its potential contribution to the adaptation in the home communities of the migrants. The contribution of the diaspora to the host communities, as well as the concerns with the integration within that community has been studied extensively throughout the years, particularly when there are frequent movements, such as from Algeria to France (Samers, 1997; Collyer, 2003) or from Egypt to oil-exporting Arab countries (Sell, 1988). In the context of climate change, migration can be either seen as an adaptive strategy or as a result of a failure to adapt (Warner, 2010). Migration as an adaptive strategy can take several forms: by migrating, the population reduces its exposure to specific risks (climatic, environmental, socio-economic and political) but also migration can act as a means to increase the socio-economic level, and through remittances, support the home community in their adaptive efforts to climate change. Nevertheless, in Yemen, the role of remittances is not more pronounced when the home community is more exposed to environmental risks, which suggests that remittances are part of the income of the community, but they are not proportional to the needs associated with the adaptation to climate change (Joseph et al., 2014).

Furthermore, it is crucial to consider the population that lacks the necessary means or refuses to migrate, which in extreme situations might place them at a higher level of vulnerability and exposure to risk. When large numbers of population migrate, the socio-economic situation of the declining population might actually deteriorate (McLeman, 2011). Nevertheless, the literature on the region does not systematically address this issue.

The mixed flows that occur throughout the region are not sufficiently addressed in the published documents. In fact, no other works have attempted to address the issues of environmental-induced migrants' protection who originate outside the region except for Kolmannskog (2015) which focused on the protection of East African migrants in Yemen.

Finally, several studies attempted to provide future projections and scenarios which reinforce the assumption that stronger impacts of climate change will further increase the intensity of migratory movements to and from the MENA region (Population and Development Review, 2010; Foresight, 2011). However, de Haas (2011) claims that it will be more prevalent on internal rather than on international migration. One such example is related to the projections of sea-level rise and its impacts on the coastal population, particularly in Egypt (Neumann et al., 2015). Regarding the increment of water scarcity, its impact on the populations will depend largely on the political and institutional actions (de Haas, 2011). The projections in terms of increased vulnerability of population could serve as a basis for preventive measures of displacement as signaled by Baumert and Kloos (2017) in their study for Egypt. Preventive displacement nevertheless poses serious concerns in terms of rights of the displaced population, since the evidence suggests that resettled populations often face socio-economic impoverishment (De Wet, 2006).

3.2 *The Specificity of the Moroccan Case*

It is difficult to establish a direct attribution of the reasons to migrate in a context in which climate change impacts are multifaceted, personal motivations are complex and environmental conditions and altered ecosystem services usually interact with a range of other economic, political, social, cultural, and demographic factors (UNFCCC, 2016b). In Morocco, the studies that focus on migration issues often deal with the international aspect of migration that includes Moroccans who migrate to other countries, or other populations that migrate to Morocco. In this sense, the country becomes a destination or a transit location for migrants who want to reach Europe. Despite being less studied, internal migration, namely from rural areas to urban centers, is more frequent, and cannot be simply characterized as an attempt to escape poverty but also as a way to seek new opportunities and to improve the quality of life.

The studies in Morocco that focus on environmentally-induced migration are concentrated in three main geographic areas: Drâa Valley (Sobczak-Szelc, 2008; Ait Hamza et al., 2009), mountainous regions (Kollmair & Banerjee, 2011), and urban centers (Tribak et al., 2019; Fernández et al., 2019).

Concerning the Drâa Valley, Ait Hamza et al. (2009) and Sobczak-Szelc (2008) identified several reasons that contributed to the deterioration of the environmental situation: desertification (advancement of the sand dunes of the Sahara desert); salinization of the water sources; reduction of the available water (partially due to the construction of the dams upstream); fungus plagues that destroyed one of the most profitable and frequent crops (dates); and also the destruction caused by locust invasions.

The research conducted by Sobczak-Szelc (2008) focused on the impact of the reduction of water availability caused by the construction of upstream dams on the migration of the community in the Drâa Valley. It should be mentioned that the dams led to a decrease of the agricultural output and an increase in the desertification process which ultimately affected the overall living conditions of the population. Aditionally, the socio-economic activities connected with the agriculture and animal husbandry were reduced and redirected to other crops and animals more resistant to the reduced water availability, while tourism related activities were increased due to the new possibility to explore visits to the dune camps. This new rearrangement of the socio-economic activities created new migratory dynamics: push factors, from agriculture and animal husbandry; and pull factors, from the tourism sector. Despite the growth in the tourism sector, Ait Hamza et al. (2009) state that it could not prevent the out-migration of the local population. The possible destinations for the migrants consist of the main urban centres around the country, and the movement tends to be temporary, even though there is a growing tendency for international and more permanent migration.

Within the frame of the Foresight project (2011), Kollmair and Banerjee (2011) produced a report on the drivers of migration in the mountainous regions of developing countries. In the case of Morocco, the authors identified various drivers, including environmental ones, which have contributed to migratory movements. Among the environmental changes, the authors identify "deforestation, overgrazing, declining soil fertility, soil erosion and periodic drought" (Kollmair & Banerjee, 2011:11). Among the contextual changes, the authors identified the change in the patterns of transhumance, which was a trend that Ait Hamza et al. (2009) also associated with the oases in the Drâa Valley in parallel with the sedentarisation of former nomad groups.

The studies presented previously for Morocco mostly focus on the drivers in the local communities that trigger migration responses, however Tribak et al. (2019) approach migration from the perspective of the urban centers that receive the new arrivals, mostly internal migrants, and the ways in which the fast rate of urbanization creates new vulnerabilities for the migrants.

Generally, the impact of sea-level rise is not analyzed in general but Wodon et al. (2013:121) anticipate that "an increasing number of floods and as well as a rising sea level may also induce future migration movements".

The authors Sow et al. (2015) provided an analysis of how migratory movements are actively supporting home communities in adapting to the impacts of climate change, by focusing not only on financial remittances from the diaspora, but also on non-financial remittances, such as the transfer of knowledge and technology through partnerships with non-governmental organizations.

Finally, other studies on environmental migration in Morocco focused on the country's role in the migratory movements that originate from sub-Saharan Africa to reach the Northern countries (Wodon et al., 2013; Sow et al., 2015), including from a security perspetive (Werz &Conley, 2012).

4 Discussion

Overall, the analysis of the various studies for the region confirms that there is still an insufficient knowledge regarding the diversity of challenges that environmental change will pose in terms of its contribution to the complex phenomenon of human mobility. This conclusion is corroborated by Piguet et al. (2018) that mapped the available literature on environmental migration around the world and concluded that the MENA region, together with South America and Central Asia, are under-represented in the studies on environmental migration despite the high risks that the populations in these regions face.

The United Nations High Commissioner for Human Rights (2018) suggests that the response to migration should specifically address gender issues, as part of a wider approach of the human rights of the migrants. For the studies in the region, however, there is an absence of a gender approach, despite the general concerns across the countries in terms of female labor and political integration (Assaad et al., 2018). In regard to human rights in general, there is also a lack of studies that address the specific situation of environmental migrants.

The studies in the region focus mainly on presenting a portrait of the historical and current situation at a national or regional scale, and with the exception of the studies that are based on national censuses; in general there is not a presentation of actual numbers. Local studies, instead of general descriptions for the full country, might be more successful in capturing the local dynamics of conflict and migration, therefore providing a more accurate view of the phenomenon, without being limited by general concerns with extrapolation of the results at a national or regional level, since these might require a simplification of the various factors, thus distorting the research. There are several countries for which there is a lack of studies, such as the countries in the Arabian Peninsula, despite being the destination for migrants from various regions in the world (United Nations, Department of Economic and Social Affairs, Population Division, 2016).

The overall impacts of climate change, pollution, and land degradation are not analysed in detail as research has focused on two specific impacts: increased water scarcity in the East Mediterranean countries (Selby, 2006; Weinthal et al., 2015) and sea-level rise in Egypt (Agrawala et al., 2004; Warner, 2010). Sudden-onset events, driven by climate change or caused by other environmental issues, are also not addressed as potential triggers for migration, with the exception of the needs assessment of the population displaced by an earthquake in Turkey (Daley et al., 2002).

Despite the polymorphic profile of MENA region in regard to migration, there is a lack of focus on migrants originating from other countries, and whose migration process might have been directly impacted by slow- and sudden-onset events. Regarding the host communities, an analysis on the impact of migration on the environment, particularly the risks faced by low-income populations in unplanned urbanizations, is another issue in the environmental migration domain that was not addressed in the MENA region except for the migrants in Fez, Morocc as advanced by Tribak et al. (2019).

In order to capture the complexities of the interlinkages between environmental change and migration, Ait Hamza et al. (2009) argue that longitudinal studies need to use questionnaires, interviews, and observation techniques to understand the evolution of the issue in time and accommodate changes of the context and their impact on the phenomena.

Furthermore, the contribution of migrants to the adaptation of their home communities to climate change is usually limited to an analysis of the role of financial remittances that compensate the loss of income with the exception of the work from Sow et al. (2015) that specifically explored the Moroccan diaspora's contributions to action directly related to climate change adaptation and non-financial remittances.

Finally, in order to avoid an increase in the exposure of the population to the various risks associated with environmental change, one of the possibilities available to the governments is preemptive relocation, which is not addressed in the literature, except for Baumert and Kloos (2017) who examined the issue in the Greater Alexandria region in Egypt.

5 Conclusion

The analysis of the studies published for the MENA region in the field of environmental migration puts in evidence the need to intensify the geographical coverage of the countries studied as well as the approaches used in order to frame the contribution of environmental change to human mobility. The presence of security constraints in several countries of the MENA region poses additional challenges in the collection of data, particularly in terms of internal migration and environmental changes. Countries like Morocco and Egypt count several studies, but the geographical coverage within those countries is also limited, with most studies focusing on particular regions and environmental phenomena such as desertification and sea-level rise.

There is a tendency to frame the climate change and migration movements in a securitization approach, without "captur[ing] the complex, mobile and interconnected nature and key challenges of climate change and migration" (Boas et al., 2019:902). A simplified analysis of these phenomena in a region that already has various situations of conflict needs to be avoided, thus the need for further studies that provide scientific evidence and demonstrate the complexity of human mobility. In this sense, considering the importance of local context to capture that same complexity, it is relevant to ensure that all countries are covered in the studies, but also that different local dynamics within the same country are analyzed.

Finally, as a region with a polymorphic migration profile, there is lack of studies in the communities that originate outside the region but whose motivations for migrating might be related to environmental change, notably migrants from the Horn of Africa in the Arabian Peninsula.

References

Afifi, T. (2009a). *Egypt: Case study report for the EACH-FOR project*. UNU-EHS.

Afifi, T. (2009b). Egyptian water and soil: A cause for migration and security threats? In J. L. Rubio, U. Safriel, R. Daussa, W. Blum, & F. Pedrazzini (Eds.), *Water scarcity, land degradation and desertification in the mediterranean region* (NATO science for peace and security series C: Environmental security). Springer. https://doi.org/10.1007/978-90-481-2526-5_12

African Union. (2009). *African Union convention for the protection and assistance of internally displaced persons in Africa – Kampala convention*. Kampala.

Agrawala, S., Moehner, A., El Raey, M., Conway, D., van Aalst, M., Hagenstad, M., & Smith, J. (2004). *Development and climate change in Egypt: Focus on Coastal Resources and the Nile*. OECD.

Ait Hamza, M., El Faskaoui, B., & Fermin, A. (2009). *Migration and environmental change in Morocco: The case of rural oasis villages in the Middle Drâa Valley*. Case Study Report: EACH-FOR – Environmental Change and Forced Migration Scenarios.

Ali, M. (2010). *Years of drought: A report on the effects of drought on the Syrian Peninsula*. Heinrich-Böll-Stiftung - Middle East Office.

Assaad, R.A., Hendy, R., Lassassi, M. & S. Yassin (2018) Explaining the MENA paradox: Rising educational attainment, Yet Stagnant Female Labor Force participation. IZA Discussion Paper No. 11385

Baumert, N., & Kloos, J. (2017). Anticipating emerging risks and vulnerabilities from sea level rise induced preventive resettlement in greater Alexandria, Egypt. In M. Behnassi & K. McGlade (Eds.), *Environmental change and human security in Africa and the Middle East*. Springer.

Boas, I., Farbotko, C., Adams, H., et al. (2019). Climate migration myths. *Nature Climate Change, 9*, 901–903. https://doi.org/10.1038/s41558-019-0633-3

Burke, S. M., Mulligan, M., Parks, K., & van Soesbergen, A. (2011). Migration and global environmental change – DR8a: Quantifying change in ecosystem services and exposure to hazards in the Mediterranean basin over the next 50 years that might be relevant to migration. UK Government's Foresight Project, Migration and Global Environmental Change, UK Government Office for Science.

Cândido Fleury, L., Almeida, J., & Premebida, A. (2014). O ambiente como questão sociológica: conflitos ambientais em perspectiva. *Sociologias, 16*(35), 34–82.

Collyer, M. (2003) *Explaining change in established migration systems: The movement of algerians to France and the UK*. Sussex Migration.

Daley, W. R., Karpati, A., & Sheik, M. (2002). Needs assessment of the displaced population following the August 1999 earthquake in Turkey. *Disasters, 25*(1), 67–75. https://doi.org/10.1111/1467-7717.00162

Dasgupta, S., Laplante, B., Meisner, C., Wheeler, D., & Yan, J. (2007). *The impact of sea level rise on developing countries: A comparative analysis* (Policy Research Working Paper; No. 4136). World Bank, Washington, DC. © World Bank. https://openknowledge.worldbank.org/handle/10986/7174. License: CC BY 3.0 IGO.

de Haas, H. (2011). Mediterranean migration futures: Patterns, drivers and scenarios. *Global Environmental Change, 21*(Supplement 1), S59–S69. https://doi.org/10.1016/j.gloenvcha.2011.09.003

De Wet, C. J. (Ed.). (2006). *Development-induced displacement. Problems, policies, and people* (Studies in forced migration, 18). Berghahn Books.
El-Raey, M., Dewidar, K. R., & El-Hattab, M. (1999). Adaptation to the impacts of sea level rise in Egypt. *Mitigation and Adaptation Strategies for Global Change, 4*(3–4), 343–361.
Fernández, C., Sierra-Huedo, M. L., & Chinarro, D. (2019). Climate change-induced migration in Morocco: Sub-Saharian and Moroccan migrants. In M. Paradiso (Ed.), *Mediterranean mobilities*. Springer.
Foresight. (2011). *Migration and global environmental change: Final project report*. The Government Office for Science.
Goldstone, J. A. (2002). Population and security: How demographic change can lead to violent conflict. *Journal of International Affairs, 56*(1).
Hsiang, S., Burke, M., & Miguel, E. (2013). Quantifying the influence of climate on human conflict. *Science, 341*, 1235367. https://doi.org/10.1126/science.1235367
Icduygu, A., & Sert, D. (2011). *Migration and global environmental change – PD13: International migration in the Mashriq countries: Past trends, future potentials and determinants – Possible implications for environmental migration?* Government's Foresight Project.
IOM. (2014). IOM outlook on migration, environment and climate change.
IDMC. (2018). Synthesizing the state of knowledge to better understand displacement related to slow onset events. The Warsaw international mechanism for loss and damage associated with climate change impacts. Task Force on Displacement Activity I.2.
IPCC. (2014). Climate change 2014: Impacts, adaptation, and vulnerability. Part B: Regional aspects. Contribution of working group II to the fifth assessment report of the intergovernmental panel on climate change [Barros, V.R., C.B. Field, D.J. Dokken, M.D. Mastrandrea, K.J. Mach, T.E. Bilir, M. Chatterjee, K.L. Ebi, Y.O. Estrada, R.C. Genova, B. Girma, E.S. Kissel, A.N. Levy, S. MacCracken, P.R. Mastrandrea, and L.L. White (Eds.)]. Cambridge University Press, p. 688.
IRIN News. (2009). Drought response faces funding shortfall [online]. Available at https://www.thenewhumanitarian.org/news/2009/11/24/drought-response-faces-funding-shortfall. Accessed 7 Sept 2019.
Joseph, G., & Wodon, Q. (2013). Is internal migration in Yemen driven by climate or socio-economic factors? *Review of International Economics, 21*(2), 295–310. https://doi.org/10.1111/roie.12037
Joseph, G., Wodon, Q., & Blankespoor, B. (2014). *Do remittances reach households living in unfavorable climate areas? Evidence from the Republic of Yemen, MPRA paper 56939*. University Library of Munich.
Leal Filho, W.; Azeiteiro, U. M.; Alves, F. (Eds.). (2016). Climate change and health: Improving resilience and reducing risks. Springer. Volume produced as part of the "Climate Change Management Series" published by Springer. Print ISBN 978-3-319-24658-1, E-Book ISBN 978-3-319-24660-4, https://doi.org/10.1007/978-3-319-24660-4. 532p. https://www.springer.com/gp/book/9783319246581.
Kadirbeyoglu, Z. (2008). *Turkey: Case study report for the EACH-FOR project*. UNU-EHS.
Kelley, C. P., Mohtadi, S., Cane, M. A., Seager, R., & Kushnir, Y. (2015). Climate change in the Fertile Crescent and implications of the recent Syrian drought. *PNAS*. https://doi.org/10.1073/pnas.1421533112
Kollmair, M., & Banerjee, S. (2011). Migration and global environmental change – DR9: Drivers of migration in mountainous regions of the developing world: A review. International Centre for Integrated Mountain Development, Kathmandu, Nepal & UK Government's Foresight Project.
Kolmannskog, V. (2015). Disasters and refugee protection: A socio-legal case study from Yemen. *Oslo Law Review, 3*(2).
Magnan, A., Garnaud, B., Bille, L., Gemenne, F., & Hallegatte, S. (2009). *The future of the mediterranean: From impacts of climate change to adaptation issues*. IDDRI.

McLeman, R. A. (2011). Settlement abandonment in the context of global environmental change. *Global Environmental Change, 21*(Supplement 1), S108–S120. https://doi.org/10.1016/j.gloenvcha.2011.08.004
Mulligan, M., Burke, S., & Douglas, C. (2013). Environmental change and migration between Europe and its neighbours. In E. Piguet & F. Laczko (Eds.), *People on the move in a changing climate* (Global Migration Issues) (Vol. 2). Springer.
Myers, N. (2002). Environmental refugees: A growing phenomenon of the 21st century. *Philosophical Transactions of the Royal Society of London. Series B, Biological Sciences, 357*(1420), 609–613. https://doi.org/10.1098/rstb.2001.0953
Neumann, B., Vafeidis, A. T., Zimmermann, J., & Nicholls, R. J. (2015). Future coastal population growth and exposure to sea-level rise and coastal flooding – A global assessment. *PLoS ONE, 10*(6), e0131375. https://doi.org/10.1371/journal.pone.0131375
Piguet, E., Kaenzig, R., & Guélat, J. (2018). The uneven geography of research on "environmental migration". *Population and Environment.* https://doi.org/10.1007/s11111-018-0296-4
Population and Development Review. (2010). Effects of future climate change on cross-border migration in North Africa and India. *Population and Development Review, 36*(2), 408–412. https://doi.org/10.1111/j.1728-4457.2010.00338.x
Samers, M. (1997). The production of diaspora: Algerian emigration from colonialism to Neo–Colonialism (1840–1970). *Antipode, 29*(1), 32–64. https://doi.org/10.1111/1467-8330.00034
Selby, J. (2006). The geopolitics of water in the Middle East: Fantasies and realities. *Third World Quarterly, 26*(2), 329–349. https://doi.org/10.1080/0143659042000339146
Selby, J., Dahib, O. S., Fröhlich, C., & Hulme, M. (2017). Climate change and the Syrian civil war revisited. *Political Geography, 60*, 232–244. https://doi.org/10.1016/j.polgeo.2017.05.007
Sell, R. (1988). Egyptian international labor migration and social processes: Toward regional integration. *The International Migration Review, 22*(3), 87–108. https://doi.org/10.2307/2546586
Sobczak-Szelc, K. (2008). Changes in the environment and migration in Southern Morocco – Example of the Mhamid Oasis. *Miscellanea Geographica, 13*, 239–250. https://doi.org/10.2478/mgrsd-2008-0024
Sow, P., Marmer, E., & Scheffran, J. (2015). Between the heat and the hardships. Climate change and mixed migration flows in Morocco. *Migration and Development, 5*(2). https://doi.org/10.1080/21632324.2015.1022968
Stern, N. (2014). *The economics of climate change: The stern review*. Cambridge University Press. https://doi.org/10.1017/CBO9780511817434
The Nansen Initiative (2015) *Global consultation*. Conference report, Geneva.
Tribak, A., Paradiso, M., & Azagouagh, K. (2019). Climate refugees, housing in risk areas, and vulnerability of the built environment in the fez urban area of Morocco (Case of the Medina and Outlying Districts). In M. Paradiso (Ed.), *Mediterranean mobilities*. Springer.
United Nations High Commissioner for Human Rights. (2018). *Summary of the panel discussion on human rights, climate change, migrants and persons displaced across international borders* (Human Rights Council Thirty-seventh session. A/HRC/37/35).
UNCCD. (2017). *United Nations convention to combat desertification: The future strategic framework of the convention* (ICCD/COP(13)/L.18).
UNFCCC. (2011). *Report of the conference of the parties on its sixteenth session. Decision 1/CP.16. The Cancun Agreements* (FCCC/CP/2010/7/Add.1).
UNFCCC. (2016a). *Report of the Executive Committee of the Warsaw International mechanism for loss and damage associated with climate change impacts.* (FCCC/SB/2016/3).
UNFCCC. (2016b). *Technical meeting – Action area 6: Migration, displacement and human mobility.* Executive Committee of the Warsaw international mechanism for loss and damage associated with climate change impacts.
UNFCCC. (2018). *Report of the Executive Committee of the Warsaw international mechanism for loss and damage associated with climate change impacts* (FCCC/SB/2018/L.6).
UNHCR (2017) Climate Change and Disaster Displacement - UNHCR Quick Guide [online]

UNISDR. (2015). *Sendai framework for disaster risk reduction 2015–2030*. United Nations International Strategy for Disaster Reduction.

United Nations. (2018). *Global compact for safe, orderly and regular migration*.

United Nations, Department of Economic and Social Affairs, Population Division. (2016). *International migration report 2015: Highlights* (ST/ESA/SER.A/375).

Verner, D. (2012). *Adaptation to a changing climate in the Arab Countries: A case for adaptation governance in building climate resilience*. The World Bank.

Warner, K. (2010). Global environmental change and migration: Governance challenges. *Global Environmental Change, 20*(3), 402–413. https://doi.org/10.1016/j.gloenvcha.2009.12.001

Warner, K., Afifi, T., Kälin, W., Leckie, S., Ferris, B., Martin, S. F. & Wrathal, D. (2013). *Changing climate, moving people: Framing migration, displacement and planned relocation*. UNU-EHS Publication Series. Policy Brief N° 8, June 2013. United Nations University.

Weinthal, E., Zawahri, N., & Sowers, J. (2015). Securitizing water, climate, and migration in Israel, Jordan, and Syria. *International Environmental Agreements: Politics, Law and Economics, 15*(3), 293–307.

Werz, M., & Conley, L. (2012). *Climate change, migration, and conflict in Northwest Africa: Rising dangers and policy options across the arc of tension*. Center for American Progress & Heinrich Böll Foundation.

Westing, A. H. (1994). Population, desertification, and migration. *Environmental Conservation, 21*(2), 110–114. https://doi.org/10.1017/S0376892900024528

Wodon, Q., Burger, N., Grant, A., Joseph, G., Liverani, A., & Tkacheva, O. (2013). Climate change, extreme weather events, and migration: Review of the literature for five Arab countries. In E. Piguet & F. Laczko (Eds.), *People on the move in a changing climate* (Global migration issues) (Vol. 2). Springer.

Wodon, Q., Liverani, A., Joseph, G., & Bougnoux, N. (2014). *Climate change and migration: Evidence from the Middle East and North Africa*. The World Bank.

Zickgraf, C. (2018). Immobility. In R. McLeman & F. Gemenne (Eds.), *Routledge handbook of environmental displacement*. Routledge International.

Chapter 7
The Environment-Climate-Conflict-Displacement Nexus in the Arab Region: Implications and Recommended Actions

Elhoucine Chougrani and Mohamed Behnassi

Abstract To date, the problem of environmental migration in the Arab region and its enormous impact on the national budget have received scant attention in the literature. Accordingly, this chapter aims to analyze the budgetary cost of environmental migration, taking into consideration the quality of local governance and the intersection of other significant drivers. The feasibility of integrating and incorporating the cost of environmental migration into the Gross National Product (GNP) accounting will be demonstrated. The analysis also shows a correlation between the constraints of internal and international displacement and the rapid environmental deterioration induced by natural resource depletion and scarcity, conflicts, climate change, disasters, desertification, and biodiversity loss. The extent to which environmental deterioration dynamics and internal conflicts are interlinked are, therefore, examined in this study. Additionally, the analysis shows how and to what extent such conflicts affect the political boundaries of fragile states in the context of mass migration and the competition of global powers for influence, interests, and redistribution of roles.

Keywords Arab region · Conflicts · Environmental degradation · Climate change · Displacement · Cost

---

E. Chougrani (✉)
College of Law, Economics, and Social Sciences, Cadi Ayyad University of Marrakech, Marrakesh, Morocco

M. Behnassi
International Law and Politics of Environment and Human Security, College of Law of Agadir, Ibn Zohr University, Agadir, Morocco

Center for Environment, Human Security & Governance (CERES), Agadir, Morocco
e-mail: m.behnassi@uiz.ac.ma

M. Behnassi et al. (eds.), *The Climate-Conflict-Displacement Nexus from a Human Security Perspective*, https://doi.org/10.1007/978-3-030-94144-4_7

1 Introduction

There is no conventional definition of the concept of 'environmental migration'; however, we can make use of some key elements and indicators that contribute to the understanding of such a concept. 'Environmental migration' encompasses elements related to: the natural/sudden disasters (Marthoz, 2011:145; Pentinat, 2010:322); the progressive environmental degradation; the environmental-induced conflicts; the environmental destruction caused by armed conflicts; land damages, desertification, uprooting of forests; the construction of dams; resource scarcity (especially water, land, and biodiversity); industrial disasters (e.g. Bhopal, Chernobyl, and Fukushima disasters); and the potential long-term impacts and implications induced by climate change.

The United Nations Environment Program (UNEP) defines an 'environmental refugee' as a person who, on a temporary or permanent basis, leaves his or her place of residence because of an environmental catastrophe (i.e. the impossibility of living in an area because of drought, flood or desertification) that poses a threat to his or her existence or seriously damages his or her living conditions (Beurier & Kiss, 2010:503).

The International Organization for Migration (IOM) uses the term 'environmental migrants' for "persons or groups of persons, who, for compelling reasons of sudden or progressive changes in the environment that adversely affect their lives or living conditions, are obliged to leave their habitual homes, or chose to do so, either temporarily or permanently, and who move either within their country or abroad" (IOM, 2008:15).

The definition provided by Keucheyan (2014), on the other hand, illustrates that a 'climate refugee' is a person whose decision to migrate is linked to environmental factors – the so-called environmental migration – at least partially. According to this perspective, the environmental/climate crisis produces refugees whose migration flows destabilize the areas where they settle, which may result in conflicts.

Part of the literature distinguishes between forced migration and sudden environmental degradation, but climate change, arguably, does not place limits on forced migration and sudden environmental damage. Forced migration may be linked to the gradual changes in the ecosystems' balance and state. It is essential to consider other classifications in this regard like internal and international migrations, previous-effect migrations and immediate migrations action, short-term migrations, and long-term migrations. According to Gemenne (2010:71), it is important to approach 'environmental refugees' as a cross-cutting issue. In all cases, it is difficult to establish a direct and automatic correlation between the nature of the damage, environmental capability imbalance, and the qualitative characteristics of migration. This precludes the possibility of reaching a consensus on the precise identification of environmental refugees. This is more complicated as well when it comes to counting the number of environmental migrants or predicting future quantitative trends. The more we expand the definition of the 'environmental refugee', the higher flow in number and size should be expected. On the contrary, the more we narrow the definition, the number and size of environmental refugees notably decrease.

The issue of refugees is caused by many factors such as violent conflicts, rapid environmental deterioration, climate change, natural and climate-induced disasters, desertification, and water scarcity. Political or economic asylums have also sparked legal debates on how the concepts of environmental refugee, ecological refugee or climate refugee can be incorporated into the 1951 Vienna Convention relating to the Status of Refugees, which makes no reference to environmental and climate drivers of migration and does not give any legal status to environmental/climate migrants. In addition, multiple debates have surrounded the issue of climate refugees, including political discussions (such as the case of the extreme right discourse which tends to capitalize more and more on the refugee's status in an attempt to win elections), cultural dialogues (such as the difficulty of clearly delineating the boundaries between political and ecological issues), and economic surveying.

In this perspective, Stern (2006) illustrated the 'real' cost of climate change when considering the national budgets and increased population displacement (250 million in 2050) (Gemenne, 2010:71). Furthermore, Stiglitz et al. (2009) on measuring economic effectiveness and social development for the year 2009 also advocated the importance of integrating the principle of sustainable development into the Gross National Product (GNP).

In the MENA region, the literature has covered internal conflicts, the struggle between multiple actors, the contending strategies adopted by states, and the 'anti-movements', as well as the extent to which national budgets can meet the induced costs of such dynamics. However, key indicators such as the intersection of demographic shifts, war outbreaks, failure of development strategies, fragility and weakness of public policies, climate change, environmental degradation, and depletion of natural resources can be determinants of environmental and climate-induced conflicts and migration. However, this may require efforts and skills to isolate each one of these factors/determinants and create a distinct category of migrants and/or displaced persons. In this context, the present research analyses the extent to which internal conflicts and environmental deterioration are interlinked, and how these linkages affect the political boundaries of fragile states by reference to mass displacements and the rivalry of regional and international powers for influence, interests, and redistribution of roles. In addition, the research assesses the question of the cost of migration induced by climate change, environmental degradation, and natural resource depletion for the Arab region. The impact of such costs, especially in terms of disintegration of national borders and related challenges, such as the mobility of environmental migrants within the national territory (i.e. from the village to the city with travel terms that are restricted to distance and time). Sometimes, these refugees cross national borders because they are forced to do so. That is to say, a displaced person does not desire to leave his or her country as much as he or she seeks to escape the immediate environmental stress.

Moreover, the new dynamics induced by increased globalization and interdependence have led to the dismantling of concepts such as sovereignty, national borders, and traditional sources of threat to domestic security, considered as key foundations of international law and the realist theory. Alternatively, to be in command of natural resources and peoples' abilities, liberal and neoliberal theories have developed concepts and notions that strengthen these trends, such as border opening, free trade, market liberalization, free-trade zones (FTZ), land confiscation, and 'humanitarian'intervention (Slaughter & Alvarez, 2000:246). The Arab region has, therefore, been caught in this tendency and has been unable to keep up with market mechanisms (Adam Smith's invisible hand) since the implementation of the Structural Adjustment Program (SAP) and the Washington Consensus. Conflicts and wars have spread in the context of environmental degradation and natural resource scarcity while the region is increasingly a hot-spot for climate change. There is no doubt that the costs of all these dynamics are constantly increasing. Can environmental migration flows and, thus, induced costs be halted in the context of diluting and moving sovereign borders of fragile Arab states?

2 Environmental Degradation, Climate Change, Conflicts, and Migration in the Arab Region: Links and Costs

In this section, the links between environmental degradation, climate change, and conflict are analyzed. Besides, this correlation and its costs for the Arab region are demonstrated.

2.1 The Environment, Climate, Conflict, Migration Nexus in the Arab Region

Any interference in the internal affairs of states is not justified from the perspective of international law. Accordingly, the 1992 United Nations Framework Convention on Climate Change (UNFCCC) emphasizes in its preamble the principle of state sovereignty by reference to the UN Charter and the established principles of international law, that is the sovereign right of states to exploit their resources according to their own environmental and developmental policies.

Conversely, and as indicated by several empirical studies (Gemenne, 2010:68), the most vulnerable populations cannot avoid environmental deterioration because of their insufficient capacities and the lack of adequate migration policies. The inevitable result is that these groups are the most affected by environmental degradation and disruption. For Aguilar (2019:83), natural resources are themselves a source of conflict, so vulnerable groups cannot engage in conflicts because the power seems to be the most determinant element in deciding who should have control over such natural resources.

Due to the presence of armed conflicts in the Arab region – particularly in Iraq, Syria, and Yemen – along with the danger of the so-called Islamic State (ISIS, ISIL or Daech), which took control over some dams in Iraq and Syria, there are fears that radical groups might poison water installations and dams and cut food supplies to force the population to leither leave or fight.

The cases of Israel and the United States call for more attention for several reasons. First, we cannot ignore the fact that Israel's tactics in removing and uprooting olive trees (a symbolism of Palestinian survival) were signs of a destructive war waged against the Palestinian people. Second, the construction of the separation wall in the West Bank, an illegal construction according to the International Court of Justice's decision (ICJ Wall separation, 2004), is favoring the confiscation of Palestinian property and water resources. Third, Israel's unlawful use of white phosphorus munitions during its attack on Gaza (2008, 2009 and 2014) has had damaging results. Fourth, the use of depleted uranium by the United States on Iraq, and its effects on the environment and health, increased the burden on national budgets as well as the forced displacement of civilians from the hotbeds of tension and conflict.

As for the links between climate change and violence, scholars are divided into two main groups (Gleditsch & Nordås, 2010; Scheffran et al., 2012). The first group claims that it is difficult to establish a link between environmental factors and the demise of states (given the intersection of social, political, and economic factors). The second group, however, provides evidence to prove the existence of a link between violence and land degradation, deforestation, and resource scarcity (especially water). To support this claim, Hulot (2015:200) asserts that climate change increases poverty, suffering, and inequality. Climate change is often seen as a source of conflict and, at other times, as a threat multiplier. Indeed, it might not be responsible for the rise of new threats; however, it exacerbates problems that already exist – especially in risk zones like Middle East, Africa, Asia, and Latin America – such as drought, desertification, and resource scarcity.

In light of the above discussion about climate change as a threat multiplier, the controversial correlation between the climate-induced drought that struck Syria between 2006 and 2010 and the spark of the Intifada can be given as an example (Levy et al., 2017:34–35). Between 2006 and 2011, Syria suffered from droughts and the number of wells in the country has moved from 13,500 in 1999 to 21,300 in 2007. Moreover, climate change has also contributed to the increasing desertification in Syria (Hulot, 2015: 200). Farms were destroyed, livestock was reduced, and village communities were displaced – between two million and three million Syrians fell into poverty. Drought pushed hundreds of peasants to migrate to cities in search of jobs and many felt abused by the government. Also, hundreds of millions of refugees have migrated due to climate change to areas with fragile infrastructure (Gueldry, 2013:164).

It is important to note here the lack of accurate statistics on the number and trends of refugees due to climate change, natural resource scarcity, and environmental disruption in Arab countries. Understanding the extent to which an environmental refugee has fulfilled his or her dream within an 'environmentally safe'place raises various challenges in the Arab region relating to health, security, employment, and increased pressures on natural resources (the resource curse). According to Levy et al. (2017), the major challenge is to develop models that can statistically predict the correlation between climate change and collective violence.

The environment can be a primordial and elemental factor for human displacement. Environmental conditions have been closely linked to the movements of nomadic pastoralists adapting to an appropriate pastoral situation. Such conditions are also behind the movement of the population as a result of natural disasters. The conflict that has been raised in Darfur, Soudan (PNUD, 2009:49) is described as the first conflict directly linked to climate change from the perspective of UNEP (von Schorlemer & Maus, 2014:9). Although there are numerous and interrelated sources of the clash in Darfur – especially ethnic rivalry between Arab and non-Arab tribes, increasing population rates, as well as bad governance and oil discovery in the region – the impacts of climate change, particularly the effects of drought and desertification on the northern part of Darfur, are certainly indisputable (Gemenne, 2010:87–88).

There is seemingly a correlation between climate and security challenges (Bastien & Baillat, 2018:80), which requires the implementation of policies that would meet such challenges. To that end, three key elements are suggested as a new framework for reflection and action to shape the vision, prevention, and analysis of conflicts and international relations, which include sustainability, stability, and security. The purpose of the framework is not to overlook the role of political, ethnic, religious, social, economic, and other determinants of conflicts (Bastien & Baillat, 2018:80). Rather, it attempts to highlight the significance of environmental degradation and climate change as parts of the underlying causes of a deteriorated general situation. The framework is equally informed by the environmental approach as advanced by scholars in the area of international security.

The links between conflict and environment in Sudan, for example, extend to a long history. Over the past four decades, for instance, rainfall in the Darfur region has fallen by 30% and the Sahara has advanced by more than a mile every year. The resulting tension between farmers and herders over disappearing pasture and declining water holes underpins the genesis of the Darfur conflict (Brown et al., 2007:1143). Additionally, the interdependence level and multiple factors have had significant negative effects on the environmental condition. Indirectly, it has led to human displacement, bad governance, excessive violence associated with massive resource exploitation, and the scarcity of investments. Indeed, environmental issues have contributed to conflict and continue to do so. Competition over natural reserves (oil, gas, Nile waters) and the confiscation of wood are important reasons for creating and maintaining disorder in the Arab region (UNEP, 2007:8).

In this perspective, Harrigan (2014) argues as well that the increase in food prices was an important catalyst of the 'Arab Spring' uprisings. Conversely, Tolba and Saab (2008:20) stressed that the connection between environment and conflict is not necessarily a direct one. This is complex because this connection often occurs in parallel with other social, political, and economic pressures and dynamics.

Apart from competition over natural resources, given their scarcity and limitations, climate change can create antagonism within the territory. A case in point is small island states threatened with drowning as a result of rising sea levels; a situation unique to international law in case these states request resettling in the territory, or at least for their population to inhabit a determined territory (Gemenne, 2010:83).

The estimations of the number of people forced to migrate due to climate change range between 25 million and 1 billion people by 2050 (IOM, 2008:12). Unfortunately, these figures are not scientifically strictly justified and acceptable. Research is aimed at limiting these estimations to the populations exposed to major climatic hazards but fails to take into account adaptation measures and actions on which individuals, groups, and governments rely (UNEP, 2007:50).

The negative impacts on the environment have increased due to the advancement of techniques and methods of armed conflicts. The gravity of such impacts has worsened as it has become possible to change environmental conditions using multiple techniques and methods to create artificial clouds, rain, and snow beyond natural levels, destroy irrigation systems, ravage vegetation soils and forests, devastate crops, and contaminate water (El Haiba, 1989:7–8). It should be reckoned that these practices are in full contradiction of international humanitarian law provisions, such as:

- The Art. 53 of the 1949 Geneva Convention relative to the Protection of Civilian Persons in Time of War (commonly referred to as the 4th Geneva Convention), which prohibits the destruction of fixed or movable property, or the Art. 55 which states that: a) care shall be taken in warfare to protect the natural environment against widespread, long-term, and severe damage. This protection includes a prohibition of the use of methods or means of warfare that are intended or may be expected to cause such damage to the natural environment, and, thereby, to prejudice the health or survival of the population; and b) attacks against the natural environment by way of reprisals are prohibited;
- The 1977 Protocol I & II to the 4th 1949 Geneva Conventions;
- The 1973 Convention on the Prohibition of the Development, Production, Stockpiling and Use of Chemical Weapons and on their Destruction;
- The 1976 Convention on the Prohibition of Military or any other Hostile Use of Environmental Modification Techniques (known as the ENMOD Convention); and
- The Convention on Prohibitions or Restrictions on the Use of Certain Conventional Weapons which may be deemed excessively injurious or to have indiscriminate effects as amended on 21 December 2001.

In the same vein, reference to the impact of wars and conflicts on the environment had been earlier highlighted by the international law of environment – for instance, in the Stockholm Declaration (1972), the World Charter for Nature (1980), and the Rio Declaration (1992). In 1992, the UNFCCC stressed the priority of saving the planet from global warming; however, it seems that this convention failed to protect refugees from climate change. The first time that human mobility was recognized in international climate policy was at COP16 (2010), when Parties to the UNFCCC adopted the Cancun Adaptation Framework, including para 14(f) to invite action that addressed the full range of movements people may take when faced with climate risks: "Invites all Parties to enhance action on adaptation under the Cancun

Adaptation Framework […] by undertaking, inter alia, the following: [….] (f) Measures to enhance understanding, coordination and cooperation with regard to climate change induced displacement, migration and planned relocation, where appropriate, at national, regional and international levels". This framework provides a range of measures (research, coordination, cooperation), mobility (displacement, migration, planned relocation), and levels (national, regional, international) of action. The inclusion of the above sub-paragraph on migration and displacement gave options for undertaking actions to address human mobility (Warner, 2017).

Regarding this issue, the Paris Agreement comes up short; it does not address the legal status of refugees or mandate their protection and assistance. However, it does mention migrants in the Preamble, and it calls for a task force to "develop recommendations for integrated approaches to avert, minimize and address displacement related to the adverse impacts of climate change". In addition, of the 185 intended nationally determined contributions (INDCs) submitted, only 20% of them mention migration. Overall, the Paris Agreement lacks the urgency, depth and coordinated framework necessary for addressing the immense challenges of climate-induced migration. Therefore, the rights to receive preventative assistance to avoid being displaced, to get support if one is forced to flee, and to build, live, work and integrate in new communities if one cannot return back home, remained unaddressed (Lambert, 2015).

2.2 *The Financial Cost of Conflicts in the Arab Region*

Extreme events, such as severe floods or droughts, and environmental degradation adversely affecting agricultural production, livestock, or water availability have pushed people to migrate. Migration in the Arab region encompasses different individuals and groups from a variety of social and economic backgrounds (ESCWA and IOM, 2015:15).

The economic cost of environmental degradation (COED) in Arab countries has been estimated at 5% of the GDP by Tolba and Saab (2008:228). In Morocco, for example, Croitoru and Sarraf (2018) assume that the cost has been estimated in 2014 at 3.52% of the GDP (Table 7.1). In the same vein, Riccard (2017) mentions that the hydro-meteorological events alone have generated more than 3/4 of economic losses in Morocco. Taken together, floods (44%) and forest fires (43%) account for almost US$ 458 million.

Iraq is one of the most vulnerable Arab countries which suffers from environmental damage due to many decades of war against Iran (1980–1988), the invasion of Kuwait in 1990, and the American invasion of the country in 2003. The environmental damage caused by the Gulf War II in 2003 alone – particularly for the rehabilitation of water installations and the purification of polluted water – was estimated between $6.4 and $11 billion (Ait Hmadouch, 2005:49); after 2003, 70% of Iraqi citizens had access to unhealthy water (Strategic Foresight Group, 2009:49).

Table 7.1 COED in Morocco, 2014 (US$ billion and % of the GDP)

	Lower bound	Upper bound	Average	% of GDP
Water	1.3	1.4	1.4	1.26%
Air	0.7	1.5	1.1	1.05%
Agricultural land	0.6	0.6	0.6	0.54%
Waste	0.4	0.4	0.4	0.40%
Coastal zone	0.3	0.3	0.3	0.27%
Forest	0.0	0.0	0.0	0.0
Cost to Moroccan society	***3.4***	***4.4***	***3.9***	***3.52%***
Carbon emissions	0.5	3.0	1.8	1.62%
Cost to global community	***0.5***	***3.0***	***1.8***	***1.62%***

Source: Croitoru and Sarraf (2018:262)

In Kuwait, and during the invasion of Iraq, more than 600 oil wells that produced about 500,000 metric tons of pollutants per day exploded, with 25–50 million barrels spilled over the land, as well as the largest recorded oil slick in the sea (6 to 8 million barrels), affecting the Gulf's marine environment. In the Kuwait case alone, environmental damage was estimated at $40 billion, representing 16% of the total war compensation claimed (Partow, 2008:165).

Depleted uranium weapons were used by the American and British forces in two destructive wars in 1991 and 2003. Experts estimated that at least 3000 tons were deployed (Almuqdadi, 2011:106). These deadly weapons undoubtedly threatened the life and health of humans as well as the environment.

In the occupied Palestinian territory, particularly in the Gaza strip, direct losses to agriculture amounted to more than $50 million and indirect losses reached $830 million. These are highly costing losses given the limited resources of the country (Strategic Foresight Group, 2009:103).

Israel's crimes against the environment in Lebanon are depicted in mine-laying in fields and around water sources, the burning of trees and orchards, the dredging of the soil; and the stealing and pumping of water (Hamad, 2005:155–156). More dramatically, the UN has estimated that in the Lebanon territory of the south and western Bekaa, Israel has left behind some 130 thousand mines and a huge number of unexploded bombs. In addition, the 2006 Israeli-Lebanese war caused significant environmental damage, the most serious of which was the targeting of an oil refinery, which resulted in an oil spill of 10,000 tons, that polluted and contaminated 150 km of Lebanese coastline and 50 km of Syrian coastline. The damage was also extended to the Litani River and its agricultural surroundings by air attacks (cluster bombs), where 26% of the bombs landed on the area (Ait Hmadouch, 2005:167–168). Also, Israeli bombardments of the petrochemical storage facility in July 2006 war in Lebanon led to a massive oil spill in the Mediterranean basin. Actually, the use the use of cluster bombs made many forests and grasslands in the South of Lebanon unsafe, precarious, and often inaccessible for the local population (Jardi et al., 2013:154).

There is no doubt that these different conflicts and hostilities resulted in internal and external displacements due to massive pollution, the poisoning of water, and the burning of forests and farms. This requires a census of the directions of individual and collective displacements owing to both environmental degradation and natural resource scarcity. In this context, climate change could aggravate the Arab region's vulnerability to natural disasters such as drought and flash floods in addition to water and food shortages, pest infestations, and most importantly sea-level rise in coastal areas (El-Sayed Selim, 2004:85–86).

3 Fragmentation of Cross-Border Migration Policies

3.1 The Cost of and Responses to Transnational Damage

The flow of forced migration through national borders, or the so-called transboundary or transnational environmental/climate migration, due to the entanglement of political conflicts, economic contexts, environmental degradation, and scarcity of natural resources, among others, poses complex and intersecting security pressures; accountability should be held by host countries as well as transit and receiving countries.

Wherever they are, refugees necessitate social protection, healthcare, and access to water and food. They are often a burden on the national budget of host countries. Meanwhile, they may also be a source of risk to environmentally safe areas (increasing pressure on fragile ecosystems, scarce natural resources, and already degraded soils). International aid to refugees is often insufficient and inadequate to protect refugees and their surroundings from risks and vulnerability.

The purpose here is not to monitor environmental/climate migration. More particularly, the aim is to rather to keep an eye on this phenomenon and understand its roots, foundations, and implications. There is a need to adopt and support a more creative approach to human mobility and its complex relationship with environmental and climate dynamics, in which migration is perceived not only in terms of failure to adapt but also as a legitimate adaptation strategy itself. For this particular reason, it is necessary to integrate migration policies into the perspective of adaptation strategies as recommended by the IOM (Christiansen, 2016:182).

Within the international development policies that regulate humanitarian assistance and environmental and climate policies, agencies interested in governance have responsibilities associated with different stages of the climate change reduction process: to support the adaptation, mitigation, and resilience of groups in the face of environmental risks and extreme events. For Mence and Parrinder (2017), agencies responsible for climate change policies who operate within a global dynamic to manage mobility of population across international borders can reinforce humanitarian efforts to avoid disasters, to maintain their return and resettlement, and to improve their integration within host societies. Moreover, integrating migration policies into the perspective of adaptation strategies would contribute to the reduction of the cost of transnational damage, on the one hand, and the advocacy of international cooperation regarding these policy areas on the other.

Statistics and scenarios concerned with the prediction of potential numbers designated for future migration differ. According to the report of the Office of the United Nations High Commissioner for Human Rights (Warren, 2016:2113), between 50 and 200 million people will be moving within or outside national borders in the middle of the present century. However, the Aid Christian report (2007) predicts a number between 200 and 250 million in the same period, and the Intergovernmental Panel on Climate Change (IPCC) and the Stern's Report (Global Humanitarian Forum, 2009:49) suggest that from 150 to 200 million people will be moving incessantly as a consequence of rising sea levels, floods, and droughts by 2050. In the same perspective, the Human Rights Council (2009) estimates that some 262 million people were affected by climate disasters from 2000 to 2004, 98% of whom live in developing countries. The risks of tropical cyclones alone affected about 120 million people annually, killing about 250 thousand people between 1980 and 2000.

Oftentimes, migrant populations usually move towards the most fertile or abundant land. Extreme events and disasters, such as drought in the Sahel region of Africa, are thought to accelerate such movements. Dry years in the Arab Mashreq and Maghreb countries are also increasing the rapidity of rural migration to cities and contributing to social, economic, cultural, and political repercussions in urban areas (Hashim, 2018:9). There is no doubt, therefore, that environmental and climate migrants are fleeing such extreme events and disasters such as drought or rising sea levels, including slow-onset changes like desertification, although the increase in these phenomena and their consequences have been met with limited political action (Schmid, 2018:37).

In his critique of international relations, Linklater (2007:4) proposes a new perspective called the 'Cosmopolitan Harm Conventions (CHCs)'. For Linklater, the approch he suggests draws upon the principle of non-harm as emphasized by the International Humanitarian Law on the duty not to harm individuals and ethnic groups. Chief among these are: the 1948 Convention on the Prevention and Punishment of the Crime of Genocide; the 1973 International Convention on the Suppression and Punishment of the Crime of Apartheid; and the international environmental law and soft law concerned with the extraterritorial effects of states – especially the Art. 3 of the 1992 Convention on Biological Diversity and the principle 21 of the 1972 Stockholm Declaration on Human Environment. In this perspective, it is estimated that the cost of transnational effects will be considerable and significant and that such effects will be diverse and different among states. More serious awareness-raising and risk analysis, together with the examination of transnational impacts of national sovereignty, would enhance the emergence of transnational governance regarding the adaptation to climate change (Hedlund, 2018).

In the same context, we can refer here to the so-called risk society. The basic idea, according to Ulrich (1992:13–23), is that we have shifted from the logic of benefits allocation to the distribution of risks along with the threats imposed by modernity and the uncertainty surrounding globalization. These risks led, first, to new disparities and inequalities between and among the North and the South. We may add that these disparities and inequalities arise within each country and between its constituent regions. A consequence of modernity is the increasing threats to both social and ecological systems as outcomes of both industrial and technical development.

Bastien and Baillat (2018:74) believe that we have moved from burden-sharing – that is, defining the responsibilities of each actor – and collective burden-sharing to the logic of opportunity sharing because it is in the public interest to reach an agreement between the parties, which helps increase employment opportunities, enhance growth, and improve well-being, etc.

In the 9th of July 2004, the ICJ took a decision upon the wall that separates Israel from Palestine. The decision undertaken by the ICJ holds that Israel is accordingly under the obligation to return the land, orchards, olive groves, and other immovable property seized from any natural or legal person for purposes of construction of the Wall in the Occupied Palestinian Territory. In this case, such restitution should prove to be materially impossible, Israel must compensate the persons in question for the caused damage. The Court considers that Israel also should compensate, following the applicable rules of international law, all natural or legal persons having suffered any form of material damage as a result of the Wall's construction (ICJ, 2004).

Aside from the context of the UN, the Nansen Initiative on Transnational Migration, launched by Norway and Switzerland in 2015, emerged in the context of disasters and climate change. The primary affirmed purpose of the initiative is to support states and various actors to prepare and respond to cross-border climate migration. This initiative, thus, comes as an attempt to form a league of states that would address this issue and develop effective approaches from a soft law perspective to protect transnational climate migration (Warren, 2016:2111–2112).

3.2 Future Impacts and Costs of Mass Migration on Public Policies

No doubt, the Arab region is at a critical stage politically, socially, and historically. Endorsing the choice of revolution against rulers has created political, social, and cultural dynamics that have severely affected public policies in various areas (revolution and anti-revolution). To escape ongoing massive migrations or at least mitigate their implications in the Arab region, the values of the community and loyalty to the nation and a culture of positive participation in the management of public affairs, under the sovereignty of a delicate economic situation highly characterized by dependency and conditionality, should be enhanced. Market mechanisms have noticeably accelerated the gradual degradation of the environment and the scarcity of natural resources with many social and security implications. To counter such dynamics, especially after the revolution, "states can guarantee a place for both competition and synergy through collective expansion of market access and inclusive decision-making" (Al-Bassam, 2014:61).

Environmental exhaustion is both the cause and the consequence of political pressure and military conflict. Nations have often competed to either impose or resist control over primary resources and energy provision, land, river basins, sea lanes, and other basic environmental resources. These conflicts can worsen as resources become scarcer and competition increases (WCED, 1987:291). Moreover, there is a direct link between natural resources and development, particularly with the findings of Sachets who points to the existence of a natural resource curse since 4/5 of the countries whose economies are based on natural resources have a lower standard of living than the global average, and half of them have not reached that rate yet (Serre, 2016:70–71).

There is a complex correlation between poverty, inequality, environmental degradation, climate change, resource scarcity, and conflict. In such a nexus, what concerns the international community more is the phenomenon of 'environmental/climate-induced migration'. It seems at first sight that the cause of mass migration is political unrest or armed violence, but the underlying drivers also include the deterioration of the natural resources' base and its ability to sustain the population (WCED, 1987:291). In early 1999, there were almost 22 million traditionally and internationally recognized refugees who flee political oppression, religious persecution, and ethnic troubles. Their numbers had declined from a peak of 27 million in 1995 but remained higher than the 19 million in 1993. In addition, there were large numbers of people who were characterized as environmental refugees or who could not secure their livelihood in their homelands because of drought, soil erosion, desertification, deforestation, and other environmental problems, together with the associated problems of demographic pressures and severe poverty (Myers, 2002:609). According to Pentinat (2010), the rate of people displaced (or refugees) due to desertification, deforestation, floods, land degradation, etc. is about 17,000 a day.

Since the 1990s, much attention has been paid to the security implications of massive migrations induced by environmental and climatic changes which affect the ability of populations to access natural resources essential to their livelihoods such as arable land, forest, and freshwater. As the residency duration of forced displacement expands, the burden on hosting countries to understand and integrate new arriving migrants becomes even more difficult. In many countries, adaptation is also emerging as a highly political process as poorer segments of society demand formal recognition and clarification of rights of access to land, water, and forest resources (Melvin & De Koning, 2011).

Environmental problems undoubtedly affect the collective security. The literature on population mobility due to environmental crises started in the 1970s within the context of security studies. Basically, researchers were interested in the relationship between environmental deterioration and the emergence of conflict (Vlassopoulos, 2017:104). The current literature shows as well that climate change is already affecting certain levels of migration (Martin, 2010; Kelman, 2020; Von

Soest, 2020). However, the speed of climate change impact on living conditions leads to different patterns of migration and mobility; climate change impacts such as floods often lead to immediate and temporary displacement, while slow environmental changes such as drought involve circular migration – i.e. frequent movement between different areas (Mobjörk, 2017).

Migration was also highly inter-related with adjustment to a deteriorating living environment and a response to a failed adjustment. For instance, IDMC (2018) argues that "Internal displacement, associated with climatic changes and disasters, is a complex and dynamic phenomenon". In addition, as long as transnational climate change affects the national sovereignty of each state, it is important to think about the transnational solidarity approach, focusing on cross-border policies, and re-emphasizing the environmental rights of migrants (Klepp, 2018:150). Such an approach would decrease regional and international competition and surpass states' boundaries.

3.3 International Competition and National Borders' Fragmentation

In addition to violence, UNHCR (2008:5) posits that human rights violations and oppression by governments are also key elements in mass migrations. Individuals are in a situation of mobility because of severe poverty, the decline of living conditions, gradual urbanization as well as the effects of environmental and climatic changes, which worsen competition and conflict over scarce resources. Sometimes all of these elements overlap or reinforce each other. According to Warren (2016:2109), climate change can lead to instability, including changes in rainfall patterns (as a cause of desertification) and melting snow leading to increased extreme weather (such as storms, heatwaves, and drought). Given this perspective, Warren claims – as many researchers – that forced migration can be caused by climate change, as in the case of Somalia in 1990 and Syria in 2015. Climate change is certainly not the only cause of forced migration, but it worsens such conditions and causes which trigger the decision to migrate.

Previous experience has confirmed that the existence of raw materials in a country affects other countries' policies, especially in industrialized ones. The imperial experience has proved that one of the most important drivers of colonial competition was the desire to acquire raw resources. The battle for such resources explains the many forms of world order (conflict and cooperation). Additionally, domestic situations in countries supplying these raw materials are affected by pressures from importing states (suppression or support of a revolution or rebellion, support of certain authoritarian regimes, etc.) (El Manoufi, 1987:131).

Between 1998 and 2017, poor communities and fragile countries were more vulnerable to the harmful impacts of climate change than developed countries, because of weak institutions (Eckstein et al., 2018). Over a billion people – the world's poorest and vulnerable communities – will bear the brunt of climate change (Ayers & Huq, 2009:2). These impacts can also be highlighted concerning the ability of countries to control their natural resources as the notion of national sovereignty has declined (Park, 2013:340). For its part, the environmental crisis (Keucheyan, 2014) will certainly exacerbate natural disasters that increase the fragility and speed of existing institutions, especially in developing countries. We can add that the deterioration of national economies is in itself a key factor and a catalyst for the refugee crisis. Climate change is likely to further weaken the capacity of some vulnerable countries with weak governance structures, which could lead to increased human insecurity, conflicts, and instability. Famine and epidemics in several countries – as a result of the 2008 world food crisis– such as Haiti, Cameroon, Egypt, Mexico, Ivory Coast, Madagascar, and Indonesia, are a case in point.

Some groups can also compete and struggle within the territory of countries to oversee and, therefore, control vital resources and agricultural lands. Ethnic and political conflicts can increase competition over resources. Gemenne (2010) mentions that instability can be a menace to international security, especially if these unstable states exert major influence in a particular geographical area – e.g., Brazil, Mexico, Nigeria, South Africa, Egypt, Pakistan, and South Korea. In the same framework, it was also claimed that conflict may arise if migrants, especially of different nationalities or ethnicities, move faster and intensively to neighboring countries that are already in disagreement or have limited resources and mechanisms to deal with constraints.

The national sovereignty has radically changed as a consequence of the complex interdependence in the spheres of economy, environment, and security and it is increasingly difficult to manage shared global resources at the national level (Chougrani, 2021). In addition, countries are growingly incapable to manage current threats to the commons. Indeed, according to Bell (1988), the nation-state is becoming too small for big problems and too big for small problems. Therefore, the threats to environmental security can only be addressed through joint management and the function of multilateral procedures and instruments (WCED, 1987:301). The environmental perspective is, in our view, a vital part of the overall security equation. Thus, the aspects that may put limits to these environmental threats in the Arab region should be investigated to protect national security and sovereignty. Certainly, security does not only imply the protection of national borders from military threats, but also non-military threats induced, among others, by climate, environmental, food, and water insecurity. In this sense, it can be argued that environmental/climate security has expanded the levels and nature of security on one hand, as well as the scope of actors' intervention in risk management policies on the other. The state was the most important unit of analysis in the classical realist theory of security, but with the emergence of environmental/climate threats, other actors, such as international non-governmental organizations and public opinion, intervened in this area.

By the end of the Cold War, the concept of 'environmental security' was strongly brought into the debate in a context where military risks were still perceived as the only threat to global insecurity. More attention has been increasingly paid to the following threats given their security implications: globalization of the economy; the consequences of interactions and mutual intentions; regional differences in the world; the displacement of migrants; the lack of integration of certain political systems into the international system; the depletion of the ozone layer; the steady growth of deforestation and desertification; and the extinction of certain species (Painchaud, 2000:62). In this evolving context, the cooperative security requires comprehensive cooperation and is not limited only to issues connected to the military aspect of security, but includes issues related to resource scarcity, sustainable development, and environmental concerns (Alharbi, 2008:25).

At the beginning of the current Millennium, Schwartz and Randall (2003) submitted to the Pentagon a report on *An Abrupt Climate Change Scenario and its Implications for United States National Security*. The report contains two phases that summarize the scenario of confrontation in case of extreme climate turbulence: the first from 2010 to 2020; and the second from 2020 to 2030. Accordingly, the United States must manage borders and refugees from the Caribbean and Europe. If oil prices continue to increase, the confrontation will be in the Persian Gulf and the Caspian Sea. In the case of internal unrest in Saudi Arabia during 2025, the conflict between the US Navy and China will be likely direct. One of the main issues to focus on is that "all borders are porous; long-term border protection, use of armed protectors or control systems are no longer acceptable" (Schwartz & Randall, 2003:251), especially in a globalizing context with dynamics transcending national borders.

4 Conclusion

This research attempts to examine the extent to which the patterns of conflict in the Arab region are linked to environmental degradation, climate change, and competition over scarce resources during mass migrations. The assessment of such patterns helps monitor the cost of the so-called environmental/climate migration to national economies and public policies. The analysis attempts as well to assess the damage caused by conflicts over natural resources within and outside national borders and the extent to which international and regional competition over such resources contributed to the fragmentation of these borders. Accordingly, it seems essential to draw the following conclusions:

- It is increasingly important to operationally define the concepts of 'environmental migration' or 'environmental and climate-related refugees' among others. and integrate them into the international system through the endorsement of the human rights framework. This will ease the way for policies and strategies that address forced migration (within or outside the national territory) along with its multiple aspects and dimensions.
- Arab Countries are still reluctant to include these concepts when it comes to conflict over limited resources and forced migration. Strengthening institutions

and socially effective action would, hopefully, change that perspective. Environmental or climate migrants are part of the issue of conflict over limited resources and wealth and are evidence of the ongoing social exclusion, vulnerability, and fragmentation of society.

- The overlap between demographic data, environmental and climatic changes, armed conflicts, and the failure of development plans lead to environmental imbalance and social conflicts. It, therefore, seems indispensable to integrate migration policies into the perspective of environmental and climate adaptation strategies.
- The oversight of monitoring the financial cost of conflicts and its impact on public policies contributes to the fragmentation of national borders, which requires the re-establishment of the perspective of environmental sustainability and equity, both at intra-generational as well as inter-generational levels. Equity in this respect is possible through the distribution and sustenance of wealth and the expansion of opportunities in society.
- To mitigate the financial cost of forced environmental migration in the Arab region, a link between sustainability, stability and security must be established. Imperative also is a shift of attention towards environmental risks and their consequences that inevitably exceed the sovereign borders of states.
- Regional and international competition over the Arab region is making the environmental situation and climate change impacts worse, shifting the center of focus into the importance of security in its comprehensive sense (both traditional and non-traditional) to escape the fragmentation of the region into small entities and states, therefore worsening the existing vulnerability. Therefore, addressing environmental degradation, climate change impacts, and the potential causes of conflicts in the Arab region – largely political factors – is increasingly imperative to ensure a sustainable management of such conflicts and their induced implications, especially displacements.

Acknowledgement We would like to express our gratitude and great appreciation to Dr. Redouan Saidi (University Sultan Moulay Slimane, Morocco), and Dr. Aziz Qaissi (Ibn Tofail University, Morocco) for their valuable support during the undertaking of this research.

References

Aguilar, S. L. C. (2019). Regional conflict system in Africa: An option for analysis. In S. H. Mokaddem (Ed.), *Stability and security in Africa: The role of hard and soft power*. Policy Center for the New South and Al Akhawayn University.

Aid Christian report. (2007). *Human tide: The real migration crisis*. https://reliefweb.int/sites/reliefweb.int/files/resources/47F1DFB5696BEE43C12572DB0035D16A-Full_Report.pdf

Ait Hmadouch, L. D. (2005) النزاعات في المنطقة العربية والحاجة إلى تغيير إدارتها . [Conflicts in the Arab region and the need to change their management]. In B. Korany (ed.), التنمية الانسانية العربية في القرن الحادي والعشرين [Arab human development in the 21st century: The primacy of empowerment]. CAUS and AUB.

Alharbi, S. A. (2008). (مفهوم الأمن: مستوياته وصيغه وتهديداته (دراسة نظرية فـــي المفاهيم والأطر [The concept of security: Its levels, forms, and threats – A theoretical study], CAUS (Summer).

Almuqdadi, K. (2011). المشكلات البيئية والصحية الساخنة المهملة فـــي عراق اليوم [Key environmental and health problems neglected in today's Iraq], *Al Mustaqbal Al Arabi*, N° 394, CAUS.

Al-Bassam, D. (2014). سياسات التنمية البديلة في بلدان الثورات العربية، عمران للعلوم الاجتماعية [Alternative Development Policies in Arab Spring Countries], Omran, N° 9, ACRPS.

Ayers, J., & Huq, S. (2009). Community-based adaptation to climate change: An update, International Institute for Environment and Development, June. https://pubs.iied.org/17064iied. Accessed 6 Aug 2021.

Bastien, A., & Baillat, A. (Dir.). (2018). Les lendemains de l'Accord de Paris: L'avenir de la planète se joue-t-il hors des COP, Entretien avec Pascal Canfin, *La Revue Internationale et Stratégique*, N°109. Edition Armand Colin.

Bell, D. (1988). Previewing planet earth in 2013, *Washington Post*, January 3rd. https://www.washingtonpost.com/archive/opinions/1988/01/03/previewing-planet-earth-in-2013/3e316a37-0d62-49fb-96d2-c544f2fa70d4/3/1/1988

Beurier, J. P., & Kiss, A. (2010). *Droit international de l'environnement*. Edition Pedone.

Broad, R., & Cavanagh, J. (1999). The death of the Washington consensus? *World Policy Journal, 16*(3), 79–88.

Brown, O., Hammill, A., & McLeman, R. (2007). Climate change as the 'new' security threat: Implications for Africa. *International Affairs, 83*(6), 11411154. https://www.iisd.org/system/files/publications/climate_security_threat_africa.pdf

Chougrani, E. (2021). الدولية العلاقات وخصخصة العربية الدول وهن. المشترك وتَخصيص السّوق وَثَن تَقديس في قراءة [The Decline of Arab states and the privatization of international relations: Reading into the idolatry of the market (logic) and the privatization of the commons], *Almustaqbal Alarabi*, N° 506, CAUS.

Christiansen, S. M. (2016). *Climate conflicts – A case of international environmental and humanitarian law*. Springer. https://doi.org/10.1007/978-3-319-27945-9

Croitoru, L., & Sarraf, M. (2018). How much does environmental degradation cost? The case of Morocco. *Journal of Environmental Protection, 9*, 254–265. https://doi.org/10.4236/jep.2018.93017

Eckstein, D., Hutfils, M.-L., & Winges, M. (2018). Global climate risk index 2019. Who suffers Most from extreme weather events? Weather-related loss events in 2017 and 1998 to 2017, Briefing Paper, German Watch. https://www.germanwatch.org/sites/default/files/Global%20Climate%20Risk%20Index%202019_2.pdf

El Haiba, M. (1989). نحو الحد من آثار الحرب على البيئة [Towards the reduction of environmental impacts of wars]. *Morocco Journal of Law and the Economy of Development*, N: 19.

El Manoufi, K. (1987). أصول النظم السياسية المقارنة [Origins of comparative political systems]. Alrubaiaan library.

El-Sayed Selim, M. (2004). The environment in the 2002 arab human development report: A critique. *Arab Studies Quarterly, 26*(2), 81–89. Spring.

Gemenne, F. (2010). *Géopolitique du climat. Négociations, stratégies, impacts*. Edition Armand Colin.

Gleditsch, N., & Nordås, R. (2010). *Climate change and conflict: A critical overview* (pp. 7–24). Die Friedens-Warte. http://www.jstor.org/stable/23773941

Global Humanitarian Forum. (2009). *The anatomy of a silent crisis* (Global Impact Report). Global Humanitarian Forum.

Hamad, K. (2005). القانون الدولـي الانساني وحماية التّراث والبيئة خلال النزاعات المسلحة [International humanitarian law and the protection of heritage and the environment during armed conflicts]. Elhalabi Publishing.

Harrigan, J. (2014). *The political economy of Arab food sovereignty*. Palgrave Macmillan UK. https://doi.org/10.1057/9781137339386

Hashim, N. F. (2018). مفاهيم نظرية فـي الهجرة السكانية. دراسة تحليلية مقارنة [Theoretical concepts in population migration. A comparative analytical study], Omran, N° 26, ACRPS.

Hedlund, J. (2018). *Climate risks without borders*. Stockholm Resilience Center. https://www.stockholmresilience.org/research/research-news/2018-08-14-climate-risks-without-borders.html

Hulot, N. (2015). Le climat est un facteur d'injustice sociale. *Revue Française des Affaires Sociales, 1*, 199–207.
Human Rights Council. (2009). Annual Report of the United Nations High Commissioner for Human Rights and Reports of the Office of the High Commissioner and Secretary-General. Report of the Office of the United Nations High Commissioner for Human Rights on the relationship between climate change and human rights. A/HRC/10/61.
ICJ. (2004). Advisory opinion of the International Court of Justice on the Legal Consequences of the Construction of a Wall in the Occupied Palestinian Territory. https://www.un.org/unispal/document/auto-insert-178825/. Accessed 6 Aug 2021.
Internal Displacement Monitoring center (IDMC). (2018). *No matter of choice: Displacement in a changing climate, thematic series*. http://www.internal-displacement.org/sites/default/files/publications/documents/20181213-slow-onset-intro.pdf
International Organization for Migration (IOM). (2008). *Migration and climate change*. IOM.
Jardi, M., Fayed, R., & Ezzin, A. (2013). التّدهور البيئي. التّحدّي لاستدامة الحياة [Environmental degradation: The challenge of sustaining life]. CAUS & AUB.
Kelman, I. (2020). Does climate change cause migration? In E. Fiddian-Qasmiyeh (Ed.), *Refuge in a moving world: Tracing refugee and migrant journeys across disciplines*. UCL Press. https://doi.org/10.2307/j.ctv13xprtw.15
Keucheyan, R. (2014). La nature est un champ de bataille. Essai d'écologie politique, Paris, La Découverte (Zones).
Klepp, S. K. (2018). Framing climate change adaptation from pacific Island perspective – The anthropology of emerging legal orders. *Sociologyus, 68*(2), 149–170. https://doi.org/10.3790/soc.68.2.149.2018
Lambert, K. (2015). *The Paris agreement: Spotlight on climate migrants*. Available at: https://environment.yale.edu/blog/2015/12/the-paris-agreement-spotlight-on-climate-migrants/
Levy, B. S., Sidel, V. W., & Partz, J. A. (2017). Climate change and collective violence. *Annual Review of Public Health, 38*, 241–257.
Linklater, A. (2007). Critical theory and world politics. In *Citizenship, sovereignty, and humanity*. Routledge.
Marthoz, J.-P. (2011). *Couvrir les migrations*. Edition de Boeck Supérieur.
Martin, S. (2010). Climate change, migration, and governance. *Global Governance (International Migration), 16*(3), 397–414. https://www.jstor.org/stable/29764954?refreqid=excelsior%3Ae8ccb7612d477c1337a25c392a4a3901
Melvin, N., & De Koning, R. (2011). *Resources and armed conflict*. SIPRI.
Mence, V., & Parrinder, A. (2017). Environmentally related international migration: Policy challenges. In M. Mcauliffe & K. Koser (Eds.), *Long way to go*. ANU Press.
Mobjörk, M. (2017). Exploring the links between climate change and violent conflict, SIPRI yearbook 2017. Armements, Disarmament and International Security, SIPRI & CAUS.
Myers, N. (2002). Environmental refugees: A growing phenomenon of the 21st century. *Philosophical Transactions: Biological Sciences, 357*(1420), 609–613. https://doi.org/10.1098/rstb.2001.0953
Nicholas Stern's Report. (2006). *The stern review of the economics of climate change*.
Painchaud, P. (2000). La sécurité environnementale: concept et perspectives. *La Revue Internationale et Stratégique*, N:° 39.
Park, K. G. (2013). La protection des personnes en cas de catastrophe, Académie de Droit international, Recueil des cours, Dossier: Les frontières, Limites et délimitations internationales – Quelle importance aujourd'hui ? La Haye: Tome 368.
Partow, H. (2008). Environmental impact of wars and conflicts. In M. K. Tolba & N. W. Saab (Eds.), *Arab environment* (Future challenges). AFED.
Pentinat, B. S. (2010). Environmental refugees. problems and challenges. In *Colloque d'Aix en Provence, Le droit international face aux enjeux environnementaux, Société Française pour le Droit International (SFDI)*. Éditions A. Pedone.
PNUD. (2009). Rapport mondial sur le développement humain de 2009. Lever les barrières. Mobilité et développement humains, PNUD.

Regional Initiative for the Assessment of Climate Change Impacts on Water Resources and Socio-Economic Vulnerability in the Arab Region (RICCARD). (2017). Disaster loss data and linkage to climate change impacts for the Arab Region. https://www.unescwa.org/sites/default/files/event/materials/riccar_technical_report-unisdr_2017_1.pdf

Scheffran, J., Brzoska, M., Kiminek, J., Link, P., & Schillin, J. (2012). Climate change and violent conflict. *Science, 336*(6083) New series.

Schmid, L. (2018). Le contre-pouvoir écologique, *Revue Esprit*, N°1. https://esprit.presse.fr/article/lucile-schmid/le-contre-pouvoir-ecologique-39815

Schwartz, P., & Randall, D. (2003, October). *An Abrupt climate change Scenario and its implications for United States National Security.* https://www.iatp.org/sites/default/files/An_Abrupt_Climate_Change_Scenario_and_Its_Impl.pdf

Serre, J. (2016). *Les États fragiles*. Édition Studyrama.

Slaughter, A.-M., & Alvarez, J. E. (2000). A liberal theory of international law. *American Society of International Law, 94*, 240–253.

Stiglitz, J., Sen, A., & Fitoussi, J.-P. (2009). *Richesse des nations et bien-être des individus*. https://doi.org/10.4000/developpementdurable.8712

Strategic Foresight Group. (2009). *Cost of conflict in the Middle East.* http://www.strategic-foresight.com/publication_pdf/39166Cost%20of%20Conflict%20in%20the%20Middle%20East.pdf

The Nansen Initiative. (2015). *Agenda for the protection of cross-border displaced persons in the context of disasters and climate change* (Vol. I). The Nansen Initiative.

The World Commission on Environment and Development (WCED). (1987). *Our common future*. Oxford University Press.

Tolba, M. K., & Saab, N. (Eds.) (2008). *Arab environment future: Future challenges*. Report of the Arab forum for environment and development. AFED.

Ulrich, B. (1992). *Risk society, towards a new modernity*. Sage Publishing.

UNEP. (2007). *Sudan post-conflict environmental assessment*. UNEP.

UNHCR. (2008). Changements climatiques, catastrophes naturelles et déplacement humain: une perspective du HCR. https://www.unhcr.org/fr/protection/environment/4b151c14e/changements-climatiques-catastrophes-naturelles-deplacement-humain-perspective.html

United Nations Economic and Social Commission for Western Asia (ESCWA) and International Organization for Migration (IOM). (2015). 2015 Situation report on international migration: Migration, displacement, and development in a changing Arab Region. https://www.unescwa.org/sites/default/files/pubs/pdf/l1500271_0.pdf

Vlassopoulos, C. A. (2017). Les dynamiques des sociétés humaines et les flux migratoires. In A. Euzen, B. Laville, S. Thiebault (Dir.), L'adaptation au changement climatique. Une question de sociétés, Édition du CNRS.

Von Schorlemer, S., & Maus, S. (2014). Reflections on climate change, heritage and peace, In S. Von Schorlemer & S. Maus (Eds.), *Climate change as a threat to peace. Impact on cultural heritage and cultural diversity.* Peter Lang GmbH. https://doi.org/10.3726/978-3-653-05205-3., https://www.peterlang.com/view/9783653968507/08_Chapter01.html

von Soest, C. (2020). *A heated debate: Climate change and conflict in Africa, German Institute of Global and Area Studies (GIGA).* https://www.giga-hamburg.de/en/publications/18735981-a-heated-debate-climate-change-conflict-africa/

Warner, K. (2017). Human mobility and the Paris Agreement: Contribution of climate policy to the global compact for safe, orderly and regular migration, Input to the UN Secretary-General's report on the global compact for safe, orderly and regular migration, in response to Note Verbale of 21 July 2017.

Warren, P. D. (2016). Forced migration after Paris COP 21: Evaluating the climate change displacement coordination facility. *Columbia Law Review, 116*(8), 2103–2144.

Chapter 8
The Changing Dynamics of the Arctic Ecosystem and Food Security: The Case of the Bering Sea Region

Anita Parlow

Abstract Climate change is having catastrophic environmental, economic, and cultural impacts, reshaping the Arctic region. Human-induced warming is accelerating sea-ice melt, permafrost collapse, and altering regular species migrations on land and sea. These impacts negatively disrupt not only the finely balanced ecosystems of the Arctic's flora and fauna, but also both the commercial fishers and Indigenous peoples who depend upon hunting and fishing for survival. This chapter provides a foundational context for my subsequent two chapters in this book that focus on the Arctic region, where sea ice melt is not only causing irreversible distortions in the region but it also significantly contributes to climate change. The rapid melting of the Arctic Sea ice is injecting fresh water into the world's oceans, thus causing a slowdown of the global ocean currents that circulate the equator waters to the Arctic region which, in turn, are less able to cool the world's oceans, thus causing atmospheric temperatures to rise. This circulatory slowdown generates erratic weather patterns and a 'feedback' loop, thus melting more Arctic Sea ice and creating yet more global warming. This chapter addresses the planet's complex existential crisis through the lens of the Bering Sea, contextualizing the accelerating changes in one of the richest ecosystems worldwide. The analysis places a human face on the climate threat, while also describing the impacts and actions of advocacy and resiliency by those who live on the front lines of the climate catastrophe. With its focus on St. Lawrence Island, on the Bering Sea, the cascading changes to the Bering's complex food web and its implications for Arctic coastal communities' food security reflect the most recent chapter of a 2500-year history in the harsh yet, giving region. Arctic coastal communities, such as those on St. Lawrence Island, depending upon the marine ecosystem for their livelihoods and food, contribute the least to the earth's global carbon footprint, and, retain the highest level of knowledge on how to maintain balance in the marine ecosystem. This, and the following chapters, reflecting on the depth and breadth of understanding by Arctic's coastal communities, suggests that the world might look more deeply to the St. Lawrence

A. Parlow (✉)
Fulbright Scholar Iceland, recent, Washington, DC, USA

M. Behnassi et al. (eds.), *The Climate-Conflict-Displacement Nexus from a Human Security Perspective*, https://doi.org/10.1007/978-3-030-94144-4_8

Island's Siberian Yupiks, and all Arctic Indigenous Peoples, as a guiding and self-determined force to address the existential threat that is facing humanity.

Keywords Arctic · Bering Sea · Climate change · Geopolitics · Indigenous food security

1 Introduction

Among the most iconic images of our time is a photograph of Earth shot from space on December 7, 1972, by the Apollo 17 crew as their lunar expedition's module left the Earth's orbit and traveled toward the moon (Fig. 8.1). The picture's deep black space, the round blue and white sphere, vulnerable and alone, changed forever the way we view our planet Earth. The vibrant blue and white colors show the Earth as a complex system of air, water, and land – dominated by a vast expanse of interconnected oceans and wrapped in a thin and dynamic protective atmosphere. Seeing the dominance of water – including both seas and clouds – helps us grasp how complex, interdependent, and fragile the planet Earth is. At the same time, the image also compels us to acknowledge how difficult it is to understand the Earth and its ecosystems, and serves, as well, as an invitation to protect its complex, vulnerable, and interdependent marine and terrestrial ecosystems.

More than a few authors wrote that the 'Big Blue Marble' Earth photo became a baseline for a new global consciousness. Such a beautiful and sobering photo is said to have helped galvanize a generation of environmentalists worldwide. Indeed, environmental advocates referred to the image to point out the substantial stresses to the planet's visible weather systems, including rising global temperatures and extreme weather events. By 1992, the stresses had become so pronounced that scientists began questioning whether the undeniable anthropogenic impact had irreversibly altered the planet's climate, weather patterns, and oceans. This question sparked a conversation among geologists and environmentalists, who began calling the geological age in which we live 'the Anthropocene' era (Huntington, 2014). This 'age of the human' highlights the heavy toll that that global economic growth – particularly its reliance on fossil fuel energy, mining, and natural resources – has placed on nature. Many researchers pointed out that mass extinction of animal and plant life can be tied to human activities that reflect 'destructive and shortsighted' behavior. The release of greenhouse gases is exceeding the natural ecosystem's absorption and storage capacity, thus contributing to global warming and species extinction (Wuebbles, 2012). More optimistically, Kolbert (2015) also notes that human activity can also be "forward thinking and altruistic".

The grim reality of social and environmental challenges resulting from climate change is currently visible worldwide. Sub-Saharan Africans face increasingly severe episodes of food shortages and food insecurity (Sikapizye, 2019) and their food security is being undermined. The MENA region, as a climate change 'hotspot', has started experiencing the highest recorded temperatures ever more than a decade

Fig. 8.1 Planet Earth shot from space by the Apollo 17 crew as their lunar expedition's module left the Earth's orbit
Credit: NASA. The "Blue Marble" photograph, taken 7 December 1972 by the Apollo 17 crew

and a half. Densely populated low-lying coastal cities and island nations in the Pacific are already relocating their populations. In addition, a combination of fires, volatile storms, hurricanes, tornadoes, cyclones, tsunamis, and an increasing number of extreme weather events are occurring worldwide with increased frequency (Nicholls & Kebede, 2012).

Warming twice as fast as the rest of the planet, the Arctic is losing its sea ice, which is the vital foundation of the Arctic marine ecosystem. The absence of sea ice impedes the highly integrated food chain that begins with planktons that form on the underside of the melting sea ice, and thus, the ability of people whose food sources depend upon the sea. Further, the melting in both the Arctic and Antarctic regions amplifies the Earth's warming and the changes in global weather patterns that ensue. The fresh water released from melting sea ice and glaciers alike into the heavier salty seas slows the circulation of water in the global ocean, which contributes to planetary warming (Huettmann, 2012; Berkman & Vylegzhanin, 2013).

This chapter focuses on the marine ecosystem in the Arctic's Bering Sea region, contiguous to the Arctic region, located between Russia and Alaska, and the close connections between it and the Indigenous people and communities of the Bering Sea (Fig. 8.2). The Bering Sea region, straddling the Arctic and sub-Arctic, offers a lens into the intricacies of an Arctic marine ecosystem, its food web, and the coastal communities that depend upon a healthy sea for their food supply. The primary purpose of this work is to show how the various elements of this ecosystem, essential, synchronized, and dependent upon seasonal ice, is undergoing what some scientists believe could be permanent damage to the Arctic and sub-Arctic marine ecosystem. This chapter also reflects upon the central role the Arctic plays in the warming of the planet, serving as a thermostat for global weather patterns.

Both the marine ecosystem and the humans who live in the Bering region depend upon the seasonal synchronization of essential biological events. Alaskan Siberian

Fig. 8.2 The Bering Sea on the map

Yupiks have been living on St. Lawrence Island, a landmass of 1.2 million acres, for more than two thousand years. The islanders depend upon the marine ecosystem of the Bering Sea for their food supply. The interviewed population in this study testifies a fundamental Yupik view which stipulates that humans are part of, but do not dominate, the marine life that shapes the complex Bering Sea's marine ecosystem. The hunters, fishers, and whalers abide by a sophisticated, intricate understanding of the food web and their role in its maintenance. An understanding that follows the science, philosophy, and practices that have evolved over two thousand years of Yupik tradition and experience. The cultural and spiritual centerpiece of Yupik philosophy is to understand the sophisticated relationships that shape the Bering Sea's marine life to protect its ecological balance while partaking of its bounty.

2 Backdrop: The Arctic Region

Some four million people live in the Arctic region, which comprises parts of Canada, Denmark/Greenland, the United States/Alaska, Finland, Sweden, Iceland, Russia, and Norway. Residents include hunters, herders, fishers, whalers, coastal dwellers, and urban inhabitants. Over the past century, much of the Arctic north has experienced considerable cultural, environmental, spiritual, and economic changes (Berkman et al., 2016).

In geographical terms, oceans and seas comprise most of the Arctic. The smallest of the world's five oceans is the Arctic Ocean, which, along with its adjacent seas, encompasses an area of about 5.4 million miles (The Maritime Executive, 2017) and an average depth of only 3200 feet, located mostly north of the Arctic Circle (66.33 degrees north). For much of its extent, the Arctic Ocean is covered by ice, glaciers, and ice floes in the increasingly warming winter months.

While considered a marginal sea of the Pacific Ocean, the Bering Sea, referred to as the 'gateway to the Arctic', has much in common with the Arctic region, from which it is separated by only the Bering Strait, 66.4 north latitude, just south of the Arctic Circles' 66.5 degrees north. Due to climate change, the Arctic region is warming fast and as the melting ice makes it more accessible to commerce and shipping, the region is moving from a geopolitical margin to a strategic and commercial center (Heininen & Exner-Pirot, 2020).

The Arctic's Indigenous peoples, who account for about 10% of the region's population, are among the first to feel the effects of climate change. Given the accelerating destruction of the Arctic marine ecosystem, Native peoples are finding it an increasing challenge to navigate and protect their millennia-old hunting, fishing, whaling, and gathering practices in a manner, developed over centuries, that depends upon a degree of predictability with respect to the seasonal variation of the marine ecosystem (Nuttall, 2005).

It is a cruel irony that some of the most profound effects of climate-based destruction are located in one of the world's most pristine, isolated, and fragile environments and experienced by those, such as the residents of St. Lawrence Island, who have contributed least to global warming. The islanders like many other communities on Alaska's coast, live in complete balance with the Bering Sea's marine ecosystem and are almost entirely dependent upon its bounty. Central to the Yupik philosophy, as lived in the St. Lawrence Island village of Savoonga, are three foundational principles: (1) 'take only what your need'; (2) 'the human does not dominate the other world's species, but lives in balance with them'; and (3) 'leave no footprint behind' (Parlow, 2019b). Yet, these fishers, hunters, and whalers are among those suffering the most from the seemingly limitless consumption of fossil fuels in the rest of the world.

2.1 The Bering Sea

The Bering Sea, a combination of a southern promontory of the CAO and a marginal sea of the North Pacific Ocean, is celebrated for its highly productive marine ecosystem. Its large fisheries and rich seafloor support a complex food web, from microscopic phytoplankton to migrating large mammals such as walrus, seals, and whales. In 2006, the Center for Biological Diversity reported that the Bering Sea is home to 418 species of fish, 102 migratory bird species, and 29 types of marine mammals (Greenwald, 2006).

Situated between the North Pacific Ocean and the CAO, the Bering Sea, at 770,000 square miles, is both a growing shipping transit point that links the Pacific Ocean with the increasingly accessible passage in the Arctic region. It is also an evolving transition zone that is experiencing a cascade of changes in the ecosystems between the warming waters of both Pacific and melting sea-ice CAO. The seasonal timing and extent of sea-ice formation is the foundation for the complex food web in the CAO and the Bering Sea; it begins with phytoplankton that thrives in the fresh water that forms just below the melting sea ice and is the cornerstone of the Bering Sea ecosystem (Stevenson & Lauth, 2019). Whether the changes in the Arctic and sub-Arctic ecosystems will continue to evolve synchronously through the unique species of their food webs pose critical questions for what is likely to become an integration of Arctic and sub-Arctic waters, with the Bering Sea as a centerpiece.

Fish and whales, marine mammals, and migratory birds have provided the food supply for coastal residents who have lived on either side of the Bering Strait since prehistoric times. In the eighteenth and nineteenth centuries, the rich waters of the Bering Sea drew Russian, British, and American fur trappers and whalers whose industrial-scale operations brought trade to the people who lived in the remote Bering region (Demer & Lester, 2017). But they also brought trauma. In 1878, whalers induced mass starvation among the Native peoples who lived in the Bering Sea region. All the resources of the marine ecosystem, whales, walrus, seals, and fish were stripped from the Bering Sea waters. Nearly everyone on the Bering Sea's islands, as well as coastal communities on both the Russian and American sides, died. Further, whalers and fur traders introduced infectious diseases that compounded the effects of food scarcity. Of some several thousand Siberian Yupiks who lived on St. Lawrence Island, only one village of 32 and fewer than 300 people survived (Ackerman, 1988).

The Bering Sea region is viewed by both Western scientists and Siberian Yupik oral history and science as the point of departure for population movements across the Bering Land Bridge, Beringia, from the Siberian Arctic to what is now Alaska, the state that makes the United States an Arctic nation (O'Neill, 2009; Van Pelt, 2015).

The islands on the Bering Sea are the geographical brides to the last ice age, which left evidence of Late Paleolithic human migrations. Archeological records indicate that people traversed the Beringia route between 12,000 and 25,000 years ago during what some call, perhaps apocryphally, the 'greatest mammoth hunt of all time' (Oliver, 2007; Anderson, 2009; Dickson et al., 2008). However, more likely, is evidence that crossing the Bering from Eurasia to the Americas was a complex process that involved multiple migrations by both land and sea (Erlandson et al., 2015).

As a northern extension of the Pacific Ocean, the Bering Sea covers an area of more than 700,000 square miles between the Aleutian Islands and the Bering Strait. Lodged between Alaska to its east and Russia to the west, the Bering Sea straddles the Arctic and sub-Arctic. As the 'Gateway to the Arctic', the Bering Sea, primarily sub-Arctic, is the only waterway that connects the Pacific Ocean to the frozen CAO through the Bering Strait at the Bering Sea's northern reach. It is a direct line to the North Pole, then to the Atlantic Ocean (Berkman et al., 2016).

Considered a 'marginal sea' of the Pacific Ocean, the Bering Sea, surpassed in size by the Mediterranean and the South China Sea, separates the planet's two largest landmasses: Eurasia and the Americas. The Bering Sea, noted for its persistently cold temperature ranges, is considered by many scientists as the "world's largest and most biologically productive semi-enclosed sea" (Van Pelt, 2015).

The Bering Sea's shallow waters, ranging from 100 to 165 feet deep at the Bering Strait, the shallowest point, although far deeper in other areas such as the Aleutian basin, at 12,000 feet. The Bering Sea, covered by ice for thousands of years, was defined by a broad continental shelf that stretched continuously from Siberia to Alaska, the Americas, and Eurasia. The Bering 'land bridge' was made visible during the last ice age when sea levels were about 160 feet lower than they are now. Much was covered by ice, but the Bering's southern shore was probably ice free, 'at least in parts'. The prevailing view is that people migrated from Asia to America along this ice-free region, by foot or, possibly using boats (Oliver, 2007).

2.2 *Ice Conditions: Far Outside the Norms*

The journal *Science Advances* recently published a 2020 study that examined ice conditions in the Bering Sea as far back as 5000 years, and concluded that the current ice conditions and their trajectory gives an appreciation of "how far outside the norms the recent conditions represent". Bailey (2020) noted in this research that as "temperatures continue to rise, we are entering into a realm that lacks prior analogies so we likely do not well understand the full ramifications for the ecosystem".

Time-lapse satellite imagery since the 1970s makes the accelerating Arctic Sea ice shrinkage visible. Lodged between Russia and the United States, via Alaska, the Bering Sea is showing signs of an ecosystem transformation that could forever alter its finely tuned structure, disrupt its entire food web, from the microscopic at the benthic levels to marine mammals and birds. This disruption is already creating significant food shortages for coastal inhabitants such as Inupiats, Yupiks, Siberian Yupiks, and Inuits who depend upon a healthy marine ecosystem for their food supply. The smallest changes cascade through the marine ecosystem, causing increasingly threatening disruptions (Huntington et al., 2020). For example, in 2018, the average seafloor water temperature in the Bering dropped by 2 degrees Fahrenheit impacting the hard-shelled bivalves that live on the seafloor, serving as a nutritious fatty meal for eider ducks who once sat on the ice, waiting to feed on their prey. Now, with less ice and bivalves slimmed to worm-like states, the eiders are at risk of extinction. Others in the marine ecosystem are equally at risk. Copepods, likewise, are less fatty and in declining numbers that newly hatched haddock eat. The cascading effect has left other seabirds and fish undernourished, unable to survive through the winter, and at risk of die-offs and, eventually, extinction (University of Maryland Center for Environmental Science, 2020).

The evidence of climate stress is profound. Thoman et al. (2020) reported that the ice coverage in the Bering Sea shrank to its lowest level on record since 1850. In addition, a range of widespread impacts has been documented: a rearrangement of the marine ecosystem, die-offs of several birds, a mass stranding of seal and walrus, a redistribution of thermally sensitive fish moving northward to colder waters, and an alteration of whale migration patterns (Greenwald, 2006; Noongwook, 2004).

Whether climate change-related distortions in the Bering Sea can be stopped, reversed or redirected poses an enormously challenging question. The Bering Sea Ecosystem Committee noted as far back as 1996 that it would be "difficult for human management to cause a large, complex, and climate stressed marine ecosystem to achieve and maintain a desirable balance" (Francis, 2018).

2.3 *The Bering Strait*

The Bering Strait is located at the northern end of the Bering Sea. As a direct result of global warming, the 51-mile-wide passage between the Russian Federation and the United States is becoming one of the world's commercially and strategically important straits. The Bering Strait's narrow passageway is considered a potential choke point for marine mammals and, increasingly, ice-capable vessel traffic. Massive numbers of birds and mammals 'funnel' through the Strait during their spring and fall migrations toward the Chukchi and Beaufort Seas that border the Bering Strait to the north (Berkman et al., 2016). Currently, the increases in shipping generally draw from the Russian Federation's Northern Sea Route to the

Bering Strait's west. Traffic from China's east coast to Russia's Northern Sea Route is also growing (Farré et al., 2014). Experts anticipate that sea ice melt in Canada's Northwest Passage to the east will attract even more vessel traffic and that a significant volume of shipping will soon transit between Chinese ports and Europe through the CAO (Humpert, 2020; Østreng et al., 2013).

Climate change impacts in the Bering Sea region are magnified in the Bering Strait, which is likely to become a transportation 'choke-point' (Humpert, 2020) requiring new governance and regulatory and monitoring regimes to protect both its ecosystem and vessel traffic (Rothwell, 1996).

2.4 *Bering Strait: Climate Change Magnified*

The Bering Sea and Strait constitute a feeding ground and migration corridor for hundreds of thousands of marine mammals, including species important for subsistence hunting and fishing. It hosts marine mammals such as bowhead, beluga, and gray whales, Pacific walrus, ringed, ribbon, spotted, and bearded seals, and occasionally polar bears. Fish also traverse the Strait, including species important for subsistence hunting and fishing (Raymond-Yakoubian, 2018).

Observers accept that climate-related changes combined with increased shipping will significantly degrade the marine life of the Bering Sea ecosystem. The Bering Strait draws more than 10 million migrating seabirds annually, and thousands of seals, walrus, and whales (Greenwald, 2006). Whether an effective regulatory system to protect the marine environment and ecosystem will be promulgated and enforced in the Bering Strait region is unclear. Such protections require both regulatory processes with an emphasis on science-based decision making as well as the infrastructure to monitor and launch effective responses to violations. A combination of existing and new multinational and bilateral agreements ratified or agreed upon by the United States and Russia, two strong nations with specific coastal jurisdiction, is vital for the establishment and implementation of governance arrangements (Berkman et al., 2016).

Thus far, relatively few non-native species have been introduced into the Bering Sea. However, with the projected increases in shipping traffic, the region is likely to see an increase in the number of invasive species, including the microscopic, capable of altering or even dominating the finely tuned food web. Without any natural predators or controls on their ability to displace or alter existing food webs (Molnar, 2008), the Bering Sea could be stripped of its rich ecosystem (Grebmeier, 2012).

As navigation is already increasing through the Bering Strait, there is an increased likelihood that ships may be involved in more whale strikes or upend migratory bird paths. An equally vital environmental question is the extent to which oil and gas tankers will damage the Bering Sea ecosystem. Given the nearly universal agreement that an oil spill in Arctic waters would be disastrous, in 2013, the Arctic Council drafted the Agreement on Cooperation on Marine Oil Pollution, Preparedness, and Response in the Arctic, an effort jointly led by Russia and the

United States. While the agreement is essential, the seasonally dark, icy, and remote Bering Sea region lacks the infrastructure, including safe harbors, for robust cleanup or effective search and rescue operations.

Ensuring the effectiveness and durability of the regulatory processes that will shape Bering Sea governance requires more initiatives that build upon the International Maritime Organization's Polar Code, the designation of areas to be avoided (ATBAs), and the agreements on illegal, unreported, and unregulated (IUU) fishing (Huntington, 2014; Berkman et al., 2016) and oil spill response. Of paramount importance, the Bering region's Indigenous peoples are beginning to insist that they meaningfully shape the initiatives that directly impact their lives. Siberian Yupiks and others understand the Bering Sea region and depend upon it for their food supply. Equally significant are the Indigenous philosophies, science, and interpretation of natural law that reject concepts of human domination of the marine ecosystem. Thus, Indigenous peoples in the Arctic region serve as a voice not only for themselves but also for the flora, fauna, and the sea, itself.

2.5 Sea Ice in the Polar Zone

Marine ecosystem nutrient production occurs in all world's oceans, but nowhere is it as concentrated as in the Arctic and the Antarctic – the Sea Ice Zones (Harrison & Cota, 1991). As more predictable and denser habitats for food than the open ocean, the polar seas and its ice zones draw a wide variety of species, ranging from gulls and bird colonies, ringed seals, polar bears, penguins, narwhals, and beluga and bowhead whales. These areas also provide foraging zones for other species such as capelin and polar cod, which are prey for migrating birds, mammals, and, of course, in the Arctic region, for human hunters and fishers who have traversed the sea ice for millennia in search of food (Akeya, 2014).

From top to bottom, the water column in the Bering Sea is alive. Just beneath the sea ice, and its seasonal melts that deposit a skin of fresh water above the heavier salt water, are microscopic bacteria, algae, larvae, and unicellular organisms and phytoplankton (Assmy et al., 2017). The fresh water stratifies the water column into two layers: fresh on top, just beneath the melting ice, with the dense, heavier, and salty seawater below. The stratification allows plankton, phytoplankton, and algal blooms to "burst out" and thrive in the sunlight and fresh water (Taniguchi et al., 1976). Phytoplanktons produce their own food by photosynthesis, thus requiring the sunlight months to obtain their nutrition.

The combination of fresh and saltwater on the Bering Sea's continental shelf is the engine for the complex ecosystem's evolution and growth. When sea ice melts, the fresh water beneath the sea ice serves as the incubator for the microorganisms that form the foundation of the Bering's food chain. The sustainability of the Bering Sea ecosystem depends upon a seasonally time-sensitive combination of sea ice, seasonal melting of sea ice, and the growth of microorganisms in the fresh water just beneath the melting sea ice that, in full bloom, drop to the seafloor. The rich benthic feeding grounds comprise the base of the food chain that ranges from microscopic organisms to the largest mammals. The complex and interactive web of prey and preyed upon, from zooplankton to humans, is a cornerstone of the Bering's ecosystem (Van Pelt, 2015).

As one of the most productive marine ecosystems, the Bering Sea supports some of the world's most economically important fisheries. Bering pollock is the largest single-species fishery, while fish from the Bering Sea account for nearly half of the U.S. annual commercial fish consumed (Alexandra, 2019). The fish and other marine life also provide food for thousands of Alaskan coastal and island residents who hunt and fish exclusively for themselves, their families, and communities (Sigler et al., 2010; Laidre, 2008).

2.6 *Sea Ice Loss*

The rapid melting of Arctic Sea ice and glaciers is an indicator of how fast the climate is changing and thus, pumps increasing amounts of greenhouse gases into the atmosphere, a phenomenon referred to as a 'regime shift' (Policansky, 1996). Analysis of data gathered from satellites shows the increasing pace of sea ice loss (Fig. 8.3), with record lows in 2019, the second lowest since satellite data collection began in 1979. With sea ice in accelerating retreat, the Earth's climate is becoming warmer and sea levels are rising.

The disappearing sea ice is the most visible indicator of a changing Arctic and world (Druckenmiller et al., 2019). Arctic sea ice is a 'thermostat' for the climate system. Unlike Antarctica, which is an ice-covered continent, the Arctic is mostly an ice-covered body of water (D'Arcy Wood, 2020). The ice historically grows in the fall and winter and shrinks during the spring and summer. The melting sea ice impacts the world's interconnected marine ecosystem by impeding the vital circulation of the planet's oceans, thus contributing to a warming planet. With ever-larger swathes of Arctic Sea ice melting for longer periods, the region has become most sensitive to global warming, serves, along with Antarctica, as ice-free waters absorb more sunlight thus, amplifying planetary temperatures, and climate change (Carson, 1991).

The melting sea ice drives further global warming. Scientists refer to thick annual sea ice cover as a 'kind of glue' (Mooney, 2018) that, through its relative permanence even in the summer season, holds the Arctic ecosystem together. The sea ice cover is not a flat, thick, unbroken sheet of ice, but a 'mosaic' of various ice textures and thicknesses with different ages and characteristics. In the cold winter months, the sea ice floes freeze together (Gearheard, 2013). In the summer months, the melting ice and ice cover breakup leave fresh water under the surface.

The loss of sea ice threatens not only the ecosystem, but also the people who depend upon a healthy sea for their food supply. While the sea ice retreat is diminishing the highly lucrative commercial fishing industry (Mekik, 2013), it is the

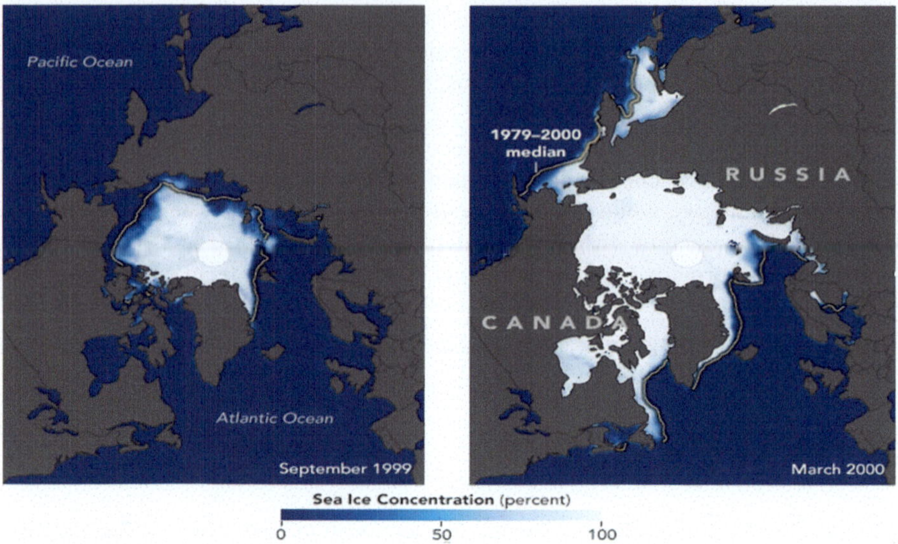

Fig. 8.3 Arctic sea ice minimum
Source: http://itbankdna.org/en/arctic-ice-reduction

subsistence hunters, fishers, and whalers, who have harvested food for their communities for at least two thousand years, that are experiencing the greatest losses (Stotts, 2015).

The National Snow and Ice Data Center (NSIDC, 2021) finds that the older, multiyear sea ice remains thicker and less likely to melt completely during the summer months. Thus, it preserves its role in maintaining a habitable global temperature range (Michon & Lindsay, 2019). With the perennial sea ice melting and diluting the oceans' salty waters, it is slowing the once faster moving currents that allow for ocean circulation which moves from the equator to the Arctic regions and return (NSIDC, 2021). The slower moving, and thus warmer oceans, are losing the ability to cool the planet.

Scientists expect Arctic-sea ice to diminish significantly sometime around mid-century or before (Meier et al., 2019). As a result, the intricate and rich Arctic marine ecosystem that depends upon sea ice is threatened (AMAP, 2017; Meier et al., 2019). NSIDC (2021) reported that 40% of international waters of the CAO in the summer months had thin, year-old ice cover. NASA has been recording a 12.85% decrease in Arctic Sea ice per decade (Lindsay & Schweiger, 2014).

The thinning sea ice impacts the ability to harvest food, as it no longer reliably holds the weight of the fishers or hunters' snowmobiles and sleds, or the weight of a harvested whale (Gerland et al., 2019; Parlow, 2019b). The early retreat of sea ice means that the blooms arrive too late to provide sufficient nutrition for the benthic aquatic life and the fish and migrating marine mammals that feed upon the microscopic plankton that grow just beneath the melting sea ice (Polyak, 2010; Huettmann, 2012).

Over the past decade, the loss of sea ice has led to increases in phytoplankton mass in the CAO that has led to an "upwelling of nutrients previously concentrated on the seafloor" driven by melting glaciers and meltwater (Cape et al., 2019). Given the combination of sea ice melt in the CAO and the newly available supply of nutrients, scientists believe that the CAO could become more productive and able to support higher trophic-level production. What this might mean for the Bering Sea as the nature of the previously distinct ecosystems become more integrated is, as yet, unclear.

Reflecting upon the magnitude of the challenge posed by the rapidly changing Bering Sea marine ecosystem, Charlie Lean, Director of Fisheries Research at the Norton Sound Fish and Game Advisory Board, resolved: "adapt or die" (Parlow, 2019a).

3 A Marine Ecosystem in Flux: Loss and Seasonal Sea Ice Melts

In 2019, the lead scientist for the Bering Sea Bottom Trawl Survey noted that the "ecosystem is in flux" (Stevenson & Lauth, 2019). Medearis (2019) reported as well that for the previous three years, the fish species "have not hit any kind of equilibrium to tell us where the fish want to be or why they are there".

The underside of melting sea ice provides the basic material on which the entire food chain depends. An estimated 1000 different species of algae flourish in an environment defined by cold and lack of light. On the underside of the sea ice, the nutrients, which form the basis for the Bering Sea's food chain, grow (Druckenmiller et al., 2019; Harvey, 2019).

Phytoplankton is the foundation upon which the Bering Sea ecosystem rests. It feeds all aquatic life from microscopic zooplankton to marine mammals. The phytoplankton blooms fall to the seafloor where they become a constituent element of the benthos, the flora and fauna found on the bottom and in the bottom sediments of the sea. As the phytoplankton blooms sink to the seafloor, they become a primary food source for the tightly knit interlinked species from other planktons to fish, seabirds, seals, whales, and humans (NOAA, 2019). Plankton is linked to the health of the entire ecosystem and its food web. Shifts in the abundance, composition, or timing of phytoplankton blooms are causally linked to the health of the food web and impact the entire ecosystem, including humans (Duffy-Anderson et al., 2019). With the seasonally timed plankton's bloom increasingly delayed due to warming, the cascade of intricate, seasonally sequenced events disrupts the entire food chain (Spies & Kristen, 2019; Grebmeier, 2012).

The melting Arctic-sea ice, an amplifier of global warming and a destabilizer of global weather patterns, serves as a "miner's canary" to warm the world (Borunda, 2020; Emmerson, 2010). Scientists hold varying views regarding how much the microscopic phytoplankton species, which release substantial amounts of oxygen into the atmosphere while consuming large amounts of carbon dioxide, impact the planet's atmosphere. But, even small changes in phytoplankton production can affect the amount of carbon dioxide absorbed, thus impacting the planet's temperature (Sugie et al., 2020). With the formation and melting of sea-ice currently delayed by some two or three months, the microscopic foundation of the Bering's food chain either doesn't bloom at all, or blooms at a time that no longer aligns with the life cycle of species that consume them, or they are smaller, less fatty, and thus less nutritious (Katz, 2019).

According to Anderson et al. (2018) and Soltwedel and Peeken (2014), the decline of phytoplankton, a foundational food source for marine ice mammals, not only is causing starvation for the copepods that move through the water columns, but it also is removing a staple of the diet of many fish, seabirds, planktonic species, and mammals, including humans.

Nutritional impoverishment at the bottom of the food chain has ripple effects starting with the sea cucumbers, snails, clams, bristle worms, crabs, and brittle stars upon which bottom-feeder mammal species, like whales, walrus, and fish feed (Wadham et al., 2019). According to walrus and whaling captain George Noongwook, "The walrus and seals have been really skinny for the last several years" (Parlow, 2019b).

3.1 *Loss of a Highly Integrated Food Web*

The food web of the Bering Sea ecosystem is well-adapted to extreme climatic conditions and is acutely sensitive to the impacts of sea ice melt (Assmy et al., 2017). The ice now forms later than in previous years, lengthening the season when sea ice no longer protects the shore from surging ocean swells and battering ocean waves. Wave damage and shoreline flooding from storms cause coastal villages like Alaska's Shishmaref, Kivalina, and dozens more, to slide into the sea (Martin, 2018).

Seemingly, small changes, such as the drop in nutrition levels of the microscopic life at the bottom of the Bering Sea's food web, can produce a domino effect with disastrous consequences (Laidre, 2008). The disruptions caused by the melting sea ice impact the entire interconnected Arctic Ocean and adjacent waters like the Bering Sea. In previous years, when the sea ice regularly moved southward through the Bering Strait from the more northerly Chukchi Sea, the ice started to melt in the spring. The melting sea ice released fresh water to its surface layer, becoming a foundational breeding ground for microscopic algae and phytoplankton.

Now, having lost at least 3 months of sea ice annually, the food system no longer matches the rhythms of the migrating fish, walrus, seals, or whales, who feed on the phytoplankton blooms that dropped to the benthic levels. As a result, the marine ecosystem suffers a cascading loss that causes food deprivation to all species, including the tightly interlinked humans who, in Savoonga, for example, have described significant changes in the location and weight of fish and mammals that have thrived in this exquisitely balanced sea for thousands of years (Overland & Stabeno, 2004).

As the sea ice recedes, plankton blooms are showing up in new areas of the Bering Sea and at different times. These changes are drawing species to new regions while native fish disappear, which is raising concern among scientists about the Bering Sea ecosystem's long-term viability (Stein et al., 2020). Thus far, it is unknown by scientists whether the Bering Sea has entered an era where the changes in the marine ecosystem will impact the complex and tightly integrated food web structure permanently. However, warming is currently changing where commercial and subsistence fishers, hunters, and whalers operate and how productive they are (Nero, 2018).

The ice-free seasons also create more significant hazards for hunters and whalers whose snowmobiles and sleds can fall through soft ice or cracks as they seek to reach their staging areas. Former Savoonga Tribal Chairman Delbert Pungowiyi observed to a group of visitors of the sea ice loss, "We used to have seven months of ice – now we have three." The sea ice once extended the island as far as the eye can see towards the Russian coast. It was consistently strong enough to hold the weight of walrus and seal hunters and their hunting and fishing gear, along with the full weight of a harvested whale to be cut and distributed to the residents of the entire island, and beyond. Savoongan whalers and walrus hunters noted that the changing wind patterns are bringing warm southerly winds that accelerate the sea ice retreat, change the temperature patterns, and cause further warming.

Global warming is also melting the permafrost, the long-frozen soil that is collapsing at accelerating rates as the ice melts. Given the substantial permafrost melt on the island, roads used to gain access to fishing and hunting camps have become impassable (Malcolm, 2009). Unable to safely traverse the island on snowmobiles or four-wheelers to reach their hunting and fishing camps, hunters and fishers are concerned about their ability to feed their families and the elders and others who are unable to hunt or fish. Because of the ice breakup and slushy ice that "comes in fast", one whaler explained, "sometimes when we go out to catch a whale, we don't know if we can get back" (Parlow, 2019b).

In anticipation of the accelerating changes to the marine ecosystem, commercial fishers plan to adjust their fishing strategies and raise prices to the consumer to maintain their lucrative and much relied upon the fishing industry. Native villagers, such as subsistence hunters and fishers who live in Savoonga, explain that Siberian Yupiks have been hunting, fishing, and whaling for more than 2000 years. They generally agreed, whatever nature presents to us, "we'll do what it takes" (Parlow, 2019b).

3.2 The Speed of Change

The speed of climate change impacts is "startling" (Huebert et al., 2005). Just over five years ago, various computer models and the Intergovernmental Panel on Climate Change (IPCC) predicted that the first sea ice-free summers in the Arctic would not arrive until the end of the century. While few, if any, models can accurately predict the future, more recent observations predict a vanishing sea ice cover very soon. NOAA (2019) and Overland and Stabeno (2004) indicate that the Arctic will be ice-free in three or four decades. Other researchers, skeptical of global climate models, suggest the Arctic's Sea ice cover could be lost well before; as early as the next decade (Maslowski et al., 2012; Thackeray & Hall, 2019; Thoman, 2020).

While the extent of sea ice is easily visible by satellite, ice thickness is more difficult to measure. This is problematic because ice thickness is the more important metric for determining the volume of ice being lost (Thompson, 2015). With data gathered from instruments as varied as submarines and satellites, Lindsay and Schweiger (2014) have calculated that ice in the CAO has thinned by 65% since 1975. Further, the authors found that the annual average ice thickness over the entire Arctic basin has been decreasing at 18 inches per decade since 2000. Wadhams (2012) reported as well that the rate of retreat and thinning has "greatly exceeded the predictions of most models". Given this accelerating rate of ice loss, the author believes that a substantial portion of the summer ice from this previously inaccessible region will soon disappear. Similarly, Witze (2019) anticipates that by mid-century the Arctic coastline and most of the previously inaccessible CAO will experience some sixty to one hundred days of open water, well beyond the current one or two months of partially open water in 2020. The Director of the Center for Marine Biodiversity at the Scripps Institution of Oceanography has advised that given the likelihood of increased oil and gas extraction, mining, and shipping, a far more precautionary, ecosystem-based approach (to fishing) is required. Otherwise, fisheries' managers are gambling with the health of our oceans and coastal communities.

3.3 Implications of Ecosystem Changes

Given the close connection that marine mammals have with sea ice, Laidre (2008) noted that that the extinction of some Arctic species particularly sensitive to warming "is likely" if current Arctic Sea ice trends continue. For example, the survival of several species of seal pups is dependent upon the timing of sea-ice break-up for suckling before weaning (Stenson, 2014). Similarly, walrus pups, dependent on sea-ice, are being crushed at the increasingly crowded haul out sites, essential for foraging for food and avoiding predators (DelViscio, 2014.) A century ago, in what became known as the 'great starvation,' industrial-scale whalers and walrus hunters depleted the Bering Sea of a foundation of its ecosystem, along with the peoples' food supply. Today, climate poses a similar threat (Fig. 8.4).

As far as the the polar cod is concerned, Wadhams (2012) believed that juvenile fish spend their first one or two years under the sea ice where they feed on shrimplike amphipods and algae-eating copepods before they swim toward the coast to join other adult cods. Without the cod, the whales and walrus lose part of their food supply.

Additionally, changes in phytoplankton productivity influence whale and walrus health, as well as the climate. Certain species of phytoplankton cause 'red tides', or poisonous algal blooms that, when consumed in sufficient amounts by mollusks, mammals, including humans, are fatal. Further, there is growing evidence that fast, unseasonal growth accelerates the growth of harmful algae blooms (HAB) that, like red tides in non-Arctic regions, threaten the ecosystem and human life in the Arctic.

Over the past several decades, paralytic shellfish poisoning (PSP) cases have been threatening humans and marine life in the area. In 2019, the Savoonga Tribal government Headquarters posted public warnings for community vigilance, warning residents to avoid consuming contaminated fish and shellfish. Whaling captain and walrus hunter George Noongwook noted that the clams that are regularly consumed by Savoongans may already be poisoned by algae toxins. "We are watching carefully," he said (Parlow, 2018).

Fig. 8.4 As sea-ice melts, tens of thousands of stranded walruses assemble on a beach near Point Lay, Alaska
Source: Photo by NOAA

4 Conclusion

In 1972, the historic view of the Apollo crew has immortalized the vast expanse of Earth's interconnected oceans wrapped in a thin and dynamic protective atmosphere. Today, this panoramic image helps us, the inhabitants of this planet, to grasp how complex, interdependent, and fragile Earth is. This understanding of the earth's fragility should compel us to more deeply understand the planet and its highly interconnected ecosystems, and improve our ability to protect its vulnerabilities. As sea ice is the foundation of the Arctic marine ecosystem, its loss, and absence, impedes the nuanced and seasonally connected orchestrations of the highly integrated food chain. The seasonal sea ice melt that begins with planktons and algae that form on the underside of the melting sea ice, interrupts the connections between algae, fish, seals, walrus, whales and, ultimately, the humans who depend upon the bounty of the sea. Further, the melting in both the Arctic, and Antarctic regions, amplifies the Earth's warming, generating steep changes in global weather patterns. With dramatic changes at both local and global levels, it is clear that a coastal catastrophe is unfolding in the Arctic, the region that contributes the least to climate change. Given their front row and close up vantage, the Arctic's Indigenous peoples are increasingly and, successfully advocating for a seat at policy tables where discussions to curb oil, gas and, mining are accelerating along with growing global movements to decarbonize. The world community will benefit from the voices of Arctic Indigenous peoples who are intricately connected to the Arctic ecosystems, understand its balances and that relationships based on ecosystem "stewardship" which, in turn, is based on environmental justice that includes the rights of all living beings, and the responsibility for collective, ecological continuance.

References

Ackerman, R. (1988). Settlements and sea mammal hunting in the Bering-Chukchi Sea Region. *Arctic Anthropology, 25*(1), 52–79. Retrieved from http://www.jstor.org/stable/40316155

Akeya, D. (2014). *White sharks of the north: The cycle of Arctic life of four generations of polar bears seen through the observant eyes of a Siberian Yupik Alaskan Eskimo*. The Catchall Press.

Alexandra, V. (2019). Why is the Bering sea important? A current view of the Bering system and climate, Bering climate. National Oceanic and Atmospheric Administration NOAA.

Anderson, A. (2009). *After the ice: Life, death and geopolitics in the new Arctic*. Harper Collins Publisher.

Anderson D. M., Richlen, M. L., & Lefebvre, K. A. (2018). Arctic report card: Harmful algal blooms in the Arctic, Arctic program. Retrieved from https://arctic.noaa.gov/Report-Card/Report-Card-2018/ArtMID/7878/ArticleID/789/Harmful-Algal-Blooms-in-the-Arctic

Arctic Monitoring & Assessment Programme (AMAP). (2017). Snow, Water, Ice and Permafrost in the Arctic (SWIPA). Arctic monitoring and assessment Programme (AMAP), xiv + 269 pp.

Assmy, P., Fernandez-Mendez, M. Duarte, P., & Granskog, M. (2017). *Leads in Arctic pack ice enable early phytoplankton blooms below snow-covered sea ice* (Scientific Reports 7). Project: Norwegian young sea-ice.

Bailey, A. (2020). Scientists document striking changes in Pacific Arctic ecosystems, University of Alaska Fairbanks, UAF News and Information, Feb 24th. https://news.uaf.edu/scientists-document-striking-changes-in-pacific-arctic-ecosystems/
Berkman, P. A., & Vylegzhanin, A. N. (2013). *Environmental security in the Arctic Ocean*. Springer.
Berkman, P. A., Vylegzhanin, A. N., & Young, O. R. (2016). Governing the Bering Strait Region: Current status, emerging issues and future options. *Ocean Development & International Law, 47*(2), 186–217. https://doi.org/10.1080/00908320.2016.1159091
Borunda, A. (2020, January 16). Weather shows evidence of climate change every single day since 1012. *Science: National geographic*.
Cape, M. R., Vernet, M., Pettit, E. C., Wellner, J., Truffer, M., Akie, G., Domack, E., Leventer, A., Smith, C. R., & Huber, B. A. (2019). Circumpolar deep water impacts glacial meltwater export and coastal biogeochemical cycling along the West Antarctic Peninsula. *Frontiers in Marine Science, 6*, 144. https://doi.org/10.3389/fmars.2019.00144
Carson, R. (1991). *The global thermostat, the sea around us*. Oxford University Press.
D'Arcy Wood, G. (2020). *Land of wondrous cold: The race to discover Antarctica and unlock the secrets of its ice*. Princeton University Press.
DelViscio, J. (2014, October 2). *Alaska, thousands of walruses take to land*. https://www.nytimes.com/2014/10/02/science/earth/in-alaska-thousands-of-walruses-take-to-land.html
Demer, L., & Lester, M. (2017, May 31). What subsistence looks like at the Apassingok family table: Platter for all with walrus, whale and more from the Bering Sea, *Anchorage Daily News*. https://www.adn.com/features/alaska-news/rural-alaska/2017/05/31/what-subsistence-looks-like-at-the-apassingok-family-dinner-table-platter-for-all-with-walrus-whale-and-more-from-the-bering-sea/
Dickson, R. R., Meincke, J., & Rhines, P. (Eds.). (2008). *Arctic-Sub-Arctic Ocean fluxes: Defining the role of the northern seas in climate* (p. 77). Springer.
Druckenmiller, M. L., Daniel, R., & Johnson, M. (2019). *Voices from the front lines of a changing Bering Sea: An indigenous perspective for the 2019 Arctic report card*. Bering Sea Elders Group, National Snow and Ice data Center and Pew Charitable Trust.
Duffy-Anderson, J. T., et al. (2019). Responses of the northern Bering Sea and southeastern Bering Sea pelagic ecosystems following record-breaking low winter sea ice. *Geophysical Research Letter*. https://doi.org/10.1029/2019GL083396
Emmerson, C. (2010). *The future history of the Arctic*. Perseus Books Group.
Erlandson, J., Braje, T. J., Gill, K. M., & Graham Michael, H. (2015). Ecology of the kelp highway: Did marine resources facilitate human dispersal from Northeast Asia to the Americas? *The Journal of Island and Coastal Archaeology, 10*(3), 1–20.
Farré, A. B., et al. (2014). Commercial Arctic shipping through the northeast passage: Routes, resources, governance, technology, and infrastructure. *Polar Geography, 37*(4), 298–324. https://www.tandfonline.com/doi/full/10.1080/1088937X.2014.965769
Francis, J. A. (2018). The Arctic is breaking climate records, altering weather worldwide: The Arctic climate is shattering record after record, altering weather worldwide. *Scientific American, 318*(4), 48–53.
Gearheard, S. F. (Ed.). (2013). *The meaning of ice: People and sea ice in three Arctic communities*. International Polar Institute.
Gerland, S., Barber, D., Meier, W., Mundy, C. J., Holland, M., Kerns, S., Li, Z., Michel, C., Perovich, D. K., & Tamura, T. (2019). *Essential gaps and uncertainties in the understanding of the roles and functions of Arctic Sea Ice* (Environmental Research Letters) (Vol. 14, No. 4). IOP Publishing Ltd..
Grebmeier, J. M. (2012). Shifting patterns of life in the Pacific Arctic and Sub-Arctic seas, annual review of marine science, Chesapeake biological laboratory, University of Maryland Center for environmental. *Science, 4*, 63–78.
Greenwald. (2006). *A biodiversity assessment of vertebrate species*. Center for Biological Diversity.

Harrison, W. G., & Cota, G. F. (1991). *Primary production in polar waters: relation to nutrient availability*, Polar Research, Wiley Online Library.

Harvey, C. (2019). As sea ice disappears, so do nutrients for wildlife: Increasingly threatened young ice is crucial in transporting nutrients to the deep, Central Arctic Ocean. *E&E News.*

Heininen, L., & Exner-Pirot, H. (Eds.). (2020). *Climate change and Arctic security – Searching for a paradigm shift*. Palgrave.

Huebert, R., Berkes, F., Fast, H., Manseau, M., & Diduck, A. (Eds.). (2005). *Breaking ice: Renewable resource and ocean Management in the Canadian North*. University of Calgary Press.

Huettmann, F. T. (2012). *Protection of the three poles*. Springer.

Humpert, M. (2020). As Arctic ice melts Russia conducts record-breaking delivery of LNG to China. *High North News*. https://www.highnorthnews.com/en/arctic-ice-melts-russia-conducts-record-breaking-delivery-lng-china

Huntington, H. (2014). *The Arctic in the Anthropocene: Emerging research questions, the National Academies Press*, committee on emerging research questions in the Arctic.

Huntington, H. P., Danielson, S. L., Wiese, F. K., et al. (2020). Evidence suggests potential transformation of the Pacific Arctic ecosystem is underway. *Nature, National Climate Change, 10*, 342–348.

Katz, C. (2019). Why is an ocean current critical to world weather losing steam? Scientists search the Arctic for answers. *National Geographic*.

Kolbert, E. (2015). *The sixth extinction: An unnatural history*. Picador Press.

Laidre, K. L. (2008). Arctic marine mammals and climate change: Quantifying the sensitivity of Arctic marine mammals to climate-induced habitat change. *Ecological Applications, 18*(2), S97–S125.

Lindsay, R., & Schweiger, A. (2014). *Arctic sea ice thickness loss determined using subsurface, aircraft, and satellite observations*. The Cryosphere.

Malcolm, J. G. (2009). *Limited progress has been made on relocating villages threatened by flooding and erosion,* report to congressional requesters: United States government accountability office (GA), GAO -09-551.

Martin, A. (2018, October 22). An Alaskan village is falling into the sea. Washington is looking the other way, the world, the big melt. https://www.pri.org/stories/2018-10-22/alaskan-village-falling-sea-washington-looking-other-way

Maslowski, W., Kinney, J. C., Higgins, M., & Roberts, A. (2012). The future of Arctic Sea ice. *Annual Review of Earth and Planetary Sciences, 40*, 625–654.

Medearis, S. (Ed.) (2019, November 15). Warm water fish migrating to northern Bering Sea are thriving.

Meier, W. N., Gerland, S., Barber, D., Mundy, C. J., Mundy, H., Marika, Z., Kern, S., Li, Z., Michel, C., Perovich, D. K., & Tamura, T. (2019, April). *Essential gaps and uncertainties in the understanding of the roles and functions of Arctic Sea ice*. IOP Publishing Ltd.

Mekik, F. (Ed.). (2013). Earth's climate: Past, present and future: Arctic Ocean circulation, the knowledge project.

Michon, S., & Lindsay, R. (2019). *Arctic Sea ice extent ties for second-lowest summer minimum on record,* NOAA, Climate.gov

Molnar, J. L. et al. (2008). *Assessing the global threat of invasive species to marine biodiversity*, ESA.

Mooney, C. (2018). The Arctic Sea ice has lost 95 percent of its oldest ice. *Washington Post.*

National Oceanic and Atmospheric Administration (NOAA/Fisheries). (2019). *Scientific teams set out to track unprecedented changes in the eastern Bering Sea*. National Oceanic and Atmospheric Administration.

National Snow and Ice Data Center (NSIDC). (2021, January 5). *Arctic Sea ice news and analysis*. https://nsidc.org/arcticseaicenews/

Nero, M. E. (2018). *Bering Sea fisheries*. Fishermen's News.

Nicholls, R. J., & Kebede, A. S. (2012). Indirect impacts of coastal climate change and sea-level rise. *Climate Policy, 12*, S28–S52. https://research.fit.edu/media/site-specific/

researchfitedu/coast-climate-adaptation-library/europe/united-kingdom-amp-ireland/Nicholls%2D%2DKebede.-2012.-UK-Indirect-Impacts-of-CC%2D%2DSLR.pdf
Noongwook, G. (2004). *Watching sea ice and weather our way.* Smithsonian Arcgtic Studies Center.
Nuttall, M. (2005). *Protecting the Arctic: Indigenous peoples and cultural survival.* Routledge, Taylor and Francis Group.
O'Neill, D. (2009, April 29). *The last Giant of Beringia* (p. 12).
Oliver, J. A. (2007). *The Bering Strait crossing: A twenty-first century frontier between east and west* (p. 25). IA Press, Information Architects Books.
Østreng, W., Eger, K. M., Fløistad, B., Jørgensen-Dahl, A., Lothe, L., Mejlaender-Larsen, M., & Wergeland, T. (2013). *Shipping in Arctic waters: A comparison of the northeast, northwest and trans polar passages.* Springer.
Overland, J., & Stabeno, P. J. (2004). Is the climate of the Bering Sea warming and affecting the ecosystem. *AGU: Advancing Earth and Space Science, 85*(33), 309–316. American Geophysical Union.
Parlow, A. (2018, July 15). Trump-Putin should find common purpose on plan for the Arctic. *Anchorage Daily News.*
Parlow, A. (2019a, June 27). Does a Russia-China alignment in the Arctic have staying power? *Arctic Today.*
Parlow, A. (2019b, November 26). Why climate policy-makers should listen more to indigenous peoples. *Arctic Today.*
Policansky, D. (1996). *Fisheries and the Bering Sea ecosystem.* The National Academy Press and PBS.
Polyak, L. (2010). *History of sea ice in the Arctic, quarternary science review* (pp. 1757–1778). Elsevier.
Raymond-Yacoubian, J. (2018). *Arctic vessel traffic and indigenous communjities in the bering strait region of alaska, Kawerak, Inc.*, Nome, AK.
Rothwell, D. R. (1996). *The polar regions and the development of international law.* Cambridge University Press.
Sigler, M. F., Harvey, R. H., & Ashjian, A. (2010). *How does climate change affect the Bering Sea ecosystem?* AGU, Advancing Earth and Space Science.
Sikapizye, L. (2019). *Severe drought threatens food security in sub-Saharan Africa.* Alliance for Science. https://allianceforscience.cornell.edu/blog/2019/10/severe-drought-threatens-food-security-in-sub-saharan-africa/
Soltwedel, Thomas and Peeken, Ilka (2014), Fact sheet: The consequences of climate change for life in the Arctic.
Spies, I., & Kristen, M. G. (2019). Genetic evidence of a northward range expansion in the eastern Bering Sea stock of Pacific Cod. *Evolutionary Applications, 17 Oct*, 362–375.
Stein, R., Macdonald, R. W., & Stent, A. (2020). *US-Russia relations – Challenging policy.* Brookings Institute.
Stenson, G. B., & Hammill M. O. (2014). Can ice breeding seals adapt to habitat loss in a time of climate change? *ICES Journal of Marine Science 71*(7), September/October.
Stevenson, D. E., & Lauth, R. R. (2019). Bottom trawl surveys in the northern Bering Sea indicate recent shifts in the distribution of marine species. *Polar Biology, 42*, 407–421. https://doi.org/10.1007/s00300-018-2431-1
Sugie, K., Fujiwara, A., Nishino, S., Kameyama, S., & Harada, N. (2020). Impacts of temperature, CO_2, and salinity on phytoplankton community composition in the Western Arctic Ocean. *Frontiers in Marine Science, 6*, 821. https://doi.org/10.3389/fmars.2019.00821
Taniguchi, A., Kazutoshi, S., Akio, K., & Fukuchi, M. (1976). Phytoplankton communities in the Bering Sea and adjacent seas. *Journal of Oceanography, 32*(3), 99–106. https://doi.org/10.1007/BF02107038
Thackeray, C. W., & Hall, A. (2019). An emergent constraint on future Arctic Sea-ice albedo feedback. *Nature Climate Change, 9*, 972–978.

The Maritime Executive. (2017, January 1). U.S. *Coast guard bering strait route study*. Available. https://www.maritime-executive.com/article/us-coast-guard-bering-strait-route-study-available

Thoman, R. L. (2020). *The record low Bering Sea ice extent in 2018: Context, impacts and an assessment of the role of anthropogenic climate change*. The Bulletin of the American Meteorological Society.

Thoman, R. L. et al. (2020). *University of Alaska*. International Arctic Research Center.

Thompson, A. (2015, March 4). *Climate central: Arctic Sea ice is getting thinner, faster*.

University of Maryland Center for Environmental Science. (2020, November 11). Late-season Arctic research cruise reveals warm ocean temperatures, active ecosystem. *ScienceDaily*. www.sciencedaily.com/releases/2020/11/201111144356.htm

Van Pelt, T. (Ed.). (2015). *Bering Sea project: Bering ecosystem study: Understanding ecosystem processes in the Bering Sea*. National Science Foundation, North Pacific Research Board.

Wadham, J. L. et al. (2019). Ice Sheets Matter for global carbon cycle, *Nature Communications, 10*(1).

Wadhams, P. (2012). *Bering strait: Marine life, what goes on beneath the floes*. Alfred Wegner Institute.

Witze, A. (2019). Dramatic sea-ice melt caps tough Arctic summer. *Nature, 573*, 320–321. https://doi.org/10.1038/d41586-019-02653-x

Wuebbles, D. J. (2012). Celebrating the 'blue marble'. *Eos, Transactions American Geophysical Union, 93*(49), 509–510. https://doi.org/10.1029/2012EO490001

Chapter 9
Arctic Geopolitics, Cross-Boundary Soft-Power, Ecosystem Protection, and Human Security in the Bering Sea and Strait

Anita Parlow

Abstract The Arctic region serves as the Earth's thermostat. A combination of increasingly open waters, melting glaciers, and changes in ocean currents is heating the Earth and changing global weather patterns. This chapter focuses on a complex and interrelated mosaic of events in the Arctic, a region, once on the periphery of global commerce and geopolitics, the region is now moving closer to its center. This research focuses on the Bering Sea region, a northern promontory of the Pacific Ocean, referred to as the 'Gateway to the Arctic'. Here, global warming is causing a cascading series of changes to the marine ecosystem, including to Indigenous and coastal communities that are tightly connected to the Bering Sea ecosystem. Further, the unfolding geopolitical and commercial interests amongst Russia, China, and United States are, together, changing the face of the Bering region. The previously ice-bound, and globally accepted 'zone of peace', with its rules-based international and domestic order, offers a challenge to the region's ecosystem. Here, increasing Russian investments combined with Beijing's growing interest as an investor and consumer of Arctic natural resources raises significant questions for both strategic cooperation and ecosystem strategies even as the United States and Russia are increasingly projecting military power in the region. The United States and Russia demonstrably agree that cooperation on marine ecosystem policies, despite unresolved contentious boundary issues, is vital. However, perhaps paradoxically, the big power cooperation also raises vital questions for coastal communities that depend upon the bounty of the sea. How the industrial scale oil, gas, mining, and shipping that are accelerating damage to the Bering raises vital questions to the use of 'soft power' by the great Arctic powers who seek to protect the marine ecosystem for the long term.

Keywords Bering Sea · Arctic · St. Lawrence Island · Marine ecosystem · Russia-American Arctic geopolitics

---

A. Parlow (✉)
Fulbright Scholar Iceland, recent, Washington, DC, USA

M. Behnassi et al. (eds.), *The Climate-Conflict-Displacement Nexus from a Human Security Perspective*, https://doi.org/10.1007/978-3-030-94144-4_9

1 Arctic Amplification: Global Ocean Circulation

Sea ice in the Bering Sea affects water temperature, salinity, and ocean currents worldwide. The formation, movement, and melting of the ice edge have played a key and predictable role in controlling and moderating the heat exchanged between the ocean and the atmosphere (Van Pelt, 2015).

For many, the melting sea ice is disrupting the Arctic Ocean's primary transport system, known as the Transpolar Drift Stream, which moves newly formed sea ice from the waters of the relatively shallow Laptov and East Siberian seas, adjacent to the Russian coast, deep into the central Arctic. The currents carry a variety of sediments, nutrients, and other forms of microscopic life with them, making them essential regulators of the ocean's biology and chemistry (Harrison & Cota, 1991; Woodgate, 2013).

The increasing likelihood of ice-free summers by mid-century is a big concern for climate stability (Granskog, 2018). As sea ice disappears, it exposes more of the ocean's surface to the sun, allowing the water to soak up more heat. Researchers suggest that this process could contribute to a catastrophic feedback cycle, in which more ocean heat causes even more sea ice to melt, which leads to more atmospheric, thus planetary, warming (Thoman et al., 2020). Wadhams (2012, 2017) claimed that "the present thinning and retreat of Arctic Sea ice is one of the most serious geophysical consequences of global warming, and is causing a major change to the face of our planet".

The ocean constitutes nearly two-thirds of the Arctic, including the CAO and the seven peripheral Russian, Nordic, and Bering Seas. All are vital to moving the complex global oceans' circulation, with sea ice cover serving as a ballast to the inflows and outflows of water from the Arctic region to the world's interconnected oceans.

The world's interconnected ocean currents and wave movements serve as a conveyor belt that transports heat from the tropics to colder, higher latitudes, transports nutrients throughout the oceans and seas, regulating the climate. The circulation is driven by differences in water density, which is caused by differences in temperature and salinity. Cold water is denser than warmer water, and salty water is denser than fresh water. Warm equatorial waters circulate to the colder Arctic regions where they cool; as this colder, denser water sinks to the deep ocean, more warm water flows in to replace them. This process maintains the temperature of both the sea and the land (Siders et al., 2016).

1.1 The Oceans' 'Conveyor Belt' Weakening

While the combination of gravity, wind, and water density drives the world's ocean 'conveyor belt' is weakening, scientists are not yet clear about whether the ocean's circulation might stabilize at new levels. Some contemporary fluctuations resonate from weakening that occurred hundreds of years ago but might be accelerated by today's relative 'sharp and sudden surges' from sea ice and glacial melting as they cause rapid weakening of the currents that drive the oceans' circulation (Berwyn, 2018a).

As the melting Arctic glaciers and sea ice disrupt the ocean's conveyor belt, the larger amounts of lighter fresh water, less able to sink, mix with the heavier seawater to form the currents that stabilize the planet's temperature. While these changes in currents are long-term processes, measured by decades, if not centuries, scientists are unsure whether the changes will 'gradually persist' in the future, or shift to an unknown state (Berwyn, 2018b).

Circulation is effectively a 'handshake' between the Arctic and the southern hemisphere. Ocean circulation plays a significant role in climate regulation as it transfers a variety of properties within the atmosphere (Dickson et al., 2008; Struzik, 2017). Without the circulation of the Arctic gyres, the oceans lose equilibrium, causing the planet to warm. This shift in equilibrium is changing not only weather patterns but also is altering how nutrients and organic materials in the ocean mix and interact, with significant implications for the entire food chain, from microscopic animals to humans (Van Pelt, 2015). Ocean circulation maintains the stability of the oceans by transporting fresh water, oxygen, and nutrients through its currents.

The Global Ocean Conveyor changes as melting sea ice and glaciers add increasing amounts of fresh water to the salty ocean that, in turn, disrupts the ocean's circulation. The Atlantic Meridional Overturning Circulation (AMOC) plays a crucial role in climate through its transport of a combination of wind, water density, and heat (Jewett & Romanou, 2017; Berwyn, 2018a, b). Fresh water and heat exchanges from the North Atlantic and the Arctic Oceans, once thought to be a 'closed system', are now understood as expanding due to sea ice and glacier melt whose fresh water is being propelled into the world's interconnected oceans (Proshutinsky et al., 2015).

As the summer sea ice continues to shrink, the howling Arctic winds are left without natural barriers. As the glacial and sea-ice melts, the layers of salt and fresh water change, impacting the water's salinity, and currents. Celine Heuze, the co-lead of the science inquiry ice-breaker, the year-long MOASIC Expedition, the Multidisciplinary Drifting Observatory for the Study of Arctic Climate, was intentionally stranded in Arctic Sea ice in 2019, to allow the climate scientists to learn how the sea-ice life cycle, the ocean under the ice and, planetary sustainability are impacted by climate change.

The changes in the dynamics between fresh and salt water as sea-ice and glaciers melt have left some degree of uncertainty about what is the dominant driver of ocean circulation needed to maintain a sea-temperature range that, in turn, maintains a habitable temperature range for the planet. The Beaufort Gyre, for example, drives the Arctic's primary circulation and stores and transports fresh water, sea ice, and heat across the Canadian Basin (Hu et al., 2019). With its clockwise rotation, the gyre moves fresh sea ice and glacial water into the Arctic seas and, subsequently, the world's oceans.

Melting sea ice speeds up the gyre, which releases greater amounts of fresh water into the oceans. If the winds were to reverse direction and pull the gyre's fresh water counterclockwise, the gyre would release yet more fresh water into the sea. It is argued in Vizza (2020) that the accelerating rates of sea ice melt and the release of large amounts of fresh water into the oceans may cause winds currents that are unimpeded by Arctic Sea ice to significantly impact climate, shift weather patterns, and redistribute ocean nutrients worldwide.

The circulation allows the oceans to absorb heat, thus moderating higher temperatures from anthropogenic greenhouse emissions released on land while altering the fundamental physical properties of the ocean (NOAA, 2019). As the sea ice and glaciers melt, the gyres, composed of larger amounts of fresh water, dramatically change the ocean's salt-fresh water balance and thus the planet's ecosystems and weather patterns (Berwyn, 2018b).

According to Meier a dynamic current system may be described as stable when changes in the system remain small and within specific parameters. But when the existing systems move outside of their predictable range, possibly irreversibly, they become unstable, unpredictable, and, perhaps, dangerous to the environment (AMAP, 2017).

1.2 Feedback Loop and a Warming Earth: The Albedo Effect

Through a heat-producing 'feedback loop', the albedo effect, the planet's heat and weather patterns increase and accelerate catastrophic climate, societal, and commercial impacts (Laidre, 2008). Albedo is a measure of how reflective a surface is so that the bright light bounces off the ocean's surface, reflecting the heat away. Thus, the sunlight reflects away from the ice, rather than being absorbed by the surface (Tenenbaum, 2017).

However, NOAA stresses that air temperatures in the Arctic region over the past 5 years have been the five highest on record since 1900. With the annual Arctic temperature rising twice as fast as the global average, Viñas (2019) cites a report by NASA which acknowledges that the Arctic's Sea ice minimum is tied for the second-lowest record since satellite records of the 1970s. Given the combination of sea ice, sea ice retreat, and open oceans, the Arctic region increasingly amplifies global warming. The Arctic Ocean is magnifying an 'unstoppable' warming feedback loop in which melting sea ice sets the stage for more melting to occur.

The Arctic-sea ice has followed an annual cycle of melting through the warm summer months and refreezing in the winter. The feedback loop phenomenon occurs when the melted sea ice leaves a darker, open ocean that absorbs twice as much of the sun's heat than bright, white sea ice, thus warming the planet at accelerated rates. Operating as a global thermostat, sea ice cover keeps global temperatures at consistent levels. A balanced albedo offers stability to the planet's climate. The polar ice reflects the sun's energy back into space. The reflective sea ice maintains temperature stability by keeping the ocean from absorbing the rays of the sun (Tenenbaum, 2017).

The feedback loop accelerates or amplifies climate factors through a "cyclical chain reaction that repeats again and again" (Goosse et al., 2018). The larger the area of ice cover, the greater the amount of heat reflected out. But, when the retreating ice exposes more ocean water, darker and more absorbent than the reflective sea ice, a circular feedback heat cycle comes into play: more melting creates more extensive open waters and thus produces more surface heat (Knies et al., 2018). Given the feedback loop, the oceans absorb twice as much of the sun's heat as bright ice deflects, thus accelerating the planet's warming.

Just beneath the sea ice and, in open surface water, polynyas, an area of open water in sea ice generally located in the same area annually, (NSIDC, 2021) form a foundation for the complex marine life. The opening of ice-free waters is distinct from polynyas, pockets of open water that form at predictable annual seasonal intervals over decades or centuries. Over time, fish and mammal life has adapted to the annually occurring polynya locations for feeding (Watts, 2019). But now, irregular and enlarging polynyas are causing epic changes not only for the marine life that depends upon regularly open sea ice to feed but also to the Earth's atmosphere itself. Monroe and IGSD (2019) warn that losing the remaining Arctic-sea ice would be the equivalent of adding one trillion tons of CO_2 to the atmosphere, on top of the 2.4 trillion tons already emitted since the Industrial Age.

2 The Southern Bering: U.S. Jurisdiction

The Arctic seas experience the world's most extreme seasonal weather patterns. These, along with the winter seasonal ice cover, are the primary influences on the Bering Sea marine ecosystem (McIntyre & McDermott, 2010). Weather patterns are not always consistent in the Bering Sea's north and south, and the Bering's geographical northern and southern ecosystems differ accordingly.

The world's most sustainable commercial fishery is located in the southeastern Bering Sea. Here, the relative stability and predictability of fishing have been a constant in the enormously productive and healthy ecosystem where nutrient-rich waters of the benthos nourish a diversity of marine life, ranging from microorganisms to walrus, seals, and whales (Stotts, 2015).

The engine behind the Bering Sea's ecosystem protection is science. Alaskan authorities make decisions concerning catch and quotas based on science and deliberative processes regarding sensitive habitats that involve many stakeholders. Since 1976, the North Pacific Fishery Management Council (NPFMC) is one of the eight Regional Councils in the USA that manage U.S. fisheries, determining the allowable annual catch in each region. In 2006, the NPFMC created a series of six marine protected areas around the magnificent coral gardens in the Aleutian Islands. Two years later, the NPFMC established the Northern Bering Sea Research Area, which excludes bottom trawling from an area the size of the state of Arizona.

As the distribution of many of the Bering Sea's southern species and fisheries move northward, seeking colder waters, a new ecosystem is being created that may straddle the Arctic, the sub-Arctic, and the Northern Pacific oceans and seas. Integration and loss of new species, could either create vast and cascading extinctions or a new ecosystem capable of supporting new life. Some scientists believe the Bering Sea is becoming fully subarctic as the CAO oceans warm and integrate with the Bering Sea. According to the Alaska-based climate scientist Thoman (2020), who advises several Bering Sea tribal governments on climate issues, "it is easier to project 50 years hence than for next year" (Parlow, 2019a, b, c, d, e). Thoman also noted that the poleward movement of various species is requiring Bering Sea policymakers to redefine sustainable management strategies as species move, change and die-off.

2.1 *Fisheries' Poleward Movement: Anomaly or Permanent Shift?*

In 2019, NOAA scientists announced that they planned a study to determine whether the massive movement of fish was an anomaly, an early stage of a transition, or a trend of fish species moving north. For instance, Spies and Kristen (2019) recognize that the permanent disappearance of fish from the Bering Sea may strongly create a significant problem for both Alaskan fishers and American consumers who fully depend upon Alaskan fish as a source of seafood protein (Spies & Kristen, 2019). The prospect of such an eventuality is of acute interest to such entities, such as the Norton Sound Economic Development Corporation (NSEDC), located on the Seward Peninsula. With respect to moving species, some Northern Bering small-scale commercial fishers have expressed concern that the commercial trawlers, currently banned from the region, would lobby for access to the northern Bering as fisheries move north. In this respect, Medred (2016) notifies that the few Nome-based fishers – some 14 boats between 30 and 51 feet long – are concerned that the massive trawlers, with their billion-dollar revenues at stake, will invade the northern waters.

A year earlier, in 2018, the U.S. and Russia were the moving forces in creating the precautionary and preventive Agreement to Prevent Unregulated Commercial Fishing on the High Seas of the Central Arctic Ocean (CAO). The Agreement, with ten signatories, agree to prohibit fishing until scientists study the effects of climate change and the CAO's ability to sustain fisheries. This collaborative approach reinforced both the Arctic's cooperative sensibility reinforced by a precautionary approach that would close any fisheries until the creation of an internationally agreed-upon regulatory system regarding commercial fishing in the high seas – once the sea ice melts (Ovando, 2018).

At this writing, the northern region of the Bering Sea shelf remains closed to commercial trawling, allaying for the time being fishers' concerns about the destruction of the seafloor habitat by trawlers chasing the northward moving fish (Oceana, 2016; Puig et al., 2012). However, if cod or other species seek northern habitats, a political fight is likely to brew regarding trawlers, the measures needed to manage fish stocks sustainably, and who will receive quotas as mandated by current Alaskan and Federal legislation (Carothers & Zanotti, 2010).

The implications are yet unknown regarding the resilience and survival factors as the Bering Sea warms, thus complicating matters for the ecosystem, the commercial fishing industry, small-scale and subsistence fishers and hunters. As climate change is shaping the pace of their interrelated transformation, scientists, policymakers, and Bering Sea regional residents are increasingly collaboratively working to develop local solutions that can be integrated into an international network that simultaneously works from the ground up and from the top down.

2.2 *Northern Bering: U.S. Jurisdiction*

The Northern Bering Sea and the Bering Strait region count among the most productive high-latitude ocean ecosystems. The northern Bering is a migration corridor for thousands of bowhead and beluga whales, hundreds of thousands of walruses and seals, and millions of migratory birds. The numbers and presence of marine life in the region vary, depending upon the timing of seasonal sea ice retreat. In 2015 and 2016, the number of migrating birds declined precipitously as the U.S. Fish and Wildlife Service reported finding more than one million bird carcasses from California to Alaska, a reported largest die-off in recent Alaskan experience (Rosen, 2019; Piatt et al., 2020).

The northern Bering's coastal communities depend upon the Bering Sea for their food supply, including nesting birds and their eggs. For Akeya (2014), the Bring Sea provides much of the marine harvest for more than 55,000 Native Alaskans who live in more than thirty coastal communities along the Bering Sea, and for whom hunting marine mammals and fishing are central to producing their annual food supply.

A nuanced understanding of Arctic ecosystems and proximity to marine food resources are defining features of life for the Arctic's Indigenous peoples, whose resilience has developed over thousands of years. The Indigenous way-of-life of life, the seasons, and their cultures are fully integrated into the marine ecosystem. "Our belief is that we don't dominate nature, we live with it," explained several elders and tribal leaders (Apassingok et al., 1985).

The disruption to the Bering food chain, and thus the food supply, is acutely evident on St. Lawrence Island, a visible reminder of Beringia, the prehistoric land bridge that connected Russia and Alaska until about 10,000 years ago. St. Lawrence Island, located off the coast of Alaska, but closer to Russia, is primarily populated by Siberian Yupiks, whose language has many words for ice, its texture, and its freezing conditions (Krupnik et al., 2010). Evidence of warming is visible everywhere on the 1.2-million-acre island – both on and offshore. As the coastal villages are no longer protected by shore ice from the often violent storms, homes are collapsing from the loss of protective sea-ice to the shore. To illustrate more, Noongwook (2004) and Nutall (2005) say that thawing permafrost, collapsed roads, tipping electricity poles, and other dramatic changes require grueling efforts to simply navigate daily life.

Yupik Whaling Captain and walrus hunter, George Noongwook, former Chairman of the Alaska Eskimo Whaling Commission, said: the reduction in sea ice, from nine-month to three-month seasons means sea mammals no longer have a platform on which to rest and feed or nurse their calves during their migrations (Parlow, 2016). This not only interrupts the mammals' life rhythms but also the hunters' access to their food supply. Noongwook emphasized that despite the critical lost sea ice that upsets the entire food chain, Yupik hunters, fishers and whalers continue to navigate climate unpredictability with agile hunting strategies that have guided them for millennia (See the Foreword by Noongwood in this book).

At the same time that a variety of fish species are moving north in search of colder waters, the Bering is experiencing a breakdown of its cold pool, a 'cold curtain' that has long separated the distinct marine ecosystems of the Bering's north and south. The cold pool determines where different species are found by providing a cold-water barrier for cold-loving species such as Arctic cod or snow crab. Persistent warming heats the cold pool, attracting species that prefer warmer waters, like commercially important Pacific cod and pollock. Previously, these remained in the southeastern Bering Sea and its outer continental shelf. With variations in sea ice cover, cold and warmer waters 'scrambles', melting away the ice curtain, and redistributing fish that are moving outside of their traditional habitat to new regions that may or may not be able to sustain them (Mueter & Litzow, 2008).

Cornwall (2019) reported that in the previous year, the once-massive cold pool was the smallest ever documented area. Research ships located fewer fish in their usual marine locations. The report surmised that given the loss of much of the cold pool that separated southern and northern fish habitats, the fish likely 'sped north' to temperatures they can thermally tolerate. As the species move to their preferred temperatures in a new marine environment, some species benefit while others suffer. In this direction, Spies and Kristen (2019) and Duffy-Anderson et al. (2019) raised key questions concerning the ability of new habitats to keep up with the demands of the new species or if the species can adapt themselves to the living context before a possible die-off.

Perhaps, curiously, the fish species' shifts in location have had, in several instances, a positive impact. Orville Toodie (as cited in Parlow, 2019b), noted that in 2019, Savoongans caught a previously unheard of 60,000 pounds of halibut: "It's the most we ever caugh […] Usually they're not in our waters, not in so many numbers". But, along with the halibut, the predator density has also increased. Salmon sharks have moved to northern waters which appear to have reduced the abundance of the typical annual catch (Fenzi, 2015).

NOAA scientists who conducted research on the Bering have questioned whether the movement of fisheries is a temporary or long-term phenomenon. It is mentioned in Parlow (2019a, b, c, d, e) that Charles Mayo, Director of the Right Whale Program at the Massachusetts-based Center for Coastal Studies, has declared in a telephone interview that he took notice of the implications for changes in management practices. Particularly, Mayo said that "oceanologists must develop alternative theories and practices to help shape new ecosystems and navigate the systemic changes that are unfolding". However, Mayo acknowledged that the combination of the massive size, scope, and scale of what appears to be a transformational catastrophe in the Bering Sea offers a challenge for the ages (Parlow, 2016).

A trifecta of headwinds is impacting the Bering Sea fishing: first, a drop in biomass and declining numbers of fish with species movement that has led to lower quotas, and thus, commercial fishers with a smaller catch through 2021; second, Covid-19 pandemic pressures have reduced the number of fishers; and third, the shorter seasons have caused shoreside fish processers to go out of business, everyone, particularly the mom and pop fishers are receiving smaller quotas, as the diminished availability of fish and mammals for commercial and subsistence consumption is becoming a defining feature.

Reflecting on the changing nature of the Bering Sea, a former Tribal Chairman, preferring anonymity, described his family's enduring connection to place, along with a philosophy that requires integration with the natural world rather than domination of it. The witness testified in Parlow (2019a, b, c, d, e) that the Yupik people have much to share with the scientists, activists, and policymakers who must conceptualize a new paradigm for the twenty-first century and beyond. Importantly, Parlow (2019b) quotes the declaration of the tribal chairman as follows: "We have a message, he said, nodding his head, taking a deep sense of responsibility from Yupik ancestors who, he said, taught us everything".

3 Commerce and Environmental Impacts: Bottom Trawling and Sustainability

The shifting species is causing Alaska's commercial fishers concern as some species are increasing in abundance and others, decreasing. Alaska enjoys a reputation as one of the world's best-managed fisheries. Its State mandated responsibility to sustainable management practices is intended to ensure that all Alaskan commercially harvested seafood species are sustainable for future generations. Fishing resources are managed sustainably and, with the significant exception of the collapse of the pollock fisheries (Bailey, 2011), they generally renew annually. Whether sustainability will continue to be a policy goal is unclear. At both Federal and state levels, science-based sustainability policies are often in tension with commercial considerations regarding revenue-generating strategies. An increasingly consuming question is whether to bring in larger short-term catches, that could deplete the marine ecosystem for generations, or deepen adherence to long-term sustainability (Oliver, 2007).

3.1 Bottom Trawlers

Critics point out that industrial-scale bottom trawling negatively impacts fish, marine mammals, and their associated habitats by scraping and plowing the seafloor, which physically alters the physical structure of seafloor habitat. Given the broad continental shelf that defines the Bering Sea; deep-sea trawling has negatively impacted the seabed's life-sustaining biomass. About 30% of groundfish are caught with bottom trawl gear in the Bering region. The National Marine Fisheries Service reported in 2001 that about 50% of the groundfish were trawl-caught in the Gulf of Alaska, 27% near the Aleutian Islands (Stevenson & Lauth, 2019).

While bottom trawling accounts for about 25% of global marine fisheries landings worldwide (Amoroso et al., 2018) in Alaska's Bering Sea, about 8% of the bottom is trawled annually (Stevenson & Lauth, 2019). The enormous, weighted trawling nets, dragged across the sensitive benthic habitats, have disturbed the foundation to the Bering's marine ecosystem while scooping up far more than the commercially salable fish (Medearis, 2019a, b). The Alaska Marine Conservation Council has endorsed and expanded best practices for bottom trawling to protect the non-fish species whose habitat is found in the benthic level seafloor, the ecosystem of which supports numerous whale species, walrus, millions of migrating shorebirds, and a variety of seals (McDermott et al., 2016; Enticknap, 2002). The diminishing and fluctuating sea-floor habitat cascades up the food chain, impacting both species distribution and variances in weight of both fish and sea-mammals (Stevenson & Lauth, 2019).

The debate over the negative impacts of industrial-scale bottom trawling is currently shifting away from whether bottom trawling impairs or reduces marine biological diversity at the benthic levels, to where and to what degree it is appropriate (Enticknap, 2002). What is clear, however, is that the biomass displacements by trawlers, no matter how ecologically sensitive, continue to raise questions from Native fishers, scientists, and regulators about trawling's long-term ecosystem impacts (Siddon & Zaddor, 2017).

3.2 Marine Protected Areas

The National Marine Fisheries Service Initiative (Federal Register, 2008) designated 180,000 square miles in the Bering Sea as an 'essential fish habitat' in 2008, prohibiting trawling in the Northern Bering region. A coalition of more than 25 Bering Sea tribal governments and Alaska Native organizations supported the initiative. Opponents of bottom trawling in the northern Bering Sea sought to protect the ecosystem and the coastal Native villages, whose culture and food supply ultimately depends on a healthy Bering ecosystem (IPCC, 2020).

This unified view of marine ecosystem protection was embraced by the North Pacific Fishery Management Council (NPFMC), the Federal body charged with managing the Bering Seas' North Pacific fisheries, which voted unanimously to prohibit bottom trawling in the northern Bering. Whether this widely used but destructive and indiscriminate form of commercial fishing in various parts of the world, including the Bering Sea, will continue to be restricted is likely to become a contentious issue as warming draws various fish species, and thus the trawlers that pursue them, northward.

In addition to the issue of ecosystem protection and curbing overfishing through a science-based quota system, there is also the contentious issue of the distribution of quotas. The smallest fishers are often priced out of costly 'catch share' programs that allow the sale and trade of quotas to those left out of quota distributions. Although declining in numbers, small-scale, mom-and-pop fishers have been advocating for an approach that includes greater equity in the distribution of quotas along the lines of Iceland's community allocations for small-scale vessels, or exchange student licenses. Competition among various stakeholders – commercial fishers, small-scale fishers, sport fishers, and subsistence fishers, hunters, and whalers – is likely to increase as climate-related warming disturbances in the Bering Sea are causing a decline in numbers and weight.

4 Fish and Geopolitics in the Bering Sea

For much of coastal Alaska, local economies are characterized by the type and scale of fishing practiced. Commercial fishing offers the principal employment opportunity for coastal community residents and beyond. It is also a significant source of revenue for both the state and local governments (McDowell Group, 2017). In this sense, Welch (2016) reported that Alaska produces more than one half of the U.S. marine harvest and national fish consumption, with an annual value of its commercial catch exceeding $4.5 billion.

Alaska's largest commercial fish catches are taken from between 3 and 200 miles from the southern shores, the regulatory limits for state and Federal jurisdiction, respectively. Shared geographically by coastal waters of Russia, the United States, and also international waters, the rich Bering Sea hosts U.S. commercial fisheries (Byers, 2010; Welch, 2016).

According to Seafood Source, Alaska has produced more than 169 billion pounds of fish since it achieved statehood in 1959 (Fisher, 2017). For a sense of scale, the McDowell Group (2017) reported that the Alaskan seafood industry brought in enough harvest annually to feed every person in the world a serving of Alaskan seafood. Alaska's commercial fishing industry is also a major supplier of commercial fish abroad. According to McDowell Group (2017), the majority of Alaska seafood is exported (66%), and Alaska seafood can be purchased in 160 countries around the world. As a result, this brings more than $3 billion into the U.S. economy, and Alaska's fishery is more valuable than all other U.S. fisheries combined (Fisher, 2017).

Geopolitics impact Alaska's commercial fisheries. Alaska and Russia harvest many of the same fish and crab species. In 2018, it was reported that the United States imported $551 million of seafood from Russia, while from China it imported $50 million of pollock taken in Russian waters. A squeeze on Alaskan commercial fishing occurred in 2014 when Russia placed an embargo on all food products, including Alaskan seafood, in retaliation for United States sanctions on Russia following its invasion of Ukraine. The United States opted not to retaliate and, instead, the following year the country increased its imports to more than 80.2 million pounds of Russian seafood, including 16 million pounds of red king crab valued at $293 million and frozen sockeye salmon valued at $16.7 million (Welch, 2016). In 2020, the Alaska Seafood Marketing Institute reported that Russia continues to undercut prices in U.S. markets for the same fish given a more lax marine regulatory regime. Additionally, in 2014, Russia banned all U.S. products, including Alaskan seafood in retaliation for U.S. sanctions that at that time accounted for $61 million in U.S. sales to Russia, thus substantially hurting Alaskan commercial fishers (Welch, 2021).

Warren (2019) claimed that the U.S. tariff dispute with China poses a risk to commercial fishing's reprocessing markets. An increased burden to Alaskan fishers emerged when, in response to the ongoing tariff wars with the United States, Beijing switched China's fish suppliers from the United States to Russia. This continues to cause distress to several Alaskan commercial fishers who spent much of the past decade patiently developing a Chinese market that is now being filled with Russian fish. "I don't know when it will come back," one life-long commercial fisher reflected (Parlow, 2019a).

4.1 Russia, China, and the United States in the Bering Sea

Nome Offshore: The Bering Sea in Winter, 2019. (Source: Anita Parlow © 2019)

Another source of anxiety to commercial and subsistence fishers, hunters, and whalers are the joint oil and gas agreements between China and Russia, which are tied to Russia's interest in increasing commerce along the Northern Sea Route and China's Polar Silk Route initiative. The expansion of oil and gas production along the Northern Sea Route is attracting growing numbers of large tankers and support vessels that ship bulk infrastructure northward and petroleum south, much of it to Chinese ports (King, 2020; Berkman et al., 2016).

The State Council Information Office of the People's Republic of China (2018) announced its intention as a 'Near Arctic State' to create a 'Polar Silk Route', thus adding an Arctic dimension to its massive, and environmentally controversial, Belt and Road Initiative, sometimes referred to as a "debt trap". As part of their mutual natural resources development interests, Presidents Vladimir Putin and Xi Jinping, in 2018, vowed to pursue economic efforts more closely. During this time, State Street Global Advisors reported that Russia and China were considering 73 joint investments worth more than 100 billion dollars.

The implications for both geopolitics and the protection of the intricate Bering Sea's marine ecosystem are enormous. Nearly 92% of the Bering Sea lies within the contiguous territorial waters and Exclusive Economic Zones (EEZs) of these two countries. Given the American and Russian interests in protecting their security and ecosystem interests in the Bering Sea, it is vital that Moscow and Washington project their soft power through increased science-based policymaking than military aerial ballet's that appear to be increasing.

As commerce and natural resource developments have expanded in recent years, NGOs, some governments and investment banks have expressed increasing interest in conservation policies regarding the marine ecosystems. Considerable interest is also emerging from non-coastal states, such as China, regarding what governance regimes, and thus non-coastal state influence, will emerge regarding international waters, in both the Bering Sea and the CAO (Berkman et al., 2016; Liu, 2020).

The perpetual existence of fish in the Mediterranean-sized CAO as an open water area remains unknown. Unprecedently, global warming continues to melt the Arctic's Sea ice which, in turn, rapidly affects the environment on many levels. No doubt, the impact is felt in the marine ecosystem and commerce. Henceforth, the need for regulatory measures to anticipates the impact of warming on development and commercial activity is becoming urgent. As a matter of fact, the regulation of shipping and the exploitation of natural resources in the Bering region require a multidimensional, multi-stakeholder, and precautionary approach to ensure low-carbon and energy-efficient shipping that minimizes risk to the rich ecosystem (Siders et al., 2016). As far as the Bering Sea is concerned, Hoag (2017) mentions that the United States and Russia are putting forwards urgent measures to reinforce their right and responsibilities over the area. More than that, Hoag adds that the two countries are showing a strong commitment to the protection and development of the Bering Sea region to ensure its environmental and political stability.

4.2 USA and Russia's Bilateral Arctic Cooperation

The bilateral cooperation between Russia and the United States in the Bering region and on the Arctic Council remains one of the best examples of cooperation between the two nations. Despite the recurrent tensions between the countries on other issues, the Arctic remains a zone of cooperation. Following the Cold War, Soviet President Mikhaïl Gorbachev called the Arctic a 'zone of peace'. His statement settled the rules of mutual respect and cooperation between the USA and the USSR. Given the coinciding interests in commercial fishing, shipping, and a healthy marine ecosystem, along with pressures by NGOs to develop an Arctic-wide standard for oil and gas development, the governance of the region appears to be moving in a cooperative direction. Notwithstanding the efforts done by the two countries to preserve the area, we believe that support and subsistence of the Indigenous communities are highly required.

Indeed, not all are competitive regarding the Bering Sea. Multilateral governance instruments such as the Arctic Council and a variety of treaties have moderated American-Russian policies, through their leadership, despite adversarial postures in other world regions (Byers, 2010). As the Bering is a prime location for illegal, unregulated, and unreported fishing, for example, Russia and the United States agreed to curb fishing by outsiders without permission. In 2015, the two countries signed a bilateral Agreement to Combat Illegal, Unreported, and Unregulated Fishing, the purpose of which is to protect the two nation's respective Bering Sea fisheries against ongoing large-scale illegal and, heretofore, unregulated fishing. As a practical matter, the agreement has much improved coordination between the relevant government agencies in the two nations.

Both the USA and Russia were the moving forces in the development of the 2018 multilateral Agreement to Prevent Unregulated High Seas Fisheries in the Central Arctic Ocean (Soyer et al., 2018; Schönfeldt, 2017). This imposed a moratorium to prevent overfishing in the rapidly warming Central Arctic Ocean (CAO) until a sustainable plan is agreed upon. Arctic coastal nations, whose EEZs will be affected by an ice-free Mediterranean-sized CAO, are concerned about the waters' vulnerability to overfishing by industrial fleets. Importantly, Highleyman (2017) noted that this agreement is the first multilateral initiative to take a legally-binding, preemptive approach to protect an area from commercial fishing before fishing has even begun.

Russian-American cooperation on fisheries management and jurisdictional issues in the Bering Sea leaves one with a meaningful degree of optimism. The sensibility of the CAO agreement supports a much needed 'precautionary' approach both in theory and practice, reinforcing a shared interest in the health and resilience of the Bering Sea's marine ecosystem (Honneland, 2004; Stein et al., 2020). Although the two coastal states in the Bering Sea are legally responsible for regulating their respective offshore activities, the Bering Sea's marine life knows no political boundaries (Gerden, 2020; Young et al., 2020).

Despite cooperation regarding the Bering ecosystem, the region is increasingly the subject of security concerns, as evidenced by episodic aerial buzzings, generally in international skies. Russia has deployed strategic 'Bear' bombers for reconnaissance and training flights, which are closely shadowed by U.S. F-22 and F-16 fighter jets. In turn, the United States has increased the tempo of its training exercises in Arctic locations, including the Sea of Okhotsk, which is surrounded on three sides by Russian territory. Within this framework, Correll et al. (2018) mention that amid context of mutual distrust, observers who are designed to protect power balance and marine boundaries, must demonstrate and express high concern for potential accidents or misinterpretations.

4.3 Marine Boundary Issues

A key issue that remains to be resolved regarding the Bering Sea, is the enforcement of the 1990–1992 agreement on the delimitation of the U.S. and Russian boundaries in the Bering Sea. When the United States purchased Alaska from Russia in 1867 (the 1867 Treaty), national exclusion zones were defined by the three-mile limit. The 1982 United Nations Convention on the Law of the Sea (UNCLOS) created EEZs out to 200 miles. Although the United States has not ratified the UNCLOS, it adheres to its provisions and it did ratify the 1992 USSR-USA Maritime Boundary Agreement, which established a line of demarcation between the two countries' EEZs related to the broad continental shelf in the Bering Sea. The United States ratified the agreement, but the Russian Federation, successor state to the Soviet Union, did not. In 2020, the Kremlin expressed interest in revising the unratified agreement regarding the maritime boundary. By 2020, the combination of Russia's growing fishing interests and the Kremlin's expressed desire to expand shipping along the Northern Sea Route and, thus, the Bering Sea suggests that the Russian government is disinclined to seek U.S. approval for Russian-related transits. For the time being, the two powers have agreed to disagree.

A widely held Russian view is that the then Soviet Foreign Minister gave up too much to "curry favor with the West" (Goble, 2020). Another commonly held belief is that such an issue should not be allowed to destabilize an essentially positive relationship between the two great powers in the Arctic 'Zone of Peace'.

The seeming paradoxical combination of the marine ecosystem and related agreements in and around the Bering Sea, and the 'peer competition' flyovers along with Russian and U.S. military exercises leave some concerned that a mishap, mistake or, crossing a red line could lead to a new Cold War in the region. However, in the 2021 Geneva-based summit between the American and Russian presidents, in a meeting designed to establish a dialogue for greater understanding and predictability for the two powers' strategic interests, the Russian President appeared to note that the Arctic is Russia's red-line, possibly offering a potential for strategic cooperation rather than muscle-flexing or jurisdictional flashpoints. Given the mutual security interests in the region, the summit offered the potential for a secure communication architecture that would not turn an adversarial relationship into an enemy in a zone of peace. A state Department official wrote to me of Arctic security concerns, "Our intent is to engage Russia in ways that advance our interests while remaining very clear-eyed about the challenge it poses" (Parlow, 2021).

4.4 *Simultaneously Ominous and Hopeful*

In 2020, two events occurred that highlight the challenges to the Arctic's ecosystems posed by oil and gas extraction, mining, and shipping. First, a fuel tank collapsed from thawing permafrost and leaked 20,000 tons of diesel oil into Russia's Ambarnaya River within the Arctic Circle, one of the largest spills in modern Russian history. According to Nechepurenko (2020), the collapse led the Russian President to declare a state of emergency. The Russian investigative committee detained the plant's manager and opened a criminal inquiry. Greenpeace Russia compared the discharge to the 1989 Exxon Valdez spill in Alaska, and one could also liken it to the BP spill off the U.S. Gulf Coast in 2010. However, unlike these two examples, which arguably resulted from corporate inattention, the fuel tank collapsed due to thawing permafrost, as one of the most visible manifestations of global warming. This, and subsequent collapses of infrastructure, along with spontaneous explosions from underground reported in the Russian Arctic caused by subterranean gases are quite likely related to global warming, offering another glimpse into a grim Arctic future (van Huissteden, 2020; Kramer, 2020).

Coincidentally, on the same day that the spill was reported to the Kremlin (albeit 2 days after it happened), a forward movement for marine ecosystem protection in the Arctic was made in a phone meeting convened by the Woodrow Wilson International Center for Scholars in cooperation with the Russian International Affairs Council (RIAC) to discuss the joint development of an ecosystem approach to protecting the Arctic's rivers, oceans, and seas (Wilson Center, 2020). Taking the measure of the political situation in the United States, the Russian and American participants agreed to small but meaningful steps to 'build trust' and hope for an improved political climate that would allow for the implementation of a science-based agreement to manage, protect, and respect the Arctic marine ecosystem.

5 Oil, Gas, and Shipping

The Arctic is, indeed, a resource-rich region. According to the U.S. Geological Survey (USGS), the Arctic region contains about 30% of the world's undiscovered gas and some 13% of the world's undiscovered oil – mostly offshore but accessible – and untold quantities of minerals both on and offshore. Russia's vast undiscovered petroleum reserves are estimated at between 66 billion tons of oil equivalent (BTOE) – accounting for about 52% of Arctic reserves. USGS estimates situate more than 87% of the resource (360 billion barrels oil equivalent) into seven Arctic EEZs, primarily the Russian Arctic, the United States, and, to a lesser extent, Canada (USGS, 2008; Kraska, 2011; King, 2020).

The Russian Federation has more at stake in the Arctic than most of the other Arctic nations. Its sovereign territory encompasses more than one half of the Arctic coastline, about 32% of its landmass, and is home to about 2 million people, roughly half of the total Arctic population. As Russia balances its renewed security concerns on an emerging and unprotected northern flank, it has also sought joint project financing from the Chinese capital. While the Kremlin is especially aware of Belt and Road 'debt traps', and the potential to be viewed as a junior partner in the relationship, its leadership is taking steps to protect its interests (The Economist, 2019).

The increasing trade in energy, despite global climate concerns, along with the growing availability and requirements for Russian icebreaker support for the projected increases in tankers and other vessels along the NSR, reflects that shipping and the growing offshore oil and gas trade are Russia's main sources of revenue. Henderson and Loe (2014) report that the Energy Agreements provide Russia with a sizeable market for its only real export. In return, Beijing, a primary investor and consumer, is gaining both energy resources and a foot in the previously inaccessible and a growing geo-strategically significant region.

With its impressive fleet of some 42 icebreakers and ice-capable ships, Russia's most visible – which some describe as 'contentious' – actions are found in its new rules to keep control over the growing commercial shipping – and naval transits – while protecting the marine ecosystem in its coastal waters. The Kremlin has indicated that China will not be exempt from its new regulatory requirements, including requirements to hire Russian icebreaker escorts, to have a Russian pilot on board, and to give advance notice of Northern Sea Route transits. Nor will Chinese vessels be offered the tax or fee reductions available to Russian ships (Solski et al., 2020).

5.1 China's Polar Silk Route Initiative

With Russia as a primary exporter and China a major producer and consumer of energy resources, the two countries' economic relationship is likely to grow in the Arctic region, at least in the short and medium term. Through its Polar Silk Route initiative, China is also becoming an increasingly dominant investor in the Arctic as it seeks to gain access and refine its influence as a global power, including in the north (Gavin, 2021). Beijing's interest in Russia's substantial oil and gas reserves along with its new investments in icebreakers and other Arctic infrastructures are indicative of China's intention to have a long-term political presence in the region. Whether Beijing's presence contains the potential to shape a new direction for the region's law and policy, as the CAO opens, raises significant questions regarding geopolitics, ecosystem protection, and law (Brady, 2017; Hsiung, 2020).

5.2 *Balancing Commercial and Marine Ecosystem Protections*

Growing choruses of environmental advocates have observed that the Arctic offshore oil and gas projects present unique challenges to an industry already under fire. These include strategies to protect marine and terrestrial ecosystem health that depends upon healthy seas for a viable commercial, subsistence future (Thorsteinson & Love, 2016). As a combination of conflicting policies, practices, and views by NGOs, governments and, commercial interests, regarding natural resource development become increasingly widespread both within and about the Arctic, virtually all stakeholders agree that the conventional approaches to the development of the Arctic's natural resources are no longer possible (Howard, 2009).

Given the growing political pressures, impacts on climate and marine ecosystems, and the fact that the development of the Arctic's oil and gas sector is risky, expensive, and controversial, some major U.S. and global investment banks have recently decided to withdraw or refrain from Arctic investments. Responding to pressures by coalitions of Indigenous and environmental groups, some leading investors have reduced their exposure in a region that many deem 'ground zero' for climate change.

Even China-critics note that given Beijing's long-range strategic interests for influence in the Arctic region, the government is currently acting within the framework of the relevant international and domestic laws of the coastal nations. But, given China's increased economic engagement on both the Atlantic and Pacific sides of the Arctic region, and its intention to build its fleet of ice-capable container ships and nuclear-powered icebreakers, the "near Arctic" nation's long-term strategy is not yet clear. Beijing's willingness to accommodate Russian law and policy that regulates and protects its sizeable northern coast, is indeed a sign of cooperation. An open question is to what extent China will want to expand its influence.

Up to now, no one is sure of China's intentions in the area despite the dominant role that the country plays in developing the new rules that are likely to be established as the sea ice retreats in the Central Arctic Ocean. It seems that some degrees of pressure put forward by China may increase global power in the Arctic, akin to the South China Seas, to seek and shape its preferred policy outcomes.

5.3 *Vessel Traffic*

The U.S. Committee on Marine Transportation Systems (CMTS), an interagency committee authorized by the U.S. Coast Guard, projected in 2019 that Arctic vessel traffic will grow somewhere between 48% and 70% from its 2008 baseline by 2030. A combination of shipping to transport oil, gas and ore, increased tourism, expansion of Arctic fleets to increase Polar security, ships carrying infrastructure supplies and seasonally rerouted vessels, are listed as the likely sources of growth. The 2009 Arctic Marine Shipping Assessment Report (AMSA), viewed as the preeminent source for understanding Arctic shipping, had previously reported that the greatest source of projected steady growth in Arctic shipping is likely to be in the oil, gas, and mining industries. According to AMSA, the "development of the rich natural resources in the Arctic is a rapidly growing industry" (Arctic Council, 2009; Farré et al., 2014). Warming has accelerated the numbers and sources of projected shipping increases.

In 2018, the International Maritime Organization approved a joint proposal by the United States and Russia to establish two-way shipping traffic corridors across the Bering Sea, as multiyear ice gives way to far thinner yearly ice, which will likely encourage increased shipping. The voluntary new shipping lanes permitted both countries to view the agreement as a step for a more comprehensive traffic-management route system that will essentially ensure the safety of navigation and assert jurisdiction and project power in the Bering region.

Beyond the planetary issue of global warming, the expansion of commercial opportunities in the Arctic is increasing the number of environmental threats to the region. For example, the combination of invasive species that often accompanies shipping and the lack of essential infrastructure to respond to oil spills have the potential to disrupt the ecosystem in ways that could accelerate climate change-driven impacts to fisheries, mammals, birds, and humans.

5.4 Icebreakers

The projected growth is indisputable. China, for example, has already sailed an icebreaker toward the North Pole, invested in Russian liquefied natural gas (LNG) projects along Russia's Northern Sea Route, and is building ice-capable container ships designed to operate on the Polar Silk Route with its two icebreakers. Its Polar Initiative is intended to connect the Pacific and Atlantic sides for its ambitions for transit shipping through the Arctic's fragile yet rugged waters (Alexeeva & Lasserre, 2018).

With its investments in natural resources, commercial fishing, infrastructure, and tourism, combined with its goal of integrating Arctic waters into its global transportation and trade strategies, China's interest opens a broader question of what operative legal and regulatory standards will come into play regarding the protection of the marine environment, climate change, and coastal and Indigenous communities (Gautier, 2019; Hsiung, 2020).

At a 2019 forum in St. Petersburg, the Russian President announced an ambitious program to expand Russia's program in the Arctic and to encourage international investment. He announced his intention to "lower profit tax rates…and reduce severance tax" to promote the development of the Arctic region with investment stability (Isachenkov & Titova, 2019). The President's invitation to accelerate oil and gas development and expand shipping capacity highlighted a sense of urgency to protect the diverse marine ecosystem of the Bering region (Ebinger et al., 2014).

Russia plans a substantial fleet of new icebreakers to help clear the way for the tankers that follow them to and from ice-free waters. Two of the new icebreakers will be nuclear powered – which makes them in some respects more eco-friendly than fuel oil-powered ships – and capable of breaking through four-foot thick ice. Russia is also building the world's most powerful diesel-engine icebreaker. It will be able to operate autonomously for 60 days in ice more than six feet thick.

The first icebreaker of this kind was built by the Russian state-owned company, Rosmorport. The Russian oil industry is likewise building more icebreakers to operate in and around Sabetta Port, home to the enormous Chinese-financed Yamal LNG project, part of China's ambitious Polar Silk Route, which will increase shipping traffic in the Bering. With the opening of Yamal LNG and the future Arctic LNG 2 facility, the U.S. Coast Guard expects more than 1000 transits through the United States' EEZ, within the next 5 years (Humpert, 2020).

In 2018, before the St. Petersburg forum, the American President opened the energy-rich Beaufort Sea, just north of the Bering Strait, to offshore drilling in federal waters. Environmental organizations quickly filed a lawsuit to block the administration's plans to "turn our oceans into oilfields", which, they said, is "really scary for the United States and the world" (Plumer & Fountain, 2020). The pending litigation claims that this offshore petroleum development violates U.S. laws, ignores the causes and effects of climate change, and has neither provisions for nor the ability to clean up an oil spill in the turbulent yet fragile Arctic waters.

Further, Savoonga Tribal Chairman and ivory carver, Ben Pungowiyi, said that in 2021, the pristine waters that surround St. Lawrence Island, have been inundated with plastic, metal, cardboard, and food containers. Pungowiyi said in a telephone interview, "there's debris washing on our shores that looks like it's coming from passing ships; more dead birds, and it looks like it's getting worse" (Parlow, 2021).

6 Conclusion

Melting Arctic Sea ice is creating a combination of problems, challenges, and opportunities in the climate-stressed High North. The core commercial activities in the region include natural resources development and its related shipping. Scholars, such as Timo Koivurova, predict that the Arctic region contains the potential to become a major global source of oil, gas, and other raw materials of critical importance to high energy-consuming nations, despite growing global initiatives to decarbonize. The Arctic is a region that is pre-eminent in building energy production balances and 'innovative approaches' to new forms of energy development. For example, Iceland's geothermal initiatives have been woven into the United Nations' clean energy training programs.

The importance of energy and shipping has moved the Arctic from a global geopolitical periphery, closer to the center of global economics and politics. With an area of ice six times the size of California already vanished, the region is drawing interest from both Arctic nations and nations with no territorial claim in the region.

Hydrocarbon resources are increasingly being shipped out of the Arctic – mainly from Russia's Northern Sea Route (NSR) to China. South-to-north vessel traffic is also likely to increase as Chinese tankers and support vessels sail to Russia's NSR to develop large-scale natural resources projects such as Yamal. This raises 'vexed' questions, both for the environment and geopolitics, regarding energy development.

Russia, currently the Chair of the Arctic Council, has launched its 2021 term under the theme 'Responsible Governance for a Sustainable Arctic'. Whether Russia, the largest Arctic nation with the greatest ice capabilities, will use this important multilateral forum to produce an Arctic-wide strategy regarding climate, ecosystems protection, and carbon fuel development remains to be seen. However, if there's a moment, despite the contentious geostrategic issues outside of the Arctic region, and within Russia itself – the time is now. And, Russia is, indeed, the power to move a human security agenda forward.

While the climate issue is an existential threat to human beings worldwide, its immediate threat weighs a lot on the Bering Sea where native communities' health and well-being depend upon what they hunt and fish. To ensure effective, durable, and sustained development in the region, the Indigenous populations must voice their suggestions and contribute massively to the policy measures undertaken by international actors when it comes to the management and governance of the region on the levels of security, political decision, and environmental governance.

References

Akeya, D. (2014). *White sharks of the north: The cycle of Arctic life of four generations of polar bears seen through the observant eyes of a Siberian Yupik Alaskan Eskimo. Village of Savoonga, St. Lawrence Island*. The Catchall Press.

Alexeeva, O. V., & Lasserre, F. (2018). An analysis on Sino-Russian cooperation in the Arctic in the BRI era. *Advances in Polar Science, 29*(4), 269–282. https://doi.org/10.13679/j.advps.2018.4.00269

Amoroso, R. O., et al. (2018). Bottom trawl fishing footprints on the world's continental shelves. *PNAS, 115*(43), E10275–E10282. https://doi.org/10.1073/pnas.1802379115

AMSA. *Identification of Arctic marine areas of heightened ecological and cultural significance: Arctic Marine Shipping Assessment (AMSA) IIc*. Arctic Monitoring and Assessment Programme (AMAP/CAFF/SDWG), Oslo, p. 36.

Apassingok, A., Walunga, W., & Tennant, E. (Eds.). (1985). *Lore of St. Lawrence Island echoes of our Eskimo elders*. Bering Strait School District.

Arctic Council. (2009). *Arctic Marine Shipping Assessment 2009 Report*. https://www.pmel.noaa.gov/arctic-zone/detect/documents/AMSA_2009_Report_2nd_print.pdf

Arctic Monitoring & Assessment Programme (AMAP). (2017). *Snow, Water, Ice and Permafrost in the Arctic (SWIPA)*. Arctic Monitoring and Assessment Programme (AMAP), Oslo, Norway, xiv + 269 pp.

Bailey, K. M. (2011). An empty donut hole: The great collapse of a North American fishery. *Ecology and Society, 16*(2), 28.

Berkman, P. A., Vylegzhanin, A. N., & Young, O. R. (2016). Governing the Bering Strait region: Current status, emerging issues and future options. *Ocean Development & International Law, 47*(2), 186–217. https://doi.org/10.1080/00908320.2016.1159091

Berwyn, B. (2018a, October 31). Global warming is messing with the jet stream. *Inside Climate News*.

Berwyn, B. (2018b, May 7). Scientists say ocean circulation is slowing. *Inside Climate News*.

Brady, A.-M. (2017). *China: As a polar great power* (p. 87). Cambridge University Press.

Byers, M. (2010). *Who owns the Arctic?* Douglas & McIntyre.

Carothers, C., & Zanotti, L. (2010). *In a climate of change: Co-producing knowledge and community-researcher relationships in the leadership and strength project in Utqiagvik, Alaska*. Purdue University, ARCUS, Arctic Research Seminar.

Corell, R. W., et al. (Eds.). (2018). *The Arctic in world affairs: A North Pacific dialogue on Arctic 2030 and beyond: Pathways to the future, 2018 North Pacific conference proceedings*.

Cornwall, W. (2019). Vanishing Sea ice poses climate puzzle. *Science, 364*(6441), 616–617.

Dickson, R. R., Meincke, J., & Rhines, P. (Eds.). (2008). *Arctic-sub-Arctic Ocean fluxes: Defining the role of the northern seas in climate* (p. 77). Springer.

Duffy-Anderson, J. T., et al. (2019). Responses of the northern Bering Sea and southeastern Bering Sea pelagic ecosystems following record-breaking low winter sea ice. *Geophysical Research Letters*. https://doi.org/10.1029/2019GL083396

Ebinger, C. K., Banks, J. P., & Schackmann, A. (2014). Offshore oil and gas governance in the Arctic: A leadership role for the U.S. *Brookings Energy Security Initiative, 14*, Policy brief No. 1.

Enticknap, B. (2002). *Trawling the North Pacific: Understanding the effects of bottom trawl fisheries on Alaska's living seafloor*. Fisheries Project Coordinator, Alaska Marine Conservation Council.

Farré, A. B., et al. (2014). Commercial Arctic shipping through the Northeast Passage: Routes, resources, governance, technology, and infrastructure. *Polar Geography, 37*(4), 298–324. https://www.tandfonline.com/doi/full/10.1080/1088937X.2014.965769

Federal Register. (2008, July 3). *Fisheries of the exclusive economic zone off Alaska; groundfish fisheries of the Bering Sea and Aleutian Islands Management Area*. A proposed rule by the National Oceanic and Atmospheric Administration, 73 FR 12357, pp. 12357–12366. https://www.federalregister.gov/documents/2008/03/07/08-988/fisheries-of-the-exclusive-economic-zone-off-alaska-groundfish-fisheries-of-the-bering-sea-and

Fenzi, F. (2015, May 12). *Chinook researchers encounter unlikely predator in Bering Sea*. AK Alaska Public Media. https://www.alaskapublic.org/2015/05/12/chinook-researchers-encounter-unlikely-predator-in-bering-sea/

Fisher, B. (2017). *Calculating the value of Alaska seafood*. Seafood Source.

Gautier, A. (2019, December 10). *Bering witness: Alaskan elders give voice to the NOAA Arctic report card, NSIDC highlights*. National Snow and Ice Data Center. https://nsidc.org/nsidc-highlights/2019/12/bering-witness-alaskan-elders-give-voice-noaa-arctic-report-card

Gavin, G. (2021, January). Russia's romance with China is all about keeping up appearances – Closer ties between Moscow and Beijing are worth more as a threat than a reality. *The Diplomat*. https://thediplomat.com/2021/01/russias-romance-with-china-is-all-about-keeping-up-appearances/

Gerden, E. (2020). *Russia unhappy with US fisheries agreements*. Russia and Bering Boundaries, Fishermen's News.

Goble, P. (2020, January 30). Moscow may soon end 'Provisional Enforcement' of 1990 Bering Strait Accord with US. *Eurasia Daily Monitor, 17*(12). https://jamestown.org/program/moscow-may-soon-end-provisional-enforcement-of-1990-bering-strait-accord-with-us/

Goosse, H., et al. (2018). Quantifying climate feedbacks in polar regions. *Nature Communications, 9*, 1919.

Granskog, M. (2018). *Exploring a more dynamic Arctic icescape*. EOS, 99 and Norwegian Polar Institute.

Harrison, W. G., & Cota, G. F. (1991). *Primary production in polar waters: Relation to nutrient availability*. Polar Res.

Henderson, J., & Julia, L. (2014). *The prospects and challenges for Arctic oil development*. The Oxford Institute for Energy Studies.

Highleyman, S. (2017). *Central Arctic Ocean: Next steps on the Arctic's newest international agreement*. See also: Nations agree to ban fishing in Arctic Ocean for at least 16 years, Oceans Conservancy.

Hoag, H. (2017). *Nations agree to ban fishing in Arctic Ocean for at least 16 years*. AAAS-Science. https://www.sciencemag.org/news/2017/12/nations-agree-ban-fishing-arctic-ocean-least-16-years

Honneland, G. (2004). Russian fisheries management. In *The precautionary approach, publications on ocean development* (Vol. 43). Martinus Nijhoff Publishers.

Howard, R. (2009, November 3). *The Arctic gold rush: The race for tomorrow's natural resources* (1st ed.). Continuum.

Hsiung, C. W. (2020, May 19). *The emergence of a Sino-Russian economic partnership in the Arctic?* The Arctic Institute. https://www.thearcticinstitute.org/emergence-sino-russian-economic-partnership-arctic/

Hu, X., Myers, P. G., & Lu, Y. (2019). Pacific water pathway in the Arctic Ocean and Beaufort Gyre in two simulations with different horizontal resolutions. *American Geophysical Union*. https://doi.org/10.1029/2019JC015111

Humpert, M. (2020). As Arctic ice melts Russia conducts record-breaking delivery of LNG to China. *High North News*. https://www.highnorthnews.com/en/arctic-ice-melts-russia-conducts-record-breaking-delivery-lng-china

IPCC. (2020). *IPCC special report on the ocean and cryosphere in a changing climate* (H.-O. Pörtner, D. C. Roberts, V. Masson-Delmotte, P. Zhai, M. Tignor, E. Poloczanska, K. Mintenbeck, A. Alegría, M. Nicolai, A. Okem, J. Petzold, B. Rama, & N. M. Weyer, Eds.). https://www.ipcc.ch/site/assets/uploads/sites/3/2019/12/SROCC_FullReport_FINAL.pdf

Isachenkov, V., & Titova, I. (2019, April 9). *Putin outlines ambitious Arctic expansion program*. Associated Press (AP). https://apnews.com/article/d0c2eb39a3b44b40ac8ddb1749ebe143

Jewett, L., & Romanou, A. (2017). *Ocean changes: Warming, stratification, circulation, acidification and deoxygenation*. Environmental Science.

King, H. (2020). *Oil and natural gas resources of the Arctic*. Geology.com.

Knies, J., Köseoğlu, D., Rise, L., et al. (2018). Nordic Seas polynyas and their role in preconditioning marine productivity during the Last Glacial Maximum. *Nature Communications, 9*, article No. 3959.

Kramer, A. E. (2020, September 5). Land in Russia's Arctic blows 'Like a Bottle of Champagne'. *New York Times*. https://www.nytimes.com/2020/09/05/world/europe/russia-arctic-eruptions.html

Kraska, J. (2011). *Energy security in the time of climate change* (pp. 130–205). Cambridge University Press.

Krupnik, I., Aporta, C., Gearheard, S., Laidler, G. J. J., & Holm, L. K. K. (Eds.). (2010). *SIKU: Knowing our ice, documenting Inuit Sea ice knowledge and use*. Springer.

Laidre, K. L. (2008). Arctic marine mammals and climate change: Quantifying the sensitivity of Arctic marine mammals to climate-induced habitat change. *Ecological Applications, 18*, S97–S125.

Liu, N. (2020). The rise of China and conservation of marine living resources in the polar regions. *Marine Policy, 121*, 104181. https://doi.org/10.1016/j.marpol.2020.104181

McDermott, S., Haist, V., & Rand, K. M. (2016). Evaluating the efficacy of trawl exclusion zones by estimating local Atka mackerel abundance and movement patterns in the central and eastern. *Marine and Coastal Fisheries, 8*(1), 334–349. https://doi.org/10.1080/19425120.2015.1135218

McDowell Group. (2017). *The economic value of Alaska's seafood industry*. Alaska Seafood Industry.

McIntyre, A. D., & McDermott, M. (Eds.). (2010). *Chapter: 10: Life in the world's oceans* (Marine life in the Arctic). Blackwell Publishing Ltd.

Medearis, S. (Ed.). (2019a, October 11). *Chaos in the Bering Sea ecosystem demands more surveys next year*.

Medearis, S. (Ed.). (2019b, November 15). *Warm water fish migrating to Northern Bering Sea are thriving*.
Medred, C. (2016, September 27). Alaska's Mafia-Style fisheries management. *Anchorage Daily News*.
Monroe, R., & IGSD. (2019, July 22). *Research highlight: Loss of Arctic's reflective sea ice will advance global warming by 25 years – Fast mitigation may still avoid complete loss of sea ice*. https://scripps.ucsd.edu/news/research-highlight-loss-arctics-reflective-sea-ice-will-advance-global-warming-25-years
Mueter, F. J., & Litzow, M. A. (2008). Sea ice retreat alters the biogeography of the Bering Sea continental shelf. *Ecological Applications, 18*(2), 309–320. https://doi.org/10.1890/07-0564.1
National Oceanic and Atmospheric Administration (NOAA/Fisheries). (2019). *Scientific teams set out to track unprecedented changes in the eastern Bering Sea*. National Oceanic and Atmospheric Administration.
National Snow and Ice Data Center (NSIDC). (2021, January 5). *Arctic Sea ice news and analysis*. https://nsidc.org/arcticseaicenews/
Nechepurenko, I. (2020, June 4). Russia declares emergency after Arctic oil spill. *New York Times*. https://www.nytimes.com/2020/06/04/world/europe/russia-oil-spill-arctic.html
Noongwook, G. (2004). *Watching sea ice and weather our way*. Smithsonian Arcgtic Studies Center.
Nuttall, M. (2005). *Protecting the Arctic: Indigenous peoples and cultural survival*. Routledge, Taylor and Francis Group.
Oceana. (2016). *Northern Bering Sea climate resilience area*. https://usa.oceana.org/northern-bering-sea-climate-resilience-area
Oliver, J. A. (2007). *The Bering Strait crossing: A twenty-first century frontier between east and west* (p. 25). IA Press, Information Architects Books.
Ovando, D. (2018). Improved fisheries management could offset many negative effects of climate change. *Science Advances Ecology, 4*(8), eaao1378.
Parlow, A. (2016, June). Shell and beyond: Toward an Arctic standard in the New North. *Alaska Daily News*.
Parlow, A. (2019a, June 27). Does a Russia-China alignment in the Arctic have staying power? *Arctic Today*.
Parlow, A. (2019b, November 26). Why Climate Policy-makers should listen more to indigenous peoples. *Arctic Today*.
Parlow, A. (2019c, October 8). The Arctic door opens a bit wider. *Stratfor*.
Parlow, A. (2019d, September 4). Reimagining US-China Arctic ties in the context of green energy. *Stratfor*.
Parlow, A. (2019e, July 16). Bluster, cooperation mark China's Arctic Summit. *Anchorage Daily News*.
Parlow, A. (2021). Clear-eyed in the Arctic: The Biden Putin Summit. *Anchorage Daily News*.
Piatt, J., Parrish, J. K., Renner, H. M., Schoen, S. K., & Jones, T. T. (2020). *Extreme mortality and reproductive failure of common murres resulting from the Northeast Pacific Marine Heatwave of 2014–2016*. USGS and National Wildlife Center, PLOS ONE.
Plumer, B., & Fountain, H. (2020, December 30). Trump administration finalizes plan to open Arctic refuge to drilling. *New York Times*. https://www.nytimes.com/2020/08/17/climate/alaska-oil-drilling-anwr.html
Proshutinsky, D. A., Timmermans, D., & Bamber, K. R. (2015). *Arctic circulation regimes*. Royal Society Publishing.
Puig, P., et al. (2012). Ploughing the deep sea floor. *Nature, 489*(7415), 286–289. https://doi.org/10.1038/nature11410
Rosen, Y. (2019). *Alaska's hottest month portends transformation into 'unfrozen state'*. Reuters.
Schönfeldt, K. (2017). *The Arctic in international law and policy*. Hart Publishing.
Siddon, E., & Zador, S. (2017). *Ecosystem considerations 2017 – Status of the Eastern Bering Sea marine ecosystem*. Alaska Fisheries Science Center, National Marine Fisheries Service, NOAA. https://archive.fisheries.noaa.gov/afsc/REFM/Docs/2017/ecosysEBS.pdf

Siders, A., Stanley, R., & Lewis, K. M. (2016). A dynamic ocean management proposal for the Bering Strait region. *Marine Policy, 74*, 177–185.
Solski, J. J., Henriksen, T., & Vylegzhanin, A. (2020). Introduction: Regulating shipping in Russian Arctic waters: Between international law, national interests and geopolitics. *The Polar Journal, 10*(2), 203–208. https://doi.org/10.1080/2154896X.2020.1818487
Soyer, B., Leloudas, G., & Miller, D. (2018). Tackling IUU fishing: Developing a holistic legal response. *Transnational Environmental Law, 7*(1), 139–163. https://doi.org/10.1017/S2047102517000267
Spies, I., & Gruenthal, K. M. (2019). Genetic evidence of a northward range expansion in the eastern Bering Sea stock of Pacific Cod. *Evolutionary Applications, 17*, 362–375.
Stein, R., Macdonald, R. W., & Stent, A. (2020). *US-Russia relations – Challenging policy*. Brookings Institute.
Stevenson, D. E., & Lauth, R. R. (2019). Bottom trawl surveys in the northern Bering Sea indicate recent shifts in the distribution of marine species. *Polar Biology, 42*, 407–421. https://doi.org/10.1007/s00300-018-2431-1
Stotts, J. (2015). *How to assess the Arctic from an Inuit perspective*. Alaska Inuit Food Security, Conceptual Framework, Inuit Circumpolar Conference – Alaska.
Struzik, E. D. (2017). *How a wayward Arctic current could cool the climate in Europe*. YaleEnvironment 360, Yale School of Forestry and Environmental Studies. https://e360.yale.edu/features/how-a-wayward-arctic-current-could-cool-the-climate-in-europe
Tenenbaun, L. F. (2017). *Ask NASA: Climate blog*.
The Economist. (2019, April 27). *A trap of one's own – China tries to calm jitters about the "Belt and Road" initiative*. https://www.economist.com/china/2019/04/25/china-tries-to-calm-jitters-about-the-belt-and-road-initiative
The State Council Information Office of the People's Republic of China. (2018, January 26). White Paper, China's Arctic Policy. *China's Arctic Policy*. https://www.alaskajournal.com/2021-09-14/fish-factor-unobserved-crab-mortality-unaddressed-survey-shows-stock-crash
Thoman, R. L., Richter-Menje, J., & Druckenmiller M. L. (Eds.). (2020). *The sustained transformation to a warmer, less frozen and biologically changed Arctic remains clear*. Actic Report Card 2020, National Oceanic and Atmospheric Administration (NOAA), US Department of Commerce. https://arctic.noaa.gov/Portals/7/ArcticReportCard/Documents/ArcticReportCard_full_report2020.pdf
Thorsteinson, L. K., & Love, M. S. (Eds.). (2016). *Alaska Arctic marine fish ecology catalog: U.S. Geological Survey Scientific Investigations Report 2016–5038* (OCS Study, BOEM 2016–048, 768 p.). https://doi.org/10.3133/sir20165038
U.S. Geological Survey (USGS). (2008). *90 billion barrels of oil and 1,670 trillion cubic feet of natural gas assessed in the Arctic*.
Van Huissteden, J. (Ed.). (2020). *Thawing permafrost – Permafrost carbon in a warming Arctic*. Springer. https://www.springer.com/gp/book/9783030313784
Van Pelt, T. (Ed.). (2015). *Bering Sea project: Bering ecosystem study: Understanding ecosystem processes in the Bering Sea*. National Science Foundation, North Pacific Research Board.
Viñas, M. J. (2019, September 23). *2019 Arctic sea ice minimum tied for second lowest on record*. https://climate.nasa.gov/news/2913/2019-arctic-sea-ice-minimum-tied-for-second-lowest-on-record/
Vizza, R. (2020). Arctic ice melt is changing ocean currents. *47 News*.
Wadhams, P. (2012). *Bering Strait: Marine life, what goes on beneath the floes*. Alfred Wegner Institute.
Wadhams, P. (2017). *A farewell to ice: A report from the Arctic*. Oxford University Press.
Warren, C. (2019). *Science: Vanishing Bering Sea ice threatens one of the richest U.S. seafood sources*. https://www.sciencemag.org/news/2019/05/vanishing-bering-sea-ice-threatens-one-richest-us-seafood-sources
Watts, J. (2019, June 7). *The end of the Arctic as we know it*. https://www.theguardian.com/environment/2019/jun/07/oceans-demise-the-end-of-the-arctic-as-we-know-it
Welch, L. (2016). Alaska ports lead nation in fish landings. *Anchorage Daily News*.

Welch, L. (2021, September 14). FISH FACTOR: Unobserved crab mortality unaddressed as survey shows stock crash. *Alaska Journal Commerce*. https://www.alaskajournal.com/2021-09-14/fish-factor-unobserved-crab-mortality-unaddressed-survey-shows-stock-crash

Wilson Center. (2020). Phone conference call regarding US-Russia relations in the Arctic region. *Polar Week*, Woodrow Wilson International Center for Scholars, accessed by the author on July 27–31, 2020.

Woodgate, R. (2013). Arctic Ocean circulation: Going around at the top of the world. *Nature Education Knowledge, 4*(8), 8.

Young, O., Berkman, P. A., & Vylegzhanin, A. N. (Eds.). (2020). *Governing Arctic seas: Regional lessons from the Bering Strait and Barents Sea* (Vol. 1). Springer.

Chapter 10
Decolonization, Food Sovereignty, and Climate Risks: The Case of St. Lawrence Island in the Bering Sea, Arctic

Anita Parlow

Abstract Climate change is having catastrophic environmental, economic, and cultural impacts on the Arctic region. Human-induced warming is increasing sea-ice melt, permafrost collapse and altering species migrations on land and sea. These impacts negatively disrupt not only the finely balanced ecosystems of the local flora and fauna, but also the Indigenous peoples who depend upon hunting and fishing for survival. But what exacerbates climate damage and impedes the ability of Indigenous peoples to protect their customary Arctic marine and terrestrial ecosystems are the impacts of past colonization– and its contemporary legacy. This chapter begins with the era of the Russian and European contact to today's spate of international, American, and other Arctic national domestic laws that continue to impede the ability of Native peoples to help shape domestic and international standards regarding ecosystem protection. Having said this, the chapter describes the ongoing developments by Native peoples to play a greater role in shaping Arctic law and policymaking. The variety of initiatives at domestic, regional, and international levels reflects an ongoing recovery of the right to self-determination, sovereignty and a relationship with traditional lands and waters. In the Arctic context, Native peoples, through a combination of advocacy, litigation and focused strategies, are becoming more sovereign equals among nations within the changing laws and administrative processes of the former colonizing and post-colonial powers. The 2500-year story that is unfolding on the Bering Sea's St. Lawrence Island, in the midst of the existential threat of climate change, offers an example of courage and resolve that is both an example to a world being engulfed in catastrophe, and also presents a story, in real time, of a people whose self-determination and sovereignty must include a meaningful seat at the table regarding decision-making processes about ecosystem protections and the limits of commerce.

Keywords Indigenous peoples · Arctic · St. Lawrence Island · Bering Sea · Siberian Yupiks · Food sovereignty

A. Parlow (✉)
Fulbright Scholar Iceland, recent, Washington, DC, USA

M. Behnassi et al. (eds.), *The Climate-Conflict-Displacement Nexus from a Human Security Perspective*, https://doi.org/10.1007/978-3-030-94144-4_10

1 The Context

Like many singularly powerful experiences that become deeply rooted in a peoples' psyche and understanding of their history, nearly all of the 1200 people who live in two villages, Savoonga and Gambell, on St. Lawrence Island in the Bering Sea, closer to Russia, are mindful of the 1878 starvation where nearly everyone died. With short summers, scant rainfall, nearly year-round frozen tundra, and sea-ice that surrounded and defined the island, planting was nearly impossible. Only a few types of greens and berries grew wild on the tundra (Ackerman, 1976). The only available meat supply was the Arctic fox or the wolf, red fox, and polar bear that occasionally traversed the ice pack formed on the sea-ice that once connected St. Lawrence Island to the Russian mainland for 9 months of the year (Apassingok, 1985, 1987).

Nearly all of the 4500 inhabitants who lived on this remote island, 40 miles southeast of the Russian Siberian coast and 120 miles south of the Bering Strait, were Siberian Yupik. At 100 miles long and twenty-seven miles at its widest, the volcanic rock island was populated by small coastal communities and fishing camps that inhabit the islands coast (Ackerman, 1976).

The Islander's primary food supply came from the sea. The Yupiks have been hunting, fishing, and whaling in the Bering waters for more than 2500 years. The major marine resources in the Bering include the Pacific walrus, for which St. Lawrence Island is well known, as 'The Walrus Capital of the World' (The Conas Project, 2020). The Bering Sea also hosts a range of bearded seals, ribbon seals, sea lions, and gray, bowhead, and finback whales. Millions of migratory birds, cormorants, ducks, geese, loons, gulls, terns, plovers and more provide an additional food resource as they rest and lay their eggs on the cliffs during their long journey south. Arctic Char and Tom Cod fish round out the human diet, along with eggs that the Islanders collect on the high rock formations that overlook (Audubon Alaska, n.d.) one of the richest peripheral seas on the planet (Alexander, 2020).

In 1867, the U.S. Senate ratified the *Treaty of Cession* with Russia. Through the Agreement, the Emperor ceded all Russian possessions in North America, including the adjacent islands, with his stated purpose to "strengthen, if possible, the good understanding which exists between them" (The Treaty of Cession, 1867).

The U.S. House of Representatives appropriated $7.2 million to carry into effect the Cession Treaty (Crowell & Oozevaseuk, 2006). The Bering residents, sovereign since 'time immemorial' and living in balance with the Arctic environment for millennia, were not included in the negotiations (Huntington & Noongwook, 2013). Nor was there any acknowledgement of territorial rights in the Agreement. The Bering Sea region is home to over 70 Indigenous communities such as the Iñupiat, Central Yup'ik, Cup'ik, St. Lawrence Island Yupik, Unangan, and Chukchi Peoples, Inuits, Aleuts, Athabascan, Alutia, Haida, Inupiat, Tlingit, Tsimshain, Siberian Yupiks and others (Druckenmiller et al., 2019).

With the United States' purchase of Alaska, all of the resident Indigenous[1] communities became, by decree of two foreign governments, a part of the United States, under the jurisdiction and control of a new, and distant, government authority (The 1867 Treaty). In his report on fur seals, the Special Agent of the Treasury Department, Henry Elliot, mentioned the St. Lawrence Islanders in passing, noting that the Islanders were of little interest to the significant fur trade industry, and indicated that evidence existed of the adequacy of Indigenous peoples' food supply (Ross, 2006: 13).

What followed was a cruel irony. For millennia, the Bering's native communities have been hunting two or three whales annually to preserve ecological harmony. This situation is no longer the same today as the American and European whaling vessels began massive exploitation of sea mammals for commercial purposes. The first time a whaler killed an Arctic whale was in 1848 when the New Bedford killed an Arctic bowhead of 65 feet, of more than 100 tons of weight, and possibly aged more than a century (Dolin, 2011). Over the following decades, Boston and West Coast whalers headed north into the Arctic, killing as many as 1000 whales in one season. By 1852, nearly 300 whalers from New Bedford, New England, and the west coast waged ceaseless hunting operations of bowheads until there remain few whales in the Arctic waters. Technological change accelerated the number of whales that could be killed at once. By 1880, a dart gun and a harpoon with an explosive projectile that could kill a whale instantly were introduced (Ross, 2006). The highly valuable bowheads that migrated between the Bering Sea and the Arctic Ocean, close to the ice packs, thereby 'ice whale', contained up to 300 barrels of oil and a possible 2500 pounds of baleen. At a high point in 1887, the year of the starvation, one vessel reportedly caught 28 whales. From 1848 to 1909, commercial whalers killed some 29,500 North Pacific bowheads (Ross, 2006). The purpose was to harness whale oil to provide light. With the oil drained, the whale meat, bones, and organs, were discarded. The industrial scale practices were in direct contrast to Yupik whalers who took only one or two whales annually to feed their entire villages. After butchering the whale on the thick sea-ice, the Yupiks returned to the sea what remained – with prayers of thanks.

It is estimated in Crowell and Oozevaseuk (2006) and Ackerman (1988) that between 1848 and 1885, more than 10,000 whales were hunted for whale oil from the Anadyr gulf to the south to Point Barrow in the high North. Once the whale population began to decline, the commercial whalers shifted in objectives and turned to slaughter more vulnerable and numerous walruses for industrial purposes. In this sense, Ackerman (1988) observes that the whaling industry killed more than 100,000 walruses between 1865 and 1885.

[1] In all instances, this chapter capitalizes the term 'Indigenous', as an Indigenous rights movement that emerged in the 1970s. In Academic and political writings, capitalization is a way to describe contemporary efforts to decolonize and recover confiscated lands. More than that, it is a form of style that permits to fight for self-representation, political, cultural sovereignty, and spiritual rights (Weeber, 2020).

It should be made clear that whale and walrus slaughter set the stage for coastal ecocide. In 1875, a year with frequent storms that exacerbated the reduction of access to food, a massive walrus kill reduced the communities' food supply to low levels. Most of the Yupik people survived this double disaster and still remember how the circumstances worsened 3 years later. In 1878, the surge of a storm prevented the sea-ice from freezing around the island refrained hunters from reaching the fewer quantity of walrus that remained in the sea. With a depleted stock, the confluence of events significantly diminished the portion of fish stock that the community took annually from the walrus hunt (Apassingok, 1985). This weakened the physical condition of the hunters. The situation was made worse by an influenza epidemic where nearly 90% of Siberian Yupiks died (Crowell & Oozavaseuk, 2006). Most of the villages disappeared, leaving behind only a few artifacts and random skeletal remains to prove the people and their civilization once existed (Crowell & Oozaaseuk, 2006; Laidler et al., 2009).

The New Bedford Standard reported in August, 1879 that on both the Russian and American sides of the Bering, more than one-third of the populations died from starvation. (Crowell & Oozavaseuk, 2006). Several Savoongan elders thought that a few Yupiks on the American side of the Bering survived by making their way across the Bering ice to Russia; then integrated into with the reindeer herders across the Russian Arctic (Parlow, 2019).[2]

In 2007, Boston.com reported that an Alaskan Native whaler had cut into a whale caught during their annual subsistence whale hunt and found embedded in its bones the tip of a nineteenth century unexploded bomb-lance used by New England whalers between 1885 and 1895 – a time consistent with the Bering Sea starvation. Whale oil lit the world at that time (Dolin, 2011) and provided Boston with whale oil revenues that made the New England city a citadel for global commerce (Nichols, 2010).

2 Backdrop: Recovery of Indigenous Sovereignty in the Arctic in the Context of Warming

To this day, St. Lawrence Islanders, a fiercely independent and sovereign Indigenous people, celebrate their ancestral way of life that has been culturally, politically, spiritually, and nutritionally connected to the Bering Sea for millennia (Alexander, 2020). Two centuries later, a global and deadly pandemic, called the Great Influenza, surged in 1918 and complicated the life of the local population. It is mentioned in Coggin (2019) that the disease was so fatal that starvation decimated more than half of the native population. The Arctic Indigenous communities, who depend upon a healthy marine ecosystem for their food supply, were on the front lines of the global climate catastrophe (Fig. 10.1).

[2] Interview with elders at Tribal Government lunch (August 2019).

Fig. 10.1 Waves break near the shore in Savoonga in April. (Source: Demer © 2017)

Consistent with the rest of the Arctic, St. Lawrence Island's Savoongan villages are experiencing warming twice as fast as the rest of the planet. This Siberian Yupik coastal community, with roughly 600 residents, has a front row seat to a wide range of global warming's most severe impacts. As fires rage across Siberia and Alaska, the permafrost melt has caused roads, rail, and onshore oil and gas infrastructure to collapse, and the caribou and reindeer are increasingly infected by tics that are migrating from lower latitudes. With an acceleration of oil, gas, mining, and shipping, the once inaccessible region and its rich resources have become increasingly available given the increasing rate of sea-ice melt (National Geographic, 2019).

Like many Indigenous peoples who live in the Arctic, the residents of St. Lawrence Island are facing pressures from at least four directions at once: loss of sovereignty and territorial rights; restrictive federal, state, and international regulatory regimes; climate change and sea-ice retreat; and pressures to develop oil, gas, and mining resources in a region increasingly accessible to outsiders. Moreover, the people of St. Lawrence Island differ from the rest of Alaskan Natives in that they fully own their 1.2-million-acre island by virtue of a decision made by the village elders after the 1971 Alaska Native Claims Settlement Act (ANCSA) was enacted, primarily intended to extinguish all Indigenous title to their ancient and historic aboriginal lands (Fig. 10.2).

The two remaining villages, after the great starvation, opted out of their portion of a billion-dollar settlement that would convey their ancestral lands to the US federal government. The Islanders decided to opt out of a cash settlement. Instead, the Yupiks chose to continue to maintain stewardship of the nation's sixth largest island

Fig. 10.2 The Bering Sea visible at the end of a row of houses in Savoonga, on Alaska's St. Lawrence Island. (Source: Parlow © 2019)

with its 900 miles of coastline, mountains, and lakes. In 1979, an interim conveyance of the land left all decisions with St. Lawrence Island's complex governance structures, with final title conveyed in 2016.

The final land conveyance was one of the largest in US history. It was a transfer of title to Native people who have tirelessly fought not only to keep their land despite strong external pressures, but also to maintain and protect their subsistence way-of-life. These same people now face the existential ravages of climate change. Of the decision not to sell their land that has seen 2500 years of Siberian Yupik occupancy, the Tribal leader Perry Pungowiyi expressed the voice of the Elders who came before him by saying that "as long you are owners of the Island, the Island will take care of you." Others noted that the "money runs out, the land is forever" (Khachatoorian, 2016).

Savoongans practice what they preach. Along with Gambell, the other St. Lawrence Island community totaling 1200 people, the Islanders refused the call of a mining operation on the lands to which they hold both surface and subsurface rights, despite the company's promises of jobs and income that would supplement their essentially non-cash economy. "We didn't want to injure our subsistence ways or break with the instructions of our ancestors to protect the ecosystems", said Bryan Rookok, President of Savoonga's Native Corporation, Kukulget, while reaffirming the Yupik philosophy of 'leave no footprint behind' (Parlow, 2019).[3]

[3] Interview undertaken by the author on August 2019.

3 Climate Change and Threats to Food Security

More than 40,000 Indigenous people live on Alaska's Arctic and sub-Arctic Bering Sea coastline, depending upon marine resources to put food on the table. The Bering Sea is a transition zone between the Pacific and Arctic oceans. The relatively narrow Bering Sea, between the coasts of the United States and Russia, hosts millions of seabirds, whales, walrus, seals, fish, and a growing number of oil and gas tankers, as well as commercial fishing vessels and subsistence fishers, hunters, and whalers (US CMTS, 2020).

The dramatic loss of winter sea-ice is having a negative cascading effect on the fine balances of the ice-dependent Bering Sea and Strait and its food chain. Moreover, the Arctic sea-ice melt and loss negatively impacts not only the primarily Indigenous residents of the Bering region, but also a growing instability in circulating currents, changing weather patterns, and air temperatures in the rest of the world (Wadhams, 2016).

Threats to human food security are not trivial. With the global spotlight on regional, national, and international agenda, small villages communities, and their residents – like Savoonga and Gambell on St. Lawrence Island – often remain overlooked. Evidence of warming is visible everywhere on the 1.2-million-acre island – both on and offshore – with visible and dramatic reductions and increasing impediments to the ability to secure food. Melting sea ice and intensifying high velocity storms are causing the collapse of coastal homes, no longer protected by coastal sea ice, into the permafrost thaw. This situation is exacerbated by shore loss and, sinking homes, and collapsed roads due to permafrost melt (Fig. 10.3).

George Noongwook, a Yupik Whaling Captain, a walrus hunter, and former Chairman of the Alaska Eskimo Whaling Commission, is quoted in Parlow (2019) saying that: the reduction in sea ice, from nine-month to three-month seasons means that sea mammals no longer have a platform on which they can rest, feed, and nurse their calves during migration times. This not only interrupts the mammals' life rhythms but also the hunters' access to their food supply. Noongwook's statement emphasizes that despite the critical lost sea ice that upsets the entire food chain, Yupik hunters, fishers, and whalers continue to navigate climate unpredictability with agile hunting strategies that have guided them for millennia.

Noongwook has traveled throughout the United States and Europe to describe how Savoonga's Yupik people are digging deeper into their ancient knowledge to address climate change. Reading the rapidly changing weather conditions is "who we are", he explained. "We take it day by day," he said, reflecting the kind of equanimity that deeply rooted knowledge brings (Parlow, 2019). For hundreds, if not thousands, of years, Yupik's have created quotas to preserve walrus, seals or whales so that the balance of the marine ecosystem is maintained. Further, Noongwook explained that the Yupik philosophy of leaving 'no footprint behind' reflects a people who maintain a "take only what you need" hunting practices, that fiercely protect the biodiversity of the Bering Sea's interconnected ecosystems (Parlow, 2019).

Fig. 10.3 Savoongan whaler and walrus hunter props up his house as it sinks into the melting permafrost. (Source: Parlow © 2019)

The changing fish and mammal migrations and newly-open waters also cause dramatic shifts in hunting and fishing. "You have to be fast", Noongwook explained as he described how the loss of sea ice impedes access to their food supply, as the walrus no longer regularly stop on nearby sea-ice to feed themselves and their calves, allowing for the annual harvest (Parlow, 2019). Of warming's impacts, another tribal leader said: "it's getting scary". Yet, most Islanders said that their ability to adapt to changing conditions remains grounded in their shared tradition of thousands of years of navigating change, a tradition transferred through oral history and song.

The irony of climate change impacts in the Arctic is that while Savoongans, like most Arctic subsistence hunters, fishers and whalers, contribute among the very least to climate change, they feel its harmful effects disproportionately. Noongwook also described an "alarming" cascade of ecological consequences that included changes in the timing and numbers of algae blooms that form the foundation for the entire food chain, northward migrating fish populations seeking colder waters, and massive bird die-offs in a migration corridor (Parlow, 2019). Further, the retreating sea-ice has left the once ice-protected cliffs vulnerable to raging storms that have significantly eroded the coastal lands leaving scores of Bering Sea and Strait communities sliding into the sea (Gricius-Abbott, 2021).

In terms of subsistence, key resources are moving away. Ice-dependent marine mammals and fish are moving to the new locations of ice edges and flows – often too far away to be safely hunted. Savoongans, like North Slope whalers, report that they must now travel farther out to hunt. Increased travel time and distances garner additional costs for fuel and maintenance, while increasing the risk of accident given choppier ice.

Further, changes in snow cover makes four-wheeler and snow-machine travel to hunting or fishing camps across the island more difficult and dangerous. This forces hunters and fishers to delay travel until later in the fall season when there is more snow, but less prey. Summer hunting has also been upset by permafrost collapse of roads that cross the Island to hunting and fishing camps. As a result, Demer (2017) argues that fewer and more dangerous opportunities are available for hunters to catch fish or whales.

Noongwook noted that whales are now sometimes hundreds of miles beyond what was once the sea-ice edge where they migrated north. This often leaves the small whaling boats with insufficient fuel to reach the changing whale migration paths. Further, a spate of Federal regulations in the U.S. – such as the Endangered Species Act and Areas to be Avoided – have restricted Native peoples whose non-commercial fishing and hunting activities are intended for their own families, communities, and Elders. The laws that prohibit walrus hunting are to regulate wanton commercial hunting practices, mainly by outsiders rather than the allocated take established by local Indigenous communities' customs and rules (Huntington et al., 2017).

To address the widely held concerns about an increasingly impaired marine ecosystem and a food supply reserve, the Food Sovereignty Summit was organized on the Alaskan mainland in 2019. the summit was attended by representatives of nearly a hundred of the Nome Alaskan Native Communities, a population that has been tangibly affected by the severe impacts of global warming. The delegates discussed the most recent signs of a broken ecosystem: birds falling off the cliffs, in part, from hunger; a collapse of the Bering Sea cold pool that served as a marine protected border for varying fish species; fish scattering northward for colder waters; "walrus and seals getting skinnier"; and whales and walrus are harder to reach. As the sea-ice melts there is no guarantee that the walrus or whales will be accessible or available. Equally concerning was discussed whether the small boats are able to return to the shore given the increasingly choppy and dangerous ice (Huntington et al., 2017). Several Savoongan whalers noted that the Elders are worried about the ice conditions that this generation of whalers and walrus hunters must traverse. "They make us take a satellite phone", one whaler noted, "so, if we get into trouble, we call", he said flatly (Parlow, 2019).

When it comes to the effects of climate change, the Arctic communities are generally overlooked, ignored, and marginalized in terms of the quest for solutions. Arctic communities offer innovative approaches to resiliency. Their philosophy, science, and theology offer a roadmap to resiliency and restoration of ecosystem balance for environmental and climate practice (Parlow, 2019).

4 Arctic Indigenous Science, Philosophy, and Natural Law

Western scientists and lawmakers must cease perceiving Indigenous contributions to climate research as less than "scientific". Savoonga, along with Gambell, the only other village on the 1.2-million-acre island that survived the starvation, count amongst the best examples, but not the only, of integrating scientific awareness with hundreds, if not thousands, of years of observation. Observation is the backbone of scientific inquiry. Firsthand experience, reliance for survival, and their deeply rooted philosophy of the region's rich biodiversity make the Islanders the most knowledgeable people to protect the highly integrated and bio-diverse ecosystems. For more than 2500 years, neither the whales nor the walrus was over-hunted and the ecosystem remained in balance unlike the several decades of ruthless kill before the "starvation" (Ackerman, 1976).

This highlights a fundamental difference between a Western perspective of land 'ownership' and that of Indigenous peoples, including the Arctic. When colonial powers take or purchase 'wild' or 'empty' lands, the perception of those environments is that of property. The 'owners' of land utilize resources for financial gain. The Indigenous peoples of the Arctic perceive themselves as stewards, servers, and preservers of the marine and terrestrial ecosystems. (Pennesi et al., 2012)

5 Subsistence: Territorial Rights, Human Rights, Mainly the Right to Food

5.1 U.S. Treaty of Cession with Russia in the Arctic High North

Following the Treaty of Cession with Russia, by which the United States gained title to what was Russian territory, Indigenous territorial rights were extinguished in a series of efforts over the next two centuries (Act, 1867). In 1868, the US Congress designated Alaska as a 'customs collection district' and extended the US law over the mainland, islands and waters of the territory of Alaska, with no consideration of Native rights.

By virtue of the Doctrine of Discovery, under principals of the international law, the 'discovering' nation acquired the exclusive right to engage Indigenous peoples with respect to matters of land ownership and government-to-government relations (Cohen, 1941). In the seminal case, Johnson v. McIntosh, Chief Justice John Marshall opined that under the doctrine of Discovery, Indigenous tribes have a "legal as well as just claim to retain possession of the (lands)" they historically occupied. That was the case until the Congress, which held the "plenary authority", choose to terminate such rights (Anderson, 2016).

On these vast lands held in the aboriginal title, Alaskan Native tribes, like all Native peoples before the discovery era, harvested, processed, distribute, and consumed marine mammals, fish, wildlife, and plants through an economy and distribution system that became known as subsistence – a legally enforceable right, but often used to further extinguished territorial rights.

Today, subsistence is an intensely political issue as non-subsistence hunters and fishers claim they lose both money and livelihoods if access to traditional Indigenous, aboriginal hunting, and fishing grounds were not available to them. Further, efforts to protect areas from overfishing or hunting often significantly diminishes Indigenous subsistence take, along with a woefully disproportionate impact on nutrition, arguably a human right often linked to territorial dispossession.

5.2 *Alaska Law and Subsistence*

Alaska law defines 'subsistence uses' as "[T]he non-commercial, customary and traditional uses of wild, renewable resources by a resident domiciled in a rural area of the state for direct personal or family consumption as food, shelter, fuel, clothing, tools, or transportation, for the making and selling of handicraft articles out of non-edible by-products of fish and wildlife resources taken for personal or family consumption, and for the customary trade, barter, or sharing for personal or family consumption" (AS 16.05.940 (31), (32), 1978).[4]

The Alaska state and federal law related to subsistence uses (respectively AS 16.05.940 [32] and Title VIII of ANILCA, section 803) define subsistence uses as the "customary and traditional" uses of wild resources for various uses including food, shelter, fuel, clothing, tools, transportation, handicrafts, sharing, barter, and customary trade.

While both Federal and State law differ in who qualifies to participate in subsistence hunting and fishing rights, both the State and Federal governmental definitions miss the core meaning and significance of subsistence to Alaska Natives. Not only has most of the land been in customary use and occupancy for millennia, but also subsistence hunting, fishing, and whaling 'from time immemorial' has been the foundation of Indigenous lives, culture, philosophy, ancient and historic sovereignty, and the ongoing ability to put food on the table. Both subsistence laws, arguably, amount to a 'taking.' In 1978, the Alaskan statute initially applied 'subsistence' to rural Alaskans. Afterward, the authorities expanded it by 1989 for the benefit of all Alaskans. The 1980 federal statute, The Alaskan National Interest Lands Conservation Act (ANILCA), applies only to federal lands but to all rural Alaskans.

Recent initiatives by Indigenous peoples at the United Nations on the development of human rights and customary law as legal and political frames of reference regarding the protection of sovereignty, territory, and customary use and occupancy, are gaining traction (IWGIA, 2010). This includes what is or was sea-ice use and occupancy. Or, as at least one Savoongan repeated an often-mentioned view of what is to be protected in marine terms, the sea is "our grocery store; our refrigerator" (Parlow, 2019) (Fig. 10.4).

[4] https://www.adfg.alaska.gov/static/home/subsistence/pdfs/subsistence_update_2017.pdf

Fig. 10.4 George Noongwook at home in Savoonga preparing a family walrus dinner, and an invited guest. (Source: Parlow © 2019)

6 External Pressures, Increasing Accessibility, and Outside Commercial Interests in Alaska

6.1 Oil, Gas, Mining, and Shipping

The oil and gas industries are well-established in Alaska, Russia, and Norway and the melting sea-ice is opening up new areas of the high North to carbon fuel production (Reiss, 2012; Horowitz et al., 2018). According to the Arctic Marine Shipping Assessment (AMSA) reported in Arctic Council (2009: 76), the "development of rich natural resources in the Arctic is a rapidly growing industry". Since 2004, several significant new bulk shipments have begun operations, such as the year-round oil shipments out of the port of Varandey in the Russian Arctic. In early 2008, an offshore-lease sale conducted by the U.S. Minerals Management Service for the U.S. Arctic totaled nearly $US2.7 billion; offshore gas appears to be a significant resource under consideration for development in this Arctic region (Arctic Council, 2009). Although, it is notable that a growing number of investment houses, insurers, shippers and oil and gas companies are withdrawing current and projected investments in the Arctic region given the growing political opposition to carbon fuels production in the High North on climate and environmental grounds.

The AMSA 2009 Report, viewed as a significant foundation for understanding Arctic shipping, reported that the greatest growth in Arctic shipping is likely to be in the oil, gas, and mining industries – with the greatest increases in bulk and break-bulk vessels generated from Russia, Canada, Norway, and the United States.

During this same period, in 2008, the U.S. Geological Survey (USGS) completed a Circum-Arctic Resource Appraisal (CARA) that evaluated the petroleum potential of all areas north of the Arctic Circle, 66.56* north latitude, for recoverable resources. The USGS concluded that the Arctic Circle, encompassing about 6% of the earth's surface, contains an estimated 90 billion barrels of oil, 1669 trillion cubic feet of natural gas, and 44 billion barrels of natural gas liquids, of which 84% is estimated to occur in the offshore. These resources account for about 22% of the undiscovered, technically recoverable, natural resources in the world (USGS, 2008).

The Russian Federation's Sea Route Administration announced in 2019 that 31.5 million tons of goods were transported through its Northern Sea Route (NSR) – a 56.7% increase over the previous year (Staalesen, 2020). The vast majority of the tonnage (20.5 million tons) was comprised of natural gas from Novatek's Yamal LNG project; a project in cooperation with China. China's growing interest as a producer and consumer in Arctic energy production and consumption, referring to itself as a "near Arctic nation," contains the potential to dwarf marine-ecosystem protections that are vital to subsistence and commercial fisheries (The State Council Information Office of the People's Republic of China, 2018).

Further in energy-and mineral-dependent Alaska, the Red Dog Mine has been embattled by lawsuits from the rural subsistence villagers in coastal Kivalina where their legal battles against a Canadian Mining Company and their native corporation to a settlement regarding destructive environmental practices. Moreover, the highly controversial and massive Bristol Bay gold mine initiative in the world's largest wild salmon run has galvanized both Indigenous and non-Indigenous fishers in opposition to a massive and controversial mine that would be located in the headwaters of the world's largest sockeye salmon run. The salmon hit record numbers in 2021, reinforcing the Indigenous and non-Indigenous opposition to the embattled Pebble Mine to seek to "forever" block an open pit gold mine planned for the headwaters of one of the world richest fisheries. The all-time high of 64.2 million fish, surpassed the previous record of 62.9 million and both exceeding the projected season total of the highly regulated fishing industry that disallows overfishing by subsistence or commercial fishers who generate more than $2 billion annually for the state of Alaska (Reynolds, 2021).

7 Loss of Territorial Land, Sovereignty, and Reduced Access to Traditional Land and Waters

To understand the power of Indigenous advocacy since post-cession Treaty contact, it is useful to understand the series of laws, codes, treaties and customs that constructed a colonial wall around Indigenous self-determination and sovereignty, even when well-meaning. As the underlying relationship between the Indigenous peoples and the colonial states that took the land, imposed political jurisdiction and, at best, established policies of assimilation, that serve the dominant culture at the expense of the region's first inhabitants is being increasingly challenged. One sees green shoots of decolonization in advocacy by Indigenous tribes, nations, organizations and Alaskan Native corporations regarding state, national and international law and policy. However, the grim history reflects how much work is to be done.

In its 1867 purchase of Alaska, the United States did not enter into any treaties or agreements with the residents who lived in this ice covered region for more than two thousand years (Swensen, 2014). Neither the Russians nor the Americans considered the status of Alaskan Indigenous sovereignty or the possession of aboriginal title as had been the case in the American lower 48 U.S. states where a domestic dependent nation status with a federal fiduciary responsibility in a controversial 'guardian-ward-trust' relationship. The Treaty divided the Alaskan population into two categories: 'inhabitants' and 'uncivilized tribes' (Treaty, March 30, 1867).[5]

The "inhabitants" were to be admitted into the United States or could return to Russia. The "uncivilized tribes" were neither acknowledged for their sovereignty over their tribal members – if not "use and occupancy" designation of their traditional, aboriginal lands – nor were they accorded any rights of citizenship (Case, 1984).

The 1867 Treaty of Cession at Article III was explicit in its characterization of the Native peoples, and the status of their sovereign rights, who inhabited the ceded territory. Non-native people would be allowed to return to Russia within 3 years or remain in the ceded territory and, "with the exception of the uncivilized tribes, shall be admitted to the enjoyment of all the rights, advantages and immunities of citizens of the United States". Of the Indigenous peoples, whose oral history reaches back as far as 2500 years on these lands, ice, and waters, "the uncivilized tribes will be subject to such laws and regulations as the United States, may, from time to time, adopt in regard to aboriginal tribes of that country".

A slow but firm development of the implementation of a colonial process followed with a combination of a restrictive interpretation of Indigenous inherent rights of sovereignty, territory, and use and occupancy of land (Case, 1984). The legal process of taking land, extinguishing title, and denying access to culturally and spiritually significant areas that supported hunting, fishing, and whaling has been organized and specific. Efforts to strip Native people of their sovereignty, and a loss of the right to maintain a government-to-government relationship with the United States to pursue political relations (Swensen, 2014), were a foundational step in stripping Native peoples of their rights to use, occupancy, and access to the Arctic lands that formed their universe for thousands of years. Sambo-Dorough (2014) stipulates that access to resources is paramount, be it for sovereignty, self-determination, or the safeguarding of the integrity of distinct peoples.

---

[5]Treaty concerning the Cession of the Russian Possessions in North America by his Majesty the Emperor of all the Russias to the United States of America, June 20, 1867, https://avalon.law.yale.edu/19th_century/treatywi.asp

The 1867 Treaty of Cession offers a useful point of departure both for analysis and understanding of the necessity for strategies that Indigenous peoples, including Alaskan Natives, have, and continue to use to develop actions to recover and exercise sovereignty over their aboriginal lands. Such efforts are increasingly in tandem with climate change responses, again linked to defending a way-of-life that Indigenous peoples have fought fiercely to protect, defend, and build upon for more than 2500 years.

7.1 The 1884 Organic Act and the Creation of District of Alaska on Aboriginal Lands

The core opposing interests regarding the legal status of Indigenous land rights, oil, gas, and mineral interests were clarified through the 1884 Organic Act,[6] which created the District of Alaska and allowed for the appointment of a bureaucracy, school system, district, and circuit courts. Perhaps the most important element was the economic driver that drew Federal: the enforcement of mining laws in the resource-rich Alaskan territory.

Arguably, given the deep understanding of ecosystem balances, if Alaskan tribes owned the land on which subsistence activities take place, or if Alaskan tribes were legally empowered to control subsistence resources and their territorial lands and waters, it might be more possible to mitigate the impacts of climate change not only on their local ecosystems, but also around the globe. By sitting at the political table as equals in diplomacy, in such bodies as the Arctic Council or the United Nations, environmental outcomes might improve, certainly for Native peoples and their particular interests (Ristroph, 2010; Cochran, 2013).

7.2 The 1971 Alaska Native Claims Settlement Act

A century after the 1884 Organic Act, the U.S Congress responded to significant Indigenous advocacy and enacted the 1971 Alaska Native Claims Settlement Act (ANCSA) (Moore, 1997). ANSCA extinguished Alaska aboriginal hunting and fishing rights, title to Alaska lands, and constructed a corporate assimilation approach that conveyed 23,040 acres of land for each of the 13-newly created regional Native Corporations. The corporate model was designed to promote commercial activity and was divorced from traditional subsistence land use or the seas where fishing or hunting locations were areas of historical, cultural, traditional, and vital to subsistence fishing, hunting, or whaling. Each of the newly created regional

[6]The 1884 Alaska Organic Act: https://explorenorth.com/library/history/alaska_organic_act-1884.html

Native Corporations received 23,040 acres of land, which gave "preference of land with a commercial purpose," primarily oil, gas, and mining. (Linxwiler, 2007).

In 1983, the Arctic Indigenous-wide Inuit Circumpolar Conference retained Thomas R. Berger, British Columbia Supreme Court Justice, who wrote his seminal review of ANCSA's impacts on Alaskan Natives. He concluded that the Congress enacted ANCSA to diminish the ability of Alaska Natives to control their land or meaningfully shape policy regarding land, subsistence or longstanding use and occupancy amongst peoples for whom ownership and private property had been foreign concepts. Berger reported that ANCSA's replacement of tribal governance with corporate structures, could further alienate any remaining land holdings by Native peoples if corporate failure, corporate takeovers or unpaid taxes led to confiscation (Berger, 1985). After conducting extensive field hearings throughout Alaska, Berger concluded that Alaska Natives believed that, if they owned their own land, they could protect Indigenous traditional economies and a village way of life. Subsistence is at the core of village life and land is the core of subsistence; you cannot protect the one unless you protect the other (Berger, 1985). ANCSA has protected neither.

7.3 *The 1980 Alaska National Interests Land Conservation Act*

Similarly, pressured by Indigenous advocacy groups, especially those who called ANCSA an encroachment on their subsistence rights, the Congress responded by passing the 1980 Alaska National Interests Land Conservation Act (ANILCA). The Act established a priority for the taking of fish and wildlife on public lands for subsistence uses. But, rather than linking subsistence priorities to Native status, ANILCA focused, generally, on the rights of all local rural residents, leaving Native Alaskans' subsistence rights again marginalized on their own traditional lands.[7]

By 1989, the Alaska Supreme Court determined that rural preference is a violation of the Alaskan Constitution.[8] ANILCA has been applied to Federal public lands, about 67% of the state. State law governs subsistence on state and private lands, including those owned by Native Corporations. Given that state law does not distinguish between Native and non-Natives or urban or rural residents, Indigenous peoples who depend upon subsistence way of life were placed as equal status to sport hunters or fishers. This made sovereign interest in food security a lower federal priority.

As the United States sought to avoid acknowledging Native land ownership, the government offered few, if any, legal or administrative tools for tribes to "officially articulate" their political will on the country. Instead, Alaskan Native people, along with the rest of the Arctic region, methodically and persistently used a combination of direct political action, litigation and the separate legal system itself created by the

[7] H. Res. No. 746, 92th Congress, 1st Session (1980).

[8] Supreme Court of Alaska, McDowell v. State, 785 P.2d 1 (1989). Can be found here: https://law.justia.com/cases/alaska/supreme-court/1989/s-2732-1.html

government to reclaim and recover the sovereignty over the lands, natural resources and way-of-life that they are denied (Moore, 1997; Ristroph, 2010).

8 Federal, State, Regional, and International Regulatory Processes: What Is Given Can Be Taken

Through a series of Federal regulatory protections within the United States, it appeared that the food supplies for Alaskan Native subsistence hunters, fishers and whalers would be protected. The U.S. Fish and Wildlife Services' (FWS) Endangered Species Act (ESA) and the National Marine Fisheries Services' Marine Mammal Protection Act (MMPA) exempt Alaskan Native subsistence hunting from prohibitions on take.[9] Additionally, the Migratory Bird Treaty Act[10] exempts Alaskan Native subsistence hunting from a prohibition on the take of migratory birds during the spring and summer seasons (Ristroph, 2010; Koivurova & Hasanat, 2015).

However, changes were relatively easy to affect. The administrative agencies are authorized to alter regulations on subsistence take if the agency finds that subsistence is "materially" and negatively affecting a threatened or endangered species.[11] With climate change warming ocean waters and altering species' habitat, it is likely that more species will be listed as threatened or depleted, leaving the Federal agencies to further restrict Indigenous subsistence food gathering to ensure species survival. Alaskan Natives have expressed concern. In 2009, after enacting regulations to protect a type of sae duck named the Eider Steller, subsistence hunters claimed that local observations of causation were disregarded regarding both the actual population status and the non-hunting factors that had changed conditions (Ristroph, 2010). The regulatory agency not only excluded the Alaskan Native hunters but, by omitting their input, missed some of the issues, such as climate-linked changes in food resources that might have more ably protected the Eiders (Ristroph, 2010). Additionally, climate change is altering the timing of seasonal migrations and the presence of the ducks, geese, and birds which, in turn, impacts the ability of subsistence hunters to fill their freezers for the long winter months. (Loring et al., 2011).

Distinctions between Native and non-Native hunters, fishers, and whalers are also significant as Native hunting strategies, developed over thousands of years, well before U.S. Federal policies, have maintained a balance within the ecosystem through their management approaches. Most restrictions on subsistence practices and a general lack of inclusion of Indigenous management strategies not only intrude on customary use, sovereignty, or, self-determination but also misses out on Indigenous best practices that allow for an improved understanding at the local level of how one small change to a marine ecosystem can cascade into a far larger dilemma.

---

[9] 16 U.S.C. #1539 (e) (2006).

[10] 16 USC #712 (2006).

[11] 16 U.S.C. #1539(e)(4) (2006).

Faced with the mounting restrictions on their right to control subsistence, the development of a joint strategy to co-manage natural resources with state and Federal governments gained traction. In 1994, an amendment to the Marine Mammals Protection Act provided for cooperative agreements between FWS and Alaska Native organizations to conserve marine mammals and co-manage subsistence use.[12] Indeed, FWS entered into agreements with Alaskan Native organizations, including the Eskimo Walrus Commission (EWC) that represents 19 villages (Raymond-Yakoubian et al., 2014). However, while FWS cooperated with funding, monitoring, and outreach, no real transfer of authority to the Alaskan Native subsistence representatives occurred, particularly in enforcement (Metcalf & Robards, 2008; Ristroph, 2010).

Co-management practices that bring together government and tribal structures often don't work because Indigenous and non-Indigenous entities differ in rationality and objectives. Some legal scholars have defined this as a 'post Westphalian resource management' (Shadian, 2010, 2013). Both in conservation and environmental protection outcomes, one would threaten to sweep people from their historic aboriginal lands; the other would establish an ecosystem balance that included subsistence hunters, fishers, and whalers (Ristroph, 2010: 73). Similarly, Alaskan Native tribes that participate in the Alaska Migratory Bird Co-Management Council (AMBCC) face a similar lack of power. The statewide management body composed of government agencies and tribes is charged to develop a recommendation for regulations of spring and summer harvest and conservation of migratory birds. But, for example, in 2008 the collective voice of Alaskan Native tribes to avoid placing further restrictions on their subsistence hunting quotas was disregarded (Fig. 10.5).

The Alaska Eskimo Whaling Commission (AEWC) is one of the few Indigenous organizations that have played a meaningful management role Within this new dynamic, the tribes are profiting from the specific Open water Season Programmatic Conflict Avoidance Agreements with the oil and gas industry that operated on and offshore during the whaling migrations. The agreements ensure that the communities' fishing activities do not interfere with whaling sonar, fish migration patterns, or other industry practices that might impede whales and subsistence whaling alike.

9 Climate, Sustainability, and Sovereignty in International Law

Drawing from a 2500-year-old history of the Island, an elder Yupik noted that the combination of federal and state restrictions on climate change hinders the protection of land and waters. In this respect, the Yupi man is quoted in Cochran et al. (2013) saying that: "Our ancestors taught us that the Arctic environment is not constant and that some years are harder than others. But they taught us that hard years

[12] 16 U.S.C. #1388 (2006).

Fig. 10.5 As sea-ice melts, tens of thousands of stranded walruses assemble on a beach near Point Lay, Alaska. (Source: Photo by NOAA © CoreyAccardo/AP, 2014)

are followed by times of greater abundance and celebration". "Almost as a message to self", she said, "as we have found with some aspects of modern changes along with climate change, we hold onto our culture's ancestral wisdom, but the enormous changes sometimes make us wonder when the good years will return".

The lack of meaningful control over subsistence management, along with marine and land use decisions directly threatens the survival of peoples who depend upon what the land, rivers, and sea offer. Climate change further exacerbates the practical and political diminishment of Native peoples' ability to put food on the table. The situation is made worse by the fact that Alaskan Natives generally no longer directly control their lands or waters. While Alaska Native tribes retain some of the inherent sovereign powers held by all tribes, they generally lack jurisdiction over activities on their aboriginal lands and waters (Ristroph, 2010). Restrictive government policies and Non-Governmental Organizations (NGOs) can curb customary use and occupancy and seek to protect the marine and terrestrial environment. Furthermore, human concern about the region has been integrated into the ecosystem as depicted in geologically ancient walrus sculpture, oral history, ceremony, and songs (Fig. 10.6).

10 Arctic Indigenous Sovereignty in Law and Policy

10.1 Shortcomings of Domestic Law

Given the shortcomings of state and Federal law regarding meaningful protection of territorial and subsistence rights, major Indigenous rights groups such as the Inuit Circumpolar Conference, the International Treaty Organization and others continue

Fig. 10.6 The melted early ice leaving walrus hunters with a disrupted hunting season as walrus had no place to rest and feed their young, thus complicating the ability of walrus to feed and hunters without a full food supply. (Source: http://www.marclesterphoto.com) (Photo © Marc Lester, Alaska Dispatch News, 2017)

to produce strategies to develop international law instruments that would produce protections for Indigenous sovereignty, self-determination, jurisdiction over aboriginal lands, and territorial rights, along with the sustainability of subsistence ways-of-life (Corntassel, 2014).

The central strategy has been to reverse, or limit, the imposition of colonialism and to challenge the principles of the legal systems that protect the colonial and neocolonial consumer societies that aggressively exploit natural resources (Whyte, 2016). Perhaps the largest battle from an Arctic perspective was to challenge federal and state support for the oil and gas industries, both on and offshore, and the commercial shipping industry that supports the transit of such resources (Watt-Cloutier, 2015). According to numerous Indigenous leaders, an objective was, and is, to build movements "powerful enough to force the policy changes" needed to stop, if not reverse, policies and processes that prevent Indigenous peoples, who depend upon subsistence living, from continuing to live in an interactive dependency with the natural world. The combination of political movements, careful diplomacy, and well-crafted lawsuits at the national and the regional levels have forced some positive actions to protect the region (Watt-Cloutier, 2015). However, it should be made clear that native oil, gas, and mining corporations can exercise power in Alaska to well-serve the native financial interests. On the complex and controversial issue of development of carbonresources, Karlin Itchoak, Director of the Alaska Branch of The Wilderness Society put it succinctly: "I may disagree with carbon fuel development practices, but I fully support the sovereign right of Indigenous peoples to generate revenues in the manner of their choosing – whether I oppose it or not, he said" (Parlow, 2021).

10.2 The Arctic Council

The Arctic Council, established in 1996, is the pre-eminent regional, intergovernmental body, whose primary members are the Arctic Coastal States with a central purpose of engaging in cooperative actions, particularly regarding matters of biodiversity, science and protection of marine ecosystems. The 1996 Ottawa Declaration established the Arctic Council, defining as member states: Canada, the Kingdom of Denmark (Greenland), Finland, Iceland, Norway, the Russian Federation, Sweden and the United States. Although the Arctic Council lacks enforcement capabilities, it has produced three Arctic Treaties. Its hard and soft laws are somewhat less bound by the primacy of the sovereign state than most regional or international entities with vigorous state-based structures regarding Indigenous interests. The Arctic Council's organizational structure includes a special membership category, Permanent Participants, for the increasingly vocal Indigenous participants who have expanded the Council's scope and vision.

The first three Indigenous Peoples' Organizations to become part of the Arctic Council were the Inuit through the Inuit Circumpolar Conference, the Saami through the Saami Council, and the Russian Indigenous Peoples through the Russian Association of the Indigenous Peoples of the North, RAIPON. The assigned status was of Permanent Participants, offering a higher status than a growing number of other Permanent Participants such as China, Singapore, and Spain. The number of Indigenous organizations as Permanent Participants has increased to the Aleut International Association (AIA), the Arctic Athabaskan Council, and the Gwich' in Council International (GCI).

While efforts of Indigenous advocates to be included in the Arctic Council succeeded to provide full consultation rights, this influencing, to an extent, state action, particularly regarding marine ecosystems, the Arctic Council contains a "possibility" (p. 691) for a right's based approach to Indigenous sovereignty, or, perhaps less quasi-sovereignty, and thus a truly more inclusive Arctic that is meaningfully responsive to the social movements seeking to amplify and realize the right to self-determination in the Arctic. The Arctic Council appears to be increasingly receptive to employ customary law, Treaties and meaningful rule making, particularly on global climate and environmental challenges that are shaping the twenty-first century (Kahn, 2019).

10.3 The United Nations Declaration on the Rights of Indigenous Peoples

In 2007, after three decades of efforts by Indigenous activists, attorneys, political and tribal leaders, and academics, the Declaration of Indigenous Peoples (UNDRIP) was adopted by the United Nations General Assembly (UNGA) with the support of a majority of member states; 143 in favor and 4 opposed: Canada, New Zealand, Australia, and the United States, all four nations with substantial Indigenous populations. Although Canada did endorse the Declaration in 2010, the government stated that UNDRIP is a "non-legally binding document that does not reflect customary international law nor changes Canadian law" (Corntassel, 2014).

UNDRIP has become one of the most important, albeit flawed, standards that protect a modicum of Indigenous interests, particularly where oil, gas, or mining are involved. Art. 26 of UNDRIP declares, for example, that "Indigenous peoples have the right to the lands, territories, and resources which they have traditionally owned, occupied or otherwise used or acquired". It intends to develop state-based accountability for ratifying states. But, the states must first, ratify.

Art. 19 requires "free, prior, and informed consent" when setting policies or preparing for commercial activities regarding oil, gas or mining, that may have the greatest potential impact on Indigenous peoples' subsistence economies. Oglala Sioux scholar, White Face (2013) observed that the current language of Art. 19 takes away the right of Indigenous peoples to devise their legislation and cedes that authority to the state. Instead of promoting reciprocal relationships on a nation-to-nation grounds, particularly those on the marginalized communities, Anaya (2009), a Special Rapporteur on the Rights of Indigenous Peoples to the United Nations, observes that the principles stressed in the declaration have political and legal force since they "are simply derived from human rights principles of equality and self-determination that are deemed of universal application". While this remains an open question in practice, it is incontrovertible that, in the Arctic, Indigenous rights and Indigenous peoples are increasingly successfully advocating to be included in the decision-making bodies that impact their lives; despite still a long way to go (Nicol, 2010).

10.4 The Inuit Circumpolar Council (ICC)

In August, 2021, the Inuit Circumpolar Council (ICC) leadership, and others – representing some 180,000 Inuit across the Arctic region including Alaska, Greenland Canada, and Russia – met to expand strategies for greater inclusion in the stable, rule-based region. Defined by cooperation and coordination as the Arctic faces dramatic and chaotic climate change, the objective is to make the region more accessible to commerce, mainly oil, gas, mining and shipping. The combination of warming and marine pollution presents an existential threat to those whose lives depend upon hunting, whaling and fishing in the Arctic's bio-diverse marine and terrestrial ecosystems. The increasing commerce, that is structured by law and practice to serve primarily outside interests, threatens the seals, walrus and whales that provides the primary food source to subsistence hunters and fishers, as well as commercial fishers. The Arctic's Native subsistence peoples,

As the Arctic Native peoples are fighting back in policy and law, their millennia of experience and observational science, knowledge, and priorities that have been largely ignored or marginalized, is seeking to mitigate losses from the commercial priorities, which serves not only the interests of the High North, but also the rest of the planet for whom the Arctic serves as its global thermostat.

It is anticipated that a final document developed from a series of ICC Arctic-wide meetings will be released to the public, national governments, and the Arctic Council at the 2022 ICC General Assembly. Jimmy Stotts, the Alaskan ICC representative, stated that in this and other contexts, that "we're rebuilding sovereignty one step at a time".[13]

11 Restoring Balance and the Crime of Ecocide in International Law

Indigenous leaders have signaled their intention to pursue an avenue of Indigenous rights throughout the circumpolar North as part of a post-colonial process that includes the widespread use and creation of international human rights instruments (Kronk Warner & Abate, 2013). Following World War II, a solid body of law transformed the scope and nature of international law by establishing standards on war crimes and crimes against humanity (Bankes, 2004: 103). Native communities are becoming part of an emerging global effort to build a regime of international environmental laws that would criminalize environmental destruction. Advocates of this trend want the crime of 'ecocide' to be included as the fifth crime against peace, which can be dealt with by the International Criminal Court (ICC).

With an existential threat to the planet, combined with the territorial and human rights claims of Indigenous peoples, emerging efforts by Indigenous and non-indigenous peoples alike to create a crime of ecocide offers not only the ability to bring criminal liability, but also a legal duty of care, and thus, justice for all living beings. These efforts contain the potential to invigorate a mutuality of interests amongst a variety of stakeholders who oppose large-scale decimations of natural resources for any reason. They could soon have an avenue to seek accountability, including animals, mammals, marine, and terrestrial life.

While this is only one step, and short of a recovery of territorial sovereignty, such a tribunal to consider crimes against the environment, would not only be good for

[13] Phone interview by Parlow (August, 2021).

the planet, its ecosystems, and its species, but would also help rebalance the wrongfully disproportionate allocations of environmental injury to the Indigenous peoples who have been stewards of the Arctic region for more than 2500 years.

When a Savoongan Yupik world class walrus ivory carver, whose work is displayed in the Smithsonian Museum in Washington D.C., was asked if he thought St. Lawrence Island might, 1 day, serve as a plaintiff regarding the crimes of ecocide, he nodded in the affirmative, and then smiled.

12 Conclusion

As Indigenous peoples throughout the Arctic assert their long-denied inherent rights to sovereignty through litigation and lobbying for legislative change, successes in advocacy, policy and international law are increasingly visible. With strategic assertions of self-determination at local, state, national and international levels, the headwinds that once appeared insurmountable are beginning to ease.

However, much work remains. Past legal and administrative decisions that still hold Native peoples in a "guardian-ward-relationship" for the "beneficial interest" of the tribe. Furthermore, the United States continues to hold a "fiduciary interest" on behalf of Native peoples, or, even 'co-management' of natural resources on customarily used lands. These policies, born of a colonial era, are worthy of a twenty-first century rethink.

In the Arctic region, Indigenous strategies are placing Native inherent rights and interests on a more equitable level. For instance, Indigenous leaders are more often seated at policymaking tables to make decisions on marine and terrestrial ecosystems and development issues. However, while large-scale Alaskan Native Corporations do help set the agenda for commercial development in the energy-rich Alaskan state, the perspectives of subsistence Indigenous leadership, who seek to raise specific questions for scientific inquiry and, ultimately, place their understanding of ecosystem protection policies into a legal, political, and human rights framework, remain restricted. Without the voices of subsistence Indigenous Peoples, there is little legal and political meaning to the inherent rights of self-determination, including its related responsibilities.

Further, the global existential crisis of climate change is upending long-held policy views regarding permissible ranges of shipping, mining, and oil production as it impacts the biodiversity of marine and terrestrial ecosystems. The Arctic Indigenous peoples, like the St. Lawrence Islanders, given their thousand years of observations, have an advanced, broad, scientific, and complex understanding of how ecosystems are integrated. The combination of a cross-political boundary approach, combined with day-to-day insights into migrations, weather patterns, and a profound understanding of the sequence and elements of sustainability in food chains, would serve a planetary interest and the species that inhabit it, to renew, or recover. Broad public support for nation-to-nation status would allow Native people to both exercise their right to self-determination and assist the global citizenry to improve the world's climate protection and resiliency practices that would, simultaneously, advance broader environmental, economic, and political objectives that positively affect our fragile planetary ecosystem.

References

Ackerman, R. E. (1976). *The Eskimo people of Savoonga* (1st ed.). Indian Tribal Series.

Ackerman, R. (1988). Settlements and Sea mammal hunting in the Bering-Chukchi Sea Region. *Arctic Anthropology, 25*(1), 52–79.

Act of July 27, 1868, ch. 273, 15 Stat. 24 0 and Alaska Law Review, vol 33:2. https://www.amazon.com/Eskimo-People-Savoonga-indian-tribal/dp/B003CVBX50

Alaska Stat. #16.05.94. https://www.touchngo.com/lglcntr/akstats/statutes/title16/chapter05/section940.htm

Alexander, V. (2020). Why is the Bering Sea important? *Bering Climate*. Available at https://www.beringclimate.noaa.gov/essays_alexander.html

Anaya, S. J. (2009). The right of indigenous peoples to self-determination in the post-Declaration era. In C. Charters, & R. Stavenhagen (Eds.), *Making the declaration work: The United Nations Declaration on the Rights of Indigenous Peoples* (IWGIA, Document No. 127, pp. 184–199). Copenhagen. https://www.iwgia.org/images/publications/making_the_declaration_work.pdf

Anderson, R. T. (2016). Sovereignty and subsistence: Native self-government and rights to hunt, fish, and gather after ANCSA. *33 Alaska Law Review*, 187–227. Available at https://scholarship.law.duke.edu/alr/vol33/iss2/3

Apassingok, A. (1985). *Lore of St. Lawrence Island: Echoes of Our Eskimo Elders, Vol. 1: Gambell, Bering Strait School District* (1st ed.).

Apassingok, A. (1987). *Lore of St. Lawrence Island: Echoes of Our Eskimo Elders, Vol. 2: Savoonga, Bering Strait School District* (1st ed.).

Arctic Council. (2009). *Arctic Marine Shipping Assessment 2009 report*. Second printing. Available at https://www.pame.is/images/03_Projects/AMSA/AMSA_2009_report/AMSA_2009_Report_2nd_print.pdf

Audubon Alaska. (n.d.). *Bering Sea: The ecological significance of the Bering Strait region is tremendous*. Available at https://ak.audubon.org/conservation/bering-sea

Bankes, N. (2004). *Kyoto, Constitutional Law and Alberta's Proposals*. Alta. L. Rev.

Berger, T. R. (1985). *Village journey: The report of the Alaska native review commission*. Hill & Wang Pub.

Case David, S. (1984). *Alaska natives and American laws*. University of Alaska Press.

Cochran, P., et al., (2013). Indigenous Fraeworks for observing and responding to climate change in Alaska. *Climate change, 120*, 557–567.

Cochran, P., Huntington, O. H., Pungowiyi, C., et al. (2013). Indigenous frameworks for observing and responding to climate change in Alaska. *Climatic Change, 120*, 557–567. https://doi.org/10.1007/s10584-013-0735-2. Available at http://ndl.ethernet.edu.et/bitstream/123456789/69226/1/Julie%20Koppel%20Maldonado.pdf#page=57

Coggin, J. D. P. (2019, October 15). *New report highlights Alaska's last five years of dramatic climate change*. Climate.gov. Available at https://www.climate.gov/news-features/understanding-climate/new-report-highlights-alaska%E2%80%99s-last-five-years-dramatic-climate

Cohen, F. S. (1941). *Handbook of Federal Indian Law*. Washington, United States Government Printing Office. Available at https://thorpe.law.ou.edu/cohen.html

Corntassel, J. (2014). Our ways will continue on: Indigenous approaches to sustainability. In T. Mitchell (Ed.), *The internationalization of indigenous rights: UNDRIP in the Canadian Context* (pp. 65–72). Centre for International Governance Innovation. Available at https://

www.jstor.org/stable/pdf/resrep05242.15.pdf?refreqid=excelsior%3Aaa41b4e0c3b5e8f7b69 0b9cffefa7d62

Crowell, A. L., & Oozevaseuk, E. (2006). The St. Lawrence Island famine and epidemic, 1878–80: A Yupik narrative in cultural and historical context. *Arctic Anthropology, 43*, 1–19. https://doi.org/10.1353/arc.2011.0105

Demer, L. (2017, May 26). For two Alaskan villages, walruses remain essential. *Anchorage Daily News*. https://www.adn.com/features/alaska-news/rural-alaska/2017/05/26/for-two-alaska-villages-walruses-remain-essential-as-sea-ice-disappears-can-it-last/

Dolin, E. J. (2011). Fur, fortune, and empire: The epic history of the fur trade in America. *Journal of American History, 98*(1), 182. https://doi.org/10.1093/jahist/jar170

Druckenmiller, M. L., Daniel, R., & Johnson, M. (2019). *Voices from the front lines of a changing Bering Sea: An Indigenous Perspective for the 2019 Arctic Report Card*. Bering Sea Elders Group, National Snow and Ice Data Center and Pew Charitable Trust.

Gricius-Abbott, G. (2021). Entire Alaskan communities are sinking into the ocean – And many can't move inland. *New America*. Available at https://www.newamerica.org/future-land-housing/blog/entire-alaskan-communities-are-sinking-into-the-oceanand-many-cant-move-inland/

Horowitz, L. S., Arn, K., Francis, L., Thierry, R., Stephan, S., & Sophie, T. (2018). Indigenous peoples' relationships to large-scale mining in post/colonial contexts: Toward multidisciplinary comparative perspectives. *The Extractive Industries and Society, 5*(3), 404–414.

Huntington, H. P., & Noongwook, G. (2013). *Traditional knowledge and the Arctic environment – How the experience of indigenous cultures can complement scientific research*. The Pew Charitable Trusts. Available at https://www.pewtrusts.org/~/media/legacy/oceans_north_legacy/attachments/arctic-traditional-knowledge-brief%2D%2Dartfinal_v4_web.pdf

Huntington, H. P., Quakenbush, L. T., & Nelson, M. (2017). Evaluating the effects of climate change on indigenous marine mammal hunting in northern and Western Alaska using traditional knowledge. *Frontiers in Marine Science, 4*, 319. https://doi.org/10.3389/fmars.2017.00319

International Work Group for Indigenous Affairs (IWGIA). (2010). *Indigenous Affairs 1-2/10 – Development and Customary Law*. Available at https://www.iwgia.org/en/resources/publications/305-books/2814-indigenous-affairs-1-22010-development-and-customary-law.html

Kahn, B. (2019). Some Arctic Sea Ice is Acting Like It's Mid-Summer. *Gizmodo*.

Khachatoorian, T. (2016). *Million-acre St. Lawrence Island land title signed over to native population*. https://www.alaskasnewssource.com/content/news/Massive-land-transfer-388618852.html

Koivurova, T., & Hasanat, W. (Eds.). (2015). *Current developments in Arctic Law* (Vol. 3). University of the Arctic Thematic Network on Arctic Law. Available at https://lauda.ulapland.fi/bitstream/handle/10024/62310/Current_Developments_in_Arctic_Law_3_pdfA.pdf?sequence=5&isAllowed=y

Kronk Warner, E. A., & Abate, R. S. (2013). International and domestic law dimensions of climate justice for Arctic indigenous peoples. *Revue générale de droit, 43*, 113–150. https://doi.org/10.7202/1021212ar

Laidler, G. J., Ford, J. D., Gough, W. A., Theo, I., Gagnon, A. S., Slawomir, K., Kevin, Q., & Celina, I. (2009). Travelling and hunting in a changing Arctic: Assessing Inuit vulnerability to sea ice change in Igloolik, Nunavut. *Climatic Change, 94*(3), 363–397. https://doi.org/10.1007/s10584-008-9512-z

Linxwiler, J. (2007). *The Alaska Native Claims Settlement Act At 35: Delivering on the promise*. Paper 12, 53rd, Annual Rocky Mountain Mineral Law Institute. https://ancsa.lbblawyers.com/wp-content/uploads/ANCSA-at-35-Delivering-on-the-Promise-Proof-10-25-07.pdf

Loring, P. A., Craig, G., Atkinson, D. E., & Murray, M. S. (2011). Ways to help and ways to hinder: Governance for effective adaptation to an uncertain climate. *Arctic, 64*(1), 73–88. Available at http://www.jstor.org/stable/23025667

Metcalf, V., & Robards, M. (2008). Sustaining a healthy human–Walrus relationship in a dynamic environment: Challenges for comanagement. *Ecological Applications, 18*(sp2), S148–S156. Available at https://esajournals.onlinelibrary.wiley.com/doi/10.1890/06-0642.1

Moore, G. A. (1997, March 6). *Tanana Chiefs, leaders need to acknowledge complexities created by ANCSA*. Arcticcircle.uconn.edu.

National Geographic. (2019). https://www.nationalgeographic.com/magazine/issue/september-2019

Nichols, P. (2010). *Oil and ice: A story of Arctic disaster and the rise and fall of America's last whaling dynasty*. Penguin Publishing Group. ISBN: 9781101460955, 1101460954.

Nicol, H. N. (2010). Reframing sovereignty: Indigenous peoples and Arctic states. *Political Geography, 29*, 78–80.

Parlow, A. (2019). Why Climate Policymakers should listen more to Indigenous peoples. *Arctic Today*.

Parlow, A. (2021). Evolving Ecosystems: Arctic Fisheries are shifting Northward. *The Economist*.

Pennesi, K., Arokium, J., & McBean, G. (2012). Integrating local and scientific weather knowledge as a strategy for adaptation to climate change in the Arctic. *Mitigation and Adaptation Strategies for Global Change, Springer, 17*(8), 897–922.

Rayond-Yakoubian, J., Khokhlov, Y., & Yarzutkina, A. (2014). *Indigenous knowldege and use of bering strait region ocean current*. Final report to the National Park Service, Shared Beringian Heritage Program for Cooperative Agreement H99111100026. Available at https://kawerak.org/wp-content/uploads/2018/04/OC-report-for-web.pdf

Reiss, B. (2012). *The Eskimo and the oil man: The battle at the top of the world for America's future*. Business Plus.

Reynolds, J. (2021). *Momentum builds for permanent end to pebble mine*. NRDC. Available at https://www.nrdc.org/experts/joel-reynolds/momentum-builds-permanent-end-pebble-mine-0

Ristroph, E. (2010). Alaska Tribe's melting subsistence rights. *Arizona Journal of Environmental Law & Policy, 1*(1) SSRN: https://ssrn.com/abstract=2648710 or https://doi.org/10.2139/ssrn.2648710

Ross, K. (2006). *Pioneering conservation in Alaska*. University Press of Colorado. http://www.jstor.org/stable/j.ctt1wn0r8r.12

Sambo-Dorough, D. (2014). *Indigenous peoples' right to self-determination and other rights related to access to justice: The normative framework*. https://academiccommons.columbia.edu/doi/10.7916/D8MC988Q/download

Shadian, J. M. (2010). From states to polities: Reconceptualizing sovereignty through Inuit governance. *European Journal of International Relations, 16*(3), 485–510. https://doi.org/10.1177/1354066109346887. Available at http://citeseerx.ist.psu.edu/viewdoc/download?doi=10.1.1.823.2876&rep=rep1&type=pdf

Shadian, J. M. (2013). Of whales and oil: Inuit resource governance and the Arctic Council. *Polar Record, 49*(251), 392–405. https://doi.org/10.1017/S0032247412000484. Available at: https://arcticextractiveindustries.files.wordpress.com/2015/04/shadian-of-whales-and-oil-final.pdf.

Staalesen, A. (2020). *Russian Arctic shipping up 430 percent in three years – A total of 31,5 million tons of goods was shipped on the Northern Sea Route in 2019*. Read in Russian | Читать по-русски. Available at https://thebarentsobserver.com/en/industry-and-energy/2020/02/russian-arctic-shipping-430-percent-three-years

Swensen, T. M. (2014). Of subjection and sovereignty: Alaska native corporations and tribal governments in the twenty-first century. *Wicazo Sa Review, 30*(1), 100–117. https://doi.org/10.5749/wicazosareview.30.1.0100

The 1867 Treaty concerning the Cession of the Russian Possessions in North America by his Majesty the Emperor of all the Russias to the United States of America; concluded on March 30; ratified by the United States on May 28; exchanged and proclaimed by the United States on June 20.

The CONAS Project. (2020). Available at https://conas-ak.org/conas-community/savoonga-alaska/

The State Council Information Office of the People's Republic of China. (2018). *China's Arctic Policy*. https://www.uaf.edu/caps/resources/policy-documents/china-arctic-policy-2018.pdf

U.S. Committee on the Marine Transportation System (CMTS). (2020). *A ten-year projection of maritime activity in the US Arctic Region 2020–2030*. Available at https://www.cmts.gov/downloads/Arctic_FAQs_for_External_Partners_2019.09.25.pdf

U.S. Geological Survey (USGS). (2008). *Circum-Arctic resource appraisal: Estimates of undiscovered oil and gas North of the Arctic Circle*. USGS Fact Sheet 2008–3049. Available at https://pubs.usgs.gov/fs/2008/3049/fs2008-3049.pdf

Wadhams, P. (2016, September 26). *The global impacts of rapidly disappearing Arctic Sea ice*. Yale Environment 360, Yale School of the Environment. Available at https://e360.yale.edu/features/as_arctic_ocean_ice_disappears_global_climate_impacts_intensify_wadhams

Watt-Cloutier, S. (2015). The Right to be Cold: One woman's story of protecting her culture, the Arctic and the whole planet. *Allen Lane*.

Weeber, C. (2020, May 19). *Why capitalize "indigenous"?* https://www.sapiens.org/language/capitalize-indigenous/

White Face Charmaine. (2013). *Indigenous nations rights in the balance: An analysis of the declaration on the rights of indigenous peoples*. Living Justice Press.

Whyte, K. P. (2016). Indigenous food sovereignty, renewal and U.S. settler colonialism. In M. Rawlinson & C. Ward (Eds.), *The Routledge handbook of food ethics*. Routledge. Available at: https://papers.ssrn.com/sol3/Delivery.cfm/SSRN_ID2770056_code1439012.pdf?abstractid=2770056&mirid=1

Chapter 11
Spatial Distribution and Geosimulation of Non-timber Forest Products for Food Security in Conflict Area

Taisser H. H. Deafalla, Elmar Csaplovics, Mustafa M. El Abbas, and Mohamad H. H. Deifalla

Abstract The intercountry conflicts affect in a variety of ways and the magnitude of effects varies according to the characteristics of the concerned area. Civilians in conflict are often deprived of their income sources and pushed to acute food insecurity, such as the case of the Nuba Mountains of Sudan. Disruption of food systems and markets resulted in higher food prices and shortages of water and fuel, or the food itself. The war also impacts the soil structure and physicochemical properties, especially in the Landmines area, where explosive remnants of war and improvised explosive devices destroy farmland, mills, storage facilities, and machinery. Besides that, increasing insecurity and roadblocks prevented humanitarian convoys from reaching the most vulnerable. Therefore, this chapter aims to explore, analyze, and predict the spatial relationships that affect the collection of Non-Timber Forest Products (NTFPs) using geostatistical analyst tools. This research uses disaggregate statistics to test and model such a relationship, making it a viable methodology for studying NTFPs, associated with household food supply in Nuba Mountains. The study utilized ESRI ArcGIS software version 10.1 for computation, exploratory analysis, mapping, and visualization. Ordinary Least Squares, Geographically Weighted Regression, and spatial autocorrelation analysis were used to map spatial patterns, test relationships, geo-visualize, and check for redundancy among the explanatory variables. These models allowed the pattern of association to be visualized on a map and all statistical values to be spatially represented on raster maps.

T. H. H. Deafalla (✉) · E. Csaplovics
Institute of Photogrammetry and Remote Sensing, University of Dresden, Dresden, Germany
e-mail: Elmar.Csaplovics@tu-dresden.de

M. M. El Abbas
Faculty of Forestry, University of Khartoum, Khartoum, Sudan

M. H. H. Deifalla
Department of Chemical Engineering, Sudan University of Science and Technology, Khartoum, Sudan

M. Behnassi et al. (eds.), *The Climate-Conflict-Displacement Nexus from a Human Security Perspective*, https://doi.org/10.1007/978-3-030-94144-4_11

The research concluded with some recommendations to guide regulators, policy-makers, and development agencies in making interventions that will have sustainable and equitable implications.

Keywords Conflict · Geostatistical techniques · Food security · NTFPs · Nuba Mountains

1 Introduction

Countries emerging from shocks, in particular, the armed conflict, face enormous challenges. Among these, the prominent ones are institution building, peace settlement, and sustainable economic recovery. Overcoming such challenges requires mobilizing massive funds to support various human, social, and economic interventions during the reconstruction phase. Unfortunately, over time, the economies of these conflict-torn societies are invariably seriously weakened and their institutional, as well as physical infrastructures, are damaged and in desperate need of repair (Behnassi, 2017; Deafalla et al., 2018).

Further complicating is the likelihood that the State is extremely weakened, lacks organization and capacity, besides budget deficits, large government debt, overvalued exchange rates and shallow financial resources to establish recovery activities without international aid. It can say, post-conflict economies are not 'normal' economies, and thus require strategies and policies for national development that are specifically tailored for such contexts. Sudan is one of these countries. The conflict added more load to ongoing problems of political instability in the country. It is not yet clear how the value chains of products will be shaped by conflict in the long term. Therefore, the potential cascading effects from rural livelihoods to markets to food security and national security are critical to understand.

In developing countries, most poverty reduction strategies are predicated on improving agricultural production and promoting market access and integration of smallholder producers in formal market exchange (Nang'ole et al., 2011). Development practitioners and researchers have utilized value chain approaches to examine the inter-relationships between diverse actors involved in all stages of the marketing channel and to capture the interactions of the increasing market dynamics and complexities in those countries (Bair & Peters, 2006; Pietrobelli & Saliola, 2008).

Kaplinsky and Morris (2001) defined the value chain as a description of the full range of activities which are required to bring a product/good or service from conception to its end use and beyond, through the different phases of production involving a combination of physical transformation and the input of various producer services, delivery to final consumers, and final disposal after use. It is simply a framework which helps understand how the world works and does not exist in the sense of having a tangible reality (Mitchell et al., 2009). It can be expressed, in a general form, as described in Fig. 11.1 below.

Fig. 11.1 The value chain description. (Sources: AIMS, 2016)

In recent years, the commercialization of Non-Timber Forest Products (NTFPs) has been widely promoted as an appropriate means of developing forest resources (Lawrence, 2003). The contribution of NTFPs to rural livelihoods is being increasingly recognized as it supports subsistence and generates financial income (Belcher, 2003), particularly in conflict areas (Deafalla, 2019). At the same time, because the harvesting of NTFPs is generally considered to be less damaging to forest resources than timber extraction, the exploitation of NTFPs is widely believed to be relatively compatible with forest conservation (Belcher, 2003; Newton et al., 2006). Thus, the commercialization of NTFPs potentially offers a means of achieving conservation and development goals concurrently (Marshall et al., 2003; Deafalla, 2011) by increasing the value of forest resources to local communities (Dickinson et al., 1996).

Nowadays, the value chain analysis has emerged on the new research agenda for NTFPs with a focus on rural studies. Increasingly, it is acknowledged that dependency and links to forests go beyond village boundaries. NTFPs contribute significantly, not only to the livelihood of rural residents (Angelsen & Wunder, 2003; Sunderlin et al., 2005), but also to the livelihood of migrants (Ambrose-Oji, 2003), national exchequers (Chamberlain et al., 2004), and the global economy (Leslie, 2005). The whole range of activities and relations associated with production, exchange, transport, distribution of a particular commodity, and the value chain approach (Kaplinsky, 2001) provide new practical insights into the markets (Gereffi & Frederick, 2010).

In developing countries, mapping the flow of input goods and services in the production chain provides a framework to analyze the nature and determinants of competitiveness in value chains in which small farmers can participate. It also provides the basic understanding needed for designing and implementing appropriate development programs and policies to support their in-market participation. Indeed, recently many development interventions utilized the value chain approach as an important entry point for engaging small farmers, especially, individually or collectively, in high value export markets (Rich et al., 2009).

The comparison of spatial patterns is a fundamental task in geography and quantitative spatial modeling has been used by analysts, scientists, and professionals on planning, managing, and estimation. Over the past decades, global methods of accessing and studying the relationships and spatial associations were weak and slow. For example, the traditional regression method, used by researchers, seriously suffers from a lack of information about location and attribute data and weightage values that have not been taken into account. Indeed, this weakness has stemmed from the advancement and development of strings of local spatial statistical models, which are often referred to as disaggregate statistics. Recently, with the growth of data being collected with a geospatial element, there is an increased interest in analyses requiring spatial pattern comparisons (Long & Robertson, 2017). Local spatial statistical models are designed to capture both spatial association and diversity (heterogeneity) simultaneously. Indeed, the 'one model fits all' syndrome that characterized global statistical techniques has motivated modern geographers and other spatial analysts to model and explore the local pattern of relationships that exist between variables.

Unfortunately, till now, the academic literature has not provided satisfactory answers. To address this shortfall of data, the objectives of this study are to explore, analyze, and predict the spatial relationships that affect NTFPs' value chains. It uses geostatistical analyst tools and reveals the gaps in knowledge, particularly in African countries emerging from armed conflicts, focusing on economic impacts and consequences. It also discusses the policies that can help promote the security of supply and ensure a sustainable economic recovery. The above gives practical insights into how the economic reforms and development of the market system must be adapted to reduce poverty through the use of value chains, besides strengthening regional value chains and reducing vulnerability to external shocks. This research also identifies the available options and guidelines for sustainable economic development.

The research takes a holistic approach to determine people's (mainly targeted communities) knowledge, attitudes, and perceptionsrelating to conflict, their needs, suitable interventions, and the effectiveness of competent authorities. The findings will inform adaptations of ongoing programmes and the design of suitable policy responses to strengthen regional value chains, reduce vulnerability to external shocks, and build economic resilience against future crises. It will help simulate and predict the dependency scenarios of NTFPs and forest changes to evaluate their impact on poverty alleviation and uses for the prediction of livelihood security. Specifically, the research focuses on the most important NTFPs according to local knowledge, namely, *Ziziphus spina-christi*, *Balantitesaegyptiaca*, *Adansonia digitata*, *Tamarindus indica*, *Acacia nilotica*, *Grewia tenax,* and *Acacia senegal*.

2 Study Site

The study site,Nuba Mountains of Sudan, is located in Southern Kordofan State (Fig. 11.2), bordered by Darfur in the west, Abyei and Republic of South Sudan in the south, White Nile and North Kordofan state in the north, with five districts Dilling, Kadugli, Rashad, Talodi and Abujubayha (Deafalla et al., 2018). The study

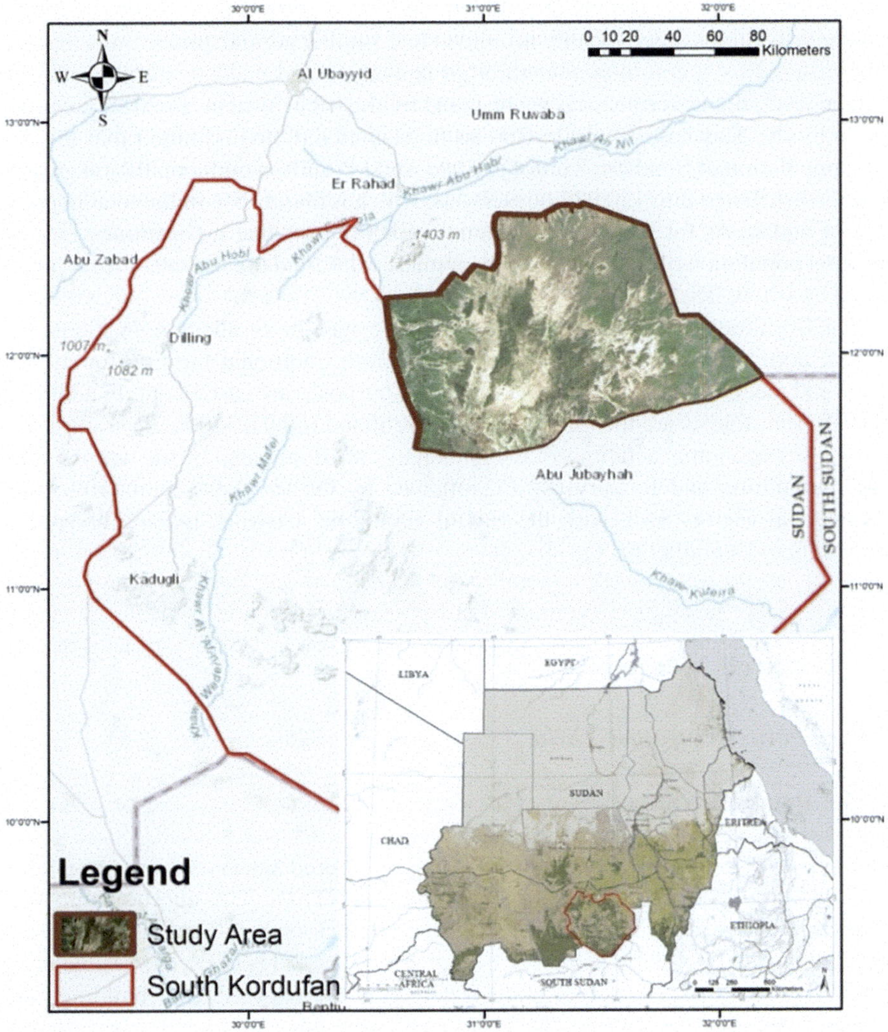

Fig. 11.2 Location of the Study Area. (Source: DIVA-GIS, developed by the authors)

area lies in the Rashad district of the State. It is biodiversity rich with various natural tree species. However, the existing risks and vulnerabilities resulting from the impacts of environmental and climatic changes, high poverty rate, ethnic and religious fractionalization, ethnic conflicts between nomads and sedentary farmers, and the socio-political conflicts, are destabilizing the critical areas in the western and southern parts of the State. The interaction of these factors compounds the risk to a higher level, compared with other states of the country (Deafalla, 2019).

With a total area of 135,696 km^2, the study site lies between latitudes 11° and 12° N, and longitudes 30° and 32° E (Deafalla, 2019). Three different locations in the

Eastern Nuba Mountains were studied, namely, Elabbassia, Rashad, and Abu Karshola localities. Each locality is divided into smaller administrative units that are administered by a commissioner through legislative and executive bodies. At the village level, the government is represented by the tribal system. In 2008, a census made by the Sudanese Centrale Bureau of Statistics (CBS) admitted that the the total population of Southern Kordofan State was 1.3 million with population growth (2.4%) distributed into 120,986 households, which is about 15% of the total population of Sudan. As for the urban population, CBS reports that it constitutes 21% of the total population of South Kordofan, nomads 1.1%, and the sedentary rural population 76.9% (CBS, 2009).

The livelihood activities found in the area are agro-pastoralism, nomadic pastoralism, horticulture, and rain-fed agriculture. Both traditional farming for subsistence and mechanized farming for commercial purposes are carried out. In addition, NTFPs from forests are the third source of livelihood (UNDP, 2006; Deafalla et al., 2018). Key economic activities of the inhabitants are dominated by the work in fishing, agriculture, and forestry (82%), followed by the services sector (10%), the commercial sector (4%), with the rest of economic activities making about 2% according to Deafalla (2011).

3 Research Methods

3.1 *Socio-Economic Data*

3.1.1 Data Collocation

The household survey was conducted using structured interviews among households who collect or trade each of the selected NTFPs. The questionnaire was divided into two parts. The first one captured the factors that affect the decision of an individual or household to collect, consume, and market NTFPs. Factors included gender, family size, education, main occupation, types of material building, and energy types used by households. The second part aimed to identify the different cultural and technological aspects of NTFPs adopted by different communities in utilizing, harvesting, and storing these products. In total, 224 household collectors and 50 household traders were interviewed in villages, local markets as well as markets in North Kordofan, White Nile and Khartoum States, where the collection and trading of the selected NTFPs are concentrated. Furthermore, a market survey, based on it, was used to obtain quantitative data on prices, taxes and quantities sold. This data was integrated with other data from 2008 (322 questionnaires) to help make a comparison between the past and the current situation of NTFPs, as well as to be used in simulation model.

The specific trade data was obtained from statistics of local and international trade and through purposive interviews at different levels of the trading chain (merchants, agents, FNC, Ministry of Environment, Forestry and Physical Development, and Ministry of Foreign Trade, Sudan). The interviews were also carried out in different areas which covered value chains of these products (based on Deafalla, 2011) namely: South Kordofan (Rashad and Elabbassia localities); North Kordofan (Um Rawaba and Alrahad localities); White Nile (Kostiy locality); and Khartoum States (Omdurman, Bahri, and Khartoum localities).

3.1.2 Data Analysis

Descriptive statistical analyses were applied to analyze data concerning social characteristics and respondents' perspectives about different aspects of the NTFPs value chain activities. Furthermore, Spearman Rank-Order (SRO) correlation and Pearson Correlation Coefficient (PCC) were applied to test the relationship between these factors and to estimate the probability of participation of households in the collection of NTFPs, besides calculating the relative magnitudes of such probabilities. This was based on different socio-economic/ecological factors (ethnicity, gender, education, age, main occupation, duration of the main occupation, war, migration, markets, etc.) that affect such activities.

3.2 Remotely Sensed Data

Earth observation data provide an immense amount of information on LU/LC data (Phiri & Morgenroth, 2017). Accordingly, it is used to study the land cover (LC) classification in the study area. Landsat (2014) imagery, in addition to high-resolution Digital Elevation Model (DEM) data from Advanced Spaceborne Thermal Emission and Reflection Radiometer (ASTER) and meteorological information, were used to gather the spatial characteristics of the region such as: vegetation cover (density and type); terrain features; and surface temperature among others. To describe and verify the ground-types adequately, several ground-truthing methods were used. The first method involved the field points or ground-truthing. A total of 325 points were gathered during the field visit randomly from the secure areas in the study site representing different LC types, and to create a 'test set' for the classification accuracy. The ground reference points were collected by using the GARMIN eTrex Venture HC GPS device to record the coordinates of each visited point.

3.2.1 Data Pre-processing

Pre-processing is necessary for extracting and quantifying meaningful information from remotely sensed data (Iqbal & Khan, 2014; Butt et al., 2015). The typical pre-processing operations in the present study included geometric correction, radiometric correction and topographic normalization and image enhancement by using

Environmental Visualization (ENVI), Earth Resources Data Analysis System (ERDAS) Imagine and ArcGIS software.

3.2.2 Image Classification

(a) *Image Segmentation*

Multiresolution segmentation strategy was applied to the dataset in the present study as a preliminary step of Geographic Object-Based Classification (GEOBIA) approaches to create meaningful unclassified image objects (object primitives) and to generate different abstraction levels. This segmentation algorithm is characterized as a bottom-up region-merging technique, applying a threshold optimization procedure for image segments, which categorizes the image objects according to a maximum acceptable heterogeneity based on a defined threshold to maximize their respective similarity (Baatz et al., 2004).

(b) *GEOBIA*

For this study, the object-based approach has been proposed for discriminating different LC classes based on group pixels with the analogous spectral and spatial response, based on predefined criteria to extract features of interest. The GEOBIA-hierarchical system was proposed in this study to attain LC classes. The rule-based classification of the study area was built on three types of knowledge. Firstly, the spectral domain knowledge was applied to construct the hierarchical structure of LC classes. Secondly, spectral classification rules were adopted based on training data, where the training data helps generate thresholds to be used later as rules for discriminating and classifying LC categories more accurately. Finally, to support spectral knowledge in the classification of LC, spatial rules based on user experiences were used to increase accuracy (Fig. 11.3). Trimble eCognitionTM Developer 8.7 software was used to analyze the images.

3.3 GIS Data

The study utilized ESRI ArcGIS software version 10.1 for computation, exploratory analysis, mapping, and visualization. It was chosen because it presents numerous extensions for spatial statistical and geostatistical modeling. Ordinary Least Squares (OLS) regression, Geographically Weighted Regression (GWR), and spatial autocorrelation analysis were used to map spatial patterns, test relationships, geovisualize, and check for redundancy among the explanatory variables. These models allow the pattern of association to be visualized on a map and all statistical values to be spatially represented on raster maps.

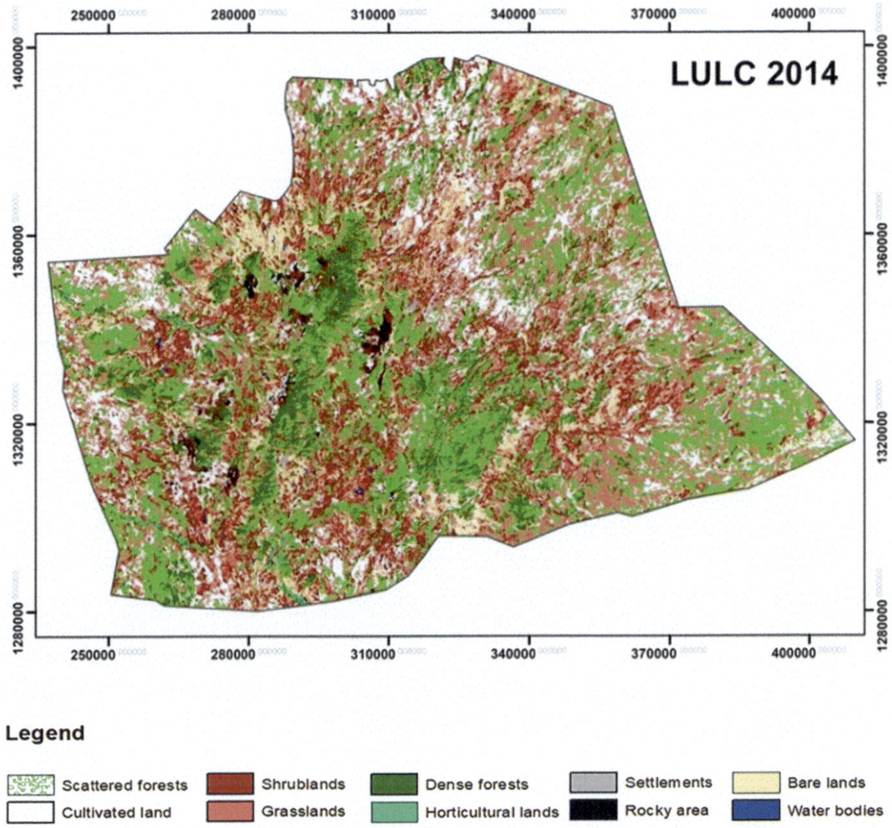

Fig. 11.3 LU/LC classes in study site

3.3.1 OLS

OLS is a global statistical model for testing and examining relationships between variables. It is defined by Hutcheson (2011) as a "generalized linear modeling technique that may be used to model a single response variable which has been recorded on at least an interval scale". The model relies on determining the dependent variable (Y) by producing an unbiased minimum sum of error square in Y in regard to the independent variable X (Shrestha, 2006). For unbiased observations, the prediction should be equal to the expected value of the dependent variable for a particular set of data (Fernandes & Leblanc, 2005). OLS is based on a set of assumptions: normality, homogeneity and independence of residuals (Montgomery et al., 2001). It uses a single equation to estimate the relationships between the dependent variable and the explanatory variable(s) and assumes a stationarity or static relationship across the study region. Based on Nkeki and Osirike (2013), the OLS model's equation for this analysis is presented as:

$$Y = \beta_0 + \beta_1 X + \varepsilon$$

Where: Y refers to the dependent variable, X are the explanatory variables, β_0 is the intercept, meanwhile β_1 is Coefficients and ε: Random error term.

The technique may be applied to single or multiple explanatory variables and also categorical explanatory variables that have been appropriately coded. In ecological applications, for example, it is (a far and away) the most widely-used regression technique for forest monitoring (Zhang & Shi, 2004; Berterretche et al., 2005).

In this study, the OLS was used as a diagnostic tool and for selecting the appropriate predictors with respect to their strength of correlation with the criterion variable for the GWR model. It can automatically check for multicollinearity (redundancy among predictors). Feature datasets were added as location data and were arranged as dependent variables and factors of influence on the dependent variables, which were fixed as explanatory variables. The result will be attributed with spatial pattern map, coefficient value and R2 value. The coefficient value is described to better understand the location of samples as statistically significant, while the R2 value explains the significant fraction between 0.0 and 1.0. The analysis used locations of NTFPs demand as dependent variable and heterogeneity such as: area of village, risk level, income, elevation, census, distance to market, distance to road, and precipitation and temperature as independent variables. The multicollinearity was assessed with the variance Probability and Robust Probability of the OLS. If the Probability and Robust Probability values are greater than 0.05, it therefore indicates the existence of multicollinearity among the predictors.

3.3.2 Moran's Index (Moran's I) and Spatial Regression

For more than 20 years, the theory of spatial autocorrelation has been a key element of geographical analysis (Chen, 2013). Spatial Autoregression (SAR) is one type of commonly used spatial models that take into account the effects of spatial autocorrelation by assuming it as an intrinsic part of the ecological process (Anselin, 2005). It plays an important role in geographical analysis and can predict a value of an outcome variable based on values of explanatory variables (Stieve, 2012). Although there are various correlation measurements, two are widely used. One is Moran's I (Moran, 1948), and the other is Geary's coefficient (Geary, 1954). The former is a generalization of Pearson's correlation coefficient, and the latter is analogous to the Durbin-Watson statistic of regression analysis. Compared with Geary's coefficient, Moran's index is more significant to spatial analysis (Chen, 2013). Stieve (2012) defined Moran's I as a "test for spatial autocorrelation, which examines whether a phenomenon is clustered or not". Its values range from +1 (positive autocorrelation) and −1 (negative autocorrelation).

In the current study, the autocorrelation statistic technique was used to see whether there exists a spatial auto-correlation or clustering of the residuals that violates the assumption of OLS. Progressively, the spatial independency of the NTFPs was assessed by using the global spatial autocorrelation coefficient Moran's I. This is defined by the equation according to Nkeki and Osirike (2013):

$$1 = \frac{n\sum_{i=1}^{n}\sum_{j=1}^{n} w_{ij}\left(y_i - y^{-}\right)\left(y_i - y^{-}\right)}{\left(\sum_{i\neq j}\sum w_{ij}\right)\sum_{i=1}^{n}\left(y_1 - y^{-}\right)^2}$$

As depicted in the equation above, (n) represents the total number of villages (polygons), (i) and (j) depict the various villages, (y) and (y_i) is the residuals of location (i) and (j) respectively, while (y) stands for the mean of the residual and (w_{ij}) represents a spatial weight matrix for measuring spatial proximity between (i) and (j) locations.

3.3.3 GWR

The global modeling techniques, such as OLS linear and other nonlinear models, cannot detect spatial variation and relationships within geographic entities. As a result, intrinsic relationships may be obscured and spatial association between variables in a region is concealed (Nkeki & Osirike, 2013). Such incomplete information derived from global statistics, when adopted for addressing policy issues, may be counterproductive. To strengthen this weakness, statistical geographers (Brunsdon et al., 1996; Fotheringham et al., 2002) recently came up with GWR, which has been introduced to solve such problems related to spatial non-stationary. It is a technique designed to explore spatial non-stationarity or heterogeneity in geographic dataset which global statistical models cannot explain the relationship between sets of variables (Nkeki & Osirike, 2013). The regression method models the local relationships between the predictors and an outcome of interest. It is capable of handling various relationships of variables in local spatial patterns for modeling, examining, monitoring, and decision making (Fotheringham et al., 2002; Shrestha, 2006). Such an analysis is conducted within a single framework, where the final outputs of the analysis provide a variation of variables in local spatial patterns and a map of the spatial variation in relationships (Fotheringham et al., 2002; Jamhuri et al., 2016).

The recent integration of GWR into ESRI ArcGIS has further increased the quality of output. For example, GIS-based GWR has the spatial capability of displaying the parameter estimates and coefficient of determination regarding all variables in a raster surface and vector map respectively for easy and quick visual interpretation of relationships and the detected spatial patterns. GWR has an immense application in almost every domain, including ecological and RS studies (Foody et al., 2003). For instance, it analyzes the forest structural attributes collected from the field measurements and spectral response extracted from remote coincided and sensed imagery (Foody et al., 2003; Lu et al., 2004). Khamis (2012) and Netrdová and Nosek (2016) used it to analyze the regional unemployment rate in Europe. Shi et al. (2006) engaged GWR to investigate the effects of local spatial heterogeneity on multivariate relationships of white-tailed deer distribution, using LC patch metrics and climate factors. In addition, Hu et al. (2015) used it to improve the ecosystem service in Fuzhou City, China.

On the other hand, Gao and Li (2011) applied GWR to examine spatially non-stationery and scale-dependent relationships between urban landscape fragmentation and related factors. While Clement et al. (2009) used GWR to analyze the factors that drive afforestation in Northern Vietnam, a study by Wang et al. (2005) employed it to estimate the Net Primary Production (NPP) of the Chinese forest ecosystems. Besides this, Propastin et al. (2008) applied it to investigate the impact of scales on the prediction of rainfall uncertainty. The underlying tenet of GWR is that parameters are estimated anywhere in the study region given a criterion variable and one or a set of explanatory variables which have been measured in a known location (Charlton & Fotheringham, 2009). It generates an equation for every component in the dataset by calibrating each variable using the target feature and its neighbors. In this respect, nearby features produce a higher weight in the calibration than distant features (Scott & Janikas, 2010). This approach may likely uncover spatial relationships or associations neglected by OLS (Nkeki & Osirike, 2013). The model multiplies geographically weighted spatial matrix consisting of geo-referenced data. The matrix defines the neighborhood spatial relationships between villages and aids the detection of spatial variation in the relationship among the variables. The traditional global regression framework can be represented in the following equation:

$$Y = \beta_0 + \beta_1 X_1 + \beta_2 X_2 + \ldots \beta_n X_n + \varepsilon \tag{11.1}$$

Where: (Y): Dependent variable, (X): Explanatory variables, while (β_0): Intercept, (β_n) Coefficients and (ε) Residuals.

The basic GWR model as developed in Eq. (11.2) by Fotheringham et al. (2002) is estimated as:

$$Y_i = \beta_0 \left(u_{i,} v_i\right) + \sum_k \beta_k \left(u_{i,} v_i\right) X_{ik} + \varepsilon_i \tag{11.2}$$

Besides standard regression notation, the term (ui, vi) denotes the geographic coordinates of the ith point in space. Hence, Eq. (11.2) yields i local regressions over each sample point, and a total of *ik* regression coefficients.

If the relationships do not exhibit spatial variability, (2) coincides with (1). Under standard regression assumptions, both equations can be solved by the OLS method, which includes a spatial weight for (2). The major output from GWR model for each observation (settlements) is a set of parameter estimates (local coefficients for each explanatory variable) and associated diagnostics (standard errors, Cook's D statistics, local R, statistic, and local standard deviation) that can be visualized within a GIS platform (Charlton & Fotheringham, 2009; Fernandez et al., 2013). The series of maps often generated are vital tools for understanding the level of spatial relationship and show locations of each predictor exhibiting a stronger influence on the dependent variable.

GWR was originally developed to analyze spatial point data to allow for the interpolation of values that are not included in the data set. It is applied under the assumption that the strength and direction of the relationship between a dependent variable and its predictors may be modified by contextual factors. In this application, spatial units (settlements) vary in size and shape over the study area. Consequently, an adaptive kernel was preferred to allow the automatic specification of an appropriate distance or number of nearest neighbors. This allows the spatial context (Gaussian kernel) to vary in extent as a function of feature density. It constructs a smaller spatial context, where the feature distribution is denser and larger on the one hand and distribution is sparse on the other. Shahid and Bertazzon (2015) defined the bandwidth of adaptive kernels as the number of nearest neighbors forming each local area. Based on Brunsdon et al. (1996), the weight inside the bandwidth reaches a monotonical zero as the distance increases. The study used Akaike information criterion to select the optimal bandwidth according to Weisent et al. (2012), using the bandwidths of 50 nearest neighbors.

3.4 Simulation Modeling

Simulation modeling methods are used to understand dynamic complex systems and to predict the outcome of change (Shahidand & Bertazzon, 2015). In this study, a simulation system dynamics model was specified in ArcGIS software version 10.1. The simulation model was structured around the GWR NTFPs model. The study used the scatter plot matrix graph to compare the datasets to look for patterns and relationships to be used in the simulation.

After testing variables, three major categories resulting from GWR analysis were identified and selected as explanatory variables. These include risk level, the sum of the forest area, and the average distance.

The dependent variable in this model is the quantities of NTFPs collected in the years 2008 and 2014. These statistical values were entered into the prepared GIS vector polygon map as non-spatial data. To visualize the spatial distribution of such data, hotspots were selected to calculate the elevation and density of forest (Dense and Scattered forest) based on the class of forest which resulted from the analysis of image 2014 for the selected villages in the study area (Fig. 11.4).

Once the simulation model was tested in the equilibrium village, interventions were introduced. The GWR model results were inspected to identify neighborhoods where communities' forest dependency is high, but the collection of NTFPs is low and significantly associated with the selected variables. The local NTFPs quantities collected interventions were simulated for the 19 selected settlements, based on two alternative scenarios: 0–10 points (low/negative); and a 10–20 points (high/positive) increase in risk level. All the two explanatory variables are modifiable at some scale. Based on the quantities of NTFPs collection from 2008 and 2014, the model is

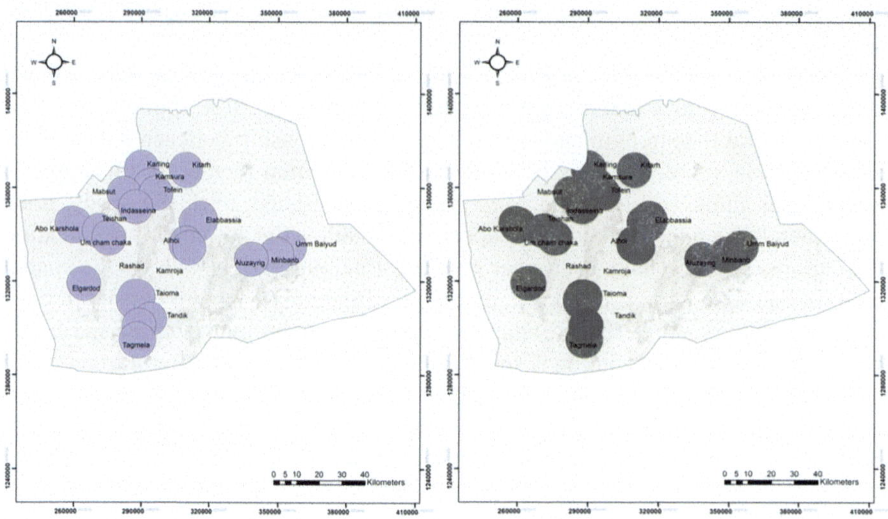

Fig. 11.4 Buffer zones created to calculate the forest density (right) and the elevation (left) within specific distance to settlements

running a simulation process that ranges from 2014 to 2030. After this initial intervention, GWR models were re-run on the newly collected NTFPs values, and the quantities collected were still significantly associated with the risk level. The same steps were taken to simulate the local forest density. Based on the area of forest in 2014, the model ran a simulation from 2014 to 2030. After this initial intervention, GWR models were re-run on the new density forest values, and as well the density was still significantly associated with the average distance for each village. Figure 11.5 shows the complete simulation model with all the GWR coefficients and variables.

4 Results and Discussion

Modeling the Heterogeneous Factors Affecting NTFPs

Recently, socio-economic benefits of service roles of forests have been recognized but have not been properly integrated into forest management for sustainable development (FAO, 2005). However, in most developing countries emerging from shocks like the armed conflict, estimates of socio-economic values of NTFPs are not available. Where they are available, they are normally underestimated. There is a vast array of NTFPs being traded locally and nationally. In Africa, for example, an assortment of fruits, spices, and medicines from various open-air markets for poverty alleviation are traded, though there is not much information to demonstrate this. However, when local people are well organized, they can often generate more income from NTFPs activities (Royal Swedish Academy of Agriculture and Forestry

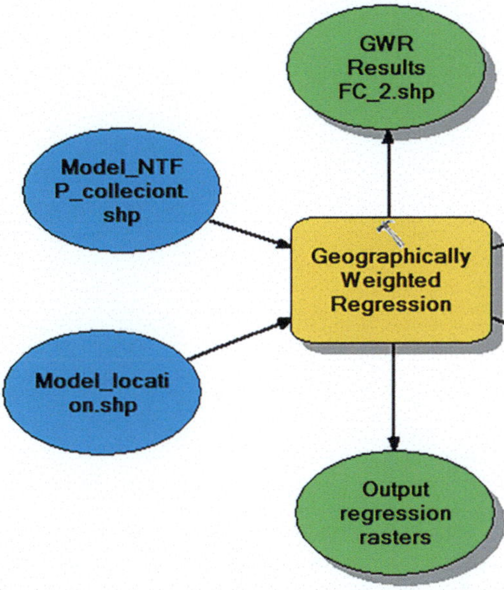

Fig. 11.5 The model's framework

et al., 2005). In Sudan, research on NTFPs is limited. Socio-economic research is even scarcer and is usually related to Gum Arabic and some resins (El Abass, 2009; Mohammed, 2011). Other studies such as El Amin and Ballal (1996), Deafalla et al. (2014) provided description data on the production and consumption of NTFPs. This study examined many factors that could affect NTFPs collection. Both models, global OLS and local GWR, were able to capture and detect prominent factors (variables) that influence NTFPs collection in the study area. However, in this study, only useful predictors without bias were entered into the local model. In the exploratory analysis using OLS, as shown in Fig. 11.6, many predictors were entered into the models based on their effects, such as income, distance to market, distance to road, average distance, sum area, and risk level that is calculated by descriptive analysis.

The OLS model was calibrated to diagnose multicollinearity among the explanatory variables and the result shows that the census, distance to market, distance to road, precipitation and temperature variables, returned Probability and Robust Probability (Robust_Pr [b]) values were higher than the set redundancy threshold of 10. Therefore, the variables were removed from the model. The final result of the OLS model is presented in Table 11.1, which shows, however, that all the predictors returned Large Variance Inflation Factor (VIF) values fairly greater than 1.0, indicating that none of the variables are redundant (Deafalla, 2019).

Among these explanatory variables, three are statistically significant, and these are: average of the distance from villages to forest, sum of the area, and the risk level (Fig. 11.7).

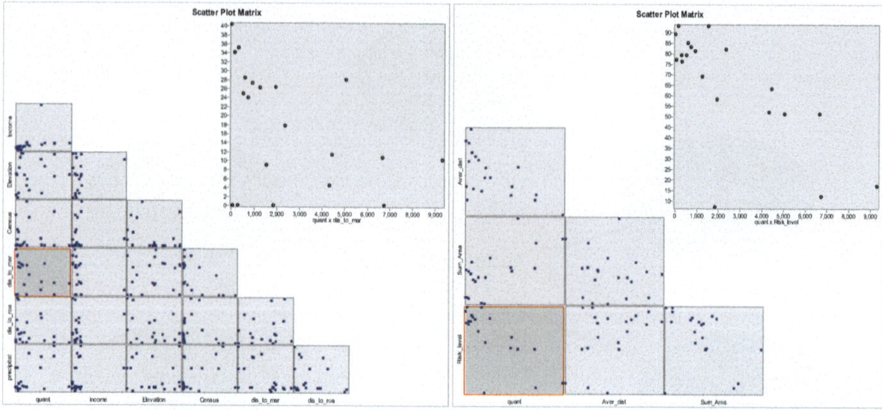

Fig. 11.6 Test of different explanatory variables

These variables are the most important concerning the distribution of NTFPs quantities collected. The average of the distance and the risk level returned negative relationships (Table 11.1). This finding implies that as the risk level and distance to forests increase, the quantities collected of NTFPs decrease. This result coincided with the current situation in Nuba Mountains (Deafalla, 2019).

The explanatory variables Average distance, Sum_area of the forest and risk_level returned significant t-values of −3.2, 3.7 and −3.1, respectively. Due to the current situation in the study area, this is probably the reason why it fails to return a significant t-value of the risk level and the average distance to the forest. This result, therefore, suggests that security may be an important variable for increasing the quantities collected of NTFPs. The present finding agreed with Deafalla et al. (2018) who noted that the security situation has a direct impact on the local livelihood. The OLS global model revealed that it explained about 78% (adjusted $R^2 = 0.788$) of the variation in the quantities collected of NTFPs and forest density (Table 11.2).

The ANOVA returned a significant F-value = 24.68 and the Wald statistic has a significant chi-squared value = 116.17. This means that the model proves to be statistically significant (Deafalla, 2019). Jarque-Bera statistic returned a non-significant chi-squared value = 0.08 (Table 11.2) indicating that the model's prediction is free from bias (i.e. the residuals are normally distributed). The chi-squared value (2.9) of the Koenker test is statistically significant. Importantly, it indicates the relationship between some, or perhaps all, of the explanatory variables and the criterion variable is non-stationary or consistent across the region. The explanation for this is that some independent variables may be important with respect to predicting the outcome of the quantities of NTFPs collected and forest density in some settlements, but in others, they may demonstrate weak predictive capability. It is evident that the model's fitness will likely be improved with GWR, since the Koenker statistic detected non-stationarity in the relationship. This is due to the fact that GWR assumes that relationships across space are non-static. However, the result was further confirmed statistically by applying spatial autocorrelation statistic (global

Table 11.1 Summary of OLS results – model variables

Variable	Coefficient [a]	StdError	t-Statistic	Probability [b]	Robust_SE	Robust_t	Robust_Pr [b]	VIF [c]
Intercept	5328.104017	1377.817630	3.867060	0.001367*	1325.904556	4.018467	0.000994*	–
Aver_dist	−863.257811	264.127995	−3.268331	0.004831*	241.277104	−3.577869	0.002515*	1.216477
Sum_area	0.928128	0.248665	3.732435	0.001815*	0.211186	4.394827	0.000453*	1.203993
Risk_level	−41.100535	12.873501	−3.192646	0.005664*	13.493969	−3.045845	0.007705*	1.411924

*Significant parameter at 0.05 level

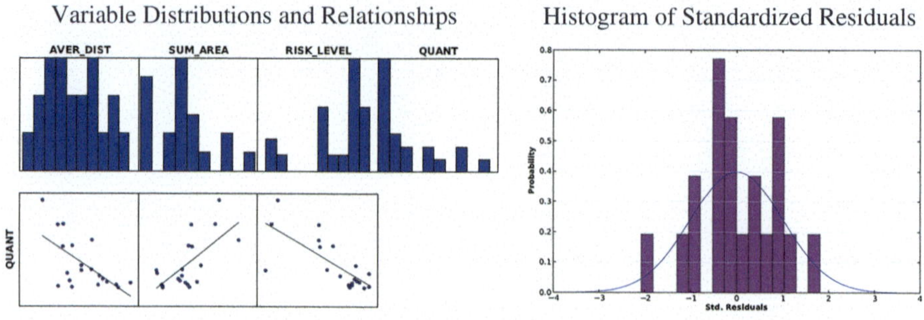

Fig. 11.7 Model NTFPs collection

Table 11.2 OLS Diagnostics

Parameters	Values	P-value
Joint F-statistic	24.680578	0.000003*
Joint wald statistic	116.172734	0.000000*
Koenker (BP) statistic	2.913721	0.405120
Jarque-Bera statistic	0.081070	0.960276

R^2 = 0.822305; Adjusted R^2 = 0.788987; AICc = 351.612035 *Significant parameter at 0.05 level

Moran's I). This will automatically detect significant clustering or random patterns in the residuals. The Moran's I report (Fig. 11.8) revealed that the pattern of the residuals is significantly different from random, with a Moran's index value = 0.05 and z-score value = 1.25. That is, the residuals have no statistically significant spatial autocorrelation. In this case, all empirical evidence points to the fact that the OLS residuals fit properly.

Some predictors exhibited high spatial variability in the resultant parameter estimates of GWR model. These predictors are sum of forest area, level of risk and the average distance. They are the same that resulted from OLS model. All reflected a combination of negative and positive coefficients across the study area whereas the OLS global coefficients for the sum of forest area and the level of risk returned negative values. This provides evidence that the relationship between the criterion variable and the explanatory variables, captured by OLS, is more complex and, for a reliable result, it needs a local model. The calibrated GWR results suggest that it is a significant improvement on the global model, comparing both models with the AICc values, show that the value is reduced from 351.612035 * for OLS model to 299.28 for GWR model. The difference is roughly 52.33, based on Nkeki and Osirike (2013) in these cases (that implying to the local models), fitness is higher when explaining spatial datasets such as the quantities collected of NTFPs and forest density (Table 11.3).

As expected, the GWR model enhanced the explaining power of the OLS model by about 11% (Table 11.2). Mapping the residuals of GWR indicates that itis randomly distributed (Fig. 11.10). This means the model is properly specified. Verifying with autocorrelation statistic (Moran's I) returned randomly distributed residuals

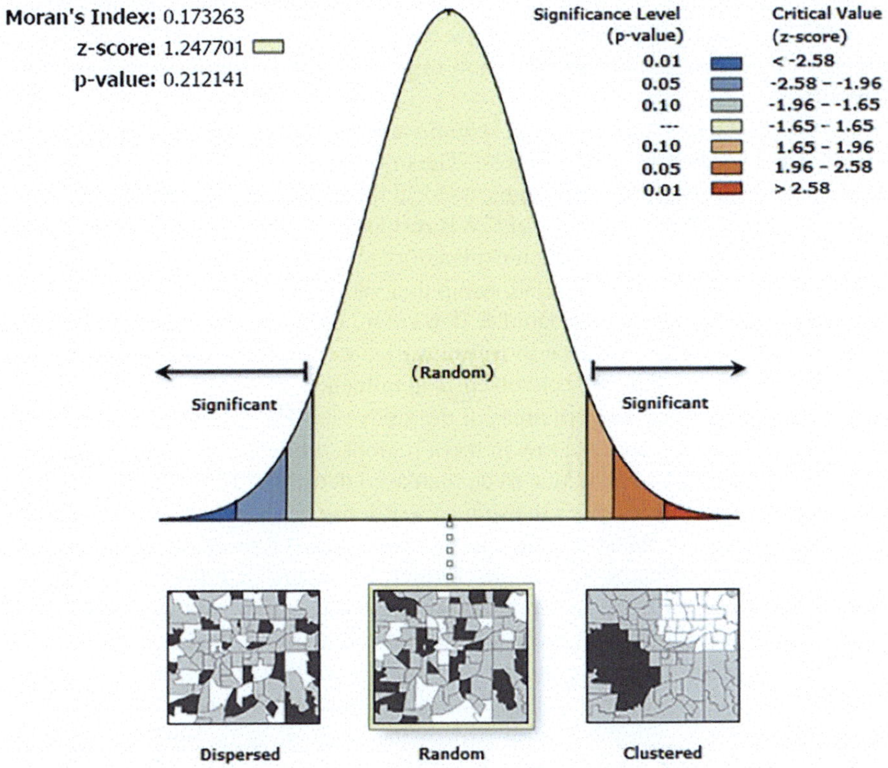

Given the z-score of 1.25, the pattern does not appear to be significantly different than random

Moran's Index	Expected Index	Variance	z-score	p-value
0.173263	-0.052632	0.032779	1.247701	0.212141

Fig. 11.8 Spatial autocorrelation regression report

Table 11.3 Comparison between OLS and GWR results

Fitness parameter	OLS	GWR
AICc	351.612035*	299.28
R^2	0.8	0.9
R^2 adjusted	0.78	0.85

*Significant parameter at 0.05 level

with a z-score = 1.24 and Moran index = 0.17. This means the model is properly specified. Verifying with autocorrelation statistic (Moran's I) returned randomly distributed residuals with a z-score = 1.24 and Moran index = 0.17. Figure 11.10

displays the R^2 value as a spatial smoothing of GWR model, showing the area where the model's prediction and strength of the relationship are improved. Furthermore, they show that there is regional variation in the strength of the relationship in the study region.

At the regional level, the resultant spatial variation shows that the strength of the relationship decreases from east to west. The only exception was in sum area of the forest variable. Thus, this pattern suggests local fluctuation in the relationship (non-stationarity). A fundamental merit of GWR resides in its ability to display and visualize the parameter estimate of each explanatory variable on a raster surface, which leads to rendering the complex relationship that varies over space easier to comprehend (Nkeki & Osirike, 2013; Shahid & Bertazzon, 2015). As shown by GWR local coefficients, the average distance to forests variable is an important factor for estimating the quantities of NTFPs collected. The influence of this predictor is stronger in the north, eastern and western areas of the study site (Fig. 11.9a). This is reasonable because the forest cover is low in these regions, while its influence in the central and south margins is weak due to an increased density of forests in these areas.

Another important variable is the sum area of forests, which has a high influence in the eastern parts of the study area, as shown in Fig. 11.9b. Most of the LU/LC represents Cultivated lands. Unlike the former, its sphere of influence is smaller. This predictor proved to be less relevant in the western and southern parts, even though there is a high concentration of households that utilize the lands under cultivation. Risk level, as a very significant predictor as well, exhibits a strong negative influence over the dependent variable in the northwestern part of the region as shown in Fig. 11.9c. On the central part of the study area, the influence is very strong and continues to the east and down to the south. The inverse relationship that risk level seems to reflect on the quantities of NTFPs collected, especially in the central parts, shows that it is a factor that should be looked at with respect to policy development and, as well, to conducting other management investigations.

The resultant raster surface for the predictors shows that there is spatial variation in the relationship between the selected explanatory variables and the quantities of

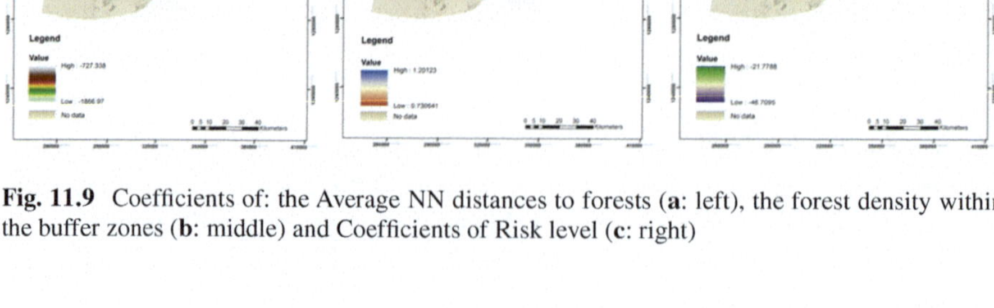

Fig. 11.9 Coefficients of: the Average NN distances to forests (**a**: left), the forest density within the buffer zones (**b**: middle) and Coefficients of Risk level (**c**: right)

NTFPs collected, as well as with forest density across the study area (Figs. 11.10 and 11.11). Positive and negative relationships were manifested in the result of GWR, where the positive relationship means that, as the forest area increases, the quantities of NTFPs collected will equally increase. On the other hand, the negative relationship means that, as the level of risk and the average distance increases, the quantities of NTFPs collected will, and in contrast, decrease. Local coefficient estimates for each explanatory variable are presented in Figs. 11.10 and 11.11.

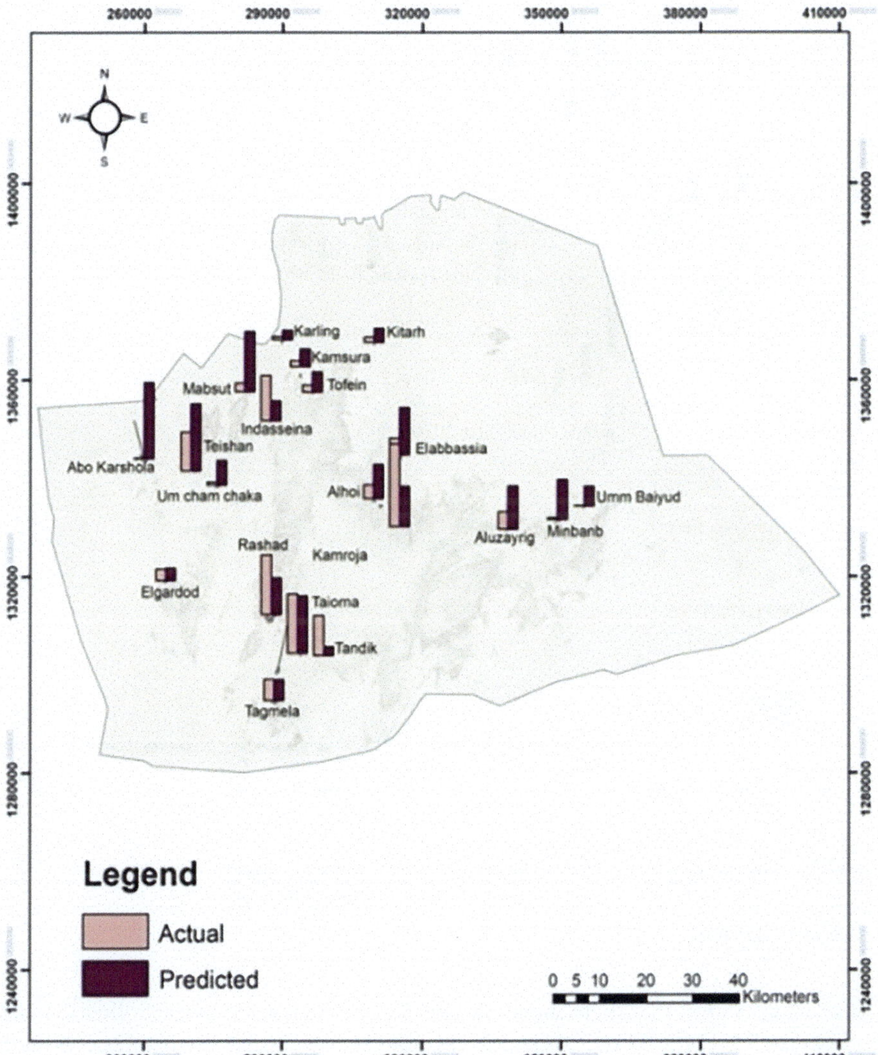

Fig. 11.10 The current (2014) and predicted (2030) quantities of NTFPs collection (left)

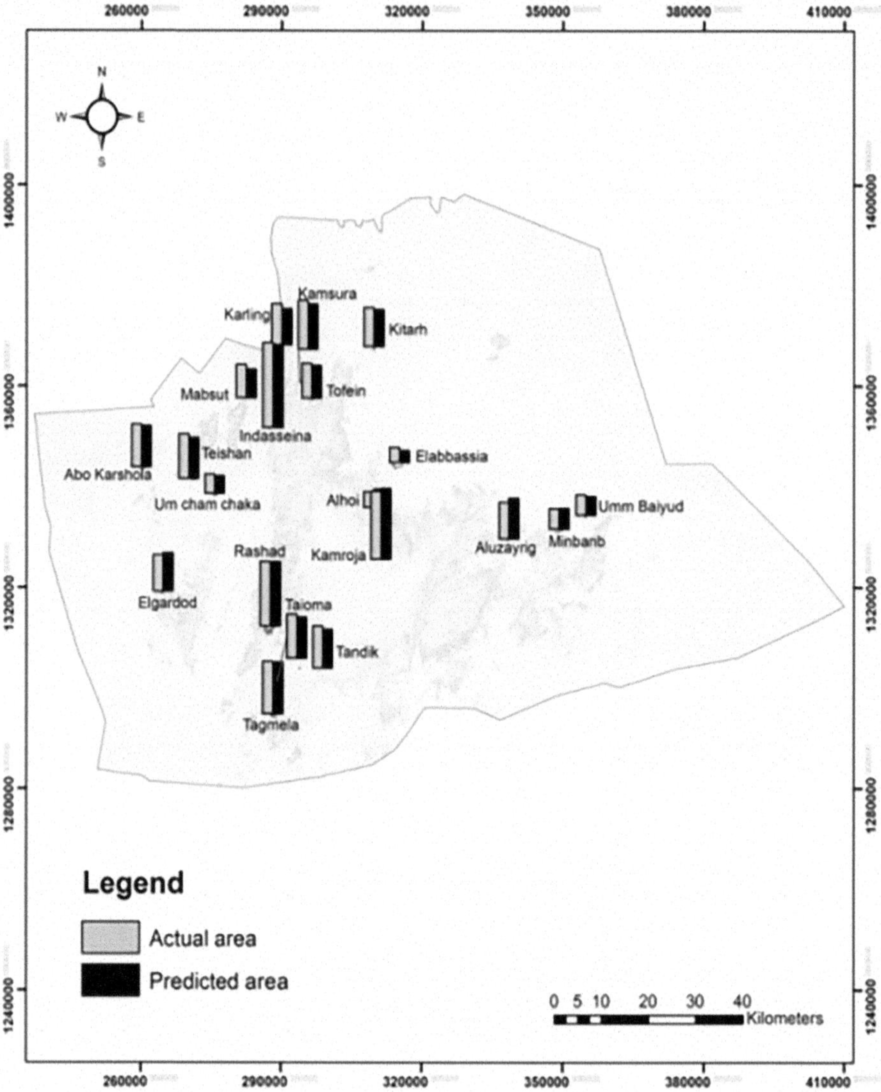

Fig. 11.11 The current (2014) and predicted (2030) forest density (right)

5 Conclusion

The research shows that there are some obstacles which restrain the collection and marketing of NTFPs. These obstacles are related to ecological, social, economic, and institutional factors. The study also detected and extracted key information concerning stationarity and non-stationarity in spatial data by using GIS-based local model and global statistics. This information was used to explore the relationship between NTFPs collection and risk level, forest areas, and average distance to a forest in Nuba Mountains. Global statistical models often assume homogeneity of relationships between variables across space. However, this exploratory analysis explains the spatial variation in the relationship between geographic datasets and across geographic regions. It statistically demonstrated that local models exhibit better fitness than global models when modeling spatial data.

Finally, this study is a contribution to the field of GIS, spatial statistics, forest modeling, and simulation. It shows that the development of modeling tools incorporating RS and GIS data could improve forest management systems. It would be interesting to integrate socio-economic/ecological mechanisms into these models by linking them to environmental data derived from satellite images. This should further improve the predictability and usefulness of RS applications in NTFPs management. It is important to secure sustainable NTFPs supply by conserving identified species, improving their market value through pre to post processing, expanding their marketing opportunities, and protecting identified NTFPs uses and users' knowledge. This would lay foundation stones towards improving their environmental benefits in a sustainable manner that leads to ecosystem stability, strengthening the regional value chains, and reducing the vulnerability to external shocks.

References

AIMS Education – UK. (2016). *What is value chain? Value chain definition, its management and analysis | AIMS lecture.*

Ambrose-Oji, B. (2003). The contribution of NTFPs to the livelihoods of the 'forest poor': Evidence from the tropical forest zone of south-west Cameroon. *International Forestry Review, 5*(2), 106–117.

Angelsen, A., & Wunder, S. (2003). *Exploring the forest-poverty link: Key concepts, issues and research implications* (CIFOR occasional paper no. 40) (70 pages).

Anselin, L. (2005). *Exploring spatial data with GeoDaTM: A workbook* (244 pages). Center for Spatially Integrated Social Science, Department of Geography, University of Illinois.

Baatz, M., Benz, U., Dehghani, S., Heynen, M., Oltje, A. H., Hofmann, P., & Lingenfelder, I. (2004). *eCognition professional user guide, version 4.0.* Definiens Imaging GmbH.

Bair, J., & Peters, E. D. (2006). Global commodity chains and endogenous growth: Export dynamism and development in Mexico and Honduras. *World Development, 34*(2), 203–221.

Behnassi, M. (2017). Climate security as a framework for climate policy and governance. In B. Mohamed & K. McGlade (Eds.), *Environmental change and human security in Africa and the Middle East* (pp. 3–24). Springer.

Belcher, B. M. (2003). What isn't an NTFP? *International Forestry Review, 5*(2), 161–168.

Berterretche, M., Hudak, A. T., Cohen, W. B., Maiersperger, T. K., Gower, S. T., & Dungan, J. (2005). Comparison of regression and geostatistical methods for mapping Leaf Area Index (LAI) with Landsat ETM+ data over a boreal forest. *Remote Sensing of Environment, 96*, 49–61.

Brunsdon, C., Fotheringham, A. S., & Charlton, M. E. (1996). Geographically weighted regression: A method for exploring spatial nonstationarity. *Geographical Analysis, 28*(4), 281–298.

Butt, A., Shabbir, R., Ahmad, S. S., & Aziz, N. (2015). Land use change mapping and analysis using remote sensing and GIS: A case study of Simly watershed, Islamabad, Pakistan. *The Egyptian Journal of Remote Sensing and Space Sciences (EJRS), 18*, 251–259.

Central Bureau of Statistics (CBS). (2009). Fifth Sudan National Population and Housing Census. *Census Tables*. Available from: www.cbs.gov.sd. Accessed 1 Apr 2014.

Charlton, M., & Fotheringham, A. S. (2009). *Geographically weighted regression white paper* (pp. 1–17). White Paper for Science Foundation, Ireland. Available from: https://www.geos.ed.ac.uk/~gisteac/fspat/gwr/gwr_arcgis/GWR_WhitePaper.pdf

Chamberlain, J. R., Schwarze, U., Wang, P. R., Hirata, R. K., Hankenson, K. D., Pace, J. M., Underwood, R. A., Song, K. M., Sussman, M., Byers, P. H., & Russell, D. W. (2004). Gene targeting in stem cells from individuals with osteogenesis imperfecta. *Science, 303*(5661), 1198–1201.

Chen, Y. (2013). New approaches for calculating Moran's index of spatial autocorrelation. *PLoS One, 8*(7), e68336.

Clement, F., Orange, D., Williams, M., Mulley, C., & Epprecht, M. (2009). Drivers of afforestation in northern Vietnam: Assessing local variations using geographically weighted regression. *Applied Geography, 29*(4), 561–576.

Deafalla, T. H. H. (2011). *Contribution of Non wood products to rural livelihood in South Kordofan State* (M.Sc. thesis). University of Khartoum, Khartoum, Sudan, pp. 6–45.

Deafalla, T. H. H. (2019). *Integrative assessment and modelling of the non timber forest products potential in Nuba Mountains of Sudan by field methods, Remote sensing and GIS* (PhD. thesis). Submitted to Faculty of Environmental Sciences, University of Dresden, June 2019, Germany. 319 pages.

Deafalla, T. H. H., Csaplovics, E., & El-Abbas, M. M. (2018). Environmental change impacts and human security in semi-arid region, the case of Nuba Mountains of Sudan. In M. Behnassi, G. Himangana, & P. Olf (Eds.), *Human and environmental security in the era of global risks* (pp. 187–203). Springer.

Deafalla, T. H. H., Dafa-Alla, D. M., & El-Abbas, M. M. (2014). Chapter 22: The importance of Non Wood Forest Products for rural livelihoods: The case of South Kordofan state, Sudan. In M. Behnassi, A. A. Shahid, & N. Mintz-Habib (Eds.), *Science, politics of modern agricultural system: Global context to local dynamics of sustainable agriculture* (pp. 323–332). Springer.

Dickinson, M. B., Dickinson, J. C., & Putz, F. E. (1996). Natural forest management as a conservation tool in the tropics: Divergent views on possibilities and alternatives. *Commonwealth Forestry Review, 75*(4), 309–315.

El Amin, M., & Ballal, M. E. (1996). *Land tenure, use and usufructoryritght in traditional agrofrorestry, gum belt of western Sudan* (unpublished).

El Abass, M. A. (2009). Contribution of Non-timber Forest Products (NTFPs) in Household Livelihood in the Rural Areas of Shiekan Province, North Kordofan State, Sudan, PhD. thesis University of Khartoum, Khartoum, Sudan. 3–15. In Sharawi HA (1987). Acacia senegal in the gum belt of western Sudan: A cost benefit analysis. M.Sc thesis, Department of Forestry, University College of north Wales, Bangor. In Ishraga, (2003). The economics of Boswellia papyrisera in southern Kordofan. M.Sc. thesis, University of Khartoum, Sudan.

FAO. (2005). *Environmental impacts of forest utilization and mitigation practices*. On http://www.fao.org/forestry/environment/11787/en/. Accessed 2 Oct 2014.

Fernandes, R., & Leblanc, S. G. (2005). Parametric (modified least squares) and non-parametric (Theil–Sen) linear regressions for predicting biophysical parameters in the presence of measurement errors. *Remote Sensing of Environment, 95*, 303–316.

Fernandez, J. M., Chuvieco, E., & Koutsias, N. (2013). Modelling long-term fire occurrence factors in Spain by accounting for local variations with geographically weighted regression. *Natural Hazards and Earth System Sciences, 13*, 311–327.

Foody, G. M., Boyd, D. S., & Cutler, M. E. J. (2003). Predictive relations of tropical forest biomass from Landsat TM data and their transferability between regions. *In Remote Sensing of Environment, 85*, 463–474.

Fotheringham, A. S., Brunsdon, C., & Charlton, M. (2002). *Geographically weighted regression: The analysis of spatially varying relationships* (284 pages). Wiley.

Gao, J., & Li, S. (2011). Detecting spatially non-stationary and scale-dependent relationships between urban landscape fragmentation and related factors using Geographically Weighted Regression. *Applied Geography, 31*(1), 292–302.
Geary, R. C. (1954). The contiguity ratio and statistical mapping. *The Incorporated Statistician, 5*(3), 115–145.
Gereffi, G., & Frederick, S. (2010). The global apparel value chain, trade and the crisis: Challenges and opportunities for developing countries. In O. Cattaneo, G. Gereffi, & C. Staritz (Eds.), *Global value chains in a postcrisis world: A development perspective* (pp. 157–208). The World Bank.
Hu, X., Hong, W., Qiu, R., Hong, T., Chen, C., & Wu, C. (2015). Geographic variations of ecosystem service intensity in Fuzhou City, China. *Science of the Total Environment, 512–513*, 215–226.
Hutcheson, G. D. (2011). Ordinary least-squares regression. In L. Moutinho & G. O. M. Hutcheson (Eds.), *The SAGE dictionary of quantitative management research* (pp. 224–228). Sage.
Iqbal, M. F., & Khan, A. A. (2014). Spatiotemporal Land Use Land Cover change analysis and erosion risk mapping of Azad Jammu and Kashmir, Pakistan. *The Egyptian Journal of Remote Sensing and Space Science, 17*(2), 209–229.
Jamhuri, J., Azhar, B. M. S., Puan, C. L., & Norizah, K. (2016). GWR-PM – Spatial variation relationship analysis with Geographically Weighted Regression (GWR) – An application at Peninsular Malaysia. *8th IGRSM International Conference and Exhibition on Remote Sensing & GIS (IGRSM 2016) IOP Publishing; IOP Conf. Series: Earth and Environmental Science, 37*(2016), 012032.
Kaplinsky, R. (2001). Globalisation and economic security. *IDS Bulletin, 49*(6), 13–24.
Kaplinsky, R., & Morris, M. (2001). *A handbook for value chain research. Working paper prepared for the International Development Research Centre (IDRC)* (p. 113). Institute for Development Studies.
Khamis, F. G. (2012). Measuring the spatial correlation of unemployment in Iraq-2007. *Modern Applied Science, 6*, 17–27.
Lawrence, A. (2003). No forest without timber? *International Forestry Review, 5*(2), 87–96.
Leslie, A. D. (2005). The ecology biodiversity value of sycamore (Acer pseudoplatanus L) with particular reference to Great Britain. *Scottish Forestry, 59*(3), 19–26.
Long, J., & Robertson, C. (2017). Comparing spatial patterns. *Geography Compass, 12*(2), e12356.
Lu, D., Mausel, P., Brondízio, E., & Moran, E. (2004). Change detection techniques. *In International Journal of Remote Sensing, 25*(12), 2365–2401.
Marshall, E., Newton, A. C., & Schreckenberg, K. (2003). Commercialisation of non-timber forest products: First steps in analysing the factors influencing success. *The International Forestry Review, 5*(2), 128–137.
Mitchell, J., Keane, J., & Coles, C. (2009). *Trading up: How a value chain approach can benefit the rural poor* (92 pages). COPLA – Trade and Poverty in Latin America. COPLA Global: Overseas Development Institute.
Mohammed, H. M. (2011). *Management of Natural Stands of Acacia seyal Del. variety seyal (Brenan).* Doctoral thesis, Dresden University of Technology, Dresden, Germany, pp. 25–26.
Montgomery, D. C., Peck, E. A., & Vining, G. G. (2001). *Introduction to linear regression analysis* (3rd ed., 672 pages). Wiley.
Moran, P. A. P. (1948). The interpretation of statistical maps. *Journal of the Royal Statistical Society, Series B, 37*(2), 243–251.
Nang'ole, E., Mithöfer, D., & Franzel, S. (2011). *Review of guidelines and manuals for value chain analysis for agricultural and forest products* (ICRAF occasional paper no. 17) (30 pages). World Agroforestry Centre.
Netrdová, P., & Nosek, V. (2016). Spatial patterns of unemployment in Central Europe: Emerging development axes beyond the Blue Banana. *Journal of Maps, 12*, 701–706.

Newton, A. C., Marshall, E., Schreckenberg, K., Golicher, D., te Velde, D. W., Edouard, F., & Arancibia, E. (2006). Use of a Bayesian belief network to predict the impacts of commercializing non-timber forest products on livelihoods. *Ecology and Society, 11*(2), 33.

Nkeki, F. N., & Osirike, A. B. (2013). GIS-based local spatial statistical model of cholera occurrence: Using geographically weighted regression. *Journal of Geographic Information System, 5*(6), 531–542. https://doi.org/10.4236/jgis.2013.56050

Phiri, D., & Morgenroth, J. (2017). Developments in landsat land cover classification methods: A review. *Remote Sensing, 9*, 967.

Pietrobelli, C., & Saliola, F. (2008). Power relationships along the value chain: Multinational firms, global buyers and performance of local suppliers. *Cambridge Journal of Economics, 32*(6), 947–962.

Propastin, P., Kappas, M., & Erasmi, S. (2008). Application of geographically weighted regression to investigate the impact of scale on prediction uncertainty by modelling relationship between vegetation and climate. *International Journal of Spatial Data Infraestuctures Research, 3*, 73–94.

Rich, K. M., Baker, D., Negassa, A., & Ross, R. B. (2009). *Contributed Paper prepared for presentation at the International Association of Agricultural Economists Conference, Beijing, China.* Available from: https://ageconsearch.umn.edu/bitstream/51922/2/51922.pdf

Royal Swedish Academy of Agriculture and Forestry (KSLA), African Forest Research Network (AFORNET) at the African Academy of Sciences (AAS), and Food and Agriculture Organisation of the United Nations (FAO). (2005). *Lessons learnt on sustainable forest management in Africa* (Policy briefs no 10).

Scott, L. M., & Janikas, M. V. (2010). Spatial statistics in ArcGIS. In M. Fischer & A. Getis (Eds.), *Handbook of applied spatial analysis* (pp. 27–41). Springer.

Shahid, R., & Bertazzon, S. (2015). Local spatial analysis and dynamic simulation of childhood obesity and neighbourhood walkability in a major Canadian city. *AIMS Public Health, 2*(4), 616–637.

Shi, H., Laurent, E. J., LeBouton, J., Racevskis, L., Hall, K. R., Donovan, M., Doepker, R. V., Walters, M. B., Lupi, F., & Liu, J. (2006). Local spatial modeling of white- tailed deer distribution. *Ecological Modeling, 190*, 171–189.

Shrestha, P. M. (2006). *Comparison of ordinary least square regression, spatial autoregression, and geographically weighted regression for modeling forest structural attributes using a Geographical Information System (GIS)/Remote Sensing (RS) Approach* (204 pages). Master of Geographic Information Systems Department of Geography, Calgary, Alberta.

Stieve, T. (2012). *Moran's I and spatial regression*. Available from: http://sites.tufts.edu/../Morans-I-andSpatial-Regression.docx. Last accessed 28 June 2016.

Sunderlin, W. D., Angelsen, A., Belcher, B., Burgers, P., Nasi, R., Santoso, L., & Wunder, S. (2005). Livelihoods, forests, and conservation in developing countries: An overview. *World Development, 33*(9), 1383–1402.

United Nations Development Programme (UNDP). (2006). *Pastoral production systems in south Kordofan, Sudan (Study 2)* (pp. 8–13).

Wang, Q., Ni, J., & Tenhunen, J. (2005). Application of a geographically-weighted regression analysis to estimate net primary production of Chinese forest ecosystems. *Global Ecology and Biogeography, 14*(4), 379–393.

Weisent, J., Rohrbach, B., Dunn, J. R., & Odoi, A. (2012). Socioeconomic determinants of geographic disparities in campylobacteriosis risk: A comparison of global and local modeling approaches. *International Journal of Health Geographics, 11*(45), 1–16.

Zhang, L., & Shi, H. (2004). Local modeling of tree growth by geographically weighted regression. *Forest Science, 50*, 225–244.

Chapter 12
Ecology of Zoonotic Pathways Indicating Conflict and Mass Migration

Fred Kruidbos

Abstract This chapter illustrates how ecosystem stability theory has been explained by early models, evolving towards ideas about how biodiversity and functional overlap of species create ecological system resilience, resulting in complex ecosystem services that provide ecological safety and security. It explains how damage to existing systems can lead to virtually irreparable ecological regime shifts that, in turn, lead to serious damage to safety and security worldwide. An issue that is further explained in Chap. 13 of this book. Through the so-called dilution effect, it explains the outbreak of the new coronavirus (SARS-CoV-2), which now causes the zoonotic pandemic disease COVID-19, as a symptom of very serious damage to the stability of ecosystems worldwide. And thus illustrates the worldwide social-ecological conflict. In addition, this chapter illustrates that biodiversity, habitat, and climate change are all interrelated, influenced by, and crucial to the human existence. Therefore, a profound understanding of ecological fundamentals is critical to any strategy involved in maintaining long-term viability between humans and nature. For decades, the loss of species and habitat has increased to such an extent that it has now even affected the climate. As climate change affects resource availability, it therefore influences conflict behavior at different organizational levels, from individuals to intra- and international relations. Therefore, this chapter serves as both a theoretical background and an introduction to possible solutions to human-induced conflict and mass migration, which are already underway and which are certain to continue for the foreseeable future.

Keywords Ecological stability · Zoonosis · Dilution effect · Regime shift · Mass migration · Conflict

F. Kruidbos (✉)
Kruidbos Ecological Research and Consultancy, K-SN Ecological Services B.V, LC, Helmond, The Netherlands
e-mail: info@kruidbos.com

M. Behnassi et al. (eds.), *The Climate-Conflict-Displacement Nexus from a Human Security Perspective*, https://doi.org/10.1007/978-3-030-94144-4_12

1 Introduction

During the writing of this chapter, the whole world is confronted with one of the most severe pathogenic outbreaks since the 1918–1919 influenza pandemic – called the Spanish flu – that killed approximately 50 million people (Taubenberger & Morens, 2006). The novel coronavirus (SARS-CoV-2),[1] which is now causing the pandemic disease COVID-19, is likely to kill hundreds of thousands more in the near future (Baud et al., 2020).

Like the novel coronavirus, the *influenza* pandemic of 1918–1919 was caused by a zoonotic virus with mixed backgrounds in both domesticated and wild animals[2] (Anhlan et al., 2011). In the past few decades, accelerating global changes linked to an expanding global population have led to the emergence of a striking number of newly described zoonoses, including hantavirus pulmonary syndrome, monkeypox, SARS, and simian immunodeficiency virus (the animal precursor to HIV), Ebola, the Nipah virus, West Nile Virus, Chikungunya virus and others (Jones et al., 2008; Karesh et al., 2012; Wang & Crameri, 2014, European Union, 2019). So, infectious diseases are emerging at an unprecedented rate with significant impacts on global economy and public health. The social and environmental conditions that give rise to disease emergence are thus of particular interest, as are management approaches that might reduce the risk of emergence or re-emergence (Rohr et al., 2019).

The signs are revealing an alarming longer-term concern about the world and raise several questions. Are we in a social-ecological conflict with nature? If so, what is the status of this conflict and will we be able to turn the tide? What ends, ways and means do we need to improve in order to come up with adequate measures to restore the ecological balance between man and nature?

In order to be able to answer these questions, some of the basic principles of ecological system stability and resilience will be explained. Then, the analysis explains how biodiversity enhances ecosystems stability – biodiversity-stability hypothesis – and how species richness can be seen as an ecosystem service against diseases from nature. In context with this, the recent Corona-pandemic is described, as a possible symptom of social-ecological imbalance, by linking it to the loss of biodiversity. The analysis explains, accordingly, how human-induced disruption of ecological stability, like species loss and habitat destruction, has contributed to a social-ecological separation between man and nature, resulting in serious synergistic negative consequences for both humans and nature at different scale and function. Human-induced climate change and ecological regime shifts are clear manifestations of such dynamics.

---

[1] Phylogenetic analysis by Zhu et al. (2020) revealed that 2019-nCoV falls into the genus betacoronavirus, which includes coronaviruses (SARS-CoV, bat SARS-like CoV, and others) discovered in humans, bats, and other wild animals.

[2] The phylogenetic studies of all eight RNA gene segments of influenza A viruses may indicate that the 1918 pandemic strain originated from a H1N1 swine virus, which itself might be derived from a H1N1 avian precursor, which was separated from the bulk of other avian viruses a long time ago.

This work prepares for the Chap. 13 of this book in which some implications for the military are postulated. Based on both ecological arguments as well as a comprehensive harm principle, this research intends to explore whether and how military capabilities can be used in mitigating the damage already done and how to restore coviability by preventing worsening of the alleged social-ecological separation. This may be done by symptom management – e.g. Humanitarian Aid and Disaster Relief – on one end of the conflict spectrum, and preventing escalation – e.g. suppressing criminal and terrorist development related to green crime – on the other.

2 Biodiversity-Stability and Threats from Nature

Decades prior to the novel Corona-pandemic, it was already clearly visible that on global basis massive habitat loss and species extinction have been going on at an unprecedented rate and scale. At the same time, our own species population has increased at the fastest rate in human history, leading to a current world population size of no less than 7.8 billion people.[3] It makes sense to assume that this has severe negative consequences on the natural world as well as on human beings.

In order to better understand this, the sections below will explain some of the ecological principles that are fundamental to ecosystem stability and form the basis of the most accepted theory on modern outbreaks of diseases such as COVID-19, the dilution-effect-hypothesis. It is important to understand that ecosystems are made out of populations of different species that together form a system of functionalities. The way in which these functionalities are cooperating to create ecosystem stability and resilience can be explained by different theories.

2.1 Diversity-Stability

The idea that biological diversity enhances ecological stability has inspired a huge body of scientific research. Since the 1950s, the emphasis has shifted from the stability of individual populations to that of entire communities or ecosystems (McCann, 2000; Mikkelson, 2009). The synthesis of all these ideas is that diversity can be expected, on average, to give rise to ecosystem stability. Evidence indicates that diversity is not the driver of this relationship; rather, ecosystem stability depends on the ability for communities to contain species, or functional groups that are

[3] According to the most recent United Nations estimates elaborated by *Worldometer*, https://www.worldometers.info/world-population (September 2020).

capable of differential response (McCann, 2000; Botton et al., 2006). Without being complete and technical, the historic development of these ideas will be explained below.

2.1.1 Stability

In 1859, Darwin proposed that an area is more ecologically stable if it is occupied by a large number of species than if it is occupied by a small number. In 1955, this idea was formalized by MacArthur (Fig. 12.1). He stated that communities are more stable as the number of links increase and that the population of each species tends to a specific constant, independent of the initial population of the species (compare below with category 1 McCann). MacArthur reasoned as follows: If a given species preys on several others, its population will fluctuate less in response to environmental variations affecting one of its prey than it would if the species in question ate fewer prey species. Similarly, if a species has many predators, its population will vary less in response to exogenous changes in one predator's population size (Mikkelson, 2009). So, less diverse communities would be expected to be more prone to destructive oscillations in populations and invasions of other species. This was later confirmed by several researchers (Elton, 1958; Holling, 1973, 1992; Frank & McNaugton, 1991; Tilman et al., 2006; Richardson & Pyšek, 2007).

Similarly, Yodzis (1981) showed that models structured from compiled food-web relationships, with plausible interaction strengths, were generally more stable than randomly constructed food webs. So, if Yodzis' results indicated that interaction strength was probably crucial to stability (compare below with category 2 McCann), more had to be happening since empirical evidence had already proven a positive correlation between diversity and stability (McCann, 2000). Ehrlich and Ehrlich (1981) and Schindler (1990) showed that the effects of adding and removing species from an ecosystem varies. Their models propose that the ecological functions of

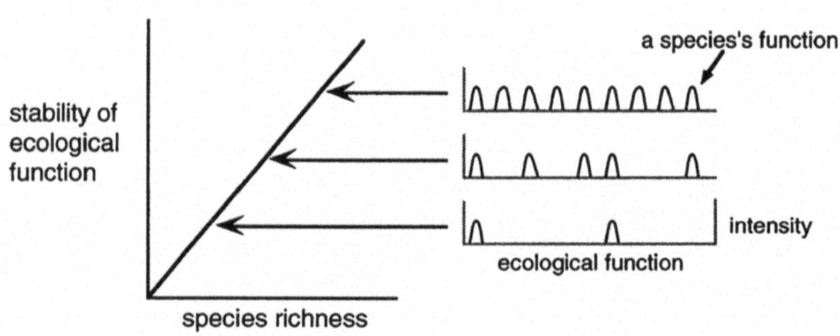

Fig. 12.1 Representation of the Darwin/MacArthur-model on how increasing species richness increases ecosystem stability
As species accumulate, they fill this space. The width and height dimensions of the inset diagrams represent the breadth and intensity of a species' ecological function
Source: Peterson et al. (1998)

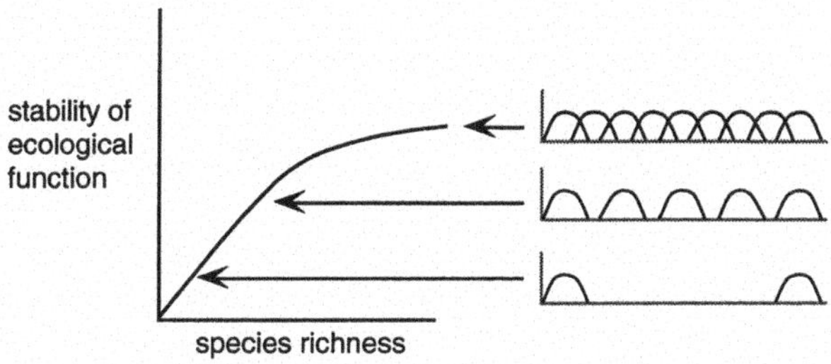

Fig. 12.2 Rivet-model of Ehrlich and Ehrlich
If species are added to an ecosystem, their functions begin to overlap or complement one another, therefore they become rivets. This overlap allows ecological function to persist despite the loss of a limited number of species, since species with similar functions can compensate for the elimination or decline of other species. However, the increase of stability decreases as species richness increases
Source: Peterson et al. (1998)

different species overlap, so that even if a species is removed, ecological function may persist because of compensation by other species with similar functions. Just like several rivets attached to an airplane wing may fall of before the plane loses its wing, so may several species be lost before the system collapses (Peterson et al., 1998). This idea is illustrated in Fig. 12.2.

Holling (1992) started his work with the following: "this paper tests the proposition that a small set of plant, animal, and abiotic processes structure ecosystems across scale in time and space. Earlier studies have suggested that these key structuring processes establish a small number of dominant temporal frequencies that entrain other processes. These frequencies often differ from each other by at least an order of magnitude. If true, ecosystems therefore will have a few dominant frequencies that are endogenously driven and that are discontinuously distributed". His idea is called the 'extended-keystone-hypothesis' and is very similar to the 'driver-passenger-hypothesis' of Walker (1992) which model is illustrated in Fig. 12.3.

In comparison with the previous ideas (Figs. 12.1 and 12.2), the relation between species richness and system stability has changed from continuity to discontinuity. Walker (1992, 1995) discriminates between *driver* and *passenger* species. He defines a driver as a species that has a strong ecological function. Such species significantly structure the ecosystems in which they and passenger species exist – compare with extended keystone hypothesis of Holling (1992). Passenger species are those species that have minor ecological impact. Driver species can take many forms: they may be 'ecological engineers', such as beavers or gopher tortoises which physically structure their environments; or 'keystone species', such as sea otters or asynchronously fruiting trees that have strong interactions with other species (Peterson et al., 1998; Paine, 1969; Mills et al., 1993). Walker (1995) proposes that since most ecological function resides in the strong influence of driver species, it is their presence or absence that determines the stability of an ecosystem's ecological function.

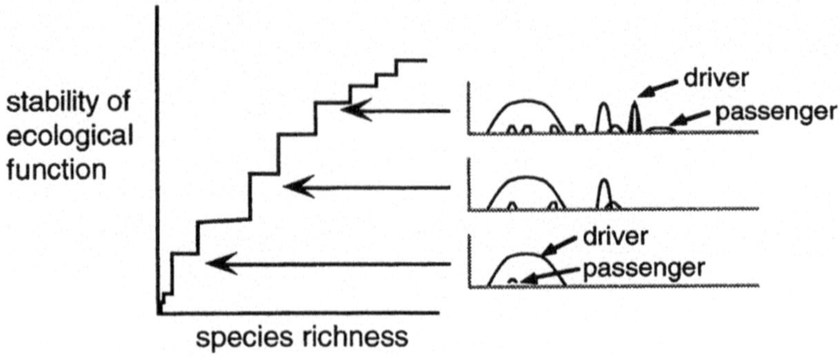

Fig. 12.3 Walker's 'drivers and passengers' model of redundant ecological function (1992, 1995) *The model proposes that ecological function is unevenly distributed among species. Drivers have a large ecological impact, while passengers have a minimal impact*
Source: Peterson et al. (1998)

In 1991, it was for the first time documented that stability of plant community species composition – in grasslands of Yellowstone National – increases with diversity; Frank and McNaugton (1991) showed that, after drought, more diverse grassland communities recovered faster than less diverse communities. In line with this, Tilman et al. (2006) conducted a long-term study on plant communities (1996–2005), and found that greater numbers of plant species led to greater temporal stability of ecosystem annual aboveground plant production. Despite the lower-temporal stability of individual species, the stability of the ecosystem increased with diversity; showing that the stability of species richness increases at the community level because plants respond differentially to variable background processes according to the redundancy of functions as explained above. Proulx et al. (2010) also found that diversity promotes temporal stability across levels of ecosystem organization. They demonstrated that in a grassland, diversity increases stability: (i) across trophic levels (producer, consumer); (ii) at both the system (community, ecosystem) and the component levels (population, functional group, phylogenetic clade); and (iii) primarily for aboveground processes. Besides these studies, others have demonstrated that the stability of many ecological processes, but not all, increases with species richness.

McCann (2000) suggests that these findings would better been interpreted as indications that ecosystem function and stability are more related to functional diversity – e.g. the grazing-tolerance of certain plant species preventing herbivores from dramatically reducing plant biomass on the Serengeti, grasses, nitrogen fixing legumes, and other herbs. The author claims that "as real populations are variable, it is possible that the persistence of complex communities depends to some degree on population fluxes. Such background population variability, whether driven by biotic or abiotic processes, can provide species with the opportunity to respond differentially to their environment (which comes close to the findings of both Yodzis and Tilman). In turn, these differential species responses weaken the destructive potential of competitive exclusion". So, within the community, species may stabilize the system by facilitating each other's presence by a number of different ecological processes.

According to Isbell (2010), stabilizing species interactions, which cause a species to limit itself more than it limits other species, are predicted to promote biodiversity, ecosystem stability, and ecosystem functioning and occur when interspecific interactions – (between individuals from different species) are more favorable than intraspecific interactions (between individuals of the same species). Resulting, therefore, in a rare species advantage, common species disadvantage, or both. Isbell explains this with the following example: when species consume different resources or consume the same resources at different times or places, resource competition will be stronger between two individuals from the same species than between two individuals from different species. Consequently, species have an advantage when they are rare because competition is relatively weak. Also, they have a common disadvantage when competition is relatively strong. So, this can maintain biodiversity because it prevents any particular species from competitively excluding all other species.

McGrady-Steed and Morin (2000) showed – in controlled aquatic systems – that, in contrast to results for individual species, temporal variation of entire functional groups composed of multiple species decreased as species richness increased. Therefore, the stability of collective community attributes, such as functional groups, may be greater in more complex communities. Two ideas have been advanced in explanation of these findings. One explanation is that increasing diversity increases the odds that at least some species will respond differentially to variable conditions and perturbations. The second is that greater diversity increases the odds that an ecosystem has functional redundancy by containing species that are capable of functionally replacing important species. Taken together, these two notions have been called the 'insurance hypotheses". This view has been extended to suggest that the greater the variance in species' responses contained in a community, the lower the species richness required to insure the ecosystem (McCann, 2000).

Peterson et al. (1998) combined the existing models into their model. They reasoned that the model that best describes an ecosystem appears to depend upon: (i) the variety of functional roles that are occupied in that system; and (ii) the evenness of the distribution of ecological function among species. This relationship is illustrated in Fig. 12.4. They reasoned that: (i) ecosystems consisting of species that each performs different ecological functions will be less redundant – show less or no overlap, less flattening of the curve – than ecosystems consisting of the same number of species that each performs a wide variety of ecological functions – and therefore show some overlap (compare flattening between graphs on the left and right-hand site in Fig. 12.4). Similarly, (ii) if there is little difference between the ecological impact of different species, there is little point in differentiating driver and passenger species and they can all be considered rivets (compare discontinuity in the graphs between top and bottom in Fig. 12.4). So, by varying the degree of functional overlap and the degree of variation in ecological function, different versions of the existing models can be made. The authors conclude that: "since no

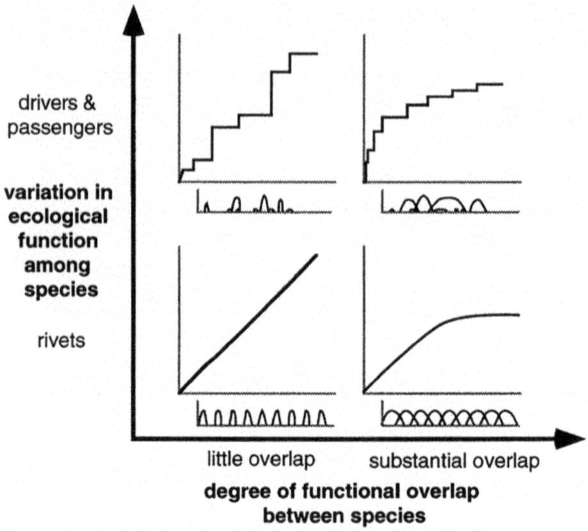

Fig. 12.4 The relationship between stability and species richness
This relationship varies with the degree of overlap that exists among the ecological function of different species and the amount of variation in the ecological impact of species ecological function. Overlap in ecological function leads to ecological redundancy (right hand graphs). If the ecological impact of different species is similar, they are 'rivets', whereas if some species have relatively large ecological impact, they are 'drivers' and others are 'passengers'
Source: Peterson et al. (1998)

species are identical, redundancy does not reside in groups of species, but rather it emerges from the interactions of species (compare with e.g. Yodzis, 1981). Therefore, it is not possible to substitute species for one another; rather, there are many possible combinations and organizations of species that can produce similar ecological functions".

2.1.2 Resilience

The findings that interaction strength was probably crucial to stability are in line with Holling's (1973) ideas. In 1996, Holling proposed that the concept of 'ecosystem stability' – the ability to return to an equilibrium after a temporarily disturbance – does actually present two contrasting aspects: one that focuses on maintaining *efficiency of functions*, which corresponds to engineering resilience; and one that focuses on *maintaining the existence of function* which was named ecological resilience (Botton et al., 2006). This engineering resilience can be seen as a combination of resistance to disturbance and speed to return to the equilibrium, while the ecological resilience can be seen as the magnitude of disturbance that can be absorbed before the system flips into another regime of behavior, another

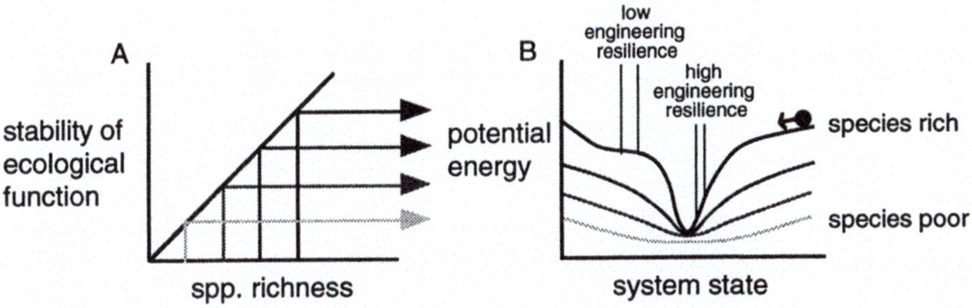

Fig. 12.5 The relationship between stability and species richness represented by a set of stability landscapes
The dynamics of a system are expressed by a landscape, and its 'state' is represented by a ball that is pulled into pits. Different landscape topographies may exist at different levels of species richness. In this model, the stability of a state increases with the depth of a pit. Zones of the stability surface that have low slopes have less engineering resilience than do areas that have steep slopes
Source: Peterson et al. (1998)

stability domain (Botton et al., 2006; McCann, 2000). The conceptual differences between these two are clearly illustrated by Peterson et al. (1998). They visualize these abstract processes by different landscapes with pitfalls and a ball. In this concept, the landscape topographies are related to species richness while the pits in the landscape represent stable states of the system. The deeper the pits, the more stable the state; the wider the pit, the more resilient the system is at that state (see Fig. 12.5 and Fig. 12.6).

Since ecosystems do not only consist of species but also of many other (a)biotic factors, they are exposed to gradual changes in e.g. climate, nutrient loading, habitat fragmentation, and biotic exploitation. Therefore, nature is usually assumed to respond to a gradual change in a smooth but certain way. However, sometimes ecosystems can change state very rapidly. Scheffer et al. (2001) studied large-scale shifts in major ecosystems like lakes, coral reefs, oceans, forests and arid lands. They showed, although diverse events can trigger such shifts, that a loss of resilience (wide of the valley in Fig. 12.6) usually paves the way for a switch to an alternative state; Table 12.1 shows different examples that they studied. Here we discuss only one example as given by Scheffer et al. (2001). Landscapes can be kept open by herbivores – often in combination with fires – because seedlings of woody plants, unlike adult trees, are easily eliminated by herbivores. Conversely, woodlands, once established, are stable because adult trees cannot be destroyed by herbivores and shading reduces grass cover so that fires cannot spread. Well analyzed examples are African woodland dynamics in Botswana and Tanzania, where regeneration of woodlands occurred for a few decades from the 1890s because of low herbivore numbers due to a combination of rinderpest epidemic and elephant hunting. Once established, these woodlands could not be eliminated by grazers (Scheffer et al., 2001).

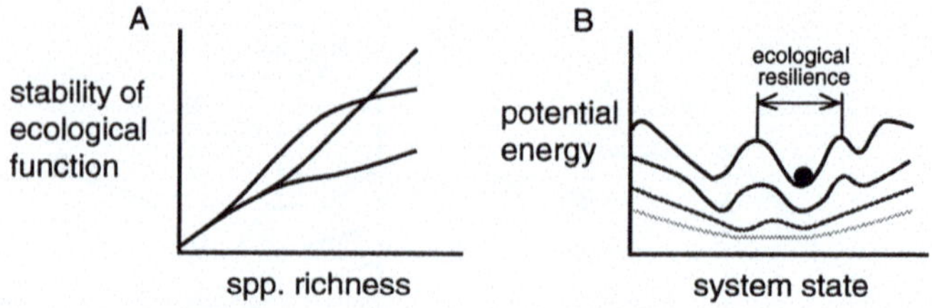

Fig. 12.6 A system may be locally stable in a number of different states
Disturbance that moves the system across the landscape and slow systemic changes that alter the shape of the landscape both drive the movement of a system between states. The stability of a state is a local measure. It is determined by the slope of the landscape at its present position. The resilience of a state is a large-scale measure, as it corresponds to the width of the pit the system is currently within
Source: Peterson et al. (1998)

Table 12.1 Characteristics of some major ecosystem state shifts and their causes

Ecosystem	State I	State II	Events inducing shift from I to II	Events inducing shift from II to I	Suggested main causes of hysteresis	Factors affecting resilience
Lakes	Clear with submerged vegetation	Turbid with phytoplankton	Killing of plants by herbicide Killing of *Daphnia* by pesticide High water level	Killing of fish Low water level	Positive feedback of plant growth Trophic feedbacks	Nutrient accumulation
Coral reefs	Corals	Fleshy brown macroalgae	Killing of coral by hurricane Killing of sea urchins by pathogen	Unknown	Prevention of coral recolonization by unpalatable adult algae	Nutrient accumulation Climate change Fishing
Woodlands	Herbaceous vegetation	Woodlands	Fires Tree cutting	Killing of grazers by pathogen Hunting of grazers	Positive feedback of plant growth Inedibility of adult trees	Overgrazing Climate change
Deserts	Perennial vegetation	Bare soil with ephemeral plants	Climatic events Overgrazing by cattle	Climatic events	Positive feedback of plant growth	Climate change
Oceans	Various	Various	Climatic events	Climatic events	Physical	Fishing Climate change

Source: Scheffer et al. (2001)

2.1.3 Scale

Since biodiversity is one of the most important factors influencing ecological stability, the ecological mechanism – e.g. ecological niche[4] – behind biodiversity are of utterly importance. Ritchie and Olff (1999) found that niche differentiation is both related to body size and the availability of food resources on different scales. In their model, larger species exclusively use large patches that contain food with low-resource concentrations, whereas smaller species exclusively use small patches that contain food with high resource concentration. Due to this, each patch size contains a certain amount of available resources. By operating at different scale – in both

[4]Ecological niche is the match of a species to a specific environmental condition. It describes how an organism or population responds to the distribution of resources and competitors (for example, by growing when resources are abundant, and when predators, parasites and pathogens are scarce) and how, in turn, it alters those same factors (for example, limiting access to resources by other organisms, acting as a food source for predators and a consumer of prey). For more details: https://en.wikipedia.org/wiki/Ecological_niche

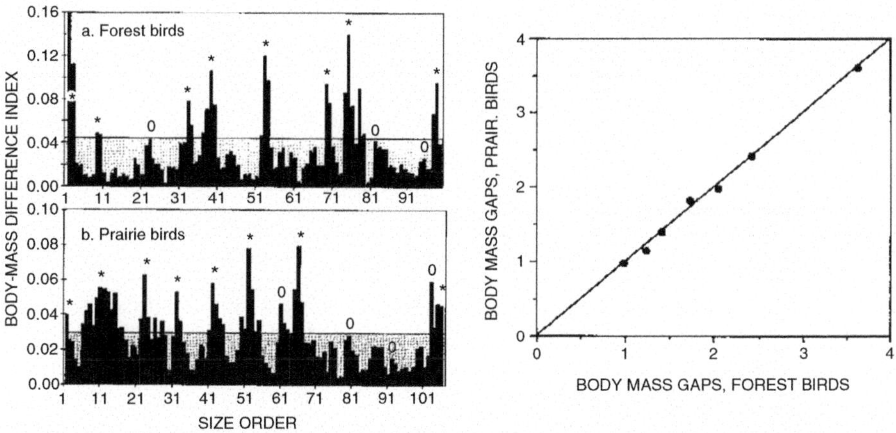

Fig. 12.7 Holling's distribution of body-mass gaps for birds of the boreal forest and prairie region as their size increases
Both the diagrams show a pattern which illustrates that body mass – representing body size – comes in classes. So, animals sharing the same environmental region show a similar discrete distribution of body mass
Source: Holling (1992)

space and time – the niches of smaller species are separated from that of larger species. So, both can exist.

Already in 1992, Holling presented the discrete appearance of body size within nature. His findings show a distribution of body-mass gaps for different groups of species. He showed that the body-mass of birds in boreal forest and prairie region of North-Amerika are clustered with 'gaps' in between (Fig. 12.7) in more or less the same pattern (Fig. 12.8). These findings can be seen as an example of the result of spatial scaling laws as found by Ritchie and Olff (1999).

Even though species operate on different scales – in both time and space – and occupy different niches, they interact with each other. This can easily be understood if you compare for example the consumption of a leaf by an aphid, a rabbit, or an elk. All three are feeding on – parts of – a plant but operating at different scales. Another example might be the feeding on a plant by an aphid that is eaten by a great tit – a small songbird that feeds on invertebrates in its territory – that is on its turn eaten by a sparrow hawk which holds a large territory to be able to feed on different kinds of (mostly) small territorial songbirds.

So, species can be divided into functional groups, based upon their ecological roles, as well as on specific scales that they exploit. Therefore, Peterson et al. (1998) proposed that the resilience of ecological processes, and therefore of the ecosystems they maintain, depends upon the distribution of functional groups within and across scales. They hypothesized that "different species use resources at different spatial and temporal scales. Members of a functional group use similar resources, but species that operate at larger scales require those resources to be more aggregated in space than do species that operate at smaller scales. Within scales, the presence of different functional groups provides robust ecological functioning, whereas replication of function across scales reinforces ecological function. The combination of a diversity of ecological function at specific scales and the replication of function across a diversity of scales produces resilient ecological function".

2.1.4 Metapopulations

As explained above, species use resources at different spatial and temporal scales and, therefore, they coexist. Since resources are available within habitats, the latter can be seen as islands that differ in size and isolation. MacArthur and Wilson (1967) showed that species richness – number of species – increases with island area but decreases with isolation (Fig. 12.8; Guo, 2015). They developed their theory based on ocean islands, but the so-called species-isolation relationship, can also be applied on 'terrestrial islands' (e.g. compare Simberlof, 1976; Simberloff & Abele, 1976; with De Vries & Den Boer, 1990; Den Boer et al., 1996) and provides the basis for metapopulation dynamic models that were first developed as a pest management tool by Levins (1969). Levins realized that pests can only be understood and controlled if they are viewed from a broader perspective: "the objectives of the control program do not depend only on the biology of the pest. We may want to achieve complete extirpation over part of the region, minimize average pest population over the whole region [...] the purpose of this report is to show that the pattern of environmental variation in space and time can be utilized in the control of pests and to indicate the information which is needed for the selection of the most promising predator" (Levins, 1969:237). Within Levins' metapopulation model, habitat patches have the same size and quality and can either be occupied or not.

Metapopulations are conceived as spatially-structured populations consisting of distinct units (subpopulations), separated by space or barriers, and connected by dispersal movements. Metapopulations characteristically demonstrate a turnover of local populations (subpopulations) going extinct and becoming re-established, resulting in a distribution pattern that shifts over time (Opdam, 1991). In other words, a metapopulation is a spatially-structured population in time. In the real

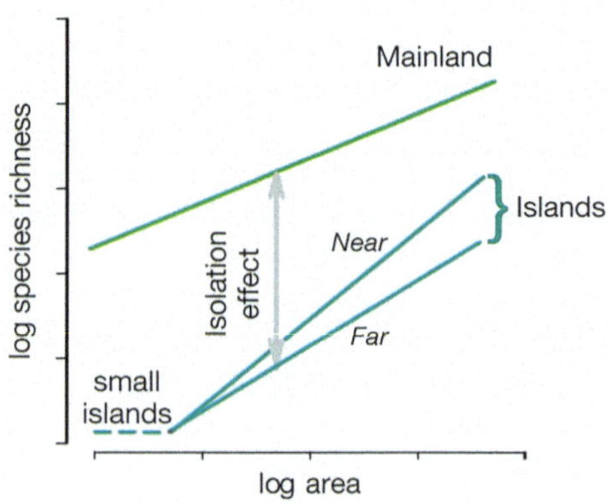

Fig. 12.8 A representation of the island biogeography theory that takes both area (habitat size) and isolation (distance to species sources) into account
The short, dashed line that is not in the original MacArthur and Wilson model indicates the 'small island effect' due to a lack of suitable habitats
Source: Guo (2015:2)

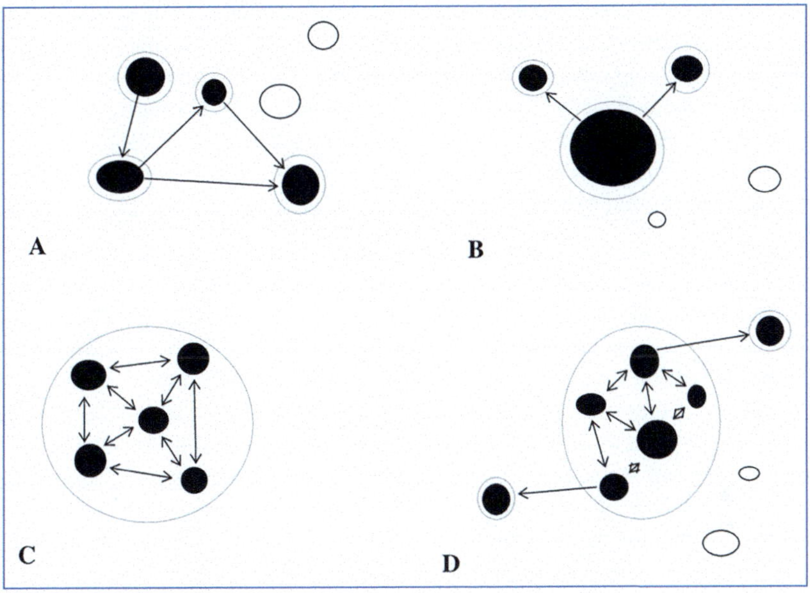

Fig. 12.9 Metapopulation types proposed by Harrison and Taylor (1997)
Closed circles represent occupied habitat patches, open circles represent unoccupied habitat patches. Dashed circles represent the boundaries of local populations. Arrows represent dispersal. A: Classical model; B: Mainland-Island model; C: Patchy population; D: Mixed metapopulation combining types B and C
Source: Cale (1999:4)

world, habitat patches differ in quality, size, and isolation and, therefore, subpopulations differ in space and time.

Harrison (1991) and Harrison and Taylor (1997) found that a few real populations fit the 'classical' metapopulation concept proposed by Levins. They argued that real metapopulations lie along a gradient of extremes, which can be described by one of the four metapopulation types shown in figure. 12.9. Alternatively, according to Cale (1999) populations had the structure of a metapopulation, but were not in equilibrium and were declining:

- Classical metapopulations have habitat patches with similar probabilities of extinction, and the persistence of these metapopulations is dependent on the recolonization of local populations which go extinct (Fig. 12.9a).
- Mainland-island or Source-sink metapopulations have a local population which is extinction resistant – i.e. Mainland or Source – and other local populations which have much higher extinction probabilities – i.e. Island or Sink – but are maintained by dispersal from Mainlands or Sources. The difference between Mainland-Island and Source-Sink metapopulations is that in the former extinction resistance is the result of large population size, while in the latter this resistance is the result of qualitative differences in the habitat (Fig. 12.9b).

- In patchy populations dispersal between habitat patches is so common that individuals from different patches mix freely. Therefore, these patches are effectively a single population. Harrison and Taylor (1997) argued that in patchy populations the distances or rates of movement among patches are less important to population persistence than in true metapopulations (Fig. 12.9c).
- Some species have mixed metapopulation structures which contain elements of several different metapopulation types (Fig. 12.9d).

When patches have a high rate of dispersal and a high rate of reproduction, and therefore have a genetically perfect connectivity, the metapopulation is called panmictic. But when habitat size, quality or dispersal opportunities are altered, metapopulations are likely to experience differences in extinction risks. As a consequence, some species will be in favor while others will suffer and disappear, altering community species' composition, and generating a cascade of possible effects. One of these effects is the facilitation of zoonoses which is explained in the next section.

Since ecological stability seems to depend upon species diversity across space and time, recent loss of many species and habitat implicates severe stability loss (vice versa) within and across ecosystems. Some of these effects and the underlying mechanism will be discussed as well in the next section.

2.2 *Zoonoses and the Dilution-Effect*

2.2.1 Direct and Indirect Transmission of Zoonoses

Zoonoses are infectious diseases caused by pathogens (viruses, bacteria, fungi, parasites or unconventional agents like prions) that are naturally transmissible from – mostly vertebrate – animals to humans.[5] Zoonoses can be transmitted from animals to humans either directly[6] (non-vector born) or indirectly via an invertebrate vector (vector born); both differing in ecological complexity. According to WHO,[7] zoonoses can be divided into three classes: (i) endemic zoonoses, which are present in many places and affect many people and animals; (ii) epidemic zoonoses, which are sporadic in temporal and spatial distribution; and (iii) emerging and re-emerging zoonoses, which are newly appearing in a population or have existed previously but are rapidly increasing in incidence or geographical range.

Examples of non-vector zoonotic disease and their organisms are anthrax (Bacillus anthracis; hosts: grazing herbivores); bird flu (Influenza A virus subtype H5N1; hosts: wild and domesticated birds); capillariasis (Capillaria spp.; hosts:

[5] https://en.wikipedia.org/wiki/Zoonosis; https://www.who.int/topics/zoonoses/en; https://www.merriam-webster.com/dictionary/zoonosis

[6] Infection via hosts can occur via: (i) direct contact, e.g., saliva, blood, urine, feces, etc.; and (ii) indirect contact, e.g. by contaminated food, contaminated water sources, or contaminated habitats like plants and soil.

[7] http://www.emro.who.int/fr/pdf/about-who/rc61/zoonotic-diseases.html

rodents, birds foxes); Ebola virus disease (Ebolavirus spp.; hosts: chimpanzee, gorilla, fruit bats, and other vertebrates); echinococcosis (Echinococcus spp.; hosts: wild and domestic canine such as wolf, fox, jackal as well as sheep, pigs and rodents); glanders (Burkholderia mallei.; hosts: beavers, other rodents, raccoons, deer, goats, sheep, dogs and cats); and Influenza (Influenza A virus; hosts: horses, pigs, domestic and wild birds, wild aquatic mammals such as seals and whales, minks and farmed carnivores).

Examples of vector-borne zoonotic disease and their organisms are: African sleeping sickness (Trypanosoma brucei rhodesiense; hosts: both wild as domestic vertebrates – vector: tsetse fly); Babesiosis (Babesia spp.; hosts: mice-vector: ticks); Barmah Forest fever (Barmah Forest virus; hosts: kangaroos, wallabies and opossums-vector: mosquito); Japanese encephalitis (Japanese encephalitis virus; hosts: pigs, water birds-vector: mosquitos); Lyme disease (Borrelia burgdorferi; hosts: deer, wolves, dogs, birds, rodents, rabbits, hares, reptiles-vector: ticks); Pneumonic plague (Yersinia pestis; hosts: rabbits, hares, rodents, ferrets, goats, sheep, camels-vector: flea); Rocky Mountain spotted fever (Rickettsia rickettsii; hosts: dogs and rodents-vector: ticks); and West Nile fever (Flavivirus; hosts: birds, horse-vectors: mosquito). Many of these infections are enzootic (i.e., stably established) in animal populations, and transmit from animals to people with little or no subsequent person-to-person transmission (e.g. rabies). Other zoonotic pathogens can spread efficiently between people once introduced from an animal reservoir, leading to localized outbreaks (e.g. Ebola virus) or global spread (e.g. pandemic influenza) (Karesh et al., 2012; WHO, as stated above). Most of them have been transmitted from domesticated and wild animals to humans for thousands of years, but some of these diseases have only emerged recently (Marano & Pappaioanou, 2004); SARS-CoV-2 is clearly an example of the latter one.

2.2.2 Host Competence, Community Composition, and Infection Risk

So, many infectious diseases of humans that are caused by pathogens that reside in non-human animal reservoirs are transmitted to humans: via the bite of an arthropod vector (e.g. Schmidt & Ostfeld, 2001; LoGiudice et al., 2003); others directly from the host reservoir itself (e.g. Khalil et al., 2016). Since many pathogens are able to infect multiple host species that vary in their competence, host community composition – including vector composition – can influence disease outcomes (e.g. Smit et al., 2003; Hofmeester, 2016; Hofmeester et al., 2017; Keven et al., 2019; Gildenhard et al., 2019). If more diverse assemblages support a greater fraction of low-competency hosts, biodiversity losses have the potential to increase disease risk (Johnson et al., 2013). In other words, the presence of hosts with a low capacity to infect feeding vectors – such as ticks – dilutes the effect of highly competent reservoirs and, thus, reduces the risk of disease. This idea, suggested by Van Buskirk and Ostfeld (1995, 1998) and called the 'dilution effect' by Ostfeld and Keesing (2000), was further developed by Schmidt and Ostfeld (2001). Since then, the dilution effect

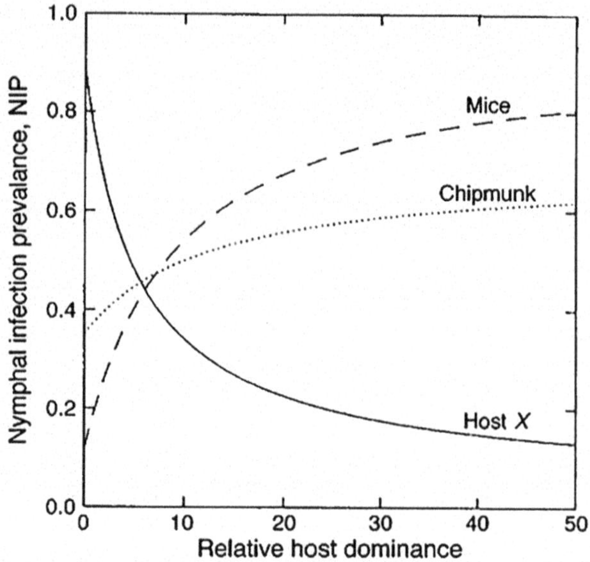

Fig. 12.10 Nymphal infection prevalence (NIP; proportion of nymphal ticks infected by Borrelia burgdorferi) as a function of the relative dominance of each host species
Source: Schmidt and Ostfeld (2001:614)

has been tested by many scientists (e.g. LoGiudice et al., 2003; Johnson et al., 2013; Khalil et al., 2016).

Schmidt and Ostfeld (2001) used the prevalence of the spirochete *Borrelia burgdorferi* – causing Lyme disease in humans – to demonstrate the presence and potential magnitude of the dilution effect. When they compared their model on infection prevalence with actual field data, they found that the prevalence of the spirochete in field-collected ticks *Ixodes scapularis* was much lower than expected if the ticks would have predominantly fed on the highly competent reservoirs white-footed mice *Peromyscus leucopus* and eastern chipmunks *Tamias striatus*. By adjusting their modeling, they found that increased species richness – but not evenness – reduces the disease risk. This is illustrated by Fig. 12.10.

Figure 12.10 shows that the chance of a nymph[8] getting infected with *B. burgdorferi* declines with the presence of bad competence hosts (Host x). These findings were confirmed by LoGiudice et al. (2003) who demonstrated that the most common non-mouse hosts of North America are relatively poor reservoirs for *B. burgdorferi* and so Nymphal Infection Prevalence (NIP) declines as host diversity increases. Figure 12.11 shows the differences in dilution potential – defined as the difference in percentage points between the expected NIP in a two-host community consisting of mice plus the focal species and a community in which mice are the only possible host – of these most common non-mouse hosts, as tested by LoGiudice et al. (2003).

[8] Tick development comes in four different stages: egg, larva, nymph, and adult. Each stage – with the exception of the egg – needs a sufficient blood meal in order to reach the next stage of development. The larvae predominantly feed on mice.

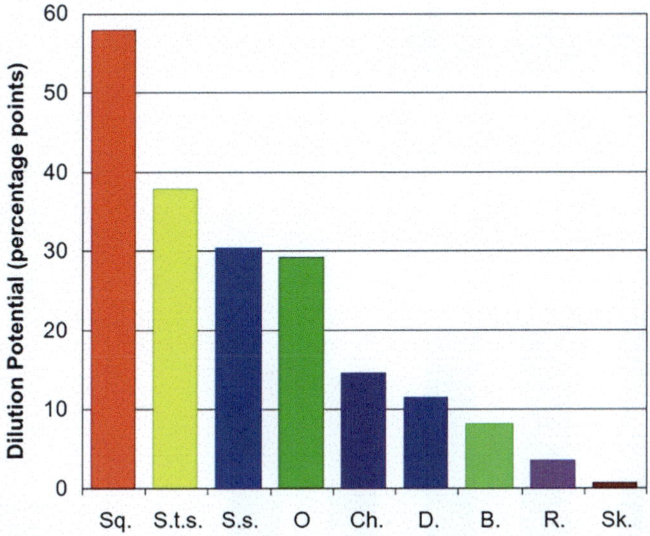

Fig. 12.11 The ability of each species to reduce the effect of white-footed mice (the most competent reservoir) on NIP
Dilution potential is the difference (in percentage points) between the expected NIP in a two-host community consisting of mice plus the focal species and a community in which mice are the only possible host. Sq., squirrel; S.t.s., short-tailed shrew; S.s., Sorex shrew; O, opossum; Ch., chipmunk; D, deer; B, birds; R, raccoon; Sk, skunk
Source: LoGiudice et al. (2003:570)

In line with the dilution effect, Johnson et al. (2013) showed that host diversity inhibits transmission of the virulent pathogen *Ribeiroia ondatrae*[9] and reduces amphibian disease as a result of consistent linkages among species richness, host composition and community competence. Figure 12.12 shows the results of amphibian richness and composition within 345 wetlands across a 758-hectare region in California, USA. Johnson et al. (2013) found that the most-competent amphibian host *Pseudacris regilla* (Fig. 12.12c) was also the most common species with progressive decreases in the fraction of highly competent hosts in more diverse assemblages (Fig. 12.12a), resulting in a decrease in community competence over the observed species richness gradient (Fig. 12.12b).

So, the 'dilution effect' implies that where species vary in susceptibility to infection by a pathogen, higher diversity often leads to lower infection prevalence in hosts as proven by the above examples. This effect comes not only from species richness alone. Also, the way in which the species interact is of importance. For instance, Ostfeld and Holt (2004), Levi et al. (2012), Khalil et al. (2016), Hofmeester (2016), and Hofmeester et al. (2017) state that an infection reduction may be resulting from

[9] *Ribeiroia ondatrae* is a parasite (trematode) in the genus Ribeiroia, which is believed to be responsible for many of the recent increases in amphibian limb malformations; particularly missing, malformed, and additional hind legs. See https://en.wikipedia.org/wiki/Ribeiroia_ondatrae (19-04-2020).

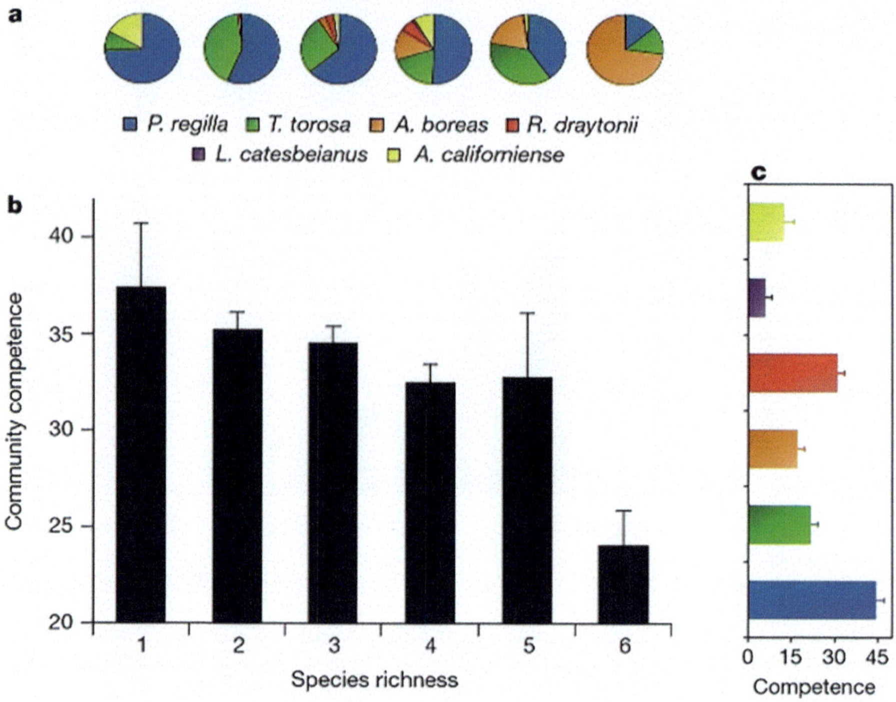

Fig. 12.12 Influence of amphibian species richness on the capacity of communities to support parasite infection in naturally occurring wetlands
*Relative abundance of each host species (**a**), mean community competence (**b**) (+1 s.e.) as a function of amphibian species richness, and (**c**) species competence as a host for the trematode pathogen Ribeiroia ondatrae*
Source: Johnson et al. (2013)

the presence of both competitors and predators altering host behavior and reducing host density; in particular when pathogens are directly transmitted to humans. Khalil et al. (2016) describe such a mechanism for *Puumala Hantavirus* (PUUV) that is transmitted by the bank vole *Myodes glareolus.* They found that both field voles *Microtus agrestic* – severe competitor of bank voles – and common shrews *Sorex araneus* – competitor and facultative predator of bank voles – dilute PUUV infection in bank voles. Assumingly, through changing their movement patterns and reducing contact rates and duration between infected and susceptible individuals – i.e. dilution effect via 'encounter reduction', as well as competition between field and bank voles – and so suppressing bank vole densities. They also found a negative correlation between nest box occupancy of Tengmalm's owl *Aegolius funereus* – main predator of bank vole – and PUUV prevalence in bank vole; assuming both direct and indirect increased infection by a decrease in predation risk.

These findings strongly reflect the importance of the main principles of ecosystem stability as described by Peterson et al. (1998). The extent to which biodiversity and community composition of ecosystems affect their functions is an issue that grows ever more compelling as human impacts on ecosystems increase (LoGiudice, 2003).

The accelerating rates of species extinctions and disease emergence underscore the importance of understanding how changes in biodiversity affect disease (Johnson et al., 2013). As the loss of habitat and biodiversity increase globally, the coronavirus outbreak – following Ebola, SARS, MERS and Bird flu – may be just the beginning of mass pandemics (Vidal, 2020). Therefore, we will zoom in on different human-induced mechanisms that are responsible for ecosystem degradation and form drivers of both ancient and novel diseases as well as other threats.

3 Socio-Ecological Coviability and Conflict

3.1 Novel Zoonoses as Symptoms of Social-Ecological Conflict

3.1.1 Zoonotic Emergence

It is estimated that, globally, about one billion cases of illness and millions of deaths occur every year from zoonoses (WHO EMRO, 2020). This means that more than 60% of human infectious diseases are caused by pathogens shared with wild or domestic animals (Jones et al., 2008; Karesh et al., 2012). During the last three decades, over 30 new human pathogens have been detected (Katz, 2019; Fig. 12.13),

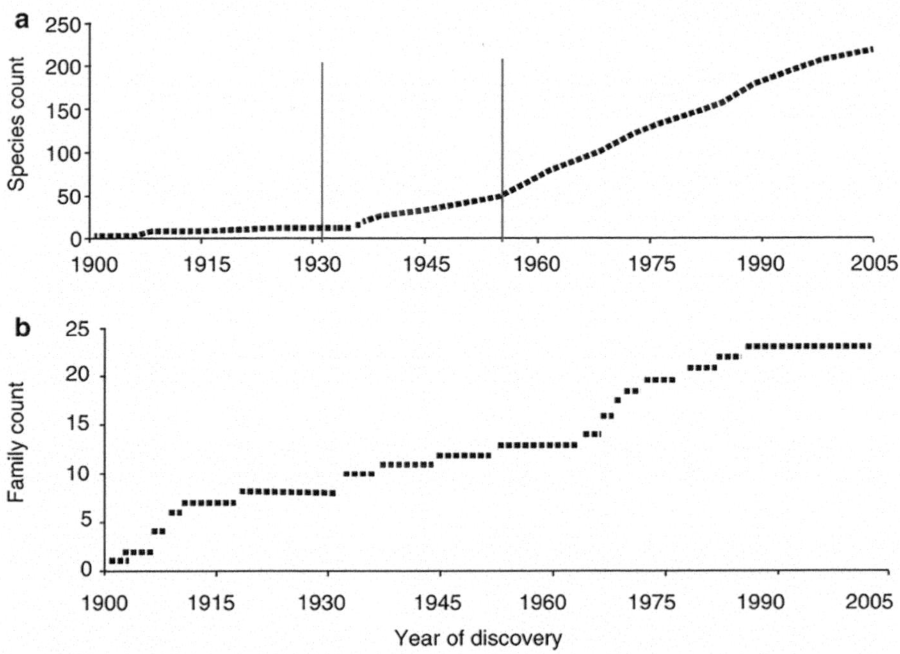

Fig. 12.13 New viruses infecting humans. Discovery curves for human viruses
*(**a**) Virus discovery curve by species. Cumulative number of species reported to infect humans. Statistically significant upward breakpoints are shown (vertical lines). (**b**) Virus discovery curve by family. Cumulative number of families containing species reported to infect humans*
Source: Data from Woolhouse et al. (2008); adapted from Katz (2019:191)

75% of which have originated in animals. Not just now, but also in the recent past infectious diseases: i.e. West Nile virus infection and monkeypox have appeared in other places than which they originated from and diseases – like HIV SARS, MERS and Avian Influenza – have emerged on a global scale (Marano & Pappaioanou, 2004; WHO EMRO, 2020). So, emerging zoonoses, both vector-borne and non-vector borne are a growing threat to global health and have caused hundreds of billions of US dollars of economic damage in the past decades. The recent COVID-19 pandemic will even cost tremendously more in the very near future. It is also likely to become an even greater threat since it has already been reported that the virus has spilled over from humans back to feline and mustelids; for the time being only in captivity (Shi et al., 2020; McGreevy, 2020; Beasley & Paul, 2020).

If we look at how zoonoses develop, taking into account what is already discussed, we can characterize them in different ways. Morse et al. (2012) roughly distinguish three different development stages (Fig. 12.14):

- Stage 1: a pre-emergence state in which naturally occurring microbes are transmitted between their animal reservoirs. Disturbance to the ecology of these populations – e.g. due to changes in land use – changes the dynamics of microbial transmission and can lead to a risk of pathogen spillover to other non-human wildlife or livestock hosts (but not people);
- Stage 2: localized emergence either through self-limiting spillover events or large-scale spillover, which leads to the person-to-person transmission for a few pathogen generations.
- Stage 3: some spillover events might lead to indefinitely sustained person-to-person outbreaks, international or global spread, and the emergence of a true pandemic.

The size, spread, and potential effect of events increase from stage 1 to stage 3, but the frequency falls so that full stage 3 pandemics are quite rare (Morse et al. 2012). Unfortunately, the outbreaks of past decades are signaling that the properties of the outcome of stage 3, as defined by Morse et al. (2012), are now quickly getting outdated.

In addition to the organization of our societies, new methods of land use also play a crucial role in the risks of both the emergence and spread of (novel) zoonoses. Many zoonoses can be linked to large-scale changes in land use that affect biodiversity and relations between animal hosts, people, and pathogens. Land modification (e.g. agricultural encroachment, deforestation, road construction, dam building, irrigation, wetland modification, mining, the concentration or expansion of urban environments, coastal zone degradation, and other activities) cause a cascade of factors that aggravate infectious disease outbreak such as changes in vegetation patterns, vector and host species dynamics (e.g. abundance, distribution and demographics), microclimates, and human contact with domestic and wild animals (Patz et al., 2004; Karesh et al., 2012). Such changes are particularly intense in, but not restricted to, tropical regions where the primary forest is opened up to mining, logging, plantation development, and oil and gas extraction. This deforestation poses a threat to

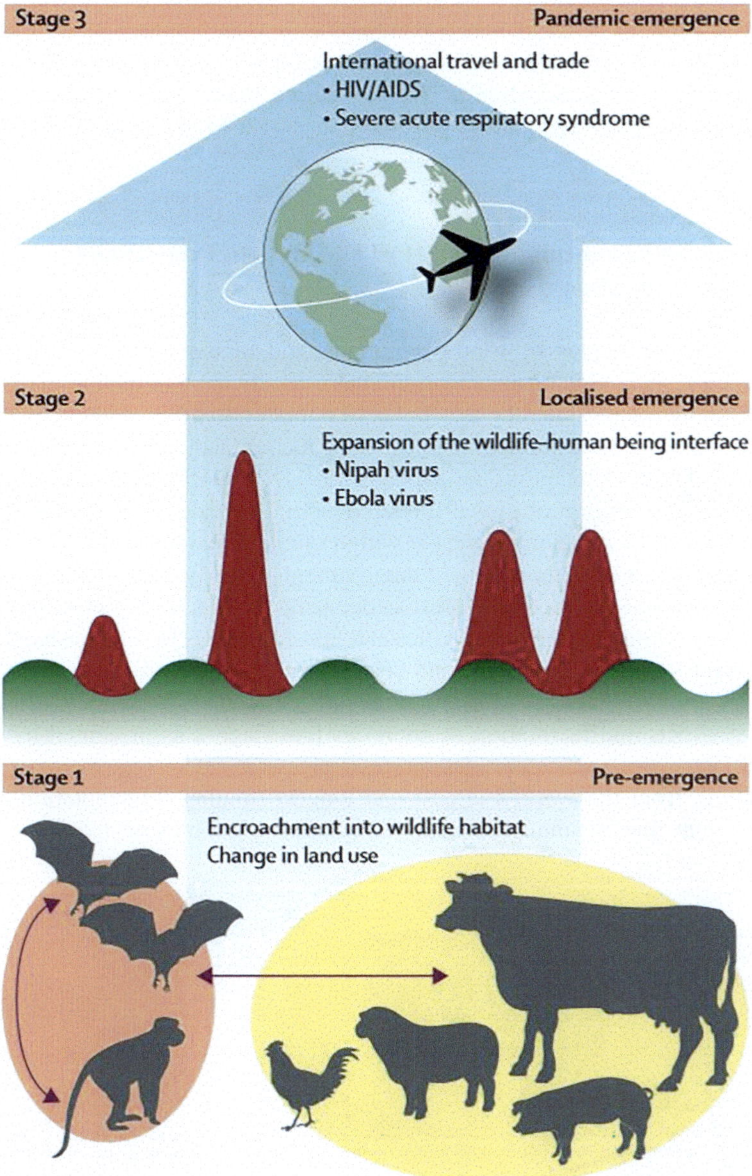

Fig. 12.14 Emergence of pandemic zoonotic disease
Source: Morse et al. (2012:1958)

global health because many of these regions are emerging disease hotspots – rich in wildlife biodiversity and probably rich in the diversity of microbes, many of which have not yet been encountered by people (Jones et al., 2008; Karesh et al., 2012).

Also, habitat fragmentation due to urbanization is of great significance since here the level of changes of contacts between host, vector, and humans is at its highest. It is for this reason that many new – vector-borne – pathogens have appeared in new places in recent decades, while at the same time many endemic diseases have increased in incidence (Kilpatrick & Randolph, 2012; Morse et al., 2012; Karesh et al., 2012).

3.1.2 Habitat Fragmentation and Host Facilitation: From Dilution to Amplification

Based on the principles of island biography and metapopulation dynamics, it becomes clear that habitat fragmentation, that is the division of habitat into smaller isolated fragments separated by a complex of human-transformed land cover, leads to long-term changes in habitat structure and function and loss of species (see the Sect. 2.1.4 on metapopulations and Opdam, 1991). Similarly, Haddad et al. (2015) analyzed the distribution of global forest cover and revealed that, in 2015, nearly 20% of the world's remaining forest is within one hundred meters of an edge near agricultural, urban or other modified environments. Nearly 70% of the remaining forest is within one kilometer of forest's edge (Fig. 12.15a, b). So, most forests are within the proximity of human activities and non-forest species like husbandry animals. Therefore, most forests are in a constant interaction with humans and their husbandry animals leading to all sorts of negative effects on both sides – e.g. negative edge effects and facilitation of zoonoses like stage 2 as presented by Morse et al. (2012).

Resulting from the world's largest and longest running fragmentation experiments (at that time spanning 35 years and disparate biomes on five continents),

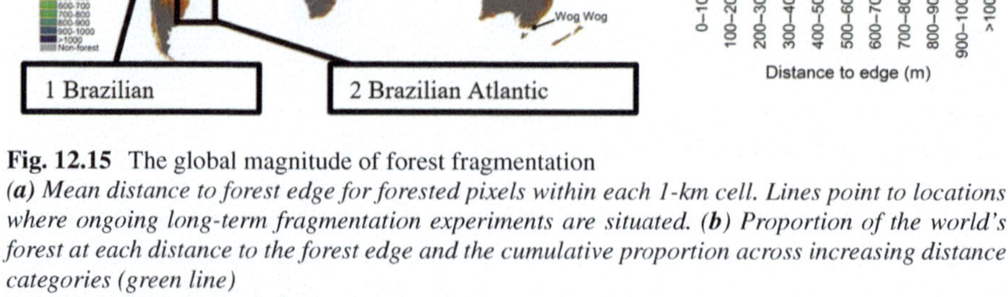

Fig. 12.15 The global magnitude of forest fragmentation
(a) *Mean distance to forest edge for forested pixels within each 1-km cell. Lines point to locations where ongoing long-term fragmentation experiments are situated.* ***(b)*** *Proportion of the world's forest at each distance to the forest edge and the cumulative proportion across increasing distance categories (green line)*
Source: Haddad et al. (2015:2)

Haddad et al. (2015) deduced evidence of unexpected long-term ecological changes caused by habitat fragmentation. Figure 12.16a–c show the effect of fragmentation over time as the proportional change in fragmented relative to non- or less fragmented treatments. As illustrated in the graphs, the species richness changes – in this case: arthropod, bird, butterfly, and plant species – are negative in time, showing a delayed loss of species due to fragmentation. Hence, extinction debt by ecosystem degradation. Besides the prognoses by theory, the authors report to be "struck by the persistence of degradation to biodiversity and ecosystem processes and by the increase in many of the effects over time. For example, extreme rainfall events at Wog Wog appeared to delay the decline in plant species richness for 5 years after fragmentation. In the Kansas Experiment, a lag of 12 years occurred before fragmentation effects on plant succession were detected". Similar effects were found for immigration debt and ecosystem function debt – represented by: biomass, dissolved carbon, total nitrogen, and total organic carbon.

Next to species decline, habitat degradation will change species compositions within habitats. Ostfeld and LoGiudice (2003) refer to several studies (e.g. Nupp & Swihart, 1996, 2000; Rosenblatt et al., 1999) which reveal that, in highly fragmented landscapes, species that are poor reservoirs for zoonotic pathogens and compete with or predate on highly competent rodents are disappearing, resulting in weakening of the dilution effect. In general, it can be said that the so-called r-strategists,[10] species with many offspring but with a lower survival rate (like mice and rats), thrive in degraded habitats. In more stable habitats, the K-strategists[11] dominate. These are species with long-term investment in their offspring and often have a poor competence for zoonoses. If habitat loss leads to a reduction of the dilution effect due to the facilitation of vertebrate r-strategists, that are competent hosts for zoonoses, it actually amplifies the infection prevalence (Faust et al., 2016). Habitat damage – in quantity and in quality – therefore has a facilitating effect on the transmission of zoonoses to humans.

Taking into account the three developmental stages of zoonotic diseases, it is now simple to understand that increased human population and habitat destruction lead to more close encounters with zoonoses (e.g. stage 2 as defined by Morse et al., 2012). Jones et al. (2008) have made an analysis of the global distribution of relative risk[12] of emerging infectious diseases. Based on the dataset that was available in 2008 they produced a presentation of four different types of pathogens of which

[10] r-selected species are those that emphasize high growth rates, typically exploit less-crowded ecological niches and produce many offspring, each of which has a relatively low probability of surviving to adulthood. https://en.wikipedia.org/wiki/R/K_selection_theory

[11] By contrast, K-selected species display traits associated with living at densities close to carrying capacity and typically are strong competitors in such crowded niches, that invest more heavily in fewer offspring, each of which has a relatively high probability of surviving to adulthood. In the literature, R-selected species are occasionally referred to as 'opportunistic', whereas K-selected species are described as 'equilibrium'. https://en.wikipedia.org/wiki/R/K_selection_theory

[12] The relative risk is calculated from regression coefficients and variable values (in Table 1, page 992 of Jones et al. 2008) categorized by standard deviation from the mean and mapped on a linear scale.

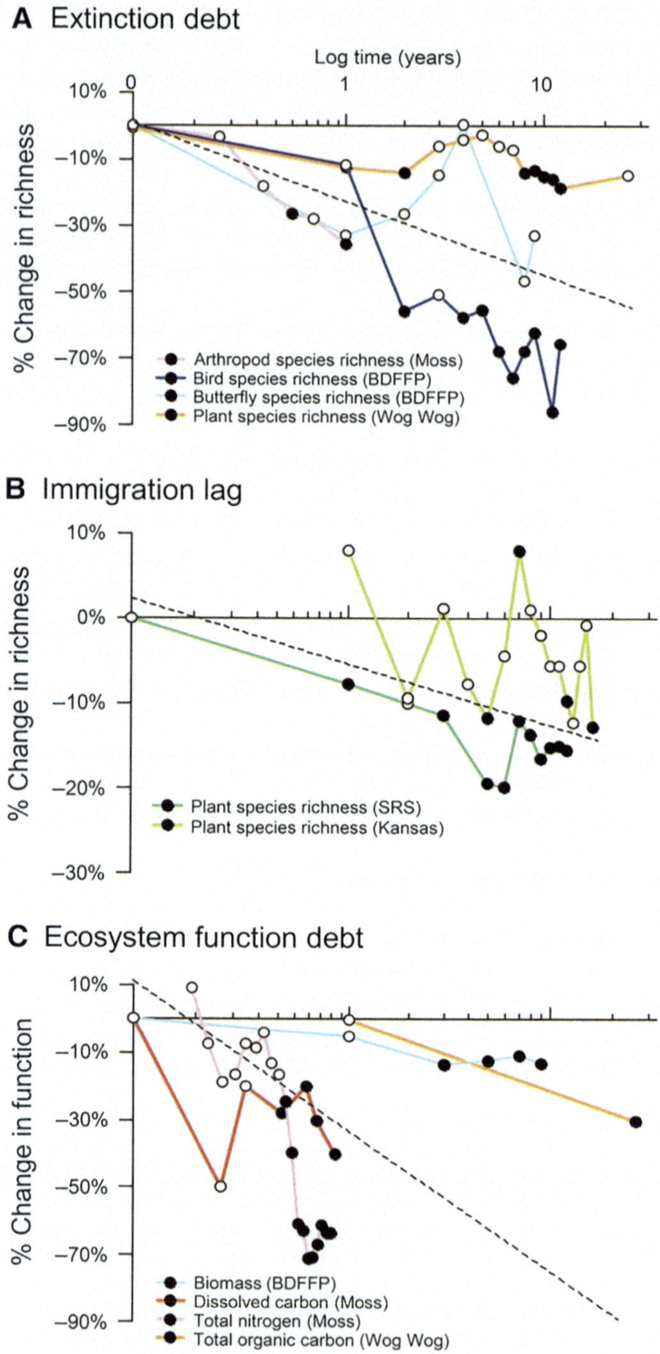

Fig. 12.16 Delayed effects of fragmentation on ecosystem degradation
*(**a**) The extinction dept represents a delayed loss of species due to fragmentation. (**b**) The immigration lag represents differences in species richness caused by smaller fragment area or increased*

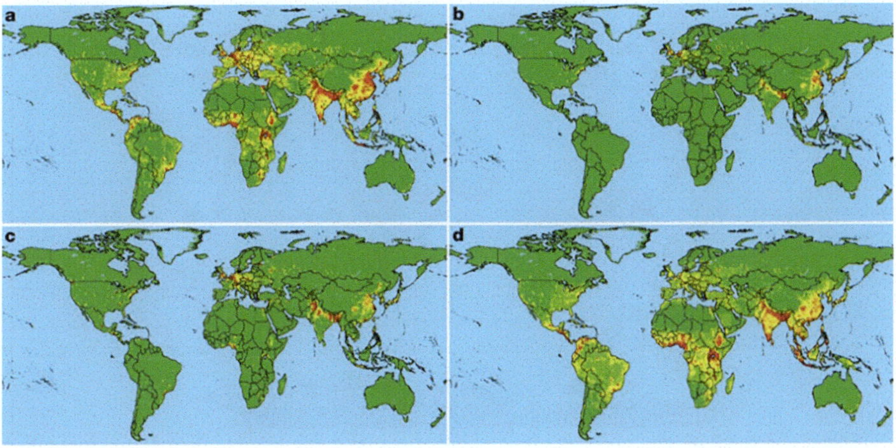

Fig. 12.17 Global distribution of relative risk of an Emerging Infectious Disease event
*(**a**) zoonotic pathogens from wildlife. (**b**) zoonotic pathogens from non-wildlife. (**c**) drug-resistant pathogens. (**d**) vector-borne pathogens. The relative risk varies from green (lower values) to red (higher values)*
Source: Adapted from Jones et al. (2008:993)

both vector-borne and zoonotic pathogens from wildlife showed a much higher risk presentation than zoonotic pathogens from non-wildlife and drug resistance (Fig. 12.17). Based on a series of drivers, including human population density, change in human population density, and wildlife diversity – mammalian species richness – Morse et al. (2012)[13] found similar results.

To illustrate the relationship between the distribution of the relative risk of emerging infectious diseases as shown in Fig. 12.17, the relative richness of terrestrial mammal species by ecoregion, according to Olson et al. (2001), is presented by Fig. 12.18. The figures show significant overlap, particularly of large parts of Asia, Africa and America. Compared with the presentation of habitat fragmentation (Fig. 12.15), the similarities between degraded land, possibilities of host presence and the relative risk of emerging infectious disease also seem to be great.

[13] See Morse et al., 2012, figure 1 on page 1957.

Fig. 12.16 (continued) *isolation during fragment succession. (**c**) The ecosystem function debt represents delayed changes in ecosystem function due to reduced fragment size or increased isolation. The percent loss is calculated as proportional change in fragmented treatments – i.e. (no of species in fragment – no of species in control)/(no of species in control) x 100. Fragments and controls were either the same area before and after fragmentation, fragments compared to unfragmented controls, or small compared to large fragments.*

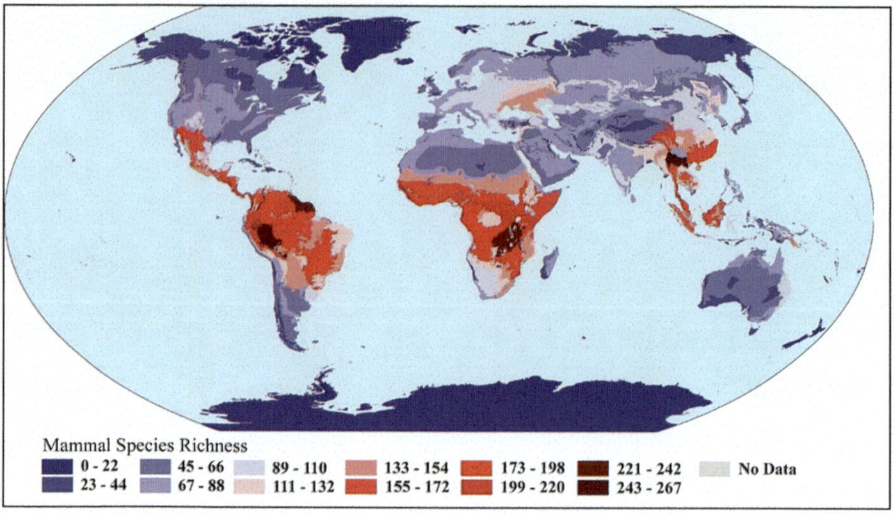

Fig. 12.18 The relative richness of terrestrial mammal species by ecoregion depicted by coloration
Warmer colors denote ecoregions containing richer assemblages
Note that this presentation is already two decades old
Source: Adapted from Olson et al. (2001:936)

3.1.3 Dispersion

One of the most important factors in the life history of a zoonosis is its ability to transmit from one host individual to the other and spread within the population. As a modern society, most of us live in large cities and are provided with excellent and extensive international logistic networks for trade and travel. Seen from the perspective of the virus – or other pathogens – our way of life meets one of the most important requirements it places on its ability to spread optimally (e.g. stage 3 as defined by Morse et al., 2012). Hufnagel et al. (2004) already showed this for the SARS outbreak of 2003 by demonstrating that the rapid worldwide spread of severe acute respiratory syndrome was directly dependent on national and international civil aviation. Figs. 12.19 and 12.20 show a similar case for the SARS-CoV-2. Unfortunately for humans, but fortunately for these viruses and other zoonotic pathogens, healthcare in many parts of the world cannot cope with major disease outbreaks and thus provides other important conditions for their spread. Furthermore, scholars (Kilpatrick & Randolph, 2012; Akiner et al., 2016; Fălcuţă et al., 2020; Wilke et al., 2020) mention that good logistic networks for international trade and travel facilitate exotic pathogens and the trade of exotic animals that may serve as vectors for zoonoses.

Now that long-term and firmly established economic systems are faltering under the current COVID-19 pandemic, most of us agree we are in conflict with nature. Looking at the source and dispersion of the virus and comparing it to human behavior, we can immediately understand that this pandemic is intimately intertwined with our modern way of life. The comparison between the two illustrations (Figs. 12.19 and 12.20) is not an artistic expression, but a clear reflection of the entanglement of the dispersion behavior of the virus with one of the humans.

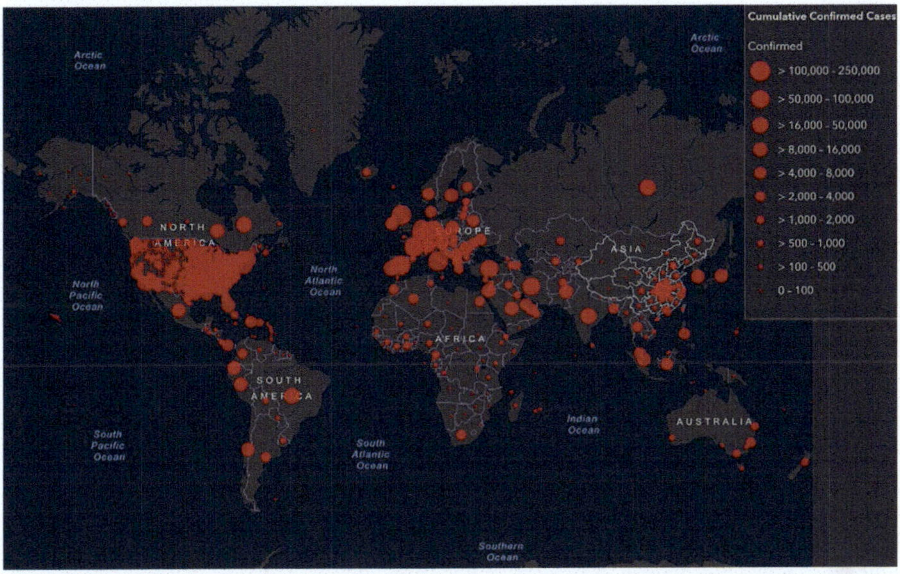

Fig. 12.19 COVID-19 distribution on 20–4-20 as presented by the Center for Systems Science and Engineering (CSSE) at Johns Hopkins University (JHU)
Source: https://systems.jhu.edu

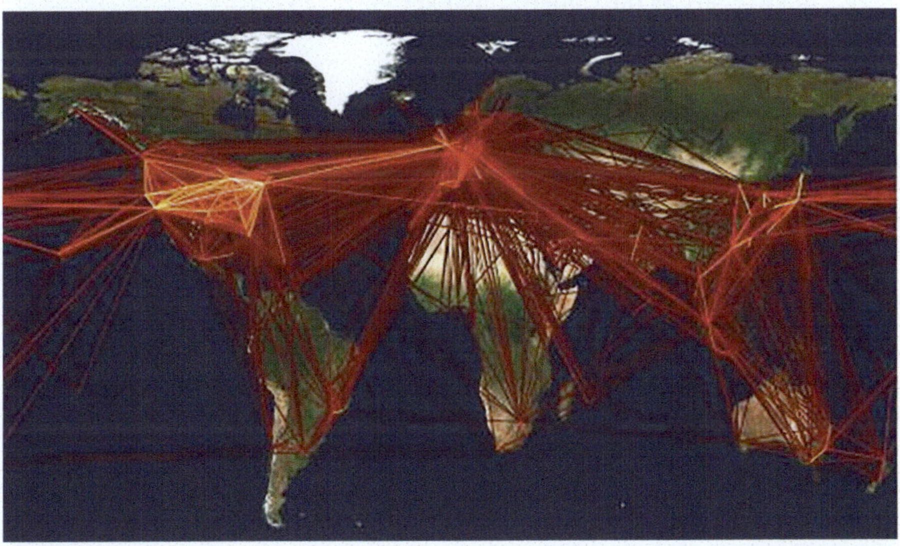

Fig. 12.20 Global aviation network
A geographical representation of the civil aviation traffic among the 500 largest international airports in > 100 different countries is shown. Each line represents a direct connection between airports. The color encodes the number of passengers per day (lighter means more) traveling between two airports. Recent figures show the same flight distribution
Source: Adapted from Hufnagel et al. (2004)

3.2 Degradation and Depletion of Ecosystem Services

Socio-ecological coviability is an emerging paradigm for rethinking the human-nature nexus by going beyond the dominant naturalistic framework based on the nature/culture dichotomy to better respond to global ecological issues, such as global warming, environmental degradation, loss of biodiversity, and desertification. Socio-ecological coviability is defined as a property of mutual dependence of the interactions between 'human' and 'non-human' systems establishing a link of viability making it possible to maintain living together (Barrière et al., 2019). Until now, this relationship quite obviously remains under pressure.

The perception of how biological diversity enhances ecological stability has already been explained in Sect. 2.1. As previously shown, the presence of modern societies has created a cascade of negative effects on the landscape, including serious damage to ecosystem services. As a result, the reciprocal interdependent relationship between humans and nature has also been damaged. As stated by Mikkelson (2009), there is reason to hope that knowledge of diversity-stability relations will help in mitigating contemporary human-induced mass extinctions. In order to stimulate this hope, the below section will focus briefly on some of the main issues of socio-ecological disruption in modern times and offer some food for thought on unconventional solutions as proposed in the last section.

3.2.1 Biodiversity as the Underestimated Centre of Gravity of Ecosystem Services

Biodiversity is affected by different drivers of change and is also a factor modifying ecosystem function. It contributes directly and indirectly to the provision of ecosystem goods and services and can be divided into different categories. The Millennium Ecosystem Assessment (2005) discriminates four main categories: *goods* (provisioning services) which are the products obtained from ecosystems; *cultural services* that represent non-material benefits delivered by ecosystems, both directly related to human well-being; *Regulating services* creating benefits obtained from regulating ecosystem processes; and *supporting services* necessary for the production of all other ecosystem services (Secretariat of CBD, 2006). These services as presented in Fig. 12.21 show the interdependence, in particular the central role of biodiversity.

Figure 12.22 emphasizes the role of (in)direct drivers of change on the loss of biodiversity. Because human presence and behavior lead to high demands for food and energy, this often results in overexploitation, habitat change, nutrient loading and pollution, as well as facilitation of invasive species and climate change. According to this figure, the role of habitat change plays a dominant role in the loss of biodiversity – a relatively thick arrow that points towards biodiversity loss. It is hopefully clear that habitat change itself is also influenced by factors identified as direct drivers of biodiversity loss in Fig. 12.22. The factors resulting from current

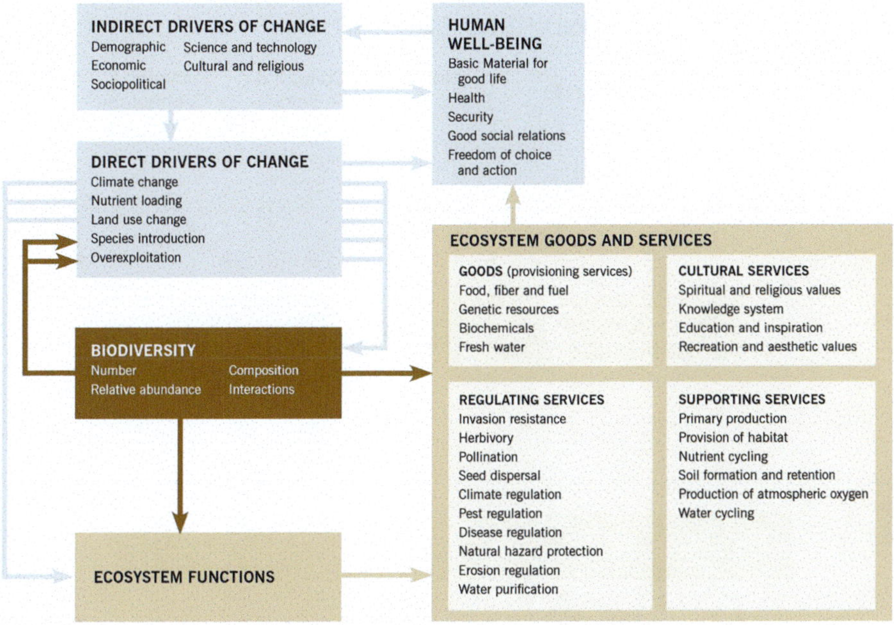

Fig. 12.21 Biodiversity, ecosystems functioning, ecoservices and drivers of change
Source: Adapted from Secretariat CBD (2006:14)

food and energy demand interaction can, directly and indirectly, act as catalysts of habitat change, and hence, strengthen each other's influence on biodiversity loss and vice versa. Furthermore, these factors can strongly degrade the properties of the mutual dependence of interactions between 'human' and 'non-human' systems.

All terrestrial, freshwater and marine ecosystems provide multiple ecosystem services. However, some ecosystems are particularly important in that they provide services that directly contribute to human health and wellbeing by providing services and goods to fulfill the daily physical, material, cultural, and spiritual needs (Secretariat of CBD, 2014). Particularly, ecosystems that directly provide human needs are of utter importance to humans; but since these systems do have critical relations with other systems, all systems are of importance. Unfortunately, Fig. 12.23 reveals that in all three distinct ecosystems there is a significant decline in species over time – green curve, see next alinea. These trends are based on the Living Planet Index (LPI)[14] that acts as a global species indicator index.

Indexes – such as the LPI – use different sources of data such as government reports, scientific papers, and research programs. Because of this, these indicator indexes automatically suffer from a variety of publication biases such as lack of resources or infrastructure for monitoring, difficulties in accessing sites as well as

[14] https://livingplanetindex.org/home/index

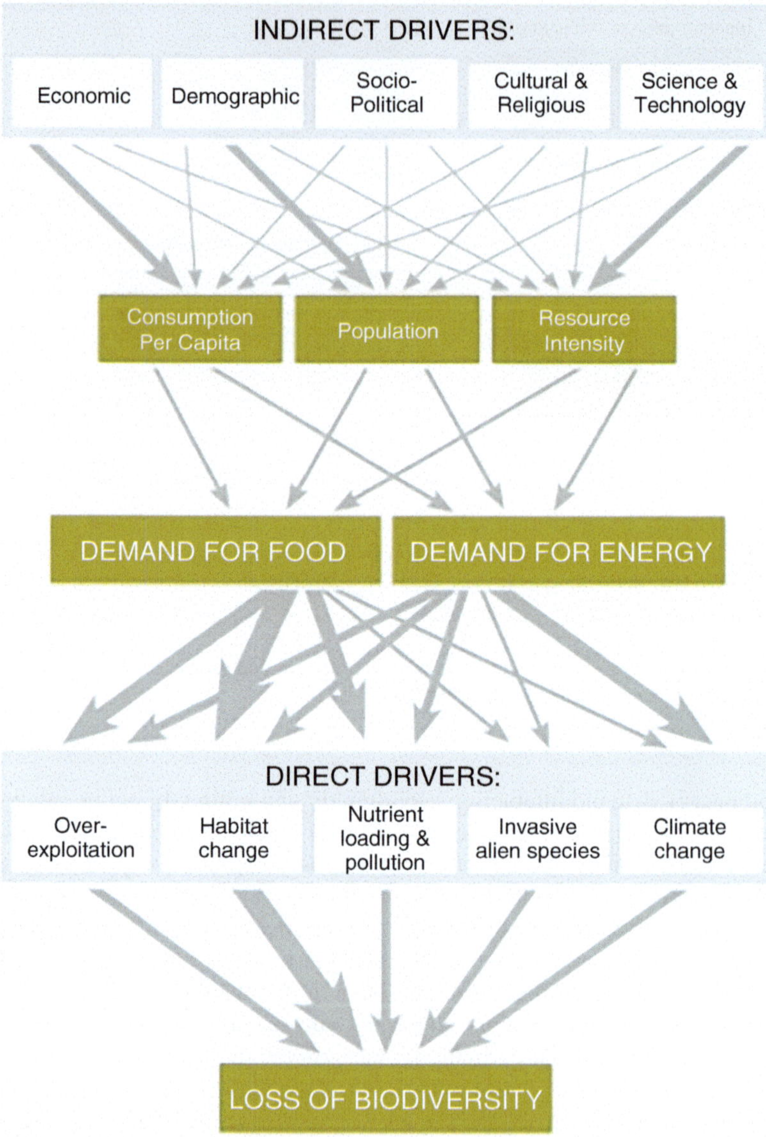

Fig. 12.22 Links between food, energy, and biodiversity loss
Source: Adapted from Secretariat CBD (2006:65)

the tendency for monitoring to occur in areas where scientists live and work or what people like to watch – for example bird watching – (Collen & Nicholson, 2014; McRae et al., 2017). Therefore, the data accuracy differs per geographic region and taxonomic group.

In order to be able to take adequate measures on species decline, it is necessary to understand the impacts of threats on biodiversity and ecosystem services.

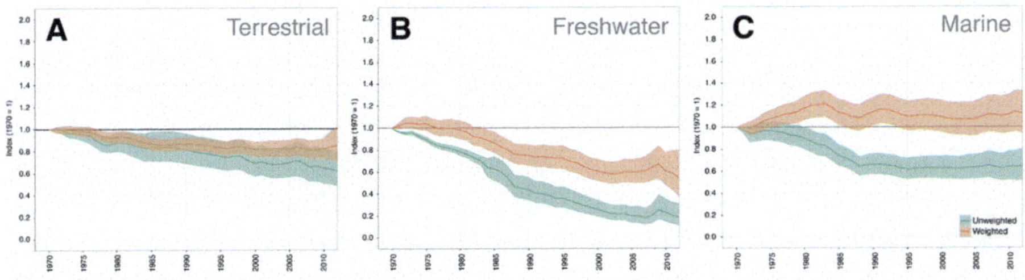

Fig. 12.23 Comparison of unweighted (red) and diversity weighted (green) Living Planet Index (LPI) for each system
(a) *Terrestrial;* ***(b)*** *Freshwater;* ***(c)*** *Marine. Solid colored lines show the average trend and shaded regions show the 95% -confidence interval of the trend*
Source: Adapted from McRae et al. (2017)

This can only be done if based on accurate trends. According to the analysis of McRae et al. (2017), the trends in population decline, as presented by the LPI, are even stronger[15] if corrected (green lines in Fig. 12.23) for several biases involved in the use of collected data, such as explained in the previous alinea. Not only do the results show steeper declines than previously estimated, they also suggest that those species for which there is poorer data coverage may be declining more rapidly. So, just like the extinction debt mentioned in Sect. 3.1.2 on metapopulations, this information should ring even more alarm bells.

3.2.2 Climate Change as a Complex Driver of Biodiversity Loss

Climate change may be defined by changing of the average weather or climate on the long term. Climate on Earth is basically formed by the interaction of the atmo-, hydro-, cryo-, litho- and biosphere, resulting in a complex of currents and patterns of water and air that are mainly driven by differences in the distribution of solar energy. These differences in energy result from the way in which the energy of the sun reaches the Earth and is distributed from the equator towards the poles. This process is influenced by many factors of which atmospheric humidity and the accumulation of greenhouse gasses – such as carbon dioxide – are most important. Since carbon dioxide is a strong greenhouse gas, its accumulation due to fossil fuel combustion on the one hand and deforestation on the other plays a crucial role in global

[15] McRae et al. (2017) corrected the original LPI 1970–2012 for diversity gaps across geographical region and species. Based on a proportional weighted index, they estimated a global population decline in vertebrate species between 1970–2012 of 58% rather than 20% from an index with no proportional weighting. From this data set, comprising 14,152 populations of 3706 species from 3095 data sources, they also found that freshwater populations have declined by 81%, marine populations by 36%, and terrestrial populations by 38% when using proportional weighting – compared to trends of −46%, +12% and +15% respectively.

warming. Climate change also involves extreme weather events like hurricane incidence, flooding and drought (Knutson et al., 2020; Kossin et al., 2020), sea-level rise, and secondary effects such as desertification, famine, and human conflicts.

Since this chapter focuses on biodiversity-stability, an extensive treatise on climate change is beyond its scope. More precisely, we will mainly focus on two major aspects of global warming: extreme temperature and sea-level rise. These two factors are significant driving forces of biodiversity loss and ecological regime shifts, resulting indirectly from human presence and behavior. Since habitat destruction and specific species loss act as drivers of ecological regime shifts, even affecting climate change, these are also briefly discussed. So, anthropogenic climate change may simultaneously be explained by top-down as well as bottom-up processes.

A. *Global Warming as a Top-Down Driver of Eco-Social Conflict*

Based on population dynamics, global warming, and physiology, Xu et al. (2020) and Raymond et al. (2020) indicate that in a business-as-usual climate scenario, the geographical position of the current temperature niche that humans, their crop and livestock have occupied for millennia will shift more over the coming 50 years than it has done over the past 6000 years. According to Xu et al. (2020), without migration one third of the global population is projected to experience a mean annual temperature over 29 °C. Such temperatures are currently found in only 0.8% of Earth's land surface, mostly concentrated in the Sahara (Fig. 12.24). Under these circumstances, the upper physiological limit of humans' ability to efficiently shed heat is reached.

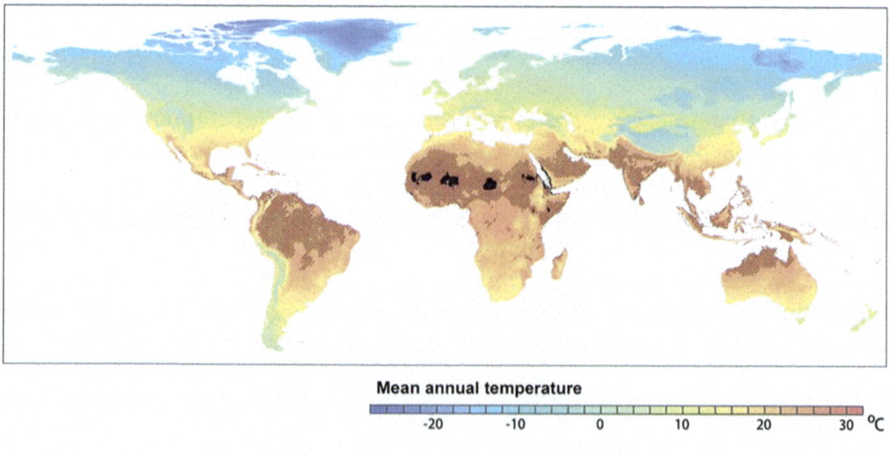

Fig. 12.24 Expansion of extremely hot regions in a business-as-usual climate scenario
In the current climate, Mean Average Temperatures (MAT's) >29 °C are restricted to the small dark areas in the Sahara region. In 2070, such conditions are projected to occur throughout the shaded area. Absent migration that area would be home to 3.5 billion people in 2070. Background colors represent the current MATs
Source: Adapted from Xu et al. (2020)

To better understand the temperature niche, a normal internal human body temperature of 36.8° ± 0.5 °C requires skin temperatures of around 35 °C to maintain a gradient directing heat outward from the core. Once the air temperature rises above this threshold, metabolic heat can only be shed via sweat-based latent cooling, and at about 35 °C, this cooling mechanism loses its effectiveness altogether (Raymond et al., 2020). Although different organisms develop different thermoregulation strategies, their physiology shares the same chemical basis on many levels. For instance, enzyme (which are made of protein) misfunctioning starts at a specific critical temperature due to the denaturation of the enzyme. Therefore, each organism is limited to a certain temperature-humidity range (Taiz & Zeiger, 1991; Rendall et al., 1997; Chapman, 1998; Yahav, 2015). Reaching the upper critical temperature is lethal and, since ideal physiological and behavioral conditions are seldom met, severe mortality already occurs at much lower temperatures (Raymond et al., 2020).

Not only will inhospitality of Earth's surface increase due to the extension of hot arid areas, also parts of hot humid areas will be inhospitable for man. This is shown in Fig. 12.24. Logically, the emergence of heat and humidity will not be too severe just for human tolerance. It may lead, however, to species loss due to migration and – local – extinction. On top of this, it is likely that the spatial-temporal dynamics of vector and reservoir species will also be influenced, affecting human and animal health (Behnassi et al., 2017; Kahime et al., 2016; Takken, 2016).

Next to temperature rise, global warming also leads to sea-level rise and will, therefore, directly affect both island and mainland ecology. Since islands are completely surrounded by water, they are highly prone to changes in sea-level rise. Islands are home to more than 20% of terrestrial plant and vertebrate species of the world with an endemism approximately eight to nine and a half times higher than on continents (Kier et al., 2009). Although different scenarios exist (Mengel et al., 2016; NOAA, 2017), even in the most optimistic scenarios, island biodiversity is likely to be affected since very low-lying islands are presumed to become partly or completely flooded. Besides this, sea-level rise will damage or destroy non-saline habitat parts of islands forcing species to leave, adapt or die (Courchamp et al., 2014). On top of that, according to Island biography rules (MacArthur & Wilson, 1967), a decrease in island surface and greater distance to the next source or mainland will lead to species decline and higher extinction rates (see Sect. 2.1.4 on meta-populations). In the coastal zones of mainland, comparable effects are likely to occur. Greater tidal ranges will lead to periodic floods that will destroy non-saline habitats. Increased frequency and amplitude of seawater floods are also expected to be more common as well as the increase of coastal erosion and saline-water intrusion (Courchamp et al., 2014; Idier et al., 2017). Taking all other possible effects of climate change into account, habitats are likely to be altered even more. Therefore, a further degradation of ecological-social coviability lies ahead.

B. *Ecological Regime Shifts: Bottom-Up and Top-Down Aspects Affecting Desirable Ecological States*

Ecological and eco-social regime shifts may more easily occur if resilience (see Sect. 2.1) has been reduced as a consequence of human actions. According to Folke et al. (2004), these actions can be defined into the following categories:

- Removal of functional groups of species and their response diversity, such as the loss of whole trophic levels (top-down effects);
- Impact on ecosystems via emissions of waste and pollutants (bottom-up effects) and climate change (top-down effects); and
- Alteration of the magnitude, frequency, and duration of disturbance regimes to which the biota is adapted.

Climate change is part of this complex of factors that influence ecological resilience and trigger ecosystems to shift to other states, from desirable to undesirable ones. However, the presence of biota and its organization at the system level also has an impact on climate. So, climate change caused primarily by anthropogenic influence should be approached as both a bottom-up and top-down process.

According to for example Fearnside and Laurance (2004), the effects of tropical deforestation on greenhouse gas emission and global warming are substantial. The rapid destruction of tropical forests is considered a major source of carbon dioxide, methane, and nitrous oxide, simultaneously creating an absorption capacity deficiency. Next to a reduced capacity to absorb carbon dioxide, deforestation causes changes in evaporation and rainfall cycle. This is of significance, in particular in the tropical regions where forest mitigates warming through evaporation cooling. So, deforestation amplifies anthropogenic effects on climate change (Fearnside & Laurance, 2004; Bonan, 2008; Staver et al., 2011; Wright et al., 2017; Cooper et al., 2020).

Wright et al. (2017) provide insights into the mechanism by which interactions among land surface processes, atmospheric convection, and biomass burning may alter the timing of wet season onset. In the Southern parts of Amazonia for example, this can lead to the extension of the dry season which enhances regional vulnerability (Wright et al., 2017). Also, further extension of the dry season may lead to a regime shift from tropical rainforest to savanna (Staver et al., 2011), if not desert. This may only take years to decades since each additional unit area of an ecosystem provides an increasingly smaller unit in time taken for that system to collapse. This means that large systems, like the Amazon rainforest, but also marine systems like the Caribbean coral reefs, tend to shift more slowly than small systems, but disproportionately faster (Cooper et al., 2020).

These kinds of climate-driven regime shifts, whether or not initially induced by habitat destruction (both legal and illegal) have serious consequences for Earth's biodiversity. Also, species loss itself can be a significant driver of regime shifts. For example, the loss of important major herbivores such as elephants, rhinoceroses, and apex predators like lion, leopard, and hyena due to poisoning, (il)legal wildlife trade and land-use change can trigger regime shifts. Therefore, (il)legal wildlife poisoning and trade as well as tropical forest loss, directly related to e.g. narcotic trafficking, form major threats to biodiversity-stability (Folke et al., 2004; Shannon et al., 2011; Pellegrini et al., 2017; Sesnie et al., 2017; Van Uhm 2016).

Concerning all this, and with respect to human well-being, utterly attention towards the need for (new) policies to focus specifically on restoring and safeguarding ecosystems is needed since the capacities for self-repair of ecosystems can no longer be taken for granted. Thus, linking biodiversity conservation with goals related to sustainable and adaptive development, and the needs of the poor, the indigenous, the local communities on one hand, and modernized urban life on the other, policies may enable societies to survive in the long term (Folke et al., 2004; Secretariat of CBD, 2014).

4 Conclusions

In the above analysis, the ecosystem stability theory has been explained by early models, evolving towards ideas about how biodiversity and functional overlap of species create ecological system resilience, resulting in complex ecosystem services that provide ecological safety and security. In addition, it has been shown that biodiversity, habitat, and climate change are all interrelated, influenced by, and crucial to human existence. Therefore, a profound understanding of ecological fundamentals is critical to any strategy involved in maintaining long-term viability between humans and nature. Against this background, the current COVID-19 pandemic simply seems to acts as an – early – indicator of conflict and mass migration.

For decades, the loss of species and habitat has increased to such an extent that it has even affected the climate system. As climate change affects resource availability, it therefore influences conflict behavior at different organizational levels, from individuals to intra- and international relations. According to Barthos and Wehr (2002), conflict behavior can occur for six main reasons: the parties may have (or think they have) incompatible goals; they may each have achieved a high degree of solidarity; have organized themselves for conflict; mobilized their conflict resources; be hostile to their adversaries; and may have sufficient material resources. One of the prominent examples can be found in conflicts regarding the availability of fresh water. Since water is a primary necessity of life, physiological stress will influence the conflict behavior of the individuals involved. Conflict behavior can manifest itself in aggression and fear.

According to Lorenz (1966), aggression and fear are both developed to act as specific cues for communicating emotional states, reducing the risk of intraspecific damage. However, supported by the complex of factors mentioned above, this communication system sometimes fails, resulting in local and regional intra- or international conflicts. In those situations, aggression and fear can lead to both internal and external movements of people. From a biological point of view, migration – as a form of movement – can be seen as an adaptation to resources that fluctuate spatiotemporally either seasonally or less predictably. Migration plays a central role in the spatial dynamics of mobile (animal) populations and differs largely in both form and function from the within-population mixing resulting from postnatal dispersal and from the inter-patch movements that characterize metapopulations. Since dispersal and migration both hold proximate stimuli and ultimate consequences, specifically related to the level of individuals ('behaviorally') or populations ('ecologically') (Dingle & Drake, 2007), what is the ecological relevance of conflict and migration?

This chapter provides ecological principles to make the reader deduce that resource and migration-driven conflicts generate new and direct environmental issues. Not only do these impacts have uncontrolled effects on the displaced individuals, but they also affect the socio-ecological systems they have left and those to which they have migrated. For Laundré et al. (2010), the long-term consequence of these issues is a cascade of restless ecological and socio-political effects.

In addition to climate-driven conflict and migration, terrorist and criminal activities can be the drivers of conflict and migration. On the one hand, stimulating migration by spreading fear and facilitating human trafficking; on the other by facilitating socio-ecological regime shifts through their impact on the environment. Known effects include habitat destruction by coca plantations and more complex ecological effects resulting from the illegal hunting of ecological engineering species such as rhinoceroses, elephants, and chimpanzees. Indeed, organized crime is responsible for millions of dollars in illicit wildlife trade, the fourth largest form of illegal activity in the world. It, therefore, weakens the socio-ecological quality of life and causes various forms of conflict. This makes it a serious threat to the long-term resilience and survival of many socio-ecological systems. Since global organizations have not been able to stop the phenomenon of illegal species and habitat loss (it has even accelerated), it may have long been time to combine existing methods with more radical but well-organized countermeasures.

Acknowledgments I wish to thank the following persons for providing insightful critiques on an early edition of this chapter: Piet Wit, Victor Spaan, Gerard Lucius, and Dr. Himangana Gupta. Last but not least, I would like to thank Dr. Mohamed Behnassi for his pertinent advice and above all for allowing me the opportunity to write and publish this chapter.

References

Akiner, M. M., Demirci, B., Babuadze, G., Robert, V., & Schaffner, F. (2016). Spread of the invasive mosquitoes *Aedes aegypti* and *Aedes albopictus* in the Black Sea region increases risk of chikungunya, dengue, and Zika outbreaks in Europe. *PLoS Neglected Tropical Diseases, 10*(4), e0004664.

Anhlan, D., Grunmann, N., Makalowski, W., Ludwig, S., & Scholtissek, C. (2011). Origin of the 1918 pandemic H1N1 influenza A virus as studied by codon usage patterns and phylogenetic analysis. *RNA, 17*, 64–73. Cold Spring Harbor Laboratory Press. RNA Society.

Barthos and Wehr. (2002). *Using conflict theory*. Cambridge Univerity Press.

Barrière, O., Behnassi, M., David, G., et al. (2019). *Coviability of social and ecological systems: Reconnecting mankind to the biosphere in an era of global change. Vol 2: Coviability questioned by a diversity of situations*. Springer.

Baud, D., Qi, X., Nielsen-Saines, K., Musso, D., Pomar, L., & Favre, G. (2020). Real estimates of mortality following COVID-19 infection. *The Lancet Infectious Diseases. March, 12*, 2020. https://doi.org/10.1016/S1473-3099(20)30195-X

Beasley, D., & Paul, K. (2020). Two New York cats become first US pets to test positive for COVID-19. *Reuters. Health News*, 23 April 2020.

Behnassi, M., et al. (2017). Assessing health vulnerability to climate change in Morocco: Governance and adaptation options. In M. Bouzid (Ed.), *Examining the role of environmental change on emerging infectious diseases and pandemics* (pp. 138–162). IGI Global, , ISBN: 9781522505532.

Bonan, G. (2008). Forests and climate change: Forcings, feedbacks, and the climate benefits of forests. *Science, 320*, 1444–1449.

Botton, S., Van Heusden, M., Parson, J. R., Smidt, H., & Van Straalen, N. (2006). Resilience of microbial systems towards disturbances. *Critical Reviews in Microbiology, 32*, 101–112.

Cale, P. G. (1999). *The spatial dynamics of White-browed Babbler groups in a fragmented agricultural landscape*. PhD thesis. University of New England.

Chapman, R. F. (1998). Chapter 19: Thermal relations. In *The insects: Structure and function* (4th ed., pp. 509–528). Cambridge University Press.

Collen, B., & Nicholson, E. (2014). Taking the measure of change. *Science, 346*, 166–167.

Cooper, G. S., Willcock, S., & Dearing, J. A. (2020). Regime shifts occur disproportionately faster in larger ecosystems. *Nature Communications, 11*, 1175. https://doi.org/10.1038/s41467-020-15029-x

Courchamp, F., Hoffmann, B. D., Russel, J. C., Leclerc, C., & Bellard, C. (2014). Climate change, sea-level rise, and conservation: keeping island biodiversity afloat. *Trends in Ecology and Evolution, 3*, 127–130. Elsevier.

Darwin, C. (1859). *On the origin of species by means of natural selection, or the preservation of favoured races in the struggle for life*. John Murray.

Den Boer, P. J., & De Vries, H. H. (1990). Survival of populations of Agonum ericeti panz. (Col., Carabidae) in relation to fragmentation of habitats. *Netherlands Journal of Zoology, 40*(3), 483–498.

Den Boer, P. J., De Vries, H. H., & van Dijk, T. S. (1996). Ground beetle species in heathland fragments in relation to survival, dispersal, and habitat preference. *Oecologia, 107*, 332–342.

Dingle, H., & Drake, A. (2007). What is migration? *Bioscience, 37*(2), 113–121.

Ehrlich, P. R., & Ehrlich, A. H. (1981). *Extinction: The causes and consequences of the disappearance of species*. Random House.

Elton, C. S. (1958). *The ecology of invasions by animals and plants*. Methuem and Co. Ltd.

European Union. (2019). *The European Union one health 2018 Zoonoses report*. European Food Safety Authority and European Centre for Disease Prevention and Control.

Fălcuţă, E., Prioteasa, L. F., Horváth, C., et al. (2020). The invasive Asian tiger mosquito *Aedes albopictus* in Romania: Towards a country-wide colonization? *Arthropods and Medical Entomology, 119*, 841–845.

Faust, C. L., Dobson, A. P., Gottdenker, N., et al. (2016). Null expectations for disease dynamics in shrinking habitat: Dilution or amplification? *Philosophical Transactions B, 327*, 20160173.

Fearnside, P. M., & Laurance, W. F. (2004). Tropical deforestation and greenhouse-gas emissions. *Ecological Applications, 14*(4), 982–986.

Folke, C., Carpenter, S., Walker, B., Sheffer, M., Elmqvist, T., Gunderson, L., & Holling, C. S. (2004). Regime shifts, resilience, and biodiversity in ecosystem management. *Annual Review of Ecology, Evolution, and Systematics, 2004*(35), 55–81.

Frank, D. A., & McNaughton, S. J. (1991). Stability increases with diversity in plant communities: Empirical evidence from the 1988 Yellowstone drought. *Oikos, 62*(3), 360–362.

Gildenhard, M., Rono, E. K., Diarra, A., et al. (2019). Mosquito microevolution drives plasmodium falciparum dynamics. *Nature Microbiology, 4*, 941–947.

Guo, Q. (2015). *Island biogeography theory: Emerging patterns and human effects. Reference module in earth systems and environmental sciences*. Elsevier.

Haddad, N. M., Brudvig, L. A., Clobert, J., et al. (2015). Habitat fragmentation and its lasting impact on Earth's ecosystems. *Science Advances, 1*(2), e15000052.

Harrison, S. (1991, January). Local extinction in a metapopulation context: An empirical evaluation. *Biological Journal of the Linnean Society, 42*(1–2), 73–88.

Harrison, S., & Taylor, A. D. (1997). Empirical evidence for metapopulation dynamics. In I. Hanski & E. Gilpin (Eds.), *Metapopulation biology: Ecology, genetics and evolution* (pp. 27–42). Academic Press. Elsevier.

Hofmeester, T. R. (2016). The role of host diversity in Borrelia burgdorferi s.s. dynamics. In M. A. H. Braks, S. E. van Wieren, W. Takken, & H. Sprong (Eds.), *Ecology and prevention of Lyme borreliosis* (Ecology and control of vector-borne diseases) (Vol. 4, pp. 173–186). Wageningen Academic Publishers.

Hofmeester, T. R., Jansen, P. A., Wijnen, H. J., Coipan, E. C., Fonville, M., Prins, H. H. T., Sprong, H., & Wieren van, H. S. (2017). Cascading effects of predator activity on tick-borne disease risk. *Proceedings of the Royal Society B Biological Sciences, 284*, 20170453.

Holling, C. S. (1973). Resilience and stability of ecological systems. *Annual Review of Ecology and Systematics, 4*(1973), 1–23.

Holling, C. S. (1992). Cross-scale morphology, geometry, and dynamics of ecosystems. *Ecological Monographs, 62*(4), 447–502.

Holling, C. S. (1996). Engineering resilience versus ecological resilience. In P. Schulze (Ed.), *Engineering within ecological constrains* (pp. 31–44). National Academy.

Hufnagel, L., Brockmann, D., & Geisel, T. (2004). Forecast and control of epidemics in a globalized world. *PNAS, 101*(42), 15124–15129.

Idier, D., Paris, F., Le Cozannet, G., Boulahya, F., & Dumas, F. (2017). Sea-level rise impacts on the tides of the European shelf. *Continental Shelf Research, 137*, 56–71.

Isbell, F. (2010). Causes and consequences of biodiversity declines. *Nature Education Knowledge, 3*(10), 54.

Johnson, P. T. J., Preston, D. L., Hoverman, J. T., & Richgels, K. L. D. (2013). Biodiversity decreases disease through predictable changes in host community competence. *Nature, 494*, 230–233.

Jones, K. E., Patel, N., Levy, M., et al. (2008). Global trends in emerging infectious diseases. *Nature, 2008*(451), 990–994.

Karesh, W. B., Dobson, A., Lloyd-Smith, J. O., et al. (2012). Ecology of zoonoses: Natural and unnatural histories. *The Lancet, 2012*(380), 1936–1945.

Katz, L. M. (2019). The emergence of zoonotic pathogens as agents of concern in transfusion medicine. In H. Shan & R. Dodd (Eds.), *Blood safety* (pp. 189–206). Springer.

Keven, J. B., Katusele, M., Vinit, R., et al. (2019). Species abundance, composition, and nocturnal activity of female Anopheles (Diptera: Culicidae) in malaria-endemic villages of Papua New Guinea: Assessment with barrier screen sampling. *Malaria Journal, 18*, 96.

Khalil, H., Ecke, F., Evander, M., Magnusson, M., & Hornfeldt, B. (2016). Declining ecosystem health and the dilution effect. *Scientific Reports, 6*, article number 31314. https://doi.org/10.1038/srep31314

Kahime, K., Behnassi, M., Mohammed, M., Samia, B., & Boumezzough, A. (2016). Infectious diseases and climate: Case of Morocco. *Journal of Coastal Life Medicine, 4*(8), 659–662.

Kier, G., Kreft, H., Lee, T. M., Jetz, W., Ibisch, P., Nowicki, C., Mutke, J., & Barthlott, W. (2009). A global assessment of endemism and species richness across island and mainland regions. *PNAS, 106*(23), 9322–9327.

Kilpatrick, A. M., & Randolph, S. E. (2012). Drivers, dynamics, and control of emerging vector-borne zoonotic diseases. *The Lancer, 380*, 1946–1955.

Knutson, T., et al. (2020). Tropical cyclones and climate change assessment: Part 2. Projection response to anthropogenic warming. *American Meteorological Society*, 303–322.

Kossin, J. P., Knapp, K. R., Olander, T. L., & Velden, C. S. (2020). Global increase in major tropical cyclone exceedance probability over the past four decades. *PNAS*. https://doi.org/10.1073/pnas.1920849117

Laundré, J. W., Hernández, L., & Ripple, W. J. (2010). The landscape of fear: Ecological implications of being afraid. *The Open Ecology Journal, 3*, 1–7.

Levi, T., Kilpatrick, A. M., Mangel, M., & Wilmers, C. C. (2012). Deer, predators, and the emergence of Lyme disease. *PNAS. USA., 109*(27), 10942–10947.

Levins, R. (1969). Some demographic and genetic consequences of environmental heterogeneity for biological control. *Bulletin of the Entomological Society of America, 15*(3), 237–240.

LoGiudice, K., Ostfeld, R. S., Schmidt, K. A., & Keesing, F. (2003). The ecology of infectious disease: Effects of host diversity and community composition on Lyme disease risk. *PNAS, 100*, 567–571.

Lorenz, C. (1966). *On Aggression* (M. Kerr Wilson, Trans.). Harcourt and World.

MacArthur, R. H., & Wilson, E. O. (1967). *The theory of island biogeography*. Princeton University Press.

Marano, N., & Pappaioanou, M. (2004). Historical, new, and reemerging links between human and animal health. *Emerging Infectious Diseases, 10*(12), 2065–2066.

McCann, K. S. (2000). The diversity-stability debate. *Nature, 2000*(405), 228–233.

McGrady-Steed, J., & Morin, P. J. (2000). Biodiversity, density compensation, and the dynamics of populations and functional groups. *Ecology, 81*, 361–373.

McGreevy, N. (2020). A tiger in the Bronx Zoo tested positive for COVID-19. *Smithsonian Magazine*. https://www.smithsonianmag.com/smart-news/tiger-bronx-zoo-tested-positive-covid-19-180974624/

McRae, L., Deinet, S., & Freeman, R. (2017). The diversity-weighted living planet index: Controlling for taxonomic bias in a global biodiversity indicator. *PLoS One, 12*(1), e0169156. https://doi.org/10.1371/journal.pone.0169156

Mengel, M., Leverman, A., Frieler, K., Robinson, A., Marzeion, B., & Winkelmann, R. (2016). Future sea level rise constrained by observations and long-term commitment. *PNAS, 13*(10), 2597–2602.

Mikkelson, G. M. (2009). Diversity-stability hypothesis. In J. B. Callicott & R. Frodeman (Eds.), *Encyclopedia of environmental ethics and philosophy*.

Millennium Ecosystem Assessment. (2005). *Ecosystems and human well-being: Synthesis*. Island Press.

Mills, L. S., Soule, M. E., & Doak, D. F. (1993). The keystone-species concept in ecology and conservation. *Bioscience, 43*(4), 219–224.

Morse, S. S., Mazet, J. A. K., Woolhouse, M., et al. (2012). Prediction and prevention of the next pandemic zoonosis. *The Lancet, 380*, 1956–1965.

National Oceanic Atmospheric Administration (NOAA). (2017). *Global and regional sea level rise scenario's for the United States*. NOAA technical report NOS CO-OPS 083. U.S. Department of Commerce. National Ocean Service. Centre for Operational Oceanographic Products and Services.

Nupp, T. E., & Swihart, R. K. (1996). Effect of forest patch area on population attributes of white-footed mice (Peromyscus leucopus) in fragmented landscapes. *Canadian Journal of Zoology, 74*, 467–472.

Nupp, T. E., & Swihart, R. K. (2000). Landscape-level correlates of small-mammal assemblages in forest fragments of farmland. *Journal of Mammalogy, 81*, 512–526.

Olson, D. M., Dinerstein, E., Wikramananyake, E. D., Burgess, N. D., et al. (2001). Terrestrial ecoregions of the world: A new map of life on earth. *Bioscience, 51*(11), 993–938.

Opdam, P. (1991). Metapopulation theory and habitat fragmentation: A review of holarctic breeding bird studies. *Landscape Ecology, 5*(2), 93–106.

Ostfeld, R. S., & Holt, R. D. (2004). Are predators good for your health? Evaluating evidence for top-down regulation of zoonotic disease reservoirs. *Frontiers Ecological Environment, 2*(1), 13–20.

Ostfeld, R. S., & Keesing, F. (2000). Biodiversity and disease risk: The case of Lyme disease. *Conservation Biology, 14*(3), 722–728.

Ostfeld, R. S., & LoGiudice, K. (2003). Community disassembly, biodiversity loss, and erosion of an ecosystem service. *Ecology, 84*(6), 1421–1427.

Patz, J. A., Daszak, P., Tabor G. M., et al., and Members of the Working Group on Land-Use Change and Disease Emergence. (2004). Unhealthy landscapes: Policy recommendations on land use change and infectious disease emergence. *Environ Health Perspectives, 112*, 1092–1098.

Paine, R. T. (1969). A note on trophic complexity and community stability. *The American Naturalist, 103*(929), 91–93.

Pellegrini, A. F. A., Pringle, R. M., Govender, N., & Hedin, L. O. (2017). Woody plant biomass and carbon exchange depend on elephant-fire interactions across a productivity gradient in African savanna. *Journal of Ecology, 105*, 111–121.

Peterson, G., Allen, G. R., & Holling, C. S. (1998). Ecological resilience, biodiversity, and scale. *Ecosystems, 1*, 6–18. Springer-Verlag.

Proulx, R., Wirth, C., Voigt, W., Weigelt, A., et al. (2010). Diversity promotes temporal stability across levels of ecosystem organization in experimental grasslands. *PLoS One, 5*(10), e13382. https://doi.org/10.1371/journal.pone.0013382

Rendall, D., Burggren, W., & French, K. (1997). Chapter 16: Using energy: Meeting environmental challenges. In *Eckert animal physiology: Mechanism and adaptations* (pp. 665–722). W.H.F. Freeman and Company.

Raymond, C., Matthews, T., & Horton, R. M. (2020). The emergence of heat and humidity too severe for human tolerance. Science. *Advances, 6*, eaaw1838.

Richardson, D. M., & Pyšek, P. (2007). Classics in physical geography revised. Elton C.S. 1958.: The ecology of invasions by animals and plants. London: Methuen. *Progress in Physical Geography, 31*(6), 659–666.

Ritchie, M. E., & Olff, H. (1999). Spatial scaling laws yield a synthetic theory of biodiversity. *Nature, 400*, 557–560.

Rohr, J. R., Barrett, C. B., Civitello, D. J., et al. (2019). Emerging human infectious diseases and the links to global food production. *Nature Sustainable, 2*, 445–456.

Rosenblatt, D. L., Heske, E. J., Nelson, S. L., Barber, D. M., Miller, M. A., & MacAllister, B. (1999). Forest fragments in east-Central Illinois: Islands or habitat patches for mammals? *American Midland Naturalist, 141*, 115–123.

Scheffer, M., Carpenter, S., Foley, J. A., Folke, C., & Walker, B. (2001). Catastrophic shifts in ecosystems. *Nature, 413*, 591–596.

Schindler, D. W. (1990). Experimental perturbations of whole lakes as tests of hypotheses concerning ecosystem structure and function. *Oikos, 57*, 25–41.

Schmidt, K. A., & Ostfeld, R. S. (2001). Biodiversity and the dilution effect of disease ecology. *Ecology, 82*(3), 609–619.

Secretariat of the Convention on Biological Diversity (CBD). (2006). *Global biodiversity outlook 2*. Montreal.

Secretariat of the Convention on Biological Diversity (CBD). (2014). *Global biodiversity outlook 4: A mid-term assessment of progress towards the implementation of the strategic plan for biodiversity 2011–2020*. Montreal.

Sesnie, E. S., Tellman, B., Wrathall, D., McSweeney, K., Nielson, E., Benessaiah, K., Wang, O., & Rey, L. (2017). A spatio-temporal analysis of forest loss related to cocaine trafficking in Central America. *Environmental Research Letters, 12*, 054015.

Shannon, G., Thaker, M., Vanak, T. A., Page, B. R., Grant, R., & Slotow, R. (2011). Relative impact of elephant and fire on large trees in a Savanna ecosystem. *Ecosystems, 14*, 1372–1381.

Shi, J., Wen, Z., Zhong, G., et al. (2020). Susceptibility of ferrets, cats, dogs and other domesticated animals to SATRS-coronavirus 2. *Science, 2020*. https://doi.org/10.1126/science.abb7015

Simberloff, D. S. (1976). Experimental zoogeography of islands: Effects of island size. *Ecology, 57*, 629–641.

Simberloff, D. S., & Abele, L. G. (1976). Island biogeography theory and conservation practice. *Science, 191*, 285–286.

Smit, R., Dijk, F., Kruidbos, F., et al. (2003). Populatiedynamiek en fenologie van teken in Nederland: effecten van geografische ligging, habitat en microklimaat op tekenpopulaties. *Infectieziekten Bulletin. RIVM, 14*(5), 167–170.

Staver, A. C., Archibald, S., & Levin, S. A. (2011). The global extent and determinants of savanna and forest as alternative biome states. *Science, 334*, 230–232.

Taiz, L., & Zeiger, E. (1991). *Plant physiology. Chapter 14: Stress physiology* (pp. 346–370). The Benjamin /Cummings Publishing Company, Inc.

Takken, W. (2016). Phenology of *Ixodes ricinus* and Lyme borreliosis risk. In M. A. H. Braks, S. E. van Wieren, W. Takken, & H. Sprong (Eds.), *Ecology and prevention of Lyme borreliosis* (Ecology and control of vector-borne diseases) (Vol. 4, pp. 149–171). Wageningen Academic Publishers.

Taubenberger, J. K., & Morens, D. M. (2006). 1918 influenza: The mother of all pandemics. *Emerging Infectious Diseases, 12*(1), 15–22.

Tilman, D., Reich, P. B., & Knop, J. M. H. (2006). Biodiversity and ecosystem stability in a decade-long grassland experiment. *Nature Letters, 441*, 629–632.

Van Uhm, D. P. (2016). *The illegal wildlife trade. Inside the world of poachers, smugglers and traders* (Studies of organized crime 15). Springer.

Van Buskirk, J., & Ostfeld, R. S. (1995). Controlling Lyme disease by modifying the density and species composition of tick hosts. *Ecological Applications, 5*, 1133–1140.

Van Buskirk, J., & Ostfeld, R. S. (1998). Habitat heterogeneity, dispersal and local risk of exposure to Lyme disease. *Ecological Applications, 8*, 365–378.

Vidal, J. (2020). Tip of the iceberg: Is our destruction for nature responsible for COVID-19? *The Guardian*, 18 March 2020.

Walker, B. (1992). Biological diversity and ecological redundancy. *Conservation Biology, 6*, 18–23.

Walker, B. (1995). Conserving biological diversity through ecosystem resilience. *Conservation Biology, 9*, 747–752.

Wang, L.-F., & Crameri, G. (2014). Emerging zoonotic viral diseases. *Revue Scientifique et Technique (International Office of Epizootics), 33*(2), 569–581.

WHO EMRO. (2020). *Zoonotic disease: Emerging public health threats in the region.* http://www.emro.who.int/fr/about-who/rc61/zoonotic-diseases.html

Wilke, A. B. B., Benelli, G., & Beier, J. C. (2020). Beyond frontiers: On invasive alien mosquito species in America and Europe. *PLoS Neglected Tropical Diseases, 14*(1), e0007864.

Woolhouse, M. E. J., Howey, R., Gaunt, E., Reilly, L., Chase-Topping, M., & Savill, N. (2008). Temporal trends in the discovery of human viruses. *Proceedings of the Royal Society B, 275*, 2111–2115.

Wright, J. S., Fu, R., Worden, J. R., Chakraborty, S., Clinton, N. E., Risi, C., Sun, Y., & Yin, L. (2017). Rainforest-initiated wet season onset over the southern Amazon. *PNAS, 1621516114.*

Xu, C., Kohler T. A., Lenton, T. M., Svenning, J-C., & Scheffer, M. (2020). *Future of the human climate niche*. PNAS.1910114117.

Yahav, S. (2015). Regulation of body temperature: Strategies and mechanism. In C. G. Scanes (Ed.), *Sturkie's avian physiology* (pp. 869–905). Elsevier.

Yodzis, P. (1981). The stability of real ecosystems. *Nature, 289*, 674–676.

Zhu, N., et al. (2020, February 7). A novel Coronavirus from patients with pneumonia in China, 2019. *The New England Journal of Medicine*, 1–7.

Chapter 13
Tree Species Classification of the Conflict Regions of Sudan Using RapidEye Satellite Imagery

Taisser H. H. Deafalla and Elmar Csaplovics

Abstract In the conflict-affected countries, increasing realization of the fact that forests not only provide multiple benefits to local communities but also help in conserving the environment and mitigating climate change has created global concern for their protection and conservation. However, if these resources have not been managed in a sustainable or equitable manner, this leads to further environmental degradation and global warming. In this direction, vegetation mapping may be a primary requirement for various management and planning activities at the landscape level. The study presented here focused on developing methods of tree species identification in conflict areas using aerial hyperspectral data. Five RapidEye scenes were acquired for that purpose in the Nuba Mountains of Sudan. The Geographic Object-Based Classification (GEOBIA) i.e., K-Nearest Neighbor classifier model and knowledge-based classifier, built on a developed model of integrated features (such as vegetation indices, DEM, and thematic layers), environmental knowledge, feature extracted from RapidEye images, and user expert knowledge, was applied to generate the species map. The overall accuracy, producer's accuracy, user's accuracy, and Kappa statistics methods were conducted. Additionally, for more accurate results of each class, the best classification result method was also applied. GEOBIA provides unprecedented opportunities to classify and detect tree species more accurately, over large areas, with diminishing costs and processing time. The study recommends that the use of GEOBIA to analyze RapidEye imagery for the identification of trees species in the semi-arid region should be integrated with ancillary data such as DEM and other levels of GIS data to improve the quality of the results.

Keywords K-nearest neighbor classifier model · Knowledge-based classifier · Rapideye imagery · Tree species · Spectral indices

T. H. H. Deafalla (✉) · E. Csaplovics
Institute of Photogrammetry and Remote Sensing, University of Dresden, Dresden, Germany
e-mail: Csaplovics@tu-dresden.de

M. Behnassi et al. (eds.), *The Climate-Conflict-Displacement Nexus from a Human Security Perspective*, https://doi.org/10.1007/978-3-030-94144-4_13

1 Introduction

Conflict/insecurity includes interstate conflicts, internal violence, regional or global instability, civil unrest, or political crises that lead to displacement (Deafalla, 2019). When we try to deeply understand the concept of conflict, we find it characterized by a close succession of events leading to a significant degree of human suffering and acutely threatening peace and security. It is a widespread phenomenon across the globe, drawing more concern and attention from scholars. Since the end of World War II and the Cold War up to date, and despite the efforts to foster peace and security made by the international community and many individual countries, the number of conflicts and the magnitude of their implications are still high political concerns at all scales. This can be explained, among others, by the fact that international responses to conflicts have often been short-term and counterproductive, while only long peace-building responses have the potential to provide lasting solutions. Indeed, the current crises have numerous causes and manifest in many different forms. They go hand in hand with a profound shift in the international order. For example, growing instability across the Middle East, Asia, and many parts of Africa has presented tremendous challenges for the international community, most notably record levels of forced displacement and terror attacks.

Since the 1990s, resource conflicts (e.g., the Angolan Civil War) became 'pure' resource conflicts and can be described as geoeconomic-geopolitical conflicts. Different geographical regions may have different frequencies of conflict intensity. Several studies such as Renner (2002) and Kurecic (2015) confirmed that geoeconomic rivalries precede geoeconomic conflicts, i.e., 'resource wars' induced by the competition for the control of resources. This can happen in certain regions of the world that possess abundant resources (e.g., the Gulf of the Guinea region, the Middle East, etc.), causing long-term instability which results in very weak opportunities for both economic and social development.

The inter-country conflicts affect in a variety of ways; this impact also varies depending on the characteristics of each area (FAO, 2000). Civilians in conflict are often deprived of their income sources and pushed to acute food insecurity. Food systems and markets are disrupted, resulting in higher food prices and sometimes shortages of water and fuel or food. FAO (2020) estimated that 77 million people faced acute food insecurity due to conflict in 2019, more than half of whom lived in the Middle East and Asia. Furthermore, the conflict prevents companies from working and weakens the national economy, reduces job opportunities, increases poverty levels, and turns government spending on war. Health systems are often damaged or destroyed, leaving communities entirely reliant on humanitarian aid for basic services. Besides that, increasing insecurity and roadblocks prevent humanitarian convoys from reaching the most vulnerable. On the other hand, the forced movement of people in those areas undermines, sometimes for decades, the economic development, sustainable livelihoods as well as the capacities of societies and nations (Deafalla et al., 2012).

Given the history of Sudan as a case study, the country is one of the most geographically and ethnically diverse countries in Africa. Two rounds of a North-South civil war have costed the lives of 1.5 million Sudanese. Similarly, the ongoing conflict in the Western region of Darfur has driven 2 million people from their homes. The Kordofan region was afflicted by conflict for over two decades, leading to death, destruction, and population displacement. This ended in 2002 with the Comprehensive Peace Agreement. After South Sudan's secession in June 2011, conflict erupted in many parts of Abyei, Blue Nile, and South Kordofan resulting inlarge-scale population displacement. In South Kordofan state alone, more than 630,000 people were displaced or severely affected, further weakening the health infrastructure of the State (Deafalla, 2019). After years of insecurity and displacements, exacerbated by drought, failed harvests, and high food prices since 2009/2010, a complex humanitarian crisis continues in most of Sudan. In 2018, in the Greater Darfur region, security improvements allowed substantial numbers of internally displaced persons (IDPs) to return home and engage in agricultural activities. Incidents of fighting had also declined in South Kordofan and Blue Nile States. However, 1.9 million IDPs who could not afford the basic food basket and 1.1 million refugees and asylum seekers displaced by conflict continued to need humanitarian assistance, both in camps and within host communities (Deafalla, 2019). While the political and security situation remained calm in the East, minimal advancement was made in the implementation of the 2006 Eastern Sudan Peace Agreement (ESPA).

The country is currently undergoing a major transition following the April 2019 overthrow of Omar al-Bashir and subsequent establishment of a civilian-military transitional government in August 2019 (Saferworld, 2020). Since then, it is calling for a new political paradigm to achieve peace, justice and freedom. The Sudanese transitional government prioritized peace efforts in conflict-affected areas. There have been renewed initiatives aimed at resolving the long-standing and deadly conflicts in Sudan, including in Darfur, Eastern Sudan, and the two areas of South Kordofan and the Blue Nile. In August, the transitional government and several major armed groups signed a peace agreement that would open humanitarian aid routes into areas cut off by conflict (Saferworld, 2020). Unfortunately, the power vacuum created at the local state level by the regime collapse led to intercommunal tensions once more escalated in some areas in Darfur, Abyei, and Eastern Sudan, with about 12,700 people newly displaced (OCHA, 2020). This is in addition to pending border issues with Egypt since 1958 in the North and with the Republic of South Sudan in the South, and the security crisis in June 2020 on the border between Sudan and Ethiopia. The crisis adds more pressure on Sudan's transitional military-civilian government, as it struggles to establish democracy and the rule of law in the country, while at the same time dealing with the coronavirus COVID-19 outbreak and a weak healthcare system in the country.

On the other hand, despite the availability of newly harvested cereal crops in January 2020, food insecurity continues for south Sudanese refugees and IDPs in SPLM-N-controlled areas of South Kordofan. It also persists for IDPs and conflict-affected households in Jebel Marra region in Darfur, and in several areas of Abyei, northern parts of North Darfur, parts of North Kordofan, and parts of Kassala and Red Sea states. From February to May 2020, acute food insecurity deteriorated further in South Kordafan, Red Sea, Gedaref, Kassala, and South Darfur states.

In recent years, forest products started to gain considerable importance in Sudan due to many reasons such as the exponential development in the use of Non Timber Forest Products (NTFPs) in many industries and medicine. Besides, the use of NTFPs, as food and drink, is becoming more familiar (Deafalla, 2019). Moreover, these products are particularly important during drought and famine (e.g. the famine of 1982 and war in 2011 when the NTFPs were the only available food). Information about the land cover, particularly the tree species, is crucial on technical, economic, and ecological grounds (Raczko & Zagajewski, 2018). In general, forested areas cover large expanses making them difficult to analyze using traditional methods of creating forest inventory, which also requires substantial financial investment (Peerbhay et al., 2013). With the advent of remote sensing (RS), the scope of effective planning and management of natural resources has considerably widened (Maurya et al., 2013). The use of satellite data permits timely and accurate information on very short repetitive cycles needed for monitoring (El-Abbas, 2015). It is observed that remotely sensed data can meet many of the information needs for proper forest management in a short time and at a low cost (Chaudhary, 2003). Moreover, RS allows us to collect data from dangerous or inaccessible areas (Pumayalli, 2012). Thematic maps are the common form of extracted information to end users in a format that is understandable and accessible based on continuously developed classification approaches (El-Abbas, 2015). It is precisely these properties that make satellite remote sensing such an important source of data for studies of the Earth's surface dynamics and atmosphere (Donoghue, 2002).

The Earth is a highly complex system formed by mutually interlinked components. Therefore, accurate, timely, and reliable data are very important to analyze and study these variations (Ramachandran et al., 2011). Using RS images, researchers in forest resource use have provided useful decision-making support and policy frameworks for their sustainable management. Methodologies have been developed towards this end by various authors (Babu et al., 2002; Sihag et al., 2015). However, these products are not adequately treated. There are no clear mechanisms and extension services to improve their quantity and quality, either through afforestation and/or development of pre and post-best practices of NTFPs collection (Deafalla, 2019). Furthermore, there is a huge lack of reliable data on the geo-location of the tree species, their production and trade, and the number of people involved, which makes it hard to assess the effective contribution of NTFPs to rural livelihoods (Chikamai & Tchatat, 2004). Therefore, the current study attempts to map the NTFPs species and investigate the role of these products under specific circumstances like armed conflicts. This study would help in designing sound NTFPs management strategy and policy guidance.

2 Study Site

The study area is located in South Kordofan State, bordered by Darfur in the west and Republic of South Sudan in the south, White Nile and North Kordofan state in the north (Fig. 13.1). The total population of Southern Kordofan State is 1,300,700 million based on the last population survey conducted in 2008 with a population growth rate of 2.4%. The urban population constitutes 21% of the total population of South Kordofan, nomads 1.1%, and the sedentary rural population 76.9% (CBS,

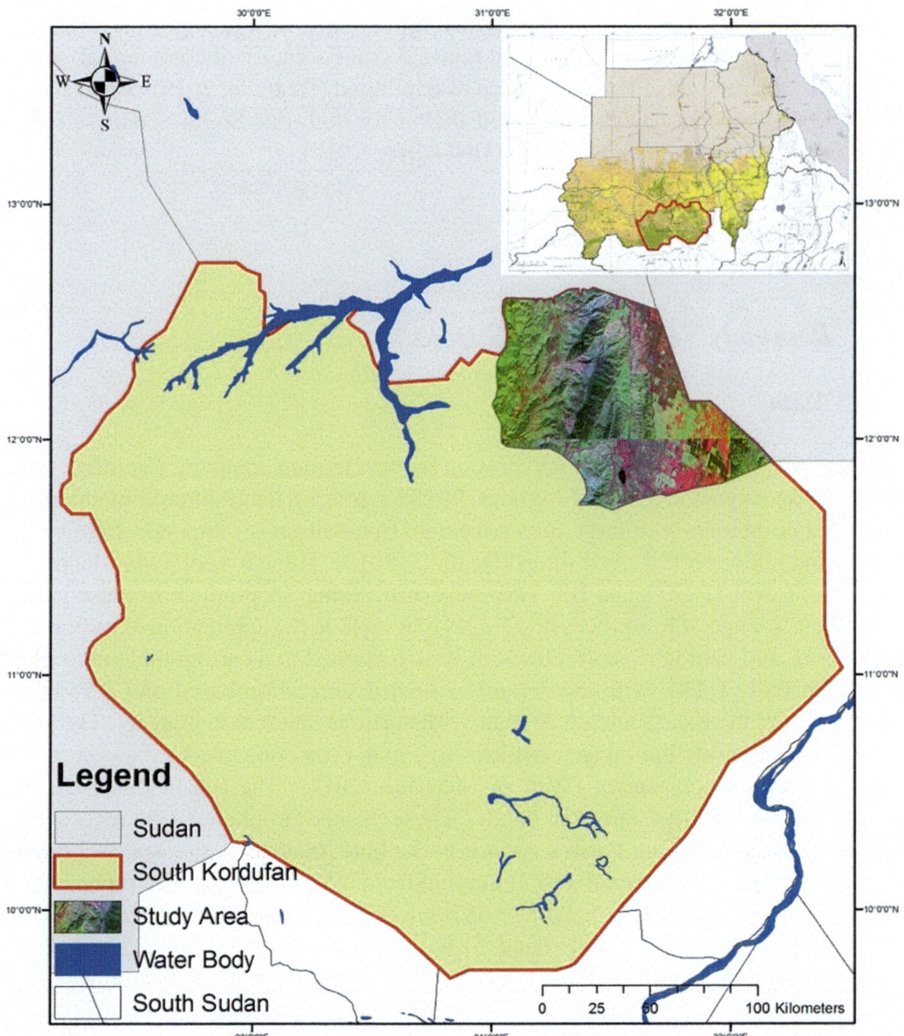

Fig. 13.1 Location of the Study Area
Source: DIVA-GIS, developed by the authors

2000). Despite the fact that some villages are overwhelmingly Islamic or Christian, and some communities practice traditional religions, the Nuba have a long tradition of religious tolerance. The region is a mountainous area where elevation reaches up to 1000 m. Around 50,000 km^2 is covered by Mountains. The high hills are mostly ridges separated by narrow V-shaped valleys. The region suffers from extensive erosion and loss of topsoil. It is classified as a sub-humid region (UNDP, 2006), with a varying climate, ranging from low-rainfall woodland savanna in the north to rich savanna in the south. Annual rainfall ranges from less than 350 mm on the northern border to more than 900 mm on the southern border (Mohammed, 2011). The average daily temperature ranges from 10 to 35 °C with an annual variation of 15 °C (El Tahir et al., 2010). The livelihood activities found in the area are agro-pastoralism, nomadic pastoralism, horticulture, and rain-fed (subsistence and commercial) agriculture. In addition, NTFPs are a source of livelihood (Deafalla, 2019). Since 2011, South Kordofan state has also provided shelter for high-rate South Sudanese refugees, nearly 43,845 refugees in 2019 (UNOCHA, 2019).

3 Research Methods

3.1 Remotely Sensed and GIS Data

3.1.1 Data

The RapidEye is a commercial RS mission by the German company RapidEye AG, launched on August 29, 2008 (El-Abbas, 2015). It offers a data source containing an unrivaled combination of large-area coverage, frequent revisit intervals, high resolution, and multispectral capabilities (Deafalla, 2019). Three RapidEye products are available: Level 1B, 3A, and B3. The study utilized the 3A product, because it has undergone a range of pre-processing stages that include the application of radiometric, sensor, and geometric corrections. It is also aligned to a cartographic map projection (usually UTM) with the default geometric correction based on GCPs. The ortho-correction process intends to remove distortions inherent in imagery. The process ensures the satellite image conforms to a map projection and includes correcting for terrain displacement (Watt & Meredith, 2011). The sensor observes the Earth's surface in five spectral bands. These bands simultaneously record the reflected energy from the Earth's surface in the blue (band 1) with a spectral range from 0.44 to 0.51 μm, green (band 2) ranging from 0.52 to 0.59 μm, red represented in (band 3) with spectral range 0.63–0.685 μm, red edge (band 4) ranging from 0.69 to 0.73 μm, and near-infrared (band 5) with spectral range 0.76–0.85 μm. It is important to note that, RapidEye is the first high-resolution multispectral satellite system incorporating red-edge band, which is sensitive to vegetation chlorophyll content (Schuster et al., 2012). That makes this system different from and privileged among the other multispectral satellite systems (Schuster et al., 2012; Ustuner et al., 2014). All these five bands shave the same calibration coefficients (Tapsall et al.,

2010). The high-resolution Rapideye imagery has been successfully utilized in several studies such as detecting changes (Zhanga et al., 2017), forest mapping (Saito & Sakaguchi, 2013), and water studies (Tetteh & Schönert, 2015).

In the current study, five RapidEye scenes, located in path/row: ranging from 174/051 to 173/052, taken on October 12, 2012, and October 11, 2013, were acquired for the identification and mapping of tree species in the Nuba Mountains. The imagery is delivered in a Geo-Tiff format, enabled to be integrated into the GIS environment and thus to be overlaid with existing GIS datasets by applying the appropriate coordinate transformation. The Geo-TIFF file was scaled from 12-Bit to 16-bit dynamic range for delivering ñ, the dynamic range, which determines the number of discrete levels of information. The 16-bit image provides 65,536 levels (Watt & Meredith, 2011). Radiometry and Color Balance were 5-Band GeoTiff; calibrated radiance-at-sensor (Watt/m 2 sr-1 μm-1) with 0% cloud cover. The pixel size was rescaled from 6.5 to 5.0 m. The five spectral bands were utilized. These bands measure discrete bandwidths (wavelengths) that are positioned to minimize atmospheric effects and are optimized to allow water penetration, discrimination of vegetation types, and plantation vigor.

3.1.2 Field Survey and Ancillary Data

Ground truthing allows image data to relate to features and materials on the ground (Thapa & Murayama, 2009). This enables calibration of RS data, as well as aids in the interpretation and analysis of what is being sensed. A dataset, constituted of 270 geo-referenced GPS points, was utilized adequately to describe and verify the species of trees across the study area, as well as to create a 'test set' to assess classification accuracy. Additional data related to climate and temperature were collected to strengthen the database. Furthermore, 65 points were randomly gathered from the study site to identify the soil types, and to create a 'test set' to assess classification accuracy.

3.1.3 Image Pre-processing

For the selected images, several pre-processing steps are required prior to analyzing satellite images, such as ortho-correction, atmospheric and topographic corrections. These processes were undertaken in ENVI, GRASS, and Erdas imagine 11 softwares and are designed to produce standardized image products from which quantitative analysis such as LC mapping, monitoring, and detection of change can be performed (Deafalla, 2019). At the completion of the pre-processing, the individual images were mosaicked to produce a single coverage (Fig. 13.2), which can then be ingested into ArcMap and used as inputs to run the customized ArcGIS and eCognition routines. Color correction algorithms (i.e.; image dodging and color balancing) were used for correcting the photometrical disparities. Seamline Generation technique (i.e., weighted seamline method) was used to avoid visual transition during the mosaicking process.

Fig. 13.2 Image mosaicking

3.1.4 Image Classification

GEOBIA

Recent advances in GEOBIA have revolutionized the processing of high to very high spatial resolution RS data such as Rapideye, IKONOS, QuickBird and Radar, providing effective computer-assisted classification techniques, for which results come close to the quality of manual image-interpretation. This is much faster, cheaper, and reproducible over large areas as well (Ma et al., 2017). Accordingly, this study used bi-temporal high spatial resolution imagery for tree species classification based on GEOBIA.

Image Segmentation

Multiresolution segmentation algorithm in Trimble eCognition TM Developer 8.7 software was applied to generate image objects from the pan-sharpened RapidEye image. This segmentation algorithm was characterized as a bottom-up region-merging technique, starting from one-pixel objects, where larger objects were

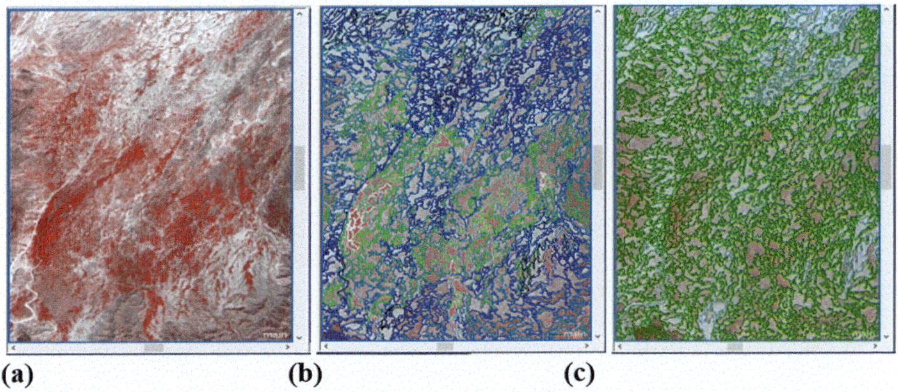

Fig. 13.3 Image segmentation results of RapidEye imagery with different scale factors; (**a**) Original image (without segmentation), (**b**) segmentation result with scale parameter of 800, shape 0.1, smoothness 0.5 and (**c**) with scale parameter of 400, shape 0.1, smoothness 0.5

generated by merging smaller ones with a series of iterative steps (Baatz et al., 2004; Li et al., 2015). Parameters that were used as input for the segmentation algorithm included: (1) weight of each input layer; (2) scale parameter; (3) color/shape weight; and (4) compactness/smoothness weight.

In this study, all of the five spectral bands of RapidEye image were used as input. The weights of color and its compactness were set as 0.1 and 0.5, respectively, in order to balance the difference of spectral/shape heterogeneity between forests and other classes. For the Forests class, the segmentation algorithm was performed at two levels, with different scales, 800 for the first and 400 for the second level (Fig. 13.3), to create objects that are connected to each other in the class hierarchy of contextual features. Each hierarchical level contained groups of objects, which were created to maximize the variability between different levels, while minimizing the variability within them (El-Abbas, 2015). As well, each level contained different information details, which have been achieved by suitable scale parameters at each level represented in Fig. 12.4. That allowed expressing each class and/or object in the respective level by one or more fuzzy sets (El-Abbas, 2015; Li et al., 2015). An image object domain has been used to link between the segmented object and the classification scheme through a defined level, objects, and fuzzy sets.

Hierarchical Fuzzy Classification

Several researchers (Benz et al., 2005; Dupuy et al., 2012; Do et al., 2016; Ma et al., 2017) proved in their studies that fuzzy classification improves class description by using understandable linguistic concepts built from expert knowledge. This also facilitates the fusion of heterogeneous multi-source information, including non-image data. Therefore, the current study proposes a method combining this kind of knowledge and feature extracted from RS images to generate tree species map. The method consists of three steps to decompose the image into a hierarchical tree structure:

- The first is to extract features, such as spectrum information, texture, and geometry shape from RapidEye images, through training samples by using KNNC.
- The second step is to derive environmental knowledge which illustrates vegetation distribution. Soil types, temperature, precipitation, and landform are chosen as the main environmental factors controlling vegetarian species. Landform indexes such as elevation, slope, and aspect are derived from high resolution DEM.
- The third step is to define tree species using the decision tree method with environmental knowledge and feature extracted from RS images, which is based on user expert knowledge, to describe forest vegetation classes by using Trimble eCognition TM Developer 8.7 software.

The study visually investigates the represented image objects to make a decision on the best features that should be considered in order to discriminate respective classes. Based on that, the values of mean layers have been selected as the main features to discriminate between most of the investigated objects (Fig. 13.4). An exception was made using subsequent rules. Geometrical features and texture values are used effectively to optimize or correct most misclassified objects. Multiple features of objects are calculated to structure data layers in RapidEye image. Spectral features are obtained based on the reflectance of the incident electromagnetic wave of different objects in each band. This includes the spectral signal of objects in each band (i.e., an average of spectral signals from all pixels within the objects), the brightness of objects (i.e., an average of spectral signals from all bands), and maximum difference (i.e., maximum variation between spectral signals of all bands). The textural features are also applied including mean, variance, homogeneity, contrast, dissimilarity, entropy, angular second moment and correlation. The shape features used consist of the geometric features of objects, including length-width ratio, compactness, density, and shape index. Normalization via min-max scaling was conducted, which is aimed at reducing the effects of different data expressions owing to various acquisitions and generation conditions (Wang et al., 2018). Additionally, feature fusion through layer stacking was applied to generate new bi-temporal images with multiple feature layers. This simplified the complex classification process and allowed intricate trees features to be effectively classified. Figure 13.4 shows the main criteria of species classification.

In the present study, the training and testing objects have been (declared) by selecting representative samples that contain meaningful spectral and spatial properties for each class. The confused training objects (critical samples) have been eliminated. Sample object has been defined by the value of 0.95, which means that 95% of an image object has to be overlapped by the sample area for a specific class category, derived by the TTA Mask, to create a class sample. The study area consists of both man-made and natural entities. Accordingly, a hierarchical system was proposed in this study to attain forest vegetation classes. The natural area is composed

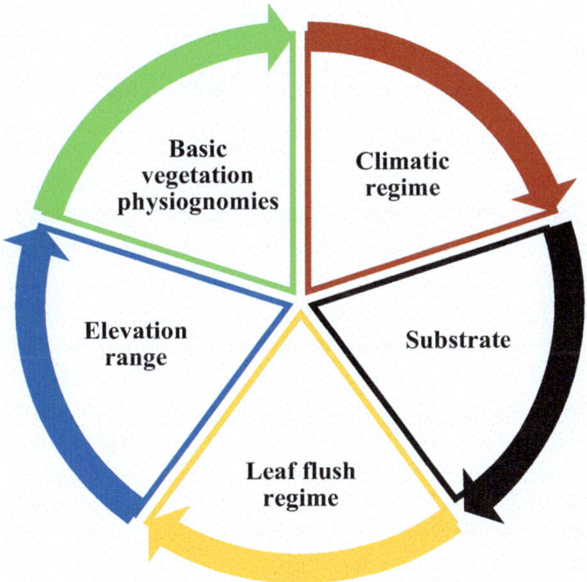

Fig. 13.4 The main criteria of species classification
Source: Adopted and developed by the authors based on Oliveira-Filho (2009, 2015)

of forests, orchards, shrubs, agricultural fields and grass. The remaining regions are covered with water, roads, and settlements. Therefore, the whole study site was divided into vegetation and non-vegetation classes at the first level due to its heterogeneous nature. This helped eliminate errors arising from mixed pixels, in non-vegetation area from further levels. Subsequently, the second level involves further division of vegetated area, which effectively reduces vegetation type-related confusions. It allows for the separation of forests class and Non-forested area that is composed of cultivated lands, Grasslands, Horticulture land and Shrublands child classes, in the third level. The forests class involves forest feature-related classifications such as forest physiognomies, trees' structure and leaf flush regime as shown in Fig. 13.4, which allow for the separation of nine species classes.

The classification was based on different hierarchical attributes:

- **NDVI and Normalized Difference Red Edge Index (NDRE)**

NDVI was used in this study; it was calculated by the following equation according to (Kross et al., 2015).

$$\text{NDVI} = \left(R_{NIR} - R_{RED}\right) / \left(R_{NIR} + R_{RED}\right)$$

A new feature in RapidEye sensor is the Red Edge band which allows a better estimation of the ground cover and chlorophyll content of the vegetation (Tapsall et al., 2010). Recently, the Red Edge band has been successfully used for the classification of vegetation, forestry, and agricultural areas (Ustuner et al., 2014). Accordingly, NDRE ratio was calculated by the following equation (Barnes & Baker, 2000) to identify and separate out species:

$$\text{NDRE} = \left(R_{band5} - R_{band4}\right) / \left(R_{band5} + R_{band4}\right)$$

The visual interpretation was based on the shape, size, and site of trees.

- **Soil**

Based on Xu and Zhuang (2007) and Wang et al. (2016), the identification of soil types is a very important aspect of forest vegetation classification because it plays an important role in the distribution of vegetation cover and in the determination of the type of vegetation that grows in a certain area (Mohammed, 2011; Peng et al., 2017). As mentioned earlier, soil samples were collected representing 65 sampled plots in the study area. At each sample plot, three soil samples were randomly collected using the cutting ring from the surface layer (0–20 cm) for assessment and determination of the soil physical properties, including the water content, bulk density, total porosity, texture, organic matter, and rock content. All of the samples were serially numbered and stored in soil bags for further analysis. For logistical reasons, the study could not make the required laboratory analysis. Therefore, the results of laboratory analysis for each class were assessed based on the analysis by Massaud (2007) and Elgubshawi (2008).

The present study utilized the RS approach for mapping the soil, which has come up with promising secondary data sources for improving digital soil mapping at all scales (Forkuor et al., 2017). The use of RS techniques in soil mapping is cost-effective and less time-consuming compared to the traditional soil mapping approaches (Mulder et al., 2011). The knowledge-based digital soil mapping, using fuzzy logic, is an accurate option for spatial prediction of soil properties (Menezes et al., 2018). Accordingly, RapidEye imagery has been used for mapping the soil types of Nuba Mountains. The visual interpretation was based on colour, shape, size, shadow, texture, pattern, free carbonates, salinity, organic matter, moisture, site and association.

Due to the complexity of soil compositions and their corresponding spectral signatures, direct connections between soil properties and their spectral responses could not be constructed (Deng et al., 2015). To address this problem, the GEOBIA approach was adopted using Trimble eCognitionTM Developer 8.7 software. In this approach, images need to be segmented before classification, for which, a multi-resolution segmentation algorithm (Baatz & Schape, 2000) was used to extract spectral, textural, and contextual information as attributes for soil classification. The criteria used to define the objects, according to the respective spatial and spectral attributes, was a scale parameter equal to 15 and shape/compactness criteria equal to 0.1/0.5. The classification was conducted based on user experience and by selecting the training objects of each class type, which were carefully selected. To further alleviate the confusion between soil and other land covers and improve the analysis, multiple features were used:

Brightness Index (BI)

The classification of bare soil areas, fallow lands, and vegetation with marked background response are enhanced using this index. In RapidEye imagery, BI uses the

Red (R), Green (G) and Blue (B) bands to represent the difference in reflectance values between soil areas as shown in the following equation based on Ray et al. (2004):

$$\mathrm{BI} = \left(\left(R^2 + G^2 + B^2\right) / 3.0\right)^{0.5}$$

Thus, it is a good indicator for separating areas of vegetation from areas of soil. In the current study, BI was employed to identify areas where soil is the dominant background or foreground material.

Redness Index (RI)

The soil characteristics are composed of organic material, granular mineral, moisture, texture, chemical components, and soil cover material etc. Based on Banerjee et al. (2014), the soil spectral reflectance depends on soil characteristics, mainly soil moisture and hydroxyl ions. So, RI has been calculated, based on Red (R), Green (G), and Blue (B) bands, to identify the hematite content of the soil using the equation by Forkuor et al. (2017):

$$\mathrm{RI} = R^2 / \left(B + G^3\right)$$

Coloration Index (CI)

Soil color has been used as one of the key attributes to differentiate and classify soil types in many areas (Forkuor et al., 2017). Therefore, CI was applied to detect the soil color using the equation below based on Ray et al. (2004):

$$\mathrm{CI} = \left(R - G\right) / \left(R + G\right)$$

Where; (R) is a Red band and (G) is a Green band

Accuracy assessment or validation is a key component of any study employing RS data, particularly for classification. This step essentially determines the quality of the information derived from remotely sensed data. In the current study, best classification result method was applied to test the soil map accuracy. The result was satisfying, where all the classes recorded high accuracy percentage as indicated in Table 13.1.

- **Digital Elevation Model (DEM)**

The classification of RS images mainly depends on the radiometric value difference of vegetations (Jing et al., 2009). In many cases, it is hard to classify the vegetation types due to the phenomenon that the different species have similar spectral values and the same species with different spectral signatures. Particularly in the mountain areas, the problem is more serious because of complex topography. According to Lee et al. (2004) and Xu and Zhuang (2007), the integration of spectrum combined with elevation information is very useful for recognizing and

Table 13.1 Accuracy assessment of soil map

Class	Objects	Mean	StdDev	Minimun	Maximum
Clay soil	9752	0.9433178	0.04642608307	0.1462936	1
Sandy soil	3006	0.9271318	0.06667027332	0.3364266	1
Rocky soil	4261	0.8797023	0.1046534	0.2274884	1
Non-cracking (more clay)	24,891	0.9326528	0.05977634427	0.2074174	1
Non-cracking (more sandy)	17,810	0.9328343	0.04682698178	0.226	1
Floodplain	32	0.8352309	0.1855172	0.1022199	1
Sandy-loam soil	348	0.9232935	0.07338409677	0.5203438	1

classifying tree species. The fusion of DEM data can not only correct the topography distortion but can also improve the classification accuracy. DEM was employed as an important criterion in classifying forest vegetations based on spectral character and its association with particular elevation. Five DEM images, provided by the USGS, derived from ASTER data with a spatial resolution of 30 m and a vertical resolution of 1 m, were used. After DEM image mosaicking, the images were projected to UTM/ zone 36 WGS 84 and resampled to match the 5-m RapidEye by using the nearest neighbor technique (Fig. 13.5).

- **Slope**

The landscape variations in vegetation are strongly related to topographic factors (Li et al., 2015). Several studies have indicated that the correlation between the diversity of species is essentially related to the slope position and altitude. For example, in Nevada, USA, the shrub land vegetation patterns frequently show a strong correlation with the slope position (Nettleton et al., 1986). In another case in Wondo Genet of Ethiopia, the tree density is negatively related to the elevation, slope, and slope aspect in the remnant moist Afromontane Forest (Kebede et al., 2013). The slope is the first derivative of elevation, and it is the rate of change of elevation in any direction. According to Alexander (2001), it is defined as the angle between a tangent to the surface and a horizontal (geoid-parallel) surface. Based on Oliveira-Filho (2009, 2015), slope leads to spectral reflectance variation of vegetation types. Therefore, the current study used the slope to help in classifying forest vegetation types. Slope units were extracted from a DEM using a break of slope rule on downslope profiles (Fig. 13.6). Each slope unit is an aggregated object of contiguous pixels and is summarized with five suites of variables: topography, shape, spectral characteristics, and variability in topographic as well as in spectral characteristics.

- **Climatic factors**

Climatic factors (temperature and precipitation) exert the largest influence on vegetation distribution and its characteristics in a global context (Prentice, 1990). A variety of climate types on Earth create deserts, tropical forests, savannas, and other vegetation types. On many time and spatial scales, climate and vegetation interact bidirectionally (Salazar et al., 2007). Climate change has affected the global

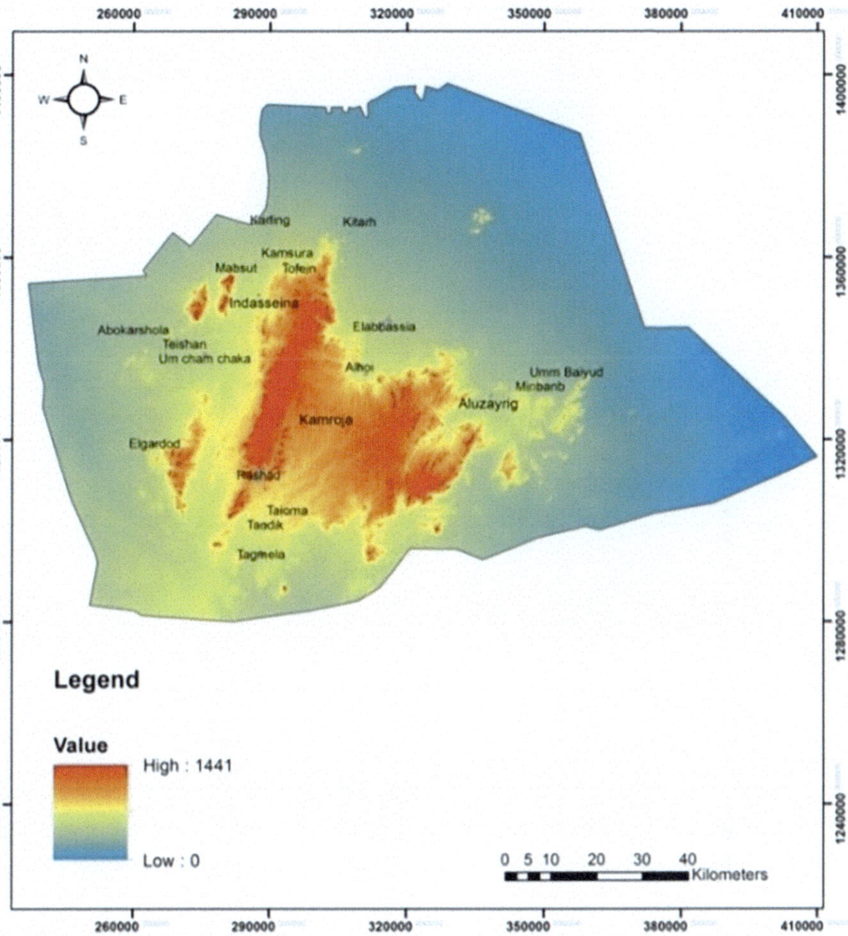

Fig. 13.5 DEM in the study area

distribution of vegetation from the distant past, and conversely, changes in the distribution and structure of the vegetation influenced climate (Nobre et al., 2007). One clear manifestation of such interaction is the global pattern of vegetative land cover and climate. Therefore, the study inserted these factors in the form of thematic layers to help in the classification of forest vegetation distribution. The climatic data were downloaded from the National Center for Environmental Information (NOAA) official website.[1]

[1] https://www.ncdc.noaa.gov/

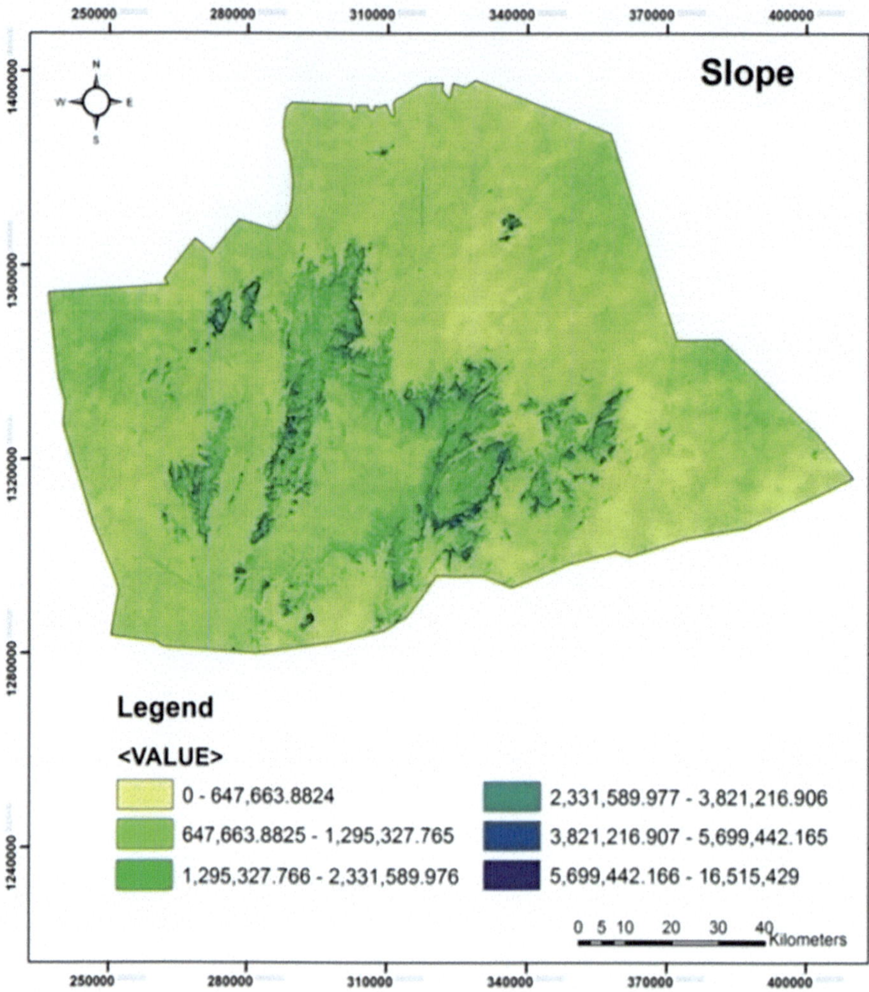

Fig. 13.6 Slope in study site

3.1.5 Map Validation

Accuracy assessment is an essential step in the classification process. In the present research, as suggested by Congalton (1991) and used in other studies (e.g. Tormos et al., 2012; Dupuy et al., 2012), ground based data is assumed to be 100% correct in accuracy assessments. Therefore, reference data was taken to assess the accuracy of using the same schemes utilized in the classification efforts. 15 samples were taken from each class to classify the image. The overall accuracy, producer's accuracy, user's accuracy and Kappa statistics methods (Manandhar et al., 2009) were conducted. Additionally, for more accurate results of each class, the best classification result method (Tiede et al., 2006) was also applied.

4 Results and Discussion

Remotely sensed EO data has revolutionized the understanding of our dynamic environment (Leeuw et al., 2010). Now, more than ever, RS is being an indispensable tool in resource management and in numerous areas of scientific research. It has given the scientists an excellent opportunity to understand the drivers, state, trends, and impacts of LU/LC on social and natural processes and helps reveal how changes happen and the resultant consequences (Verburg et al., 2009).

4.1 Trees Location

The present research uses tonal variations based on NDVI and NDRE to identify the vegetation status of the area. Vegetated areas have a relatively high reflection in the near infrared and a low reflection in the visible range of the spectrum. However, due to some similarities between classes (Table 13.2), a spectral-based classification scheme has led to ecologically senseless species vegetation classes.

Tree types, in that classification, had near or similar spectral characteristics and could not be distinguished clearly. Therefore, classification processes tackled other criteria: rainfall and temperature, topography, slope, elevation, and soil type. Based on these criteria, the spectral separability for every pair of land cover/vegetation class signatures was assessed. This helped progressively rearrange the classification scheme to guarantee higher spectral homogeneity and coherence within each final class. The soil was one of the main factors in determining the vegetation distribution and the type of vegetation that grows in the study area. The classification schemes for the study area based on user experience and spectral indices resulted in seven soil type classes, five main classes, and four sub-classes based on American systems: Clay soil, Sandy soil, Rocky soil, and Non-cracking soil that is divided into Non-cracking soil (more clay), and Non-cracking soil (more sandy). The last class was soil under water bodies: Floodplain and Sandy-loam soil as shown in Fig. 13.7.

Indeed, these different types of soil caused changes in vegetation structure and plant diversity. However, the distribution of forest vegetation, in the Nuba Mountains, is primarily governed by climatic and edaphic factors, and this is reflected in the diversity of forest types. Accordingly, the vegetation type is gradually or drastically changed and distributed from relatively poor, followed by medium and dense stands of different tree species and shrubs, as rainfall increases towards the south.

In the northern parts, the vegetation cover consists of scattered acacia trees such as *Acacia senegal* and *A. nilotica,* where the mean annual rainfall is 300–400 mm, and the slope ranges between 1,295,327.766 and 2,331,581.959. *Acacia senegal* is found at elevations up to 483 m, widely in the north and northeast parts of the study area. The current results agreed with Bown (1995) and Bein et al. (1996), who noted in their studies that the tree grows best in moist, well-drained, and neutral to acid soil. In the study area, it is found to be on non-cracking soil (more clay). The tree is

Table 13.2 Displayed NDVI and NDRE of RapidEye imagery results for the current study

Tree	NDVI	NDRE	Similar trees spectral reflectance range
Acacia senegal	0.37–0.22	0.26– 0.17	*Acacia polyacantha, Sterculia setigera, Acacia mellifera, Dalbergia melanoxylon, Azadirachta indica, Khaya senegalensis, Ailanthus excels, and Terminalia spp.*
Grewia tenax	0.39–0.22	0.25– 0.35	*Combretum hartmannianum, Anogeissus leiocarpus, Boswellia papyrifera, Acacia polyacantha,* and *Sterculia setigera*
Balanites aegyptiaca	0.30–0.16	0.22– 0.14	*Acacia oerfota, Boscia senegalensis* and *Acacia seyal*
Acacia nilotica	0.28–0.23	0.19– 0.16	*Hyphaene thebaica* and *Piliostigma reticulatum*
Adansonia digitata	0.33–0.20	0.24– 0.20	*Guiera senegalensis* and *Albizia amara*
Ziziphus spina-christi	0.24–0.21	0.20– 0.17	*Diospyros mespiliformis*
Tamarindus indica	0.36–0.30	0.22– 0.14	*Azadrachta indica, Khaya senegalensis, Ailanthus excelsa, Acacia mellifera,* and *Dalbergia melanoxylon*

an important source of income for the local people. Therefore, it is also being cultivated near villages in addition to its natural availability. *A. nilotica* was found in flooded areas on alluvial soils along streams.

The *genus Ziziphus* is known to be drought tolerant and very resistant to heat (NAS, 1980). It is found in desert areas with very low rainfall (Saied et al., 2008). In the study site, *Ziziphus spina-christi* is found, as well, in alluvial plains with deep soils at altitudes up to 400 m. It grows along the edges of ponds and *Wadis* where groundwater is available.

In central and southern part of the study site (Fig. 13.8), Balanites woodland are found, where *Balanites aegyptiaca* dominates the vegetation. It is a low-rainfall woodland savannah with annual rainfall ranging from 350 to 600 mm, located on clay soil with elevations up to 521 m with slop ranging from 2,331,589.977 to 3,821,216.906.

In hilly areas, where the elevation reaches up to 1340 m, and slope ranges between 5699.442 to 8, 542, 150, *Grewia tenax* is abundant. It is found on very dry sites, in sandy and rocky soils, while, Baobab are found on the north facing slopes, sheltered from cold southern winds with elevation ranging from 1020 to 1240 m. It tolerates poorly drained soils with heavy texture, but it does not existin deep sand and it prefers sandy topsoil overlying loamy subsoil (Huxley, 1992). It is found in sandy soils overlying loam and with a high-water table with an annual rainfall of 300–500 mm. However, Baobab was found, at its best, at approximate altitudes of 450–600 m.

Tamarindus indica is found in central, western and southern parts of the study site, with few representations in the other parts. It is located in areas that have high rainfall, with annual rainfall in the range of 600–700 mm. Based on Barwick (2004),

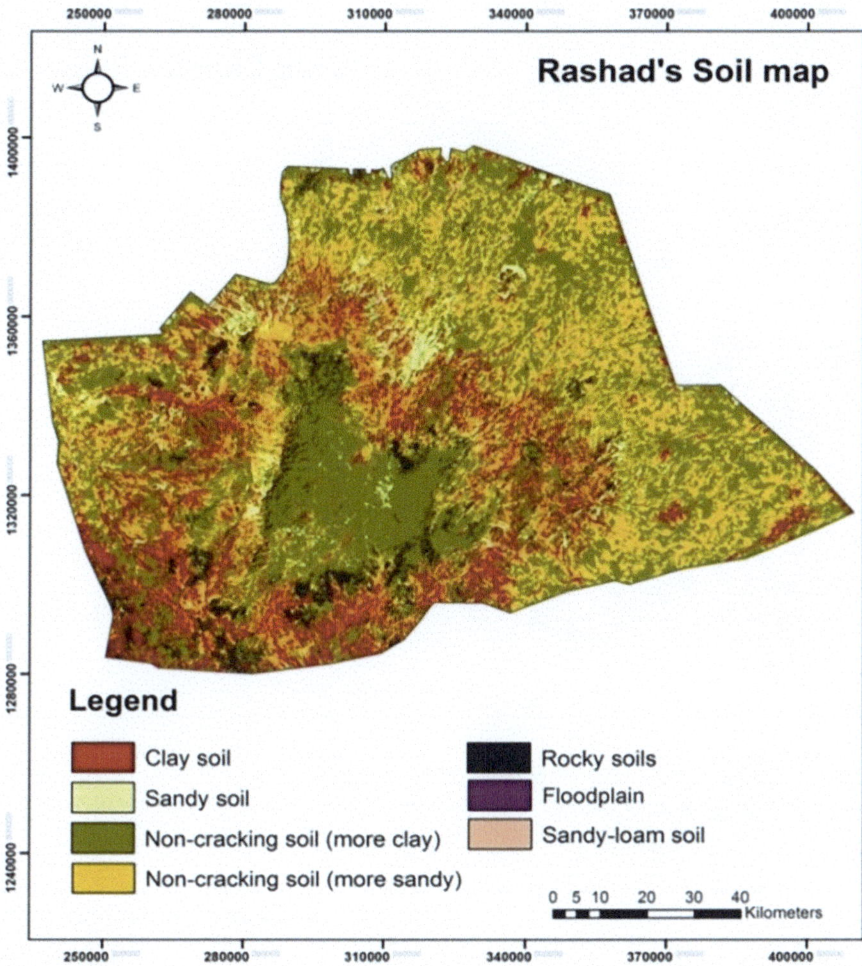

Fig. 13.7 Soil types in Nuba mountains

the tree grows in a range of soils such as clay, loam, sandy, and acidic soil types, which suggests that it is tolerant to saline conditions. It is found on non-cracking soil (more sandy) and elevations up to 1500 m.

4.2 *The Classification Accuracy Assessment*

The key element of a quantitative accuracy assessment is the creation of an error matrix (Congalton, 1991). According to Congalton (2001), "it is a square array of numbers organized in rows and columns which expresses the number of sample

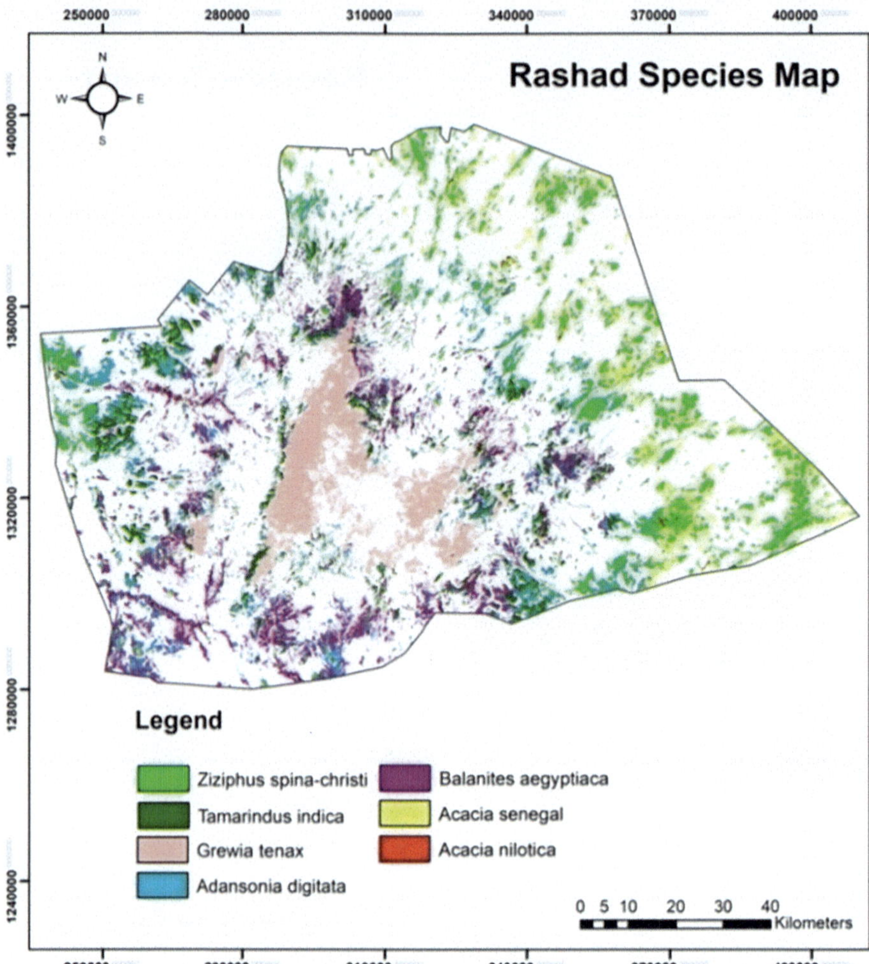

Fig. 13.8 Tree species map

units (i.e. pixels, clusters of pixels, or polygons) assigned to a particular category, relative to the actual category" as indicated by the reference data in Table 13.3. Columns of the confusion matrix correspond to the classes of objects in the validation set that belong to the ground truth samples, while rows correspond to which classes the image objects have been assigned to any of the classes in the image. The diagonal shows the objects that are correctly classified. Objects that are not assigned to the proper class do not occur in the diagonal and indicate the confusion between the different species class in the class assignment. Furthermore, the off-diagonal elements in the rows of the confusion matrix, divided by the total number of objects assigned to the RapidEye image class corresponding to the row, represent the commission errors and describe the confusion between that image class and the other species class. The commission errors describe the chance that an object, which has been assigned to a particular class, actually belongs to one of the other classes.

The total objects classified for each particular species class, and the number of objects found corrected through ground truth samples, along with the total number of reference objects from the reference data set, were computed and tabulated as presented in Table 13.4 for each classifier. Moreover, for computing the tree species class wise users and producers' accuracy for all classifiers, the following equations according to Bharatkar and Patel (2013) were used; they are considered and presented in Table 13.4 as well:

$$\text{User's accuracy}\left(\%\right)=\frac{\text{Correctly classified pixels}}{\text{Classified total pixels}} \tag{13.1}$$

$$\text{And Producer's accuracy}\left(\%\right)=\frac{\text{Correctly classified pixels}}{\text{Reference total pixels}}$$

The user's accuracy is found highest in the case of *Adansonia digitata* (100%) and only 64.8% in the case of *Ziziphus spina-christi* (lowest).Similarly, the producer's accuracy also reflects the exact classification of particular tree species classes and the matching of correctly classified objects in comparison to ground truth samples. Table 13.4 shows that this accuracy also gives better results for *Balanites aegyptiaca*. The study also noted that *Tamarindus indica* is better classified in all classification methods, while *Acacia senegal* gives very poor producer accuracy as compared to other tree species. Indeed, the result is not satisfactory.

Table 13.3 Error matrix of the classification data

User class	Acacia senegal	Tamarindus indica	Adansonia digitata	Balanites aegyptiaca	Acacia nilotica	Ziziphus spina-christi	Grewia tenax	Sum
Acacia senegal	65	0	0	0	1	0	1	67
Tamarindus indica	12	74	0	0	0	0	0	86
Adansonia digitata	0	0	48	0	0	0	0	48
Balanites aegyptiaca	2	0	1	64	2	0	0	68
Acacia nilotica	6	0	0	0	41	0	1	48
Ziziphus spina-christi	3	0	0	0	0	24	0	27
Grewia tenax	14	0	2	0	0	0	28	44
Unclassified	20	2	4	1	0	2	0	29
Sum	122	76	55	65	44	26	30	

Table 13.4 Classification accuracy assent

Accuracy	Acacia senegal	Tamarindus indica	Adansonia digitata	Balanites aegyptiaca	Acacia nilotica	Ziziphus spina-christi	Grewia tenax
Producer	0.5327869	0.9736842	0.8727273	0.9846154	0.9318182	0.923	0.933
User	0.9701493	0.961	1	0.7901235	0.8541667	0.6486486	0.875
Hellden	0.6878307	0.93673203	0.932	0.8767123	0.8913043	0.762	0.9032258
Short	0.5241935	0.9367089	0.8727273	0.7804878	0.804	0.6153846	0.8235294
KIA per class	0.4556852	0.9685673	0.8583529	0.9814364	0.9241176	0.9165490	0.9284958
Totals							
Overall accuracy	0.8245243						
KIA	0.8						

Table 13.5 Classification accuracy assessment based on best classification results method

Class	Objects	Mean	StdDev	Minimum	Maximum
Background	8360	1	0	1	1
Acacia senegal	1211	0.815	0.1098720	0.4715775	1
Tamarindus indica	2841	1	0	1	1
Adansonia digitata	2660	0.982	0.4633149	0.2528453	1
Balanites aegyptiaca	3382	1	0	1	1
Acacia nilotica	537	1	0	1	1
Ziziphus spina-christi	932	1	0	1	1
Grewia tenax	4261	1	0	1	1

This may be due to the misclassification of some training objects of *Acacia senegal,* such as *Grewia tenax* or *Tamarindus indica,* due to the similarity of trees' spectral reflection.

The overall accuracy for the species map was 82.4%, with the corresponding kappa statistics of 80%. Kappa coefficient above 75% may be interpreted as better classification than would be expected by random assignment of classes (Vieira et al., 2010). As shown in Table 13.4, the overall accuracy and kappa coefficient are comparatively more for the maximum likelihood classifier, where both represented approximately the same percentage of trees species map.

On the other hand, as in all species, the overall agreement of the best classification results method is relatively high, with most of the confusion occurring in *Adansonia digitata* and *Acacia senegal* classes as clarified in Table 13.5. This result consists of Table 13.4's outputs regarding these two products as well.

5 Conclusion

The study provides updated information about the location of these trees to help in strategy formulation and developing appropriate interventions under armed conflict conditions. It also shows great potential for the use of hyperspectral data as a tool for identifying tree species locations in a diverse mountainous forest. GEOBIA, with a multi-scale framework, has played a critical role in accurate trees species mapping and helped define tree species segmentation and the similarity of near objects, and increase interpretations of vegetation traits. NDRE of RapidEye imagery was used to examine the ability of vegetation indices derived from original spectral bands of RapidEye imagery in forest classification in semi-arid regions.

More studies regarding the monitoring of NTFPs, by the use of new techniques e.g. RS and GIS are needed. GEOBIA could be used as an efficient solution to extract and update the required information in semi-arid regions at various abstraction levels of details, utilizing optical multispectral imagery. The study recommends that the use of GEOBIA to analyze RapidEye imagery for the identification of trees species should be integrated with ancillary data such as DEM and other levels of GIS data to improve the quality of the result.

References

Alexander, N. (2001). *Automated modelling of digital elevation models for predictive ecosystem mapping in GIS*. Master thesis, University of Northern British Columbia. p. 36.

Baatz, M., & Schape, A. (2000). Multiresolution segmentation an optimization approach for high quality multi-scale image segmentation. In *Proceedings of the Angewandte geographische informations verarbeitung XII*, Karlsruhe, pp. 12–23.

Baatz, M., Benz, U., Dehghani, S., Heynen, M., oltje, A. H., Hofmann, P., & Lingenfelder, I. (2004). *eCognition professional user guide*, version 4.0. Definiens Imaging GmbH. Munchen, Germany.

Babu, T. P., Singh, P., & Nigam, R. K. (2002). Mapping of forest cover in Rohtak district, Haryana using remote sensing techniques. In Proceeding of the Geomatics 2002 conference on "IT enabled spatial data services" Tiruchirapalli, India. pp. 115–120.

Banerjee, K. B., Surajit Panda, S., Bandyopadhyay, J., & Jain, M. J. (2014). Forest canopy density mapping using advance geospatial technique. *IJISET – International Journal of Innovative Science, Engineering and Technology, 1*(7), 356–363.

Barnes, E. M., & Baker, M. G. (2000). Multispectral data for mapping soil texture: Possibilities and limitations. *Applied Engineering in Agriculture, 16*(6), 731.

Barwick, M. (2004). *Tropical and subtropical trees – A worldwide encyclopaedic guide* (484 pages). Publisher; Thames and Hudson.

Bein, E., Habte, B., Jaber, A., Birnie, A., & Tengnas, B. (1996). *Useful trees and shrubs in Eritrea* (472 pages). Publisher Regional Soil Conservation Unit.

Benz, U. C., Hofmann, P., Willhauck, G., Lingenfelder, I., & Heynen, M. (2005). Multiresolution object-oriented fuzzy analysis of remote sensing data for GIS-ready information. *ISPRS Journal of Photogrammetry and Remote Sensing, 58*, 239–258.

Bharatkar, P. S., & Patel, R. (2013). Approach to accuracy assessment for RS image classification techniques. *International Journal of Scientific and Engineering Research, 4*(12), 79–86.

Bown, D. (1995). *Encyclopaedia of herbs and their uses* (p. 244). Dorling Kindersley.

Central Bureau of Statistics (CBS). (2000). Fifth Sudan national population and housing census. Census tables. Available from: www.cbs.gov.sd. Accessed 1 Apr 2014.

Chaudhary, B. S. (2003). *Integrated land and water resources management in southern part of Haryana using remote sensing and geographical information systems (GIS)*, Ph. D. Thesis, University of Rajasthan, Jaipur, India. 78 pages.

Chikamai, B., & Tchatat, M. (2004). *Forest management for non-wood products and services in Africa*. A report prepared for the project Lessons learned on the Sustainable Forest Management in Africa, KSLA/AAS/FAO.

Congalton, R. G. (1991). A review of assessing the accuracy of classifications of remotely sensed data. *Remote Sensing of Environment, 37*, 35–46.

Congalton, R. G. (2001). Accuracy assessment and validation of remotely sensed and other spatial information. *International Journal of Wildland Fire, 10*, 321–328.

Deafalla, T. H. H. (2019). *Integrative assessment and modelling of the non timber forest products potential in Nuba mountains of Sudan by field methods, remote sensing and GIS* (PhD. thesis). Submitted to Faculty of Environmental Sciences, University of Dresden, June 2019, Germany. 319 pages.

Deafalla, T. H. H., Dafa-Alla, D. M., & Eltayeb, A. M. (2012). *Potentiality of non wood forest products to poverty alleviation and food security in Nuba Mountains, Sudan*, Symposium of Agricultural Sustainable Development and Chain-Sudan Research Collaboration, Yangshuo, China.

Deng, E., Orner, E. P., Chau, J. F., Anderson, E. M., Kadilak, A. L., Rubinstein, R. L., Bouchillon, G. M., Goodwin, R. A., Gage, D. J., & Shor, L. M. (2015). Synergistic effects of soil microstructure and bacterial EPS on drying rate in emulated soil micromodels. *Soil Biology & Biochemistry, 83*, 116–124.

Do, T. H., Raghavan, V., Vinayaraj, P., Truong, X. L., & Yonezawa, G. (2016). Pixel based and object based fuzzy LULC classification using GRASS GIS and RapidEye imagery of Lao Cai Area, Vietnam. *Geoinformatics, 27*(2), 104–105.

Donoghue, D. N. M. (2002). Remote sensing: Environmental change. *Progress in Physical Geography: Earth and Environment, 26*(1), 144–151.

Dupuy, S., Barbe, E., & Balestrat, M. (2012). An object-based image analysis method for monitoring land conversion by artificial sprawl use of RapidEye and IRS data. *Remote Sens, 4*, 404–423.

El-Abbas, M. M. (2015). *Assessing, monitoringand mapping forest resourcesin The Blue Nile Regionof Sudan usingan object-based image analysis approach*. Doctoral thesis, Faculty of Environmental Sciences, University of Dresden. Germany. 262 pages.

El Tahir, B. A., Fadl, K. M., & Fadlalmula, A. G. D. (2010). Forest biodiversity in Kordofan Region, Sudan: Effects of climate change, pests, disease and human activities. *Biodiversity Conservancy, 11*(3–4), 34–44.

Elgubshawi, A. A. I. (2008). *Soil degradation: Its aspectsand modellinga case study of Northeast Nuba Mountain South Kordofan State, Sudan*. Doctoral thesis, Faculty of Environmental Sciences, University of Dresden. Germany. 181 pages.

FAO. (2000). *Conflicts, agriculture and food security. The State of Food and Agriculture*. Food and Agriculture Organization of the United Nations. Rome, Italy. http://www.fao.org/docrep/x4400e/x4400e00.htm#TopOfPage. Accessed 12 Apr 2015.

FAO. (2020). *The state of food security and nutrition in the world (SOFI)*. Transforming Food Systems for Affordable Healthy Diets. ISBN: 978-92-5-132901-6. 320 pages.

Forkuor, G., Ozias, K. L., Welp, H. G., & Thiel, M. (2017). High resolution mapping of soil properties using remote sensing variables in South-Western Burkina Faso: A comparison of machine learning and multiple linear regression models. *PLoS One, 12*(1), e0170478.

Huxley, A. (1992). *New RHS dictionary of gardening*. Macmillan. Available from: http://www.scirp.org/(S(lz5mqp453edsnp55rrgjct55))/reference/ReferencesPapers.aspx?ReferenceID=361541

Jing, X., Wang, J., Huang, W., Liu, L., & Wang, J. (2009). Study on forest vegetation classification based on multitemporal remote sensing images. In D. Li & C. Zhao (Eds.), *Computer and computing technologies in agriculture II, Volume 1. CCTA 2008. IFIP advances in information and communication technology* (Vol. 293, pp. 115–123).

Kebede, M., Kanninen, M., Yirdaw, E., & Lemenih, M. (2013). Vegetation structural characteristics and topographic factors in the remnant moist Afromontane forest of Wondo Genet, south central Ethiopia. *Journal of Forestry Research, 24*, 419–430.

Kross, A., McNairn, H., Lapen, D., Sunohara, M., & Champagne, C. (2015). Assessment of RapidEye vegetation indices for estimation of leaf area index and biomass in corn and soybean crops. *International Journal of Applied Earth Observation and Geoinformation, 34*, 235–248.

Kurecic, P. (2015). Geoeconomic and geopolitical conflicts: Outcomes of the geopolitical economy in a contemporary world. *World Review of Political Economy, 6*(4), 523–543.

Lee, J., Shuai, Y., & Zhu, Q. (2004). Using images combined with DEM in classifying forest vegetations. 2004 IEEE international geoscience and remote sensing symposium, IGARSS '04. Proceedings.

Leeuw, J. D., Georgiadou, Y., Kerle, N., De Gier, A., Inoue, Y., Ferwerda, J., Smies, M., & Narantuya, D. (2010). The function of remote sensing in support of environmental policy. *Remote Sens, 2*(7), 1731–1750.

Li, D., Ke, Y., & Gong Hand Li, X. (2015). Object-based urban tree species classification using bi-temporal worldview-2 and worldview-3 images. *Remote Sens, 7*, 16917–16937.

Ma, L., Manchun, L. I., Ma, X., Cheng, L., Du, P., & Liu, Y. (2017). A review of supervised object-based land-cover image classification. *ISPRS Journal of Photogrammetry and Remote Sensing, 130*, 277–293.

Manandhar, R., Odeh, I. O. A., & Ancev, T. (2009). Improving the accuracy of land use and land cover classification of Landsat data using post-classification enhancement. *Remote Sens, 1*, 330–344.

Massaud, M. M. (2007). *Regeneration and propagation of Sclerocarya birrea (A. Rich.) Hochst. Tree in Rashad Area at Nuba Mountains, South Kordofan State*. Master thesis. Department of Silviculture, Faculty of Forestry, University of Khartoum. pp. 5–20.

Maurya, A. K., Tripathi, S., & Soni, S. (2013). A strategy of forest/non-forest cover mapping of Achanakmar-Amarkantak Biosphere Reserve, Central India: Based on remotely sensed imagery and GIS data. *International Journal of Remote Sensing & Geoscience (IJRSG), 2*(4), 50–54.

Menezes, M. D., SHG, S., Mello, C. R., Owens, P. R., & Curi, N. (2018). Knowledge-based digital soil mapping for predicting soil properties in two representative watersheds. *Scientia Agricola, 75*(2), 144–153.

Mohammed, H. M. (2011). *Management of Natural Stands of Acacia seyal Del. variety seyal (Brenan).* Doctoral thesis, Dresden University of Technology, Dresden, Germany. pp. 25–26.

Mulder, V. L., Bruin, S. D., Schaepman, M. E., & Mayr, T. R. (2011). The use of remote sensing in soil and terrain mapping – A review. *Geoderma, 162*(1), 1–19.

National Academy of Sciences (NAS). (1980). *Firewood crops: Shrub and tree species for energy production* (255 pages). National Academy.

Nettleton, W. D., Brasher, B. R., Spencer, E. L., Langan, L. N., & Peterson, F. F. (1986). Differentation of closely related xerolls that support different sagebrush plant-communities in NEVADA. *Soil Science Society of America Journal, 50*, 1277–1280.

Nobre, C. A., Oliveira, G. S., & Salazar, L. F. (2007). Climate and land use changes in Amazonia: Impacts on the hydrological cycle and biome distribution, in Water and Environment: Proceedings of the Workshop in the Vatican Academy of Sciences, Nov. 2005, Vatican Press, Rome. pp. 144–232.

OCHA. (2020). SUDAN, Situation Report. https://reports.unocha.org/en/country/sudan/#cf-3ko5uoWyLHUS5mXaYfOM7u

Oliveira-Filho, A. T. (2009). Classificação das Fitofisionomias da América do Sul extra-Andina: Proposta de um novo sistema – prático e flexível – ou uma injeção a mais de caos? *Rodriguésia, 60*(2), 237–258.

Oliveira-Filho, A. T. (2015). Um sistema de classificação fisionômico-ecológica da vegetação Neotropical: segunda aproximação. In J. M. Felfili, P.V. Eisenlohr, M. M. R. F. Melo. L. A. Andrade, & J. A. A Meira Neto (Eds.), *Fitossociologia no Brasil: Métodos e estudos de casos*. (Vol. 2, p. 452), Editora UFV, cap. 19.

Peerbhay, K., Mutanga, O., & Ismail, R. (2013). Commercial tree species discrimination using airborne AISA Eagle hyperspectral imagery and partial least squares discriminant analysis (PLS-DA) in KwaZulu–Natal, South Africa. *ISPRS Journal of Photogrammetry and Remote Sensing, 79*, 19–28.

Peng, J., Loew, A., Merlin, O., & NEC, V. (2017). A review of spatial downscaling of satellite remotely sensed soil moisture. *Reviews of Geophysics, 55*, 1–27.

Prentice, K. C. (1990). Bioclimatic distribution of vegetation for general circulation model studies. *Journal of Geophysical Research, 95*, 11811–11830.

Pumayalli, R. (2012). Watershed modeling by remote sensing and Agwa--Swat for Western Portion of Cusco Watershed – Peru Paperback – Import, 1 May 2012. 156 pages.

Raczko, E., & Zagajewski, B. (2018). Tree species classification of the UNESCO man and the biosphere Karkonoski National Park (Poland) using artificial neural networks and APEX hyperspectral images. *The Special Issue Hyperspectral Remote Sensing of Forest and Trees outside Forests Ecosystems, 10*(7), 1111. https://doi.org/10.3390/rs10071111

Ramachandran, B., Justice, C. O., & Abrams, M. J. (2011). Land remote sensing and global environmental change: NASA's earth observing system and the science of ASTER and MODIS. *Remote Sensing and Digital Image Processing, 2.*

Ray, S., Singh, J., Das, G., & Panigraphy, S. (2004). Use of high resolution remote sensing data for generating site specific soil management plan. *The International Archives of the Photogrammetry, Remote Sensing and Spatial Information Sciences, 35*, 127–132.

Renner, M. (2002). *The anatomy of resource wars.* Worldwatch Institute.

Saferworld. (2020). *Sudan.* Available at: https://www.saferworld.org.uk/east-africa/sudan

Saied, A. S., Gebauer, J., Hammer, K., & Buerkert, A. (2008). Ziziphus spina-christi (L.) Willd: A multipurpose fruit tree. *Genetic Resources and Crop Evolution, 55*, 929–937.

Saito, K., & Sakaguchi, H. (2013). Satellite imagery solutions for monitoring of forest and ecosystems. *Hitachi Review, 62*(3), 204–208.

Salazar, L. F., Nobre, C. A., & Oyama, M. D. (2007). Climate change consequences on the biome distribution in tropical South America. *Geophysical Research Letters, 34*, L09708.

Schuster, C., Förster, M., & Kleinschmit, B. (2012). Testing the red edge channel for improving land-use classifications based on highresolution multi-spectral satellite data. *International Journal of Remote Sensing, 33*(17), 5583–5599.

Sihag, V., Kumar, M., & Yadav Mand Bhanu, P. S. (2015). Assessment and classification of TOF (tree outside forest) using space technology in Morni Block, Haryana. *International Journal of Engineering, Management, Humanities and Social Sciences Paradigms (IJEMHS), 17*(01), 19–25.

Tapsall, B., Milenov, P., & Tasdemir, K. (2010). Analysis of rapideye imagery for annual land cover mapping as an aid to European Union (EU) COMMON AGRICULTURAL POLICY. In W. Wagner, B. Székely (Eds.), *ISPRS TC VII Symposium – 100 Years ISPRS*, Vienna, Austria, July 5–7, 2010, IAPRS, Vol. XXXVIII, Part 7B. pp. 568–573.

Tetteh, G. O., & Schönert, M. (2015). Automatic generation of water masks from RapidEye images. *Journal of Geoscience and Environment Protection, 3*, 17–23.

Thapa, R., & Murayama, Y. (2009). Urban mapping, accuracy, & image classification: A comparison of multiple approaches in Tsukuba City, Japan. *Applied Geography, 29*, 135–144.

Tiede, D., Lang, S., & Hoffmann, C. (2006). Supervised and forest type-specific multi-scale segmentation for a one-level-representation of single trees. In *1st International Conference on Object-based Image Analysis (OBIA 2006).* 5 pages. Available from: http://citeseerx.ist.psu.edu/viewdoc/download?doi=10.1.1.374.3670&rep=rep1&type=pdf

Tormos, T., Durrieu, S., Kosuth, P., Dupuy, S., Villeneuve, B., & Wasson, J. G. (2012). Object based image analysis for operational fine-scale regional mapping of land cover within river corridors from multispectral imagery and thematic data. *International Journal of Remote Sensing, 33*(14), 4603–4633.

UNDP. (2006). *Pastoral production systems in South Kordofan, Sudan (study 2)* (pp. 8–13).

UNOCHA. (2019). *Humanitarian response plan.* Available from: https://www.who.int/healthcluster/countries/sudan/sudan-2019-humanitarian-response-plan.pdf?ua=1

Ustuner, M., Sanli, F. B., Abdikan, S., Esetlili, M. T., & Kurucu, Y. (2014). Crop type classification using vegetation indices of RapidEye imagery. *ISPRS – International Archives of the Photogrammetry, Remote Sensing and Spatial Information Sciences, XL-7*, 195–198.

Verburg, P. H., Van De Steeg, J., Veldkamp, A., & Willemen, L. (2009). From land cover change to land function dynamics: A major challenge to improve land characterization. *Journal of Environmental Management, 90*(3), 1327–1335.

Vieira, S. M., Kaymak, U., & Sousa, J. M. C. (2010). *Cohen kappa coefficient as a performance measure for feature selection, pro.* IEEE International Conference.

Wang, J., Wang, H., Cao, Y., & Bai Zand Qin, Q. (2016). Effects of soil and topographic factors on vegetation restoration in opencast coal mine dumps located in a loess area. *Scientific Reports, 6*, 22058.

Wang, X., Sicong Liu, S., Peijun Du, P., Liang, H., Junshi Xia, J., & Li, Y. (2018). Object-based change detection in urban areas from high spatial resolution images based on multiple features and ensemble learning. *Remote Sens, 10*(276), 1–14.

Watt, P., & Meredith, A. (2011). Forest estate mapping using RapidEye, Theme: Radiata Management. Task No: F10404, Report No. FFR- R068, Milestone Number: 4.04.5. pp. 7–14.

Xu, Z., & Zhuang, D. (2007). The methodology of detailed vegetation classification based on environmental knowledge and remote sensing images. In *2007 IEEE International Geoscience and Remote Sensing Symposium*. pp. 2074–2077.

Zhanga, M., Zhoua, W., & Li, Y. (2017). The analysis of object-based change detection in mining area: A case study with Pingshuo Coal Mine. *The International Archives of the Photogrammetry, Remote Sensing and Spatial Information Sciences, XLII-2/W7*, 2017. ISPRS geospatial week 2017, Wuhan, China.

Chapter 14
Securitization of Human-Induced Environmental Conflict: Implications for the Military (in the Struggle for Life)

Fred Kruidbos

Abstract Many are now convinced that we are living in a self-made Anthropocene era marked by global changes and a precipitous slope of species extinction, which require unorthodox measures. In addition to the enormous increase in the global human population, several specific actors can be identified that can be held responsible for such a shift. The ecological effects generated by these actors differ in nature and manifest themselves at different organizational levels within the complex of socio-ecological interactions as they act as drivers of induced dynamics. Since these processes are likely to have a significant impact on the future international safety and security landscape, this chapter discusses the securitization of human-induced environmental conflict. It offers a vision for a restructuring of the existing multidisciplinary and multidimensional deployment of military capabilities with a specific focus on conflict prevention. It identifies wildlife crime issues such as illegal wildlife trade and habitat destruction – as one of the root causes of future international destabilization resulting in a fluid mosaic landscape of diverse state and non-state actors in peace and conflict, shifting powers, and further loss of nature – as an argument to the use of special military capacity. In doing so, new capabilities from a multidisciplinary approach, in which diplomacy-development-defense is intertwined with ecology and time (3D-ET), are created as a means to reduce human-environmental conflict. This will be underpinned by proposing Foreign Ecological Security Operations (FESO) as a new core activity for Special Operations Forces (SOF).

Keywords Securitization · Environment · Conflict prevention · Wildlife crime · Military · SOF

F. Kruidbos (✉)
Kruidbos Ecological Research and Consultancy, K-SN Ecological Services B.V, LC, Helmond, The Netherlands
e-mail: info@kruidbos.com

M. Behnassi et al. (eds.), *The Climate-Conflict-Displacement Nexus from a Human Security Perspective*, https://doi.org/10.1007/978-3-030-94144-4_14

1 Securitization of Human-Induced Environmental Conflict

Since the famous hokeystickcurve by Mann et al. (1998, 2004) on the exponential atmospheric accumulation of greenhouse gases (especially CO_2), the industry has raised much doubt as to whether climate change and induced global warming are caused by anthropogenetic influence; somehow similar to the current anti-Corona campaigns organized by some prominent leaders. But, many are now convinced that we live in the self-made Anthropocene era and that politics and policy have nothing (and all) to do with that fact. However, a policy is crucially needed to address the survival of existing social-ecological systems (SES). Most importantly, the current settings have obviously not led to a significant effective improvement in the survival of many systems. This chapter focuses on this aspect and highlights some of the most important organized disruptions of ecological systems of the recent past and present. It reflects on the disastrous consequences of the organized illicit wildlife trade that is held responsible for the destruction of significant parts of nature and its resources on which many social-ecological systems depend and focuses on the role the military might play in mitigating these effects. Therefore, it brings us to the discussion of the securitization of the environment.

Warner and Boas (2019:1473) argue that "security concerns and risk assessments tend to be about what *might* happen. We worry about them before they happen, seeking a 'way out' in anticipation of a potential crisis to retain a sense of agency. This may incite would-be securitizers to crank up the alarm to get heard, invoking apocalyptic visions of climate-induced violence and mass displacement". But, exactly the opposite can be true as well, since it is the power holders who are the beneficiaries of the current state that evolved under their regime; after all, waving the problems away facilitates the continuation of current power settings and policy. In this chapter, I will clearly demonstrate why the argument of 'fear of what might happen' has already been proven to be wrong; without any doubt, it can easily be seen from out of space, presenting the so-called apocalyptic vision to be true indeed. This 'what might happen' – for instance sea-level rise in the Netherlands – has now evolved in 'what has happened' – for instance large-scale deforestation of tropical forests across each continent. 'What will happen' still lays ahead. Since some processes take time to become visible (see Chap. 12, Sect. 3.1.2 on extinction debt), it is, therefore, legitimate to suggest a new and different approach is justified to stop the clock from ticking further away from 12.05 h. According to Warner and Boas (2019), a securitizing move involves an existential, life-and-death threat and its corollary: an extraordinary course of action as the only way out. Due to the many socio-ecological systems that have already been removed from the planet Earth, and the many that are yet to come, this now seems legitimate.

Today's world is more connected than ever before and set against the increasing security risks. As the human population and its behavior have created forces that threaten the stability of SES as they exist today, a certain group of actors play a prominent role in this. This results in self-reinforcing effects that together pose a very serious threat to safety and security, the long-term survival of SES and to the survival of much of life on Earth. This is highly undesirable and should be halted.

Since the actors of these eco-crimes have evolved into well-organized and often violent organizations, their *modus operandi* often goes beyond the capacity of regular police forces. This brings us to the question if the military may be useful to assist in prevention and mitigation of the risks and insecurity that result from modern socio-ecological conflicts. In an attempt to maintain and strengthen social-ecological coviability and resilience, this analysis explores the possible role of the military as a response mechanism against human-induced environmental conflict and induced displacement. This can only be done by broadening the horizon of the conventional military security domain, enabling military – as well as civil – doctrine to adapt to this new threat.

The signs are flashing red. This title of an article on the interplay between climate change and violent extremism in the Western Sahel (Middendorp & Bergema, 2019) sounds alarming. But when the co-author of the article and former Dutch Chief of Defense, General Tom Middendorp spoke on National TV about climate change and its relevance to the national security, he had to explain himself to the Minister of Defense and other high placed politicians. Within the military itself, he also had to explain himself more than once.[1] This illustrates the problem. Or, to speak by the words of the General himself: "climate change and the safety risks that arise from it are a global problem. However, the effects and risks of climate change are not yet experienced the same everywhere. Despite all climate summits, it is therefore difficult to achieve global action" (Middendorp, 2019). In this, lies the insidious danger of the newly emerging conflict.

From a securitization perspective, the aim of this chapter is to provide an ecological approach on conflict prevention, which brings us to the following question: How can the military help mitigate human-induced environmental conflict? To answer this question, the analysis provides solutions that can contribute to mitigating socio-ecological conflicts. Namely, military deployment aimed at conflict prevention with a special focus on ecosystem management. The context of conflicts outlined in this chapter is much narrower than in reality. However, this limitation is necessary in order to be able to indicate what the added value the deployment of military personnel can have in enhancing the sustainability of SES.

The Sect. 2 provides an example of modern connectivity that has resulted in the emergence of human-mass behavior that can be compared with the swarming behavior of ants and other social insects. This holds both beauty and threats. In Sect. 3, a future security landscape is sketched, while the Sect. 4 provides different examples of non-state actors, such as criminal and terrorist as well as militia organizations, that act as disrupters of ecosystem stability. Section 5 brings this in context with strategic solutions and explains how the military can play a role in mitigating the negative effects on SES, of courses, by non-state actors such as criminal and

[1] Personal comments during the Clingendael Institude International Webinar "Adapt to Defend. The Security Dimension of Climate Change", on March 4th, 2021.

terrorist organizations. Finally, the last paragraphs draw the conclusions which, hopefully, lead to a pragmatic discussion on how scientific and security services can effectively join as a worldwide interagency against eco-crime.

2 Emergence of Human Behaviour

As cities continue to grow in size and number, humans have shifted their original habitat (Ritchie & Roser, 2018) while developing high standards of living that may not be environment-friendly. Technology has also grown exponentially, creating new possibilities and threats. Cyberspace has already shown its power as an important influencer of national security (Arthur, 2011). Moreover, technology, combined with high population numbers, has given rise to new weapons of mass mobilization. Social networks – such as Facebook and Twitter – have already proven the potential to became the rallying forces for revolutions, social justice as well as manipulation by individuals, and criminal and terrorist organizations that are able to strategically place false information, creating deceptions, or hacking government systems using social networks (Yang & Tobun, 2007; Ethembabaoglu et al., 2010; Abdulhamid et al., 2011; Howard et al., 2011; Chukwuere & Onyebukwa, 2018).

The emergence of a kind of collective swarm intelligence is not new to nature. The roots of swarm intelligence are deeply embedded in the biological self-organized behaviors of social insects. For example, in their moving phase, the Neotropical army ants *Eciton burchelli* may organize large hunting raids, which may contain more than 200,000 workers collecting thousands of preys, be 15 meters or more wide and sweep over an area of more than 1500 square meters in a single day (Garnier et al., 2007). Also, ants of the old world show comparable collective behavior. Figure 14.1 shows two extremes in the raiding behavior of *Pheidologedon diversus* colonies. These colonies have one or two stable trunk trails which extend 5–100 m from the nest and remain in use for weeks, during which time ants move along them continuously, day and night. The raids of *P. diversus* can reach 6 m in width and contain tens of thousands of individuals (Moffet, 1984).

Swarms of social insects, like the ants above, construct trails and networks of regular traffic via a process of pheromone laying and following. These patterns constitute what is known in brain science as a cognitive map. The main difference lies in the fact that the insects write their spatial memories in the environment, while the mammalian cognitive map lies inside the brain. Therefore, the individual ant can be compared with the individual neuron in the brain, the swarming as a whole with cognitive maps in the brain. Thus, in a way these swarms form a sort of self-organized cognitive map while a good part of this map lies outside of the individual (Chialvo & Millonas, 1995).

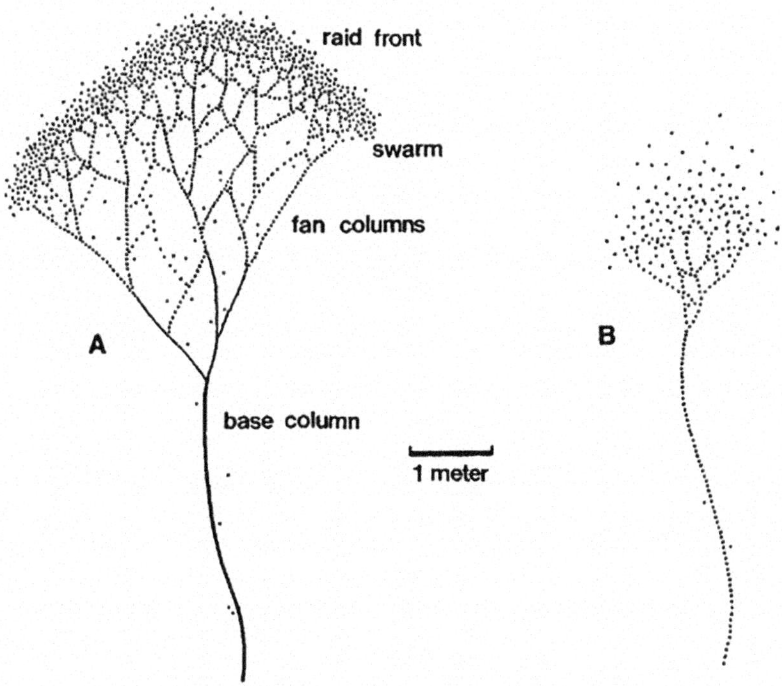

Fig. 14.1 Two extremes in the raiding pattern of *P. diversus*
In pattern A, worker density is high and concentrated at a well-defined raid front. Raids of this description vary from 2 to 6 m across. Pattern B is characteristic of smaller raids. Workers are relatively spread out; so that, in extreme cases, scattered individuals advance as much as 50 cm ahead of other ants
Source: Adapted from Moffet (1984)

With respect to the modern use of social media, the swarming behavior of individual ants might also be compared with individual social media behavior, eventually resulting in emergence of collective swarm behavior. From a security point of view, it can be imagined that monitoring swarming behavior, and controlling it in case the stability of a sovereign state is at stake, could be of the utmost importance for the sake of such a stability.

The above examples illustrate the connectivity of man and nature. Nature seems to behave in different forms at different organizational levels. All interconnected and entangled. We have seen this already in Chap. 10 of this book (Sect. 2.1.3), where spatial scaling laws determine what species might exist or not. Before focusing on how to respond to some of the security and safety aspects of these types of new phenomena, some other examples of human-induced socio-ecological conflict drivers, especially with regard to criminal and terrorist behavior, will be examined.

3 Future Security Landscape

The world is no longer the recognizable and fairly structured patchwork quilt of nations it used to be. It has become a dense and volatile network with a multitude of actors within the context of a mosaic of shifting economic and political powers driven by factors such as climate change, energy development, demographic growth, and rapidly slinking reservoirs of natural resources. So, one of the greatest challenges of the coming decades is to be prepared for difficult and unpredictable developments; a security environment with a multitude of risks and actors (Bekkers & Spiegeleire, 2010). In such an environment, conflicts are increasingly characterized by fundamental uncertainties; resulting in a strategic uncertainty, also known as deep uncertainty (Haasnoot et al., 2012; 'Hart et al., 2016; Lawrence et al., 2020). In a highly uncertain world, a high degree of adaptability to rapidly changing circumstances is a prerequisite for survival ('t Hart et al., 2016). However, this is not new to nature (Darwin, 1859; Eldredge & Could, 1972; Kutschera, 2009).

As described above, the emphasis in this research is on how to counter further degradation of SES; both neutralizing the symptoms and also preventing their occurrence. The urgency is extremely high as multiple systems have already been lost, and it is reasonable to expect that this is only the beginning of a further escalation of threats in which several non-state actors, drivers as previously mentioned, will play a crucial role.

To be effective in countering current and future threats, it is necessary to have a strong strategy, especially which it needs a global multidisciplinary approach with multiple actors involved. Referring to the remarks made by General Middendorp, as noted above, not all the effects and risks are yet experienced the same everywhere. Therein lays a deep problem that in itself acts as a multiplier of threats. In his book, *The Black Swan*, Taleb (2007) focuses on the extreme impact of rare and unpredictable outlier events and the human tendency to find simplistic explanations for these events, retrospectively. Central to Taleb's reflection is the proposition that we tend to overestimate what we know and underestimate what we do not know. Bekkers and Spiegeleire (2010) also refer to Taleb's Black Swan theory, especially concerning transition periods of fundamental change such as the transition to the post-industrial era; sudden events, processes, characteristics, and structures that we not only do not recognize in advance, but that we do not even realize we do not know until they manifest themselves.

Before answering the question on how to formulate a strategy that helps combat socio-ecological conflict, a more fundamental question should be first considered: what will the context of the near future, from which we must operate at a political-strategic level, look like?

In order to be more adaptive to future conflicts, the Dutch developed four future scenarios, which outline the most important uncertainties and, thus, contribute to the process of arriving at a realistic threat analysis and adequate strategy. The scenarios are consistent visions of the future, not by way of prediction but as a conceivable result of important developments. Future scenarios, together with the armed

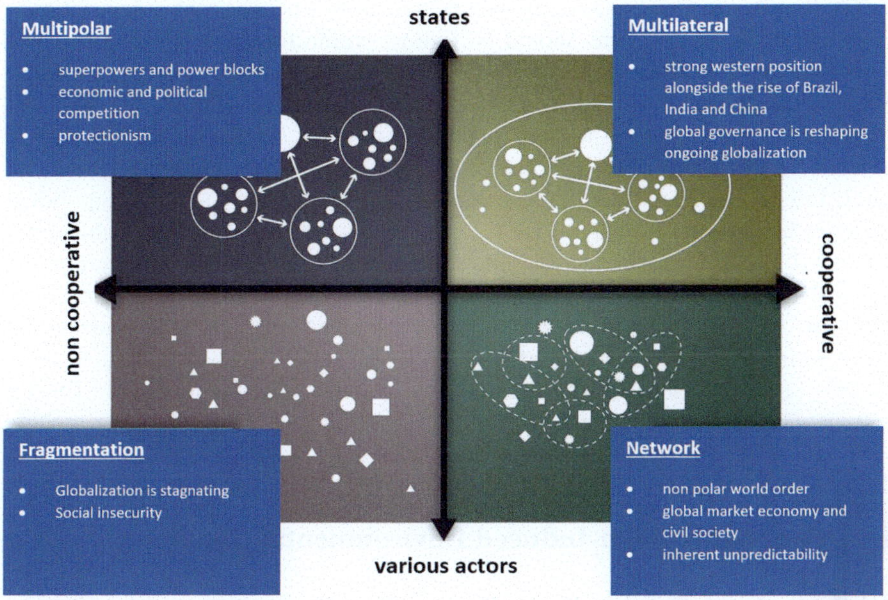

Fig. 14.2 Four future scenarios based on an analysis in which (non)state actors interact in a cooperatively or non-cooperatively way
Source: Adapted from MOD, the Netherlands (2010)

forces profiles, play a central role in determining future-proof policy options (MOD The Netherlands, 2010).

Figure 14.2 shows four scenarios based on two parameters, namely: (i) cooperativity; and (ii) actors involved. The outcome answers two strategically important questions: is the world system developing in the direction of increasing cooperation and the integration or of crumbling collaboration and fragmentation? And is (our) security mainly determined by states or by non-state actors? The combination of these questions provides a clear overview of a central issue in security politics: which actors will be affected in the future and how do they interact? The direction in which this issue develops largely determines the dynamics within the global system and, thus, has a major impact on national and international security issues (MOD The Netherlands, 2010).

As the Dutch strategic future analysis shows, the causes and effects of conflict are becoming increasingly difficult to identify. In order to mitigate the risk of 'strategic shocks' or deep uncertainty, we can argue that ecological regime shifts should be prevented at all costs, since these shifts are characterized by the loss of resilience and, therefore, increase stochasticity. This loss is highly unfavorable for strategy development. It makes things more complicated, fragile and uncertain. So, if the future landscapes of conflict will be fluid and more stochastic, how can we prepare strategically to act in a fragmented and networked landscape, a mosaic of conflict and inter-conflict where scenarios merge into one another in space and time?

The first step in strategy development is to identify where and what decisions are to be taken. These decisions should be framed within the context relevant to the organization, and their importance weighted accordingly (Krishnadas, 2015). According to Amersfoort (2016), military strategy forms a separate link between political security policy on the one hand, and military actions on the other (Von Clauzewitz 1832, *War is the continuation of politics by other means* in Howard & Paret, 1984). Therefore, in order to be able to create a military strategy, a political strategy must first exist. Thereby, the military strategy is determined from above and the armed forces, thus, have an instrumental function. Because the formulation of a political strategy goes far beyond the scope of this research, it remains here a general view on the question of how the military can make an appropriate contribution to mitigating social-ecological conflicts. To give an idea on such an approach, some drivers of human-induced environmental conflicts that cause disruption of SES, are examined below.

4 Drivers of Human-Induced Environmental Conflict

4.1 *The Vulnerability Niche Within Fragile States*

The Notre Dame Global Adaptation Initiative (ND-GAIN) Country index[2] summarizes a country's vulnerability[3] to climate change and other global challenges in combination with its readiness[4] to improve resilience. The map below shows the fragility of tropical regions, which favors some highly undesirable actors, such as terrorists, criminals, and militias; forcing local civilians to join these groups, to flee or die (Miguel et al., 2004; Jensen & Gleditch, 2009; Nillese & Verwimpt, 2009; Haenlein & Smith, 2016; van Uhm, 2016; Middendorp & Bergema, 2019), (Fig. 14.3).

The vulnerability thus created facilitates these non-state actors to expand or shift their power while luring others. The so-called Islamic State in Iraq and the Levant (ISIL), for example, is shifting its interest towards the Mali-Burkina Faso-Niger tri-border area, after their losses in Syria and Iraq (Middendorp & Bergema, 2019). As

[2]Vulnerability is composed of 36 indicators. Each component has 12 indicators, crossed with 6 sectors. Readiness is composed of 9 indicators. For index calculation method, see Chen et al., (2015: 6–7).

[3]ND-GAIN assesses the vulnerability of a country by considering 6 life-supporting sectors: food, water, health, ecosystem services, human habitat, and infrastructure. Each sector is, in turn, represented by 6 indicators that represent 3 cross-cutting components: the exposure of the sector to climate-related or climate-exacerbated hazards; the sensitivity of that sector to the impacts of the hazard; and the adaptive capacity of the sector to cope or adapt to these impacts.

[4]ND-GAIN measures readiness by considering a country's ability to leverage investments to adaptation actions. ND-GAIN measures overall readiness by considering 3 components: economic readiness, governance readiness, and social readiness.

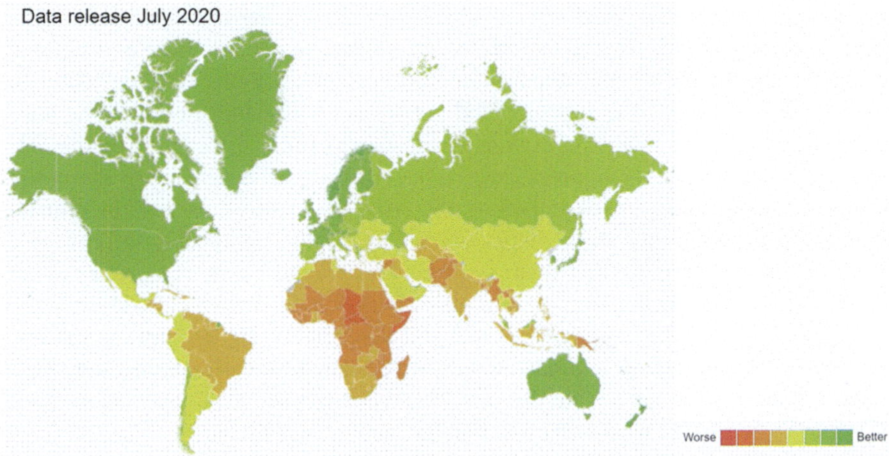

Fig. 14.3 ND-GAIN Country index showing each country vulnerability and capacity to adapt to negative effects of climate change
The more it is red, the higher the country is vulnerable and the lower is the readiness to improve resilience
Source: Adapted from *Notre Dame Univer*sity. https://gain.nd.edu/our-work/country-index/

annual temperatures rise in sub-arid environments, pressure on natural resources like water and food increases, making poor communities prone to such organizations and internal conflict. The population growth is high, coupled with rural-urban migration and fragile governance. These situations are used by terrorist organizations – like ISIL and Boko Haram – as they weaponize water and food or use scarcity as a recruitment tool (Middendorp & Bergema, 2019). Nillisse and Verwimpt (2009) showed that recruitment to militia during the civil war in Burundi, particularly in rural areas, was directly related to deficient rainfall. The lack of rain directly affects crop production and results in a negative income. Miguel et al. (2004), as well as Jensen and Gleditch (2009), showed similar findings based on economic growth shocks related to rainfall data of 41 African countries.

More terrorist organizations such as the Islamic State in the Greater Sahara (ISGS),[5] the Islamic State West Africa Province (ISWAP),[6] and the Organization of

[5] ISGS was listed on 23 February 2020 pursuant to paragraphs 2 and 4 of resolution 2368 (2017) as being associated with ISIL or Al-Qaida. https://www.un.org/securitycouncil/content/islamic-state-greater-sahara-isgs

[6] ISWAP late Tuesday 18-08-2020 overran Kukawa in the Lake Chad region, seizing people who had just returned to their homes after spending nearly 2 years in displacement camps (The Defense Post d.d. 19-08-2020: https://www.thedefensepost.com/2020/08/19/iswap-hostage-hundreds-nigeria/

Al-Qaida in the Islamic Maghreb (AQIM),[7] as well as other non-state actors like criminal organizations, are taking advantage of the socio-ecological and climate vulnerability. Therefore, the human landscape of such regions is more likely to develop into a mosaic of fragmented and networked landscapes in which the non-state actors rule. This results in regions where more or less dense and volatile networks of multiple actors suppress civil rights (Bekkers & Spiegeleire, 2010) and deteriorate biodiversity – strengthening the risks of zoonosis as well as climate change – thus triggering more displacement. Ironically enough, these unwanted non-state actors facilitate their own position. This should be an eye opener to act while there still is a chance to achieve positive outcome. So, the so-called 'civilized' world should, using the words of Bekkers and Spiegeleire (2010), show some strategic agility to anticipate a fundamentally uncertain future. Realizing that their vision of the future as seen from the year 2010 is our present day, it seems the situation can be compared to an infection that has gotten out of hand.

4.2 Drugs, Illegal Logging, and Security

On the continent of Central-South America, similar patterns occur, however, different in appearance. For example, Sesnie et al. (2017) conducted a spatio-temporal analysis of forest loss related to cocaine trafficking in Central America. Through this analysis, they provide evidence that criminal activities associated with drug trafficking networks are a progressively important driver of forest loss in the region. They estimated that cocaine trafficking could account for 15% to 30% of annual national forest loss in Guatemala, Nicaragua, and Honduras. Of these losses, 30% to 60% occurred within nationally and internationally designated protected areas.

Dávalos et al. (2011) illustrate the relationship between poor rural development, deforestation, and intertwined relationships with organized crime. Hereto, they analyzed coca production-related land cover change between 2002 and 2007 in the northern Andes, Chocó, and Amazon Forest of Colombia, which is the largest producer of coca leaf for the global cocaine market. As they quantified the impact of coca production on forest dynamics, they found that forest loss increased within proximities to coca plots. They also found that within protected areas, this relation is reduced, showing the effects of nature protection – at least concerning coca-related deforestation in these regions.

---

[7] AQIM was created in Algeria in 1998 by Hassan Hattab as the Salafist Group for Call and Combat (GSPC). GSPC was a splinter entity of the Armed Islamic Group (QDe.006), which was the largest and most extreme terrorist group in Algeria at the time. GSPC was renamed AQIM in January 2007 following the group's union with Al-Qaida (QDe.004), which was announced by Al-Qaida leader Aiman Muhammed Rabi al-Zawahiri (QDi.006) on 11 September 2006. https://www.un.org/securitycouncil/sanctions/1267/aq_sanctions_list/summaries/entity/the-organization-of-al-qaida-in-the-islamic.

Like in Middle-South America and Africa, the four main forms of serious organized crime in Southeast Asia are illicit drugs; human trafficking and migrant smuggling; environmental crimes (e.g. illegal logging and wildlife trade); and counterfed goods. Besides, factors such as lack of alternative sources of livelihood, conflict, and cultural tolerance, serious organized crime is facilitated by high levels of corruption and weak democratic traditions, often combined with weak enforcement (Idris, 2019). The literature identifies Myanmar as a particular hub of serious organized crime in Southeast Asia, especially for illegal drugs, as well as illegal logging, jade and gem mining, and wildlife trade. One of the forms in which the corruption has come to flourish is the so-called 'ceasefire capitalism', whereby the military entered into ceasefire agreements with insurgent groups, in return allowing them to carry out their illicit activities (Idris, 2019).

As a result of these kinds of habitat destructive processes, the whole region has undergone utterly dramatic ecological changes. This can be deduced from the figures presented in Table 14.1, which gives an overview of the changes in tropical forest cover for the whole Southeast Asia over the period 1990-2000-2010, as analyzed by Stibig et al. (2014). It shows that during a period of 20 years, 31,700,000 hectares of the total forest cover of Southeast Asia have been eliminated. These figures actually are an underestimation of the really removed forest surface, because parts of the clearances have been replaced by, for example, oil palm and rubber plantations and secondary forest. Moreover, they concluded, in 2014, that "in continental Southeast Asia, there is an indication of increased forest cover loss along the Annamite Mountain range (Laos, Cambodia, and Vietnam), and in the border zones of Cambodia (with Thailand and Vietnam) and of northern Myanmar". Unfortunately, this has evolved into a progressive decline which is still ongoing (Fig. 14.4).

One of the countries, already mentioned by Stibig et al. (2014), which has undergone a major decimation of its natural forest, is Cambodia. The country lost nearly 2.3 million hectares of tree cover between 2001 and 2019, and the annual rate of loss increased by almost 300% during the same period. Data from Global Forest Watch's Dashboards show that, since 2000, Cambodia has lost about 26% of its tree cover, a much higher percentage than larger, forested countries including Brazil and Indonesia. It needs to be noted here that within the territories of Cambodia, different types of tropical forests exist. To illustrate this diversity, Tani et al. (2007)

Table 14.1 Forest cover and annual change in Southeast Asia and sub-regions (areas in Mha.)

	Forest Cover				
	1990	2000[a]	2010	Annual net change 1990s	Annual net change 2000s
Sub-region	Area	Area	Area	Area (se); Change % (se)	Area (se); Change % (se)
SE-Asia[b]	268.0	250.6	236.3	1.75 (0.26); 0.67 (0.10)	1.45 (0.25); 0.59 (0.10)
Continental SE-Asia	78.7	76.5	71.7	0.21 (0.08); 0.27 (0.10)	0.48 (0.13); 0.65 (0.18)
Insular SE-Asia[b]	187.9	173.0	163.5	1.51 (0.25); 0.84 (0.14)	0.96 (0.22); 0.57 (0.13)
Indonesia[c]	123.8	112.4	104.4	1.15 (0.25); 0.98 (0.21)	0.82 (0.21); 0.76 (0.19)

PNG = Papua New Guinea

Source: Adapted from Stibig et al. (2014:253)

[a]Average from two period estimates, [b]incl. PNG & Solomon Isl., [c]including East Timor

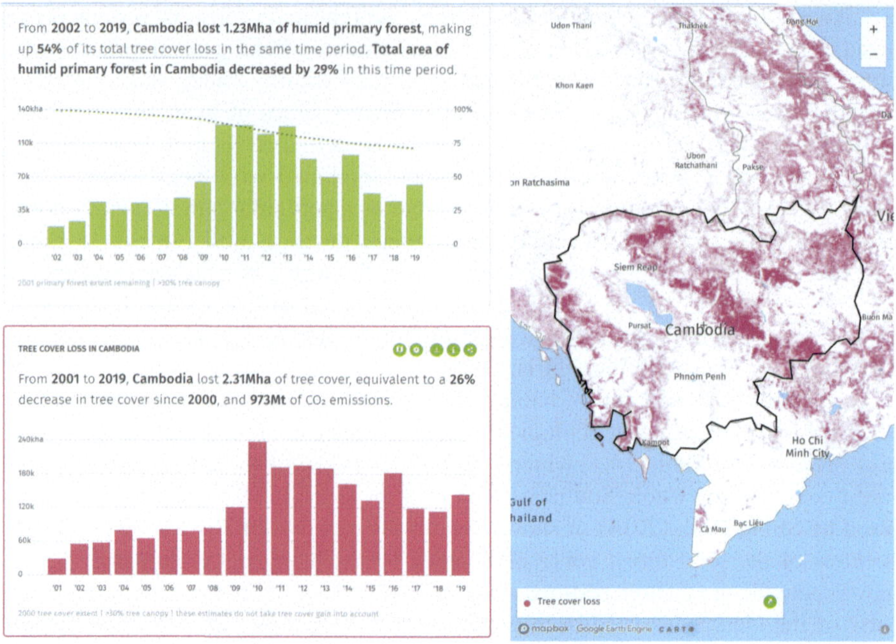

Fig. 14.4 Cambodia lost nearly 2.2 million hectares of tree cover between 2001 and 2018, and the annual rate of loss increased by almost 300% during the same period
Source: https://blog.globalforestwatch.org/data-and-research/whats-happening-in-cambodias-forests/

investigated three regions of Cambodia: *Kampong Thom, Kratie*, and *Mondolkiri*. They found four main clusters corresponding to traditional qualitative forest types known as evergreen forest, deciduous forest, hill evergreen forest, and swamp forest. The evergreen cluster was further divided into two stand types of dry evergreen forest and two stand types of secondary evergreen forest. Given the diversity of this region, one can imagine the enormous loss of biodiversity already caused by human interference and behaviour.

With organized crime operating at all levels of the supply chain, high demands on the consumer markets within the region's proximity as well as timber processing industries in nearby China, Cambodia's biodiversity has decreased very rapidly over the past decades (Humphrey, 2019; Idris, 2019; Kresek, 2019; Stibig et al., 2014). This citation from Humphrey (2019) characterizes a significant part of the situation: "as you approach Beng Per Wildlife Sanctuary, five hours north of Phnom Penh, it's difficult to tell exactly where the park begins. There is no audience of trees to greet you, no sign to welcome you. In many areas, there are no trees at all. The land is more reminiscent of parched African savannah than a Southeast Asian rainforest. Where trees do appear, they stand in uniform rows, with vessels taped to their trunks – archetypal features of rubber plantations".

The map, see Fig. 14.4, illustrates the total loss of tree cover of Cambodia and its surroundings between 2001 and 2019. The two diagrams show that trees cover loss fluctuates in time. So, the combination of maps and diagrams illustrates that deforestation fluctuates in both space and time.

4.3 Poaching, Wildlife Trafficking, and Security

Many research works have been done on terrorist organizations and militia,[8] financing part of their existence with poaching and wildlife trafficking.[9] Haenlein et al. (2016) critically address the link between poaching, wildlife trafficking, and terrorism in Africa and its implications for transnational security and domestic conflict. Hereby, they consider the assertion that Al-Shabaab, the Janjaweed, and the Lord's Resistance Army (LRA) – the three most cited organizations within this context – have become major beneficiaries, particularly in ivory. For example, Al Shabaab base of operation extends from Somalia across the border of the Horn of Africa. Similarly, the LRA, is an armed organization that operates in the northeastern Democratic Republic of Congo (DRC), southern the Central African Republic (CAR), and southwestern South Sudan. Finally, the Janjaweed, a Sudanese-based armed movement, operates from Darfur across Central Africa. They conclude that all three these terrorist groups have been proven to be involved in elephant poaching and ivory trafficking. But since their share is relatively small compared to organized crime, their role, according to the authors, should not be overestimated. From an ecological point of view, however, this seemingly modest contribution could have huge impacts on both local and regional ecosystems within their radius of action. Combined with the impact generated by organized crime groups, the ecological effects may be utterly devastating, both on nature alone as well as on SES.

4.4 Quantifying Poaching, Wildlife Trafficking, and Illegal Logging

According to Haenlein and Smith (2016), the perhaps most widely used estimate on the value of illegal trade in flora and fauna, including illegal fishing and logging, is done by United Nations Environment Program (UNEP) and Interpol.[10] In 2014, they estimated the value of illegal trade at \$70–213 billion and updated it to \$91–258 billion in 2016, resulting in a 5–7% annual rise. Thereby, illegal trade has become

[8] Depending on one's definition, we might use the word 'terrorist' instead of 'militia'.

[9] See Haenlein et al. (2016) for an impression on such a literature.

[10] https://www.interpol.int/News-and-Events/News/2016/UNEP-INTERPOL-report-value-of-environmental-crime-up-26

the fourth-largest global form of illicit activity after drug trafficking, counterfeit crimes, and human trafficking.

The illegal trade in wildlife alone is estimated to be approximately one-fourth of the legal trade (van Uhm, 2016). Deduced from an estimated legal wildlife trade value of $86 billion in 2010, van Uhm assumed the value of illegal wildlife trade in 2010 to be around $21,5 billion worldwide, and $9.5 billion for the European Union. Based on his calculations, the 2010 illegal trade of ivory alone was already over $100 million. Compared with the negative value of the current and upcoming pandemic events, the outcome of this calculation is likely to fade the benefits made by terrorists, organized crime, and corrupt government employees as well as the cost made to counter them. So, relative to the profits gained through various illegal business models on illicit flora and fauna trade, the cost of socio-ecological damage is enormous.

According to Schwab (2020) on the World Economic Forum of August 3, 2020, the achievement of a significant reduction of new diseases from tropical forests would cost globally between 22.2 and 30.7 billion each year, while the COVID-19 pandemic will likely end up costing between $8.1 and $15.8 trillion globally, referring to Dobson et al. (2020). They recently published their calculation on prevention cost with a special focus on ecology of zoonosis and investments to prevent tropical deforestation and to limit wildlife trade in order to prevent future zoonosis outbreaks. As these costs are partly generated by organized crime and terrorist organizations, actions against such organizations should, therefore, be part of the prevention tools, yet to develop.

4.5 Socio-Ecological Harm, Violent and Deadly Force of Criminal Activities

4.5.1 Socio-Ecological Harm

Responding to environmental crime involves a wide range of collaborations across many different scales and sectors. This is especially the case when addressing transnational environmental crime and its associated global environmental harms (Pink & White, 2016). From a green criminology perspective, the basic starting point is that something or someone has suffered harm (van Uhm, 2016 referring to White 2011). Depending on one's personal preference,[11] 'something' might be defined as 'the SES' and, therefore, 'someone' might be defined as 'a human or any other organism'. Van Uhm further refers to Stretesky et al. (2013) who state that a harm perspective, as in this context, is shaped by shared understanding and scientific knowledge about what constitutes ecological harm rather than being a socially constructed concept defined by politicians, as in orthodox criminology. Because of that,

[11] From a green criminology point of view, 'harm' can be approached from an anthropocentric, eco-centric or biocentric view. For more details, see van Uhm (2016: 66–69).

environmental crime is typically defined on a continuum ranging from strict legal definitions to broader harm perspectives (reference to Bricknell, 2010) such as an act committed with the intent to harm or with a potential to cause harm to ecological and/or biological systems, and for the purpose of securing business or personal advantage (referring to Clifford & Edwards, 1998). In simple terms, 'environmental harm' can be defined as a crime (White, 2011; Pink & White, 2016).

With respect to environmental harm, dramatic changes in habitat composition and species diversity as well as even in the climate have been adversely affected by both regular and criminal human behavior. Although most of these changes are directly resulting from human population growth, urbanization, other land uses, and human consumption behavior, some of these factors are related and intertwined with the illicit trade in narcotics and wildlife products. The high demand for such products and the methods used are pushing ecosystems to other, less desirable, states and thresholds. This significantly weakens the resilience of SES and is, thus, socio-ecologically harmful. As a result, it causes comprehensive harm to all involved stakeholders, with some relative short-term gains for the illegal ones.

As explained in Sect. 4.1, the influence of terrorist and criminal organizations will likely increase and expand in the near future, affecting the sustainability and resilience of SES in which they operate or which they influence indirectly. As a consequence, we may speak of socio-ecological harm, especially within the vulnerability niche of fragile tropical states. This is recognized by the United Nations as shown in the following quotes:

- "To protect people and planet in line with the Sustainable Development Goals (SDGs), and to build back better from the COVID-19 crisis, we cannot afford to ignore wildlife crime", said Ms. Waly,[12] the UN Executive Director of the United Nations Office at Vienna (UNOV)/United Nations Office on Drugs and Crime (UNODC).[13]
- The SDGs are the blueprint to achieve a better and more sustainable future for all. They address the global challenges we face, including those related to poverty, inequality, climate change, environmental degradation, peace and justice (Fugunwa & Olanbiwoninu, 2020).

These two citations together make it clear that the highest-level organization, as the United Nations is, recognizes threats and pressure on both the ecological and social systems induced by criminal acts resulting into socio-ecological problems as previously explained. In addition to the interconnectivity of SDGs, and just as Pink and White (2016) stated, this interdisciplinary problem can only be solved from a multidimensional point of view. It will, therefore, have to be solved using an interdisciplinary approach in which military components may also have a role to play. This thought matches the emphasis of the Former UN Secretary-General, Kofi

---

[12] UN News. Wildlife crime putting environment and health at risk: UN report, https://news.un.org/en/story/2020/07/1068121

[13] UN System: Chief Executive Board for Coordination, https://www.unsystem.org/content/ms-ghada-fathi-waly

Annan, on the responsibility and political commitment to end the worst forms of violence. Since the disruption of SES by organized crime and terrorism can be seen as one of the most serious threats for the near future, his repeated call to 'act in defense of common humanity' is applicable.

4.5.2 Violent and Deadly Force

Over the past decade, the global death toll from terrorism ranged from 8000 in 2010 to 44,000 in 2014 (an average of 21,000 people worldwide each year). As in most countries, terrorism accounts for less than 0.01% of deaths; in countries of high conflict this can be as much as several percent. Although media might have influenced general public on the distribution, frequency and intensity, terrorism tends to be very geographically-focused (Ritchie et al. 2019).

Terrorist presence and posture vary in space and time. Figure 14.5a shows the variation over time of terrorism as a share of total death from 1990 up and until 2017. While the trend fluctuated between 0.01% and 0.02% over the 1990s and early 2000s, it increased to 0.08% in 2014 before dropping to 0.05% in 2017. Figure 14.5b shows the highly uneven distribution of terrorism as a share of total deaths for each country, with higher death rates in the Middle East and Africa.

Table 14.2 shows a gradient of violence used in poaching by local villagers, local farmers/poachers, and professional groups (criminal organizations). While locals use simple guns and bush craft techniques, brutal organized crime groups use advanced technical field equipment and 'professionally' sedate the animals from their helicopter or four-wheel drive all-terrain vehicles. After this, depending on their methods, they heartlessly saw through the skulls of their target with their chainsaw to get their hands on the horn or tusks, without leaving any precious bits

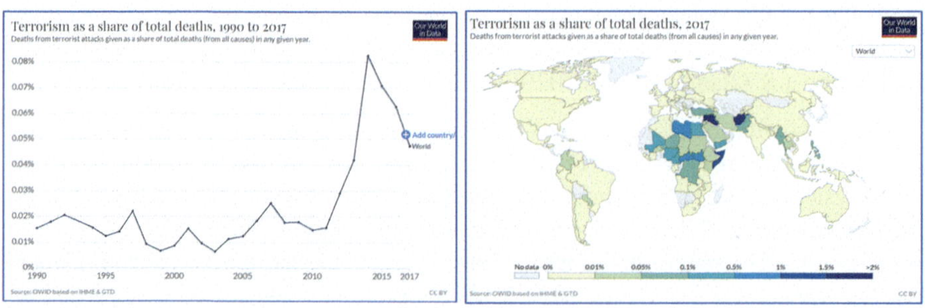

Fig. 14.5 Terrorism as a share of total deaths (1990 and 2017)
A (left side): presents worldwide deaths from terrorist attacks between 1990 and 2017. Presented as percentage of the total deaths worldwide per year. B (right side): presents the distribution per country in 2017
Source: Adapted from Ritchie et al. (2019)

Table 14.2 Types of poachers of animals for Traditional Chinese Medicine (TCM)

Type	Equipment	Method
Local villager	Guns or muskets	Poaching animals that threaten local communities and their cattle
Local farmer/ poacher	Traps, tracking signs, guns	Traditional poaching using tracking signs, traps, dogs and guns
Professional group (organized crime)	Helicopter veterinary tranquilizers, night-vision goggles, guns with silencers	Professional specialist poaching with highly sophisticated equipment in order to be efficient and effective

Source: Adapted from Van Uhm (2016:214)

behind. After taking their loot, they leave the animal in an utterly screaming pain as they come out of anesthesia, and eventually bleed to death.[14]

In addition to the use of covert weapon systems and anesthetics, poachers also use poison to kill elephants or to contaminate their carcasses specifically to eliminate vultures, whose overhead circling might otherwise reveal the poachers' presence, especially in the southern part of Africa (Ogada et al., 2016; Solly, 2019). In 2019, Solly reported on the poisoning of 530 vultures after feasting on elephant carcasses laced with poison in northern Botswana. Between 2012 and 2014, Ogada et al. (2016) recorded 11 poaching related incidents in seven African countries, in which 155 elephants and 2044 vultures were killed (Table 14.3).

4.5.3 Short Distance – Long Distance Silent Killing

In addition to the ecological effects of killing mega herbivores, the killing of vultures, which are obligate scavengers, generates many other ecological effects, some of which are likely to pose serious threats to both humans and wildlife. Because vultures have a huge home range, their killing has long-ranging effects up to hundreds to thousands of miles away. Therefore, this phenomenon can be seen as a textbook example of human-induced environmental conflicts that disrupt the eco-social balance. Figure 14.6 illustrates some of the possible routes that play a role in this.

Ogada et al. (2012) showed that the risk of infectious disease transmission is likely to increase with the absence of vultures, also affecting transmission patterns, especially among mammalian carnivores (II).[15] In addition to this, more carcasses and longer removal times can cause direct effects such as the growth of pathogen on the carcasses (I) and indirect effects like an increase in the number of scavengers (III). The latter one leads to higher transmission risks via direct physical contact between animals, both within and between species transmission (Alexander &

[14] Stroop: *Journey into the Rhino Horn War* is a 2018 South African documentary film about rhino poaching, produced by Bonné de Bod and Susan Scott. In this documentary, detailed images of the results of poachers' methods on individual rhinoceroses are presented. See https://www.stroop-film.com/

[15] The Roman figures I, II and III refer to the scheme.

Table 14.3 Details of 11 vulture poisoning incidents at elephant poached for ivory in seven African countries

Year	Month	Country	Location	No. of vultures poisoned	No. of elephants poisoned
2012	Mar.	Botswana	Kwando	200[a]	1
2012	July	Zimbabwe	Gonarezhou National Park	191	1
2013	May	Botswana	Kwando	326	3
2013	May	Mozambique	Gonarezhou Transfrontier Park	78	1
2013	July	Namibia	Bwabwata National Park	500[a]	1
2013	Oct.	Zimbabwe	Hwange National Park	219	135
2013	Oct.	Zambia	North Luangwa National Park	476	1
2013	Nov.	South Africa	Imfolozi Game Reserve	37	1
2013	Dec.	Zambia	Mfume, South Luangwa	6	1
2014	May	DRC	Virunga National Park	10	3
2014	July	Zimbabwe	Zambezi National Park	1	4

Source: Adapted from Ogada et al. (2016:594)

[a]Estimates based on the number of vulture skulls and other hones found at the scene

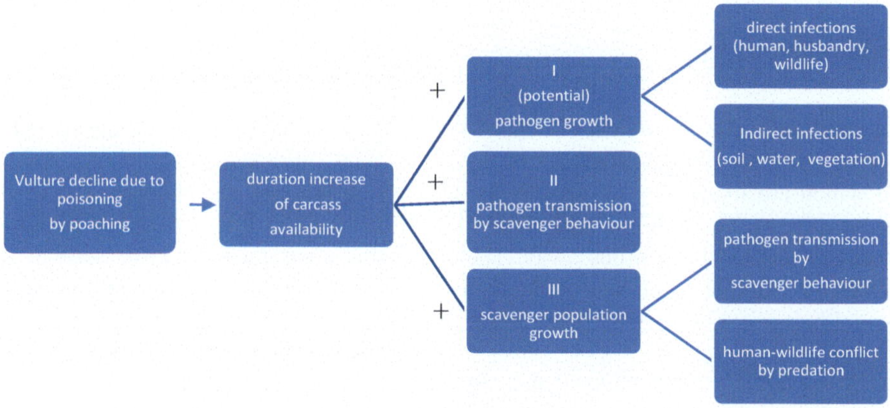

Fig. 14.6 Possible health treats resulting from vulture declines and, therefore, increased carcass exposure times

Although the main reason for the worldwide decline of vulture populations is due to the use of poison and certain medicines, active poisoning by poachers plays a significant role in disturbance of eco-social balance

Source: Deduced from Ogada et al. (2012)

Appel, 1994; Alexander et al., 1994; Craft et al., 2008; Perry et al., 2018), and also increases the chance of unwanted human-wildlife interactions such as predation on or infection of livestock (Gusset et al., 2015).

In the Serengeti ecosystem in Tanzania, for example, vultures were[16] (Pain et al., 2003 referring to Houston, 1979) the major consumers of dead ungulates, accounting for greater meat consumption than all mammalian carnivores combined. Unfortunately, this is no longer the case (Virani et al., 2010). As vultures decline, populations of many facultative scavengers are growing. Since these species are more susceptible to contract and transmit diseases among themselves and to humans, this decline can and – probably – already has resulted in both public and veterinarian health issues as well as threats to wildlife (Beucheley & Şekercioğlu, 2016; Buij et al., 2015; Green et al., 2004; Gross, 2006; Markandya et al., 2008; Pain et al. 2003; Ogada, 2014; Ogada et al., 2012; Turner et al., 2014).

Next to higher risks of disease transmission, an abundance of uneaten carcasses poses a direct threat to public health because the rotting flesh provides a breeding ground for potentially pathogenic bacteria, providing sources of disease, such as anthrax, for humans, livestock, and wildlife (Pain et al., 2003). Disease transmission can also take place via contamination of soil, water, and vegetation. Turner et al. (2014, 2016) found that anthrax-infection risk of African herbivores was highest at carcass sites and most probable at carcass sites 1–2 years of age. This is related to the fact that carcass nutrients improve soil and vegetation, and that *Bacillus anthracis* is found on grasses up to 2 years after death. They found that animal foraging responses to carcass sites shifted from avoidance to attraction, and ultimately to no preference (Turner et al., 2014).

In addition, climate change will increase the pressures on availability, accessibility, and demand for resources, especially water, adding to the risk of water-borne diseases (Chalmers, 2012; Aldeyarbi et al., 2016).

5 Strategic Solutions

5.1 Can the Military Be Part of a Solution?

In order to tackle dynamic future fields of operation, 't Hart et al. (2016) suggest a bivalent organization of the military which mixes a robust and a felexible type of orientation. Table 14.4 shows some of the dominant characteristics of the context in which military forces have to operate; eighter relatively stable – predictable[17] – or highly dynamic – unpredictable.[18] Depending on circumstances and stated goals, both types of orientation can and will be used, since landscapes of conflict are fluid in both space and time.

[16] I unfortunately have to write 'were' instead of 'are', since nowadays this is not true anymore. See for instance Virani et al. (2010).

[17] One might also use the term 'foreseeable'. Within the military community this term is used for planned periods of operation. It does not necessarily mean the theatre of operations will be predictable.

[18] These characteristics unsurprisingly show a high resemblance to both the ecological r- and K-strategists, in which the ecological r-strategist is adapted to changing environments and the K-strategist to (more) stable environments; see Sect. 3.1.2 in Chap. 10 of this book.

Table 14.4 Organizational orientation characteristics of adaptation to relative stable and, therefore, predictable (robust) and unstable or unpredictable environments (flexible)

Robust orientation (predictable environment)	**Flexible orientation** (unpredictable environment)
Aimed at stability	Aimed at flexibility
Acting from the strength of the existing organization	Adapting by the organization itself when necessary, in order to achieve the collective goals
No fundamental changes are necessary within the organization	Fundamental changes are implemented if needed to adapt
Changes in the environment are predictable as much as possible	Unpredictable changes in the environment are welcomed
The existing fixed structures (fixed SOPs[a]) are sufficient to be able to respond in a controlled manner to changes in the environment	The existing structures (SOPs) are constantly adapted to be able to respond to changes in the environment
Characteristic are the central management, fixed techniques and tactics and procedures (fixed TTPs)	Characteristic are a high degree of task maturity and autonomy of the individual. TTPs can be instantly adapted to the current environment.
Starting point: The environment is predictable and so are the procedures.	Starting point: The environment is unpredictable and so are the necessary procedures

In a constantly changing landscape of conflict, both organizational orientations seem to be a necessity

Source: Deduced from 't Hart et al. (2016)

[a]Within the military, SOP stands for Standard Operating Procedure. This is a more or less fixed procedure of how to do things within and between different levels of the organization

Table 14.4 represents a possible strategic approach of a fluid landscape in which conflict and inter-conflict are entangled both in space and time. Just like in ecology, different environments demand different approaches in strategy. This dictates different traits of participants as well as (re)allocation of scarce resources. Such differences come to expression in operational goals as well as education, training, and selection of personnel, equipment, and material. In order to be able to take advantage of a wide variety of methods, as well as to be prepared to tackle the broad scope of future landscapes of conflict, a mix of both orientations – robust and flexible – is a must. Hence, the development of a bivalent organization appears to be an adequate adaptation to the current and – near – future situation. This thought is further explored in the following sections.

5.2 Negative Feedback

Before looking at how to strategically and effectively designate scarce military resources, some of the key ecological processes involved in socio-ecological conflict dynamics should first be recalled. As already explained in Chap. 12 of this book,[19] social-ecological regime shifts may more easily occur if resilience has been

[19]See i.e., sections: 2.1 on resilience; 3.2.1 on ecosystem services, and 3.2.2.2 on ecological regime shifts.

reduced – i.e. as a consequence of human actions. Habitat destruction and loss of species and ecosystem services are drivers of such shifts that further exacerbate climate change impacts and influence future security (see also Sect. 4.1 above). For example, the loss of important major herbivores such as elephants, rhinoceroses and apex predators like lion, leopard, and hyena due to poisoning, (il)legal wildlife trade, and land-use change can trigger regime shifts. Therefore, (il)legal wildlife poisoning and trade, as well as tropical forest loss directly related to e.g. narcotic trafficking, form major threats to biodiversity-stability. The main ecological principal behind this is simple: the negative feedback or the balancing feedback. Such a principle occurs when some function of the output of a system, process, or mechanism is fed back in a manner that tends to reduce the fluctuations in the output, whether caused by changes in the input or by other disturbances.[20]

For example, keystone species play a central role in 'negative feedback loops', the loss of which needs to be prevented to avoid unpredictable socio-ecological conflict with consequences for biodiversity, ecosystem services, livelihoods, and human wellbeing (Paine, 1966; Shackleton et al., 2018; Biggs et al., 2018). Given the sensitivity of SES to the involvement of disruptive actors (criminals and terrorists) as referred to here, it is obvious to pursue an integrated approach, primarily focused on conflict prevention. It must, therefore, be recognized that the above-mentioned actors play a crucial role in the destabilization of SES, and that actions against them can only become permanently successful as a result of a systems approach, as in any other attempt to avoid, mitigate or stop complex conflicts (Rietjens et al., 2009; Olsthoorn et al., 2015; Lucius & Rietjens, 2018; Post, 2019; Rietjens & Ruffa, 2019; Faleg & Gaub, 2019; Gans, 2019).

From an economic point of view, this is also highly preferable since the direct and indirect cost of, for example, violent conflicts – i.e. related to drivers such as climate change, like in the case of the Syrian war[21] – are high and keep rising[22]. In this perspective, prevention is key. This is supported by Faleg and Gaub (2019) who

---

[20] https://en.wikipedia.org/wiki/Negative_feedback

[21] According to Kelly et al. (2015), "there is evidence that the 2007–2010 drought contributed to the conflict in Syria. It was the worst drought in the instrumental record, causing widespread crop failure and a mass migration of farming families to urban centers. The authors conclude that human influences on the climate system are implicated in the current Syrian conflict.

[22] According to Faleg and Gaub (2019), "the most visible cost is human life and health: the number of civilians dying directly from conflict doubled between 2005 and 2016 and the number of forcibly displaced persons increased fivefold between 2010 and 2016. Conflict does not just kill people, it also hurts economies. The Israeli-Palestinian conflict, for example, is estimated to have cost more than $12 trillion so far, while the destruction in Syria has by some estimates amounted to $388 billion (other aspects, such as environmental damage or indirect deaths from conflict are not included in this calculation). Some of this cost also affects the international community: in 2018, more than 135 million people were in need of humanitarian assistance and protection, predominantly due to conflict. Moreover, the number of people receiving emergency aid has risen systematically in the past 10 years, from 28 million in 2008 to 97.9 million in 2018. The gap between the financial requirements and funds available has widened: in 2008, requirements amounted to $6.3 billion, with $5.2 billion funded; in 2018, requirements were $25.2 billion, with 'only' $15.1 billion funded. Simply put, humanitarian assistance is more costly than prevention".

assume that "humanitarian assistance is more costly than conflict prevention", and Woocher (2009) who mentions that "an ounce of prevention is worth a pound of cure".

In the same line of thought, according to Faleg and Gaub (2019), internal conflicts span approximately 5–10 years risking a relapse within the first 10 years. Since post-conflict recurrences constitute only a minority of all conflict outbreaks, and more outbreaks in the near future are most likely to happen, the prevention of relapse after wars end is insufficient to prevent most new conflicts (Woocher, 2009). Therefore, early warning and early response, resilience improvement, and conflict prevention measures should be part of the strategy toolbox during both the conflict and inter-conflict phase. Now, this is where we have to look if we want to develop an effective strategy to tackle root causes by the use of scarce military resources.

In the next section, a brief exploration will be done on how to take advantage of the military as a tool to shape the landscapes of conflict and fear, discussing that the military can be of significant value during both conflict and inter-conflict phases. The analysis will be limited to the mitigation of human-induced climate change and habitat and species loss as drivers of socio-ecological conflict.

5.3 The Military

5.3.1 Regular and Special Forces

Depending on the political-strategic ends, ways, and means, the military can play a role within the human-induced environmental conflict prevention. Since the UN Security Council has recognized climate change as a 'threat multiplier' as well as a root cause of conflict in specific regions and countries, such as Lake Chad, West Africa, and Somalia (Reedt Dortland et al., 2019a, b), this issue is getting increasingly more attention. Within the military sphere, this has given rise to the question of how to prepare for future operations; new tasks in a changing environment.

The military – or armed forces – can roughly be divided into four different kinds of forces: army, navy, air, and Special Forces (for some countries). Special operations forces (SOF) are those which are not conventional (Searle, 2017). Figure 14.7 illustrates this conceptual definition.

Operations inside the box are conducted by conventional forces (CF): army, navy, and air force for national defense. Since special operations are different from conventional ones, they are, by definition, outside the conventional box (Searle, 2017). As defined by the AJP-3.5 doctrine (NATO, 2013), SOF has three main building blocks – Direct Action (DA), Special Reconnaissance (SR), and Military Assistance (MA) – on which they depend to carry out their core activities. So, SOF

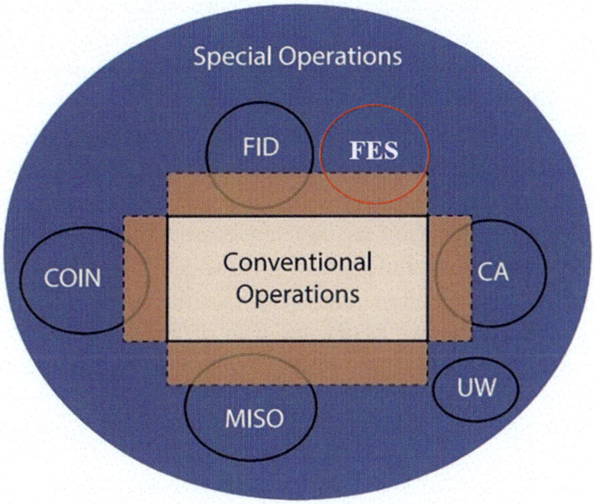

Fig. 14.7 Conceptual definition of SOF and its core activities
MISO: Military Information Support Operations; COIN: Counter Insurgency; FID: Foreign Internal Defense; CAO: Civil Affairs Operations; UW: Unconventional Warfare (See for further information: AJP 3.5, or: https://www.socom.mil/about/core-activities. MISO: are planned to convey selected information and indicators to foreign audiences to influence their emotions, motives, objective reasoning, and ultimately the behavior of foreign governments, organizations, groups, and individuals in a manner favorable to the originator's objectives. COIN: The blend of civilian and military efforts designed to end insurgent violence and facilitate a return to peaceful political processes. FID: Activities that support a Host Nations internal defense and development (IDAD) strategy and program designed to protect against subversion, lawlessness, insurgency, terrorism, and other threats to their internal security, and stability, and legitimacy. CAO: enhance the relationship between military forces and civilian authorities in localities where military forces are present. UW: actions to enable a resistance movement or insurgency to coerce, disrupt, or overthrow a government or occupying power); *and FESO: Foreign Ecological Security Operations. The edges of the military circle expand and contract as military responsibilities and authorities change over time. Also, the conventional box within that circle changes its size and shape as the definition of conventional operations evolves. Therefore, the covering edges around the conventional box symbolize the domain where special operations roles and missions have become conventional. Mostly first executed by SOF enabling forces*
Source: Adapted, but slightly adjusted, from Searle (2017)

is able to apply DA,[23] SR,[24] and MA in various contexts and for different purposes. MA basically is a way in which SOF provides improvement of local or regional security through training, mentoring or partnering with foreign military forces. This

[23] DA: Short-duration strikes and other small-scale offensive actions employing specialized military capabilities to seize, destroy, capture, exploit, recover, or damage designated targets.

[24] SR: Actions conducted in sensitive environments to collect or verify information of strategic or operational significance.

may vary from basic skills, special tactics, and procedures up and until combined operations.[25]

The scope of special operations, like regular operations, is limited by the authorities given to the military but extends to the boundaries of what the military is authorized to. However, special operations are much more likely than conventional operations to overlap with the authorities and activities of other agencies (e.g. Interpol). Furthermore, the routine overlap between special operations and the activities of both national as well as inter-governmental agencies means that the forces who conduct special operations will have special knowledge and skills to be able to perform their often independently performed tasks (Searle, 2017). This means that these kinds of forces easily adapt to different settings within the fluid landscape of (inter)conflict, extending military power as well as the strategic insight of deep within areas of operations. Therefore, the strategic utility of SOF lies in its complementary capability to partially fill the void of military force (Titulaer, 2021).

5.3.2 Foreign Ecological Security Operations and 3D-ET

As shown in Fig. 14.7, a number of core activities exist and such activities have evolved in response to specific needs. In fact, they arose as a result of past experiences. For instance, Hostage Release Operations (HRO) have developed from operations like Operation Eagle Claw[26] – also called Operation Tabas – in which the Americans tried to release their compatriot diplomats held hostage in the American Embassy in Teheran, Iran (November 4, 1978 – April 24, 1980); and Operation Nimrod[27] on May 5, 1980, in which the British rescued hostages from the Iranian Embassy in London (April 30–May 5, 1980). Like the other core businesses, also Foreign Ecological Security Operations (FESO) now appears to be needed worldwide to suppress extensive breeding grounds of conflict. Therefore, we now propose a new focus in order to develop and apply FESO as a new core activity for both SOF as well as regular forces within the context of a wide scope on socio-ecological conflicts.

One of the ways to deal with the causes of human-induced environmental conflicts is to disrupt the disrupter of the system. As a matter of fact, a military format that would fit this need, but requires further development from a holistic point of view, is called 'Swarming'. Swarming is a seemingly amorphous, but deliberately structured, coordinated, and strategic way to perform military strikes from all directions (Arquilla & Ronfeldt, 2001). It can be compared with swarming in nature, like bees and ants do (see Sect. 2). One of the major advantages of the swarming method

[25] For further details see: Allied Joint Publication (AJP)-3.5

[26] https://en.wikipedia.org/wiki/Operation_Eagle_Claw

[27] https://en.wikipedia.org/wiki/Iranian_Embassy_siege#SAS_assault

is that it also allows small countries with little capacity to contribute to both operational and strategic advantages (Haar & Haspels, 2018) significantly.

To ensure that FESO is developed from a broad socio-ecological-oriented approach, I recommend taking the Dutch 3D approach, also called the 'Comprehensive Approach' as a starting point. The Dutch used Diplomacy, Development, and Defense as a counter insurgency tool during their 'reconstruction' mission in the Province of Uruzgan, Afghanistan. The Comprehensive Approach was devised to facilitate cooperation between civilian and military actors, tackle causes of instability, and create a more safe and secure environment in areas of conflict (Olsthoorn et al., 2015). It was, therefore, a 'system-based approach'.

This work, as the Chap. 12 of this book, has shown the strong relevance of ecology to human existence. Disruption of ecological processes can lead to serious degeneration of SES causing conflicts (Fig. 14.8). The recovery of ecological systems often takes a lot of time. That is why I propose to add Ecology and Time to the existing 3D-concept (Fig. 14.9). Ecology is now positioned as a central starting point on which Diplomacy, Development and Defense form supporting controlling tools. In order to take structural account of fundamental differences in the timescale on which the intended processes take place, the time factor is added. Having said this, the Comprehensive Approach has evolved into a 3D-ET Approach. A practical first step in extending the 3D approach has been taken by people like Gallagher and Wit (2012), but should now become a fully accepted part of any major military mission.

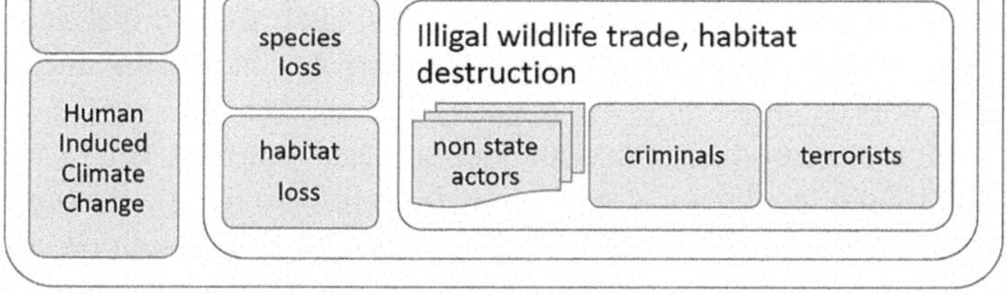

Fig. 14.8 'Trophic-context' of human-induced conflict
Drivers of this conflict are formed by different types of actors, such as criminal – and terrorist-organizations as well as other illicit non-state actors

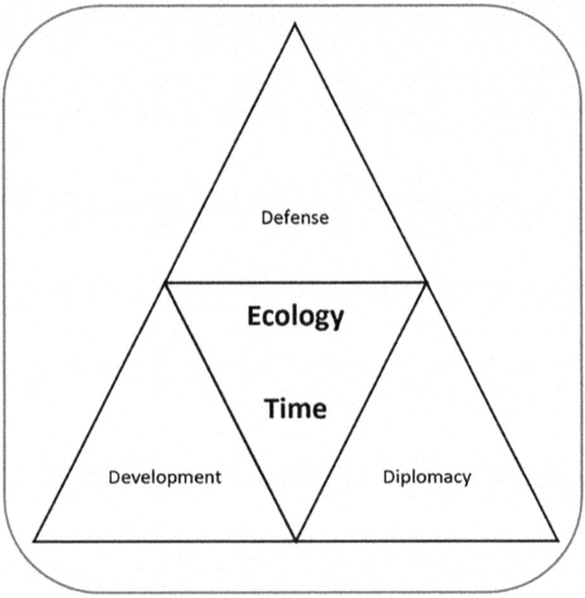

Fig. 14.9 Context of the 3D-ET Approach in which ecological security is created by incorporating Time as part of the core strategy to counter human-induced environmental conflicts, facilitated by Diplomacy, Development, and Defense

6 Conclusions and Recommendations

This chapter aims to provide some inputs to the question of how the military can help mitigate human-induced environmental conflicts. For a good part, the drivers of social ecological conflicts appear to be non-state actors such as criminal and terrorist organizations, which play significant roles in the disruption of ecological systems, eventually leading to stress, conflict, and displacement. Such disruptions are induced, for instance, by poaching of engineering species like rhinoceros, elephants, and chimpanzees or illegal logging which causes habitat loss and ecological regime shifts. This, therefore, contributes to the stochasticity of future security landscapes in which scattered or networked non-state actors, such as criminal and terrorist organizations, make a significant contribution to an ever-increasing uncertainty of safety and security.

The answer to the question of how the military can contribute is narrowed to the conclusion that both regular and special operations forces do have existing methods to counter the problems mentioned in this chapter. The only thing that needs to be done is a refocus of international political-strategic ends, ways, and means, including both factors of ecology and time. From an operational level, SOF (Special Operations Forces), in particular, can operate in the diffuse field of conflict and inter-conflict and may play a crucial role in situations that go beyond the boundaries of regular police forces. Since ecology forms the fundamental ground on which SES (Social-Ecological Systems) exist, it dictates FESO (Foreign Ecological Security Operations) to become one of the core tasks of SOF, accompanied by regular – SOF enabling – forces.

To establish a multidimensional and social-ecological approach, the 'Comprehensive Approach' – or Dutch 3D-approach – is now ready to evolve into a 3D-ET-approach. In this way, fundamental timescale differences between Diplomacy, Development, and Defense are mitigated while they all serve as social-ecological-supporting tools in mitigating conflict and induced displacement, and vice versa.

7 Epilogue

Almost immediately after completing the manuscript of this chapter, the latest insight into Amazonia's carbon budget was published on July 14, 2021 in the prestigious journal *Nature*, entitled: "Amazonia as a carbon source linked to deforestation and climate change" (Gatti et al., 2021). The present article concludes that Amazonia is now acting as a carbon source rather than a carbon sink due to deforestation and climate change. While the academic domain is frequently nourished by research on the issue, the daily news is full of reports of extreme temperatures and massive fires in Western Europe, Canada, Siberia, China, and other places on the planet where these phenomena usually do not occur.

As a climax to ongoing events, the Intergovernmental Panel on Climate Change (IPCC) issued the following press release[28]: "Geneva, August 9 – Scientists are observing changes in the Earth's climate in every region and across the whole climate system, according to the latest IPCC, released today. Many of the changes observed in the climate are unprecedented in thousands, if not hundreds of thousands of years, and some of the changes already set in motion – such as continued sea-level rise – are irreversible over hundreds to thousands of years" (IPCC, 2021). The report shows that greenhouse gas emissions from human activities are responsible for about 1.1 °C of warming since 1850–1900, and notes that average temperatures are expected to reach or exceed 1.5 °C over the next 20 years.

Do we really understand that it is 12:05 pm? May the wise words of Albert Einstein still reach us in time: "we cannot solve our problems with the same thinking we used when we created them".

Acknowledgments The author wishes to thank Dr. Victor Spaan for providing useful critiques on an early edition of this chapter. My thanks go also to Dr. Mohamed Behnassi and Dr. Himangana Gupta for providing the opportunity to write and publish this chapter, useful advice during the process, and the proofreading of the revised version.

---

[28] https://www.ipcc.ch/site/assets/uploads/2021/08/IPCC_WGI-AR6-Press-Release_en.pdf

References

Abdulhamid, S. M., Ahmad, S., Waziri, V. O., & Jibil, F. N. (2011). *International Journal of the Computer, the Internet and Management, 19*(3), 14–20.

Aldeyarbi, H. M., Abu El-Ezz, N. M. T., & Karanis, P. (2016). Cryptosporidium and cryptosporidiosis: The African perspective. *Environmental Science and Pollution Research, 23*(14). https://doi.org/10.1007/s11356-016-6746-6

Alexander, K. A., & Appel, M. J. G. (1994). African wild dogs (Lycaon pictus) endangered by a canine distemper epizootic among domestic dogs near the Masai Mara National Reserve, Kenya. *Journal of Wildlife Diseases, 30*, 481–485.

Alexander, K. A., Kat, P. W., Wayne, R. K., & Fuller, T. K. (1994). Serologic survey of selected canine pathogens among free-ranging jackals in Kenya. *Journal of Wildlife Diseases, 30*, 486–491.

Amersfoort, H. (2016). Nederland de weg kwijt. Over de teloorgang van de militaire strategie en de noodzaak van geschiedenis. *Militaire Spectator, 185*(5), 217–231.

Arthur, C. (2011). Egypt cuts off internet access. *The Guardian*, 28 January. https://www.theguardian.com/technology/2011/jan/28/egypt-cuts-off-internet-access.

Arquilla, J., & Ronfeldt, D. (2001). *Swarming and The Future of Conflict*. RAND. National Defense Research Institute. Naval Postgraduate School. California, USA. Dudley Knox Library.

Bekkers, F. F., & Spiegeleire, S. H. A. M. (2010). Strategische lenigheid. Anticiperen op een fundamenteel onzekere toekomst. *Militaire Spectator, 179*(7/8), 360–371.

Beucheley and Şekercioğlu. (2016). The avian scavenger crisis: Looming extinctions, trophic cascades, and loss of critical ecosystem functions. *Biological Conservation, 198*, 220–228.

Biggs, R., Peterson, D., & Rocha, J. (2018). The Regime Database: A framework for analysing regime shifts in social-ecological systems. *Ecology and Society, 23*(3), 9.

Bricknell, S. (2010). *Environmental Crime in Australia* (AIC Reports: Research and Public Policy Series no. 109). Australian Institute of Criminology.

Buij, R., Nikolaus, R., Whytock, D., Ingram, D. J., & Ogada, D. (2015). Trade of the threatened vultures and other raptors for fetish and bushmeat in West and Central Africa. *Oryx, 50*(4), 606–616. Flora and Fauna International.

Chalmers, R. M. (2012). Waterborne outbreaks of cryptosporidiosis. *Annali dell'Istituto Superiore di Sanità, 48*(4), 429–446.

Chen, C., Noble, I., Hellmann, J., Coffee, J., Murillo, M., & Chwla, N. (2015). *University of Notre Dame Global Adaptation Index: Country Index Technical Report*. Updated maps: July 2020. https://gain.nd.edu/our-work/country-index/

Chukwuere, J. E., & Onyebukwa, C. F. (2018). The impacts of social media on national security: A view from the Northern and South-Eastern Region of Nigeria. *International Review of Management and Marketing, 8*(5), 50–59.

Chialvo, D. R., & Millonas, M. M. (1995). How swarms build cognitive maps. In L. Steels (Ed.), *The Biology and Technology of Intelligent Autonomous Agents. NATO ASI Series* (Series F: Computer and Systems Sciences) (Vol. 144). Springer.

Clifford, M., & Edwards, T. (1998). Defining environmental crime. In M. Clifford (Ed.), *Environmental crime: Enforcement, policy and social responsibility* (pp. 5–30). Gaithersburg, MD.

Craft, M. E., Lawthorne, P. C., & Dobson, A. P. (2008). Dynamics of a multihost pathogen in a carnivore community. *Journal of Animal Ecology, 77*(6), 1257–1264.

Darwin, C. (1859). *On the origin of species by means of natural selection, or the preservation of favoured races in the struggle for life*. John Murray.

Dávalos, L. M., Bejarano, A. C., Hall, M. A., Corea, L., Corthals, A., & Espejo, O. (2011). Forest and drugs: Coca-driven deforestation in tropical hotspots. *Environmental Science and Technology, 45*(4), 1219–1227.

Eldredge, N., & Could, S. J. (1972). Punctuated equilibria: An alternative to phyletic gradualism. In T. J. M. Schopf (Ed.), *Models in Paleobiology* (pp. 82–115). Freeman Cooper.

Dobson, A. P., Pimm, S. L., Hannah, L., Kaufman, L., Ahumanda, J. A., Ando, A. W., Bernstein, A., Busch, J., Daszak, P., Engelmann, J., Kinnaird, M. F., Li, B. V., Loch-Temzelides, T., Lovejoy, T., Nowak, K., Roehrdanz, P. R., & Vale, M. M. (2020). Ecology and economics for pandemic prevention. *Science, 269*(6502), 379–381.

Ethembabaoglu, A., Sweijs, T., Boone, G., Goreham, K., & Spiegeleire De, S. (2010). *Social Networking The security shakedown of shared information*. The Hague Centre for Strategic Studies.

Faleg, G., & Gaub, F. (2019). *Iceberg Ahead! Rethinking conflict prevention* (Brief 4, d.d. 9-4-2019). European Union, ISS Institute for Security Studies. https://www.iss.europa.eu/content/iceberg-ahead-rethinking-conflict-prevention

Fugunwa, O., & Olanbiwoninu, A. A. (2020). Accelerating the sustainable development goals through microbiology: Some efforts and opportunities. *Acces Microbiology, 2020*, 2.

Gallagher, N., & Wit, P. (2012). *Society stabikization by winning the environment. Ecosystem assessment makes sense... Full situational awareness in CIMIC*. Civil-Military Co-operation Centre Of Excellence (CCOE).

Gans, B. (2019). *Stabilisation operations as complex systems – order and chaos in the interoperability continuum*. CentER, Center for Economic Research.

Gatti, L. V., Basso, L. S., Miller, J. B., et al. (2021). Amazonia as a carbon source linked to deforestation and climate change. *Nature, 595*, 388–393.

Garnier, S., Gautrais, J., & Theraulaz, G. (2007). The biological principles of swarm intelligence. *Swarm Intelligence, 1*, 3–31. https://doi.org/10.1007/s11721-007-0004-y

Green, R. E., Newton, I., Schultz, S., Cunningham, A. A., Gilbert, M., Pain, J., & Prakash, V. (2004). Diclofenac poisoning as a cause of vulture population declines across the Indian subcontinent. *Journal of Applied Ecology, 41*, 793–800.

Gross, L. (2006). Switching drugs for livestock may help save critically endangered Asian vultures. *PLoS Biology, 4*(3), 0303–0304.

Gusset, M., Swarner, M. J., Mponwane, L., Keletile, K., & McNutt, J. W. (2015). Human-wildlife conflict in northern Botwana: Livestock predation by endangered African wild dog *Lycaon pictus* and other carnivores. *Oryx, 43*(1), 67–72.

Haar, F. E., & Haspels, B. (2018). *The strategic utility of small-state special operations forces (SOF) as information collectors to support National Decision Making*. MSC thesis. Naval Postgraduate School Monterey, CA, USA.

Haasnoot, M., Kwakkel, J. H., Walker, W. E., & Laat ter, J. (2012). Dynamic adaptive policy pathways: A method for crafting robust decisions for a deeply uncertain world. *Global Environmental Change, 23*, 485–498. Elsevier.

Haenlein, C., Maguire, T., & Somerville, K. (2016). Poaching, wildlife trafficking and terrorism. In C. Haenlein & M. L. R. Smith (Eds.), *Poaching, wildlife trafficking and security in Africa. Myths and realities* (Whitehall papers) (pp. 58–76).

Haenlein C. and Smith M.L.R. (eds.) (2016). Poaching, wildlife trafficking and security in Africa. Myths and realities. . Whitehall papers.

Houston, D. C. (1979). The adaptation of scavengers. In A. R. A. Sinclair & N. Griffiths (Eds.), *Serengeti, dynamics of an ecosystem* (pp. 263–286). University of Chicago Press.

Howard, M., & Paret, P. (Eds.). (1984). *Carl von Clausewitz. On War (Vom Krieg)* (Indexed ed.). Princeton University Press.

Howard, P. N., Aiden, D., Freelon, D., Hussain, M. M., Mari, W., & Marwa, M. (2011). *Opening closed regimes: What was the role of social media during the Arab spring?*.

Humphrey, C. (2019). *Illegal logging poised to whipe Cambodian wildlife sanctuary off the map*. Forest Trackers. https://news.mongabay.com/2019/05/illegal-logging-poised-to-wipe-cambodian-wildlife-sanctuary-off-the-map/

Idris, I. (2019). *Drivers and enablers of serious organized crime in Southeast Asia. K4D: Knowledge, evidence and learning for development*. Helpdesk report.

IPCC. (2021). Climate change 2021: The physical science basis. In V. Masson-Delmotte, P. Zhai, A. Pirani, S. L. Connors, C. Péan, S. Berger, N. Caud, Y. Chen, L. Goldfarb, M. I. Gomis,

M. Huang, K. Leitzell, E. Lonnoy, J. B. R. Matthews, T. K. Maycock, T. Waterfield, O. Yelekçi, R. Yu, & B. Zhou (Eds.), *Contribution of Working Group I to the Sixth Assessment Report of the Intergovernmental Panel on Climate Change*. Cambridge University Press. In Press.
Jensen, P. S., & Gleditch, K. S. (2009). Rain, growth, and civil war: The importance of location. *Defence and Peace Economics, 20*(5), 359–372.
Kelly, C. P., Mohtadi, S., Cane, M. A. Seager, R., & Kuhnir, Y. (2015). *Climate change in the Fertile Crescent and implications of the recent Syrian drought*. PNAS Early Edition 1–6. https://doi.org/10.1073/pnas.1421533112.
Kresek K. (2019). *What's happening in Cambodia's forest? Blog on: Tree cover loss*. Accessed on 19 Aug 2020. www.globalforestwatch.org.
Krishnadas, D. (2015). *Fuse: Foresight-driven understanding, strategy and execution: Move the future*. Marshall Cavendish International (Asia) Pte Ltd.
Kutschera, U. (2009). Charles Darwin's Origin of Species, directional selection, and the evolutionary sciences today. *The Science of Nature, 96*(11), 1247–1263.
Lawrence, J., Haasnoot, M., & Lempert, R. (2020). Climate change: Making decisions in the face of deep uncertainty. *Nature, 580*, 456.
Lucius, G., & Rietjens, S. (2018). *Effective civil-military interaction in peace operations: Theory and practice*. Springer.
Mann, M. E., Bradley, R. S., & Hughes, M. K. (1998). Global-scale temperature patterns and climate forcing over the past six centuries. *Nature, 392*, 779–787.
Mann, M. E., Bradley, R. S., & Hughes, M. K. (2004). Correction: Corrigendum: Global-scale temperature patterns and climate forcing over the past six centuries. *Nature, 430*, 105.
Markandya, A., Taylor, T., Longo, A., Murty, M. N., Murty, S., & Dhavala, K. (2008). Counting the cost of vulture decline: An appraisal of the human health and other benefits of vultures in India. *Ecological Economics*. https://doi.org/10.1016/j.ecolecon.2008.04.020
Middendorp Tom. (2019). Military should fight climate change. *HagueTalks*, February 27. https://www.youtube.com/watch?v=p9VDIAhNFmw
Middendorp, T., & Bergema, R. (2019). *The warning signs are flashing red: The interplay between climate change and violent extremism in the Western Sahel* (Clingendael policy brief). Planetary Security Initiative.
Miguel, E., Satyanath, S., & Sergenti, E. (2004). Economic schoks and civil conflict: An instrumental variables approach. *Journal of Political Economy, 112*(4), 725–753.
MOD The Netherlands. (2010). *Eindrapport Verkenningen*. Houvast voor de krijgsmacht van de toekomst.
Moffet, M. W. (1984). Swarm raiding in a myrmicine ant. *Naturwissenschaften, 71*, 558–590. Springer.
NATO. (2013). *Standard AJP-3.5 allied joint doctrine for special operations, edition A, version 1*.
Nillese, E. E., & Verwimpt, P. (2009). *Grievence, Commodity Prices and Rainfall: A Village-level Analysis of Rebel Recruitment in Burundi* (MICRON research working paper 11).
Notre Dame Global Adaptation Initiative (ND-GAIN). https://gain.nd.edu/our-work/country-index
Ogada, D. L. (2014). *The power of poison: pesticides poisoning of Africa's wildlife* (pp. 1–20). Annales of the New York Academy of Sciences. ISSN 0077-8923.
Ogada, D., Botha, A., & Shaw, P. (2016). Ivory poachers and poison: Drivers of Africa's declining vulture populations. *Oryx, 50*(4), 539–596. Fauna and Flora International.
Ogada, D. L., Torchim, M. E., Kinnaird, M. F., & Ezenwa, V. O. (2012). Effects of vulture declines on facultative scavengers and potential implications for mammalian disease transmission. *Conservation Biology, 26*(3), 453–460.
Olsthoorn, P., Bollen, M., Lobbezoo, E., & Rietjens, S. (2015). The comprehensive approach and the problem of exiting civil military cooperation: Lessons from Uruzgan. In N. Noll, D. Wollenberg van den, G. Frerks, & I. Kemenade (Eds.), *Netherlands annual review of military studies. The dilemma of leaving: Political and military exit strategies* (pp. 247–270). Springer.
Paine, R. T. (1966). Food web complexity and species diversity. *American Naturalist, 100*, 65–75.
Pain, D., Cunningham, A. A., Donald, P. F., et al. (2003). Causes and effects of temperospatial declines of gyps vultures in Asia. *Conservation Biology, 17*(3), 661–671.

Perry, L. R., Marino, J., & Sillero-Zubiri, C. (2018). Going to the Dogs: Free-ranging dogs threaten an endangered wild canid through competitive interactions. *Journal of Biodiversity and Endangered Species, 6*, 1. https://doi.org/10.4172/2332-2543.1000211

Pink, G., & White, R. (2016). Collaboration in Combating Environmental Crime — Making it Matter. In G. Pink & R. White (Eds.), *Environmental crime and collaborative state intervention. Palgrave studies in green criminology*. Palgrave Macmillan. https://doi.org/10.1007/978-1-137-56257-9_1

Post, E. (2019). *Less is more conflict prevention* (Clingendael policy paper).

Reedt, D. M., Noomen, D., Kruidbos, F., & Agterhuis, S. (2019a). *Climate change and degradation of natural resources*. Implications for the military.

Reedt, D. M., Noomen, D., Kruidbos, F., & Agterhuis, S. (2019b). Climate Change and degradation of Natural resources. Implications for the military. *Atlantic Perspective, 4*, 4–11.

Rietjens, B., Bollen, M., Khalil, M., & Wahidi, S. F. (2009). Enhancing the footprint: Stakeholders in Afghan reconstruction. *The US Army War College Quaterly: Parameters, 39*(2009), 1.

Rietjens, S., & Ruffa, C. (2019). Understanding coherence in UN peacekeeping: A conceptual framework, international peacekeeping. *International Peacekeeping, 26*(4), 1–25.

Ritchie, H.., Hasell, J., Appel, C., & Roser, M. (2019). *Terrorism. Our World in Data*. https://ourworldindata.org/terrorism#which-regions-experience-the-most-terrorism

Ritchie, H., & Roser, M. (2018) Urbanization. *Published online at OurWorldInData.org*. Retrieved from: https://ourworldindata.org/urbanization

Schwab, J. (2020). *Fighting COVID-19 could cost 500 times as much as pandemic prevention measures*. World Economic Forum 3-8-2020.

Searle, T. (2017). Outside the Box: A Theory of Special Operations. In McCabe & Lieber (Eds.). Special Operations Theory, JSOU Report 17-6 The JSOU Press MacDill Air Force Base.

Sesnie, E. S., Tellman, B., Wrathall, D., McSweeney, K., Nielson, E., Benessaiah, K., Wang, O., & Rey, L. (2017). A spatio-temporal analysis of forest loss related to cocaine trafficking in Central America. *Environmental Research Letters, 12*, 054015.

Shackleton, R. T., Biggs, R., Richardson, D. M., Brandon, M. H., & Larson, M. H. (2018). Social-ecological drivers and impacts of invasion-related regime shifts: consequences for ecosystem services and human wellbeing. *Environmental Science and Policy, 89*, 300–314. Elsevier.

Solly, M. (2019). Poachers'Poison Kills 530 endangered vultures in Botswana. *Smitsonian Magazine*, June 24, 2019.

Stibig, H.-J., Achard, F., Carboni, S., & Raši, and Miettinen J. (2014). Change in tropical cover of Southeast Asia from 1990 to 2010. *Biogeosciences, 11*, 247–258.

Stretesky, P. B., Long, M. A., & Lynch, M. J. (2013). *The treadmill of crime: Political economy and green criminology*. Routledge.

Taleb, N. (2007). *The black swan: The impact of the highly improbable*. Random House. ISBN 978-1400063512.

Tani, A., et al. (2007). Principal forest types of three regions of Cambodia: Kampong Thom, Kratie, and Mondolkiri. In H. Sawada, M. Araki, N. A. Chappell, J. V. LaFrankie, & A. Shimizu (Eds.), *Forest environments in the Mekong River basin*. Springer. https://doi.org/10.1007/978-4-431-46503-4_18

Titulaer, F. (2021). Special Operations (Forces) Explained. On the nature of Western special operations and the forces that conduct them. *Militaire Spectator, 190*(2), 84–99.

Turner, W. C., et al. (2014). Fatal attraction: Vegetation response to nutrient inputs attract herbivores to infectious anthrax carcass sites. *Proceedings of the Royal Society B, 281*, 20141785.

Turner, W. C., et al. (2016). Lethal exposure: An integrated approach to pathogen transmission via environmental reservoirs. *Nature. Scientific Reports, 6*, 27311.

't Hart, M. H. E., Dekkers, P. A. P., Kamphuis, W., Sassen, J., & Vries, de T. (2016). Het vergroten van adaptiviteit bij Defensie. *Militaire Spectator, 185-7/8*, 320–335.

van Uhm, D. P. (2016). *The illegal wildlife trade* (Inside the world of poachers, smugglers and traders. Studies of organized crime 15). Springer.

Virani, M. Z., Kendall, C., Njoroge, P., & Thomsett, S. (2010). *Major declines in the abundance of vultures and other scavenging raptors in and around the Masai Mara ecosystem*. Kenya.

Warner, J., & Boas, I. (2019). Securitization of climate change: How invoking global dangers for instrumental ends can backfire. *Politics and Space, 37*(8), 1471–1488.

White, R. (2011). *Transnational environmental crime. Toward an eco-global criminology*. Routledge.

Woocher L. (2009). Preventing violent conflict: Assessing progress, Meeting Challenges, USIP Special Report 231.

Yang C. C., & Tobun D. Ng. (2007). *Terrorism and crime related weblog social network: Link, content analysis and information visualization*.

Chapter 15
Resolution of User Rights Related Conflicts in Collective Rangelands Through Negotiation: The Case of Morocco

Rabiaa Bourhim

Abstract In Morocco, rangelands cover over 53 million hectares of land. They are in most cases marginal lands for annual crops because of various reasons (drought, low soil fertility, risk of erosion, and so on). Additionally, their legal status is not favouring their development because of intensive farming. Not long ago, collective rangelands existed in a harmonious state, as there was a balance between the supply of fodder resources and livestock needs. However, in recent decades, the conditions of these areas have been increasingly deteriorating due to many factors, including overgrazing, land clearing, and uprooting. A number of changes are being noticed, including the decline in the roles and importance of traditional institutions, cumbersome administrative procedures in terms of transhumance licenses, and the non-compliance with the regulations, illiteracy, the negligence observed by local population and beneficiaries, and the contested legitimacy of the law. These changes contribute to the emergence of major conflicts whose solutions could be found in making deals or formulating charters based on compromises with the potential to organize the access to land and the exercise of user rights. The success of these measures largely depends on the approval of sedentary and transhumant parties. Customary norms could form the basis for a concerted regulation of collective user rights to ensure the sustainability and resilience of vulnerable areas. Some of the investigated Moroccan instances of customary schemes include the negotiated charters developed during colonization by the *Ouzighimt* fraction (the *Mgouna* tribe) and *Aït Moussa Oudaoud* (*Ait Zekri*) in the region of Ouarzazate as well as *Ait Rahou in Boulmane* and *Zaouiat Ahensal* in the Middle Atlas.

Keywords Collective rangelands · Transhumants · User rights · Conflicts · Customary norms · Negotiation · Charters

---

R. Bourhim (✉)
College of Law, Economics, and Social Sciences of Agadir, Ibn Zohr University, Agadir, Morocco

M. Behnassi et al. (eds.), *The Climate-Conflict-Displacement Nexus from a Human Security Perspective*, https://doi.org/10.1007/978-3-030-94144-4_15

1 Introduction

In Morocco, collectives and forest rangelands cover over 53 million ha.[1] They comprise terrestrial ecosystems made up of a group of plant species serving as a fodder resource for livestock. These plant species constitute 84% of the national territory,[2] and contribute to about 70% of the fodder needs of small ruminants (Qarro & Roose, 2010).

Arid and callous rangelands represent almost 93% of the country's surface. Almost 97% of rangelands are located in the arid and semi-arid areas of the Oriental, the Pre-Sahara, and the vast Saharan regions (Qarro & Roose, 2010). These rangelands play a vital role in feeding the livestock. They represent the main fodder source for sheep and goats in pastoral areas. They also contribute to a large degree to the diet of local cattle in these areas.

In this context, the proportion of rangelands varies between 25% and 36% in the national fodder balance, depending on climatic conditions. This contribution is significant to the eastern steppes and mountainous regions where it ranges from 70% to 90% (Naggar, 2018).

Rangelands have existed until recently in a state of balance between the supply of fodder resources and the livestock needs. Over the past few decades, the condition of rangelands has been increasingly deteriorating because of several factors, including overgrazing, land clearing, and wood uprooting. In addition, the ambiguity of the concept of beneficiaries leads to the outbreak of conflicts related to rights of access and use in collective rangelands between sedentary and transhumant herders.[3]

When analyzing the Moroccan legal system related to collective lands, namely the 2016 law on transhumance, the 1919 law on collective lands, and the 1917 law on forest conservation, or even the Islamic doctrine and endogenous norms, we found out the existence of several legal orders. This leads to questioning the nature of their relationship: is it a competition, a rivalry? In other words, who holds the law-making authority over collective tribal lands in Morocco: is it the Islamic law, the *Jmaa*, or the state?

Lastly, can the negotiation of territorial charters work as a basis for a new governance of the rights of access and use?

This chapter focusses on the negotiated rule of law and its potential role in developing the regulation of collective and forestry user rights and negotiated governance. Through an analytical approach covering both positive and endogenous

[1] www.terrecollectives.ma

[2] In Morocco, 85% of collective lands are rangelands.

[3] In the past ten years, the province of Tiznit has received 537 complaints, and the province of Chtouka has received 438 complaints (internal report of the provinces of Tiznit and Chtouka Ait Baha). During the years 2018 and 2019, the Souss Massa region has known numerous conflicts between transhumants and farmers from Ait Milk, Belfaa, Inchaden, Arbiaa Sahel, Tafrout, Idousaka Taroudant, etc.

norms, the analysis will look into the articulations between the national legality and local legitimacies. The aim is to examine the possibility of structing effective local governance and a viable management of natural resources and biodiversity susceptible of addressing conflict of use issues and supporting human security in collective rangelands. This could be done through the development of valid negotiated charters, which represent concerted acts established by actors interested in a particular territory. These charters delineate the rights and obligations and constitute the legitimate local law of the local social group with a view to co-produce legal norms. In this approach, the issue of collective and forestry user rights is investigated through the effectiveness of positive and endogenous norms (Sect. 2), and the originality of the negotiated rule of law as an essential component of regulation (Sect. 3). The analytical approach developed in this context places the regulation of collective and forestry user rights and of collective lands at the heart of socioecological issues.

2 The Regulatory Framework Governing Collective User Rights

Historically, the Moroccan society was tribal in nature with tribes organized in collective territories. These ethnic groups made use of their common heritage according to their norms (Sect. 2.1). However, as time passed, many factors – such as the pressure on renewable natural resources, the degradation of rangelands, the transhumance, and the overgrazing – have generated conflicts between sedentary populations, collectivists,[4] beneficiaries, and transhumants. The eruption of every tension leads the parties to refer to the norms that better serve their interests (Sect. 2.2). The ambiguity that characterizes the concept of the beneficiary and the collective necessity which typifies collective lands generate several issues (Sect. 2.3). However, endogenous norms have a capacity for regulation when it comes to collective and forestry user rights (Sect. 2.4).

2.1 Conceptual Framework of User Rights

From time immemorial, Moroccans have had the practice of expecting the forest and the collective rangelands to support their needs, which vary according to the nature and condition of the forest and the tribe's lifestyle. The Forestry law – under the article 21 of the Dahir[5] of October 10, 1917 related to the conservation and

[4] All members belonging to the community or ethnic group.

[5] In Moroccan legislation, this term "Dahir' designates the seal of the King affixed to the texts of laws voted in the parliament.

exploitation of forests – has recognized and maintained the main forestry user rights attributed to the Moroccan user.

Until the beginning of the twentieth century, the Moroccan society was based on an ethical organization of groups, which made use of a collectively exploited territory. At that time, appropriating collective lands has been confused with the possession of property rights. Since the adoption of the Dahir of 1917, the legislator has introduced the concept of 'user rights', which includes the possibility of using rangelands, collecting dead wood, and so on. These user rights are not unlimited; they are subject to certain conditions aiming to establish a compromise between the necessity to preserve collective and forestry rangelands and the fulfillment of the users' rights on these lands.

The concepts related to users, user rights, and the exercise of these rights seem to be poorly understood[6] both by the beneficiaries and by the public institutions responsible for regulating these rights. However, communities recognize the user rights of the transhumant parties on rangelands. This unlimited freedom could be considered as one of the main reasons leading to the deterioration of rangelands. It also creates several problems between communities with regard to the allocated grazing periods and the protection measures of rangelands.

The collective exploitation of rangelands has generated several debates. Some authors – such as Hardin (1968) – posit that when the user rights of a resource are shared, the discrepancies between the marginal costs and the benefits of individuals and groups provoke an overexploitation of the resource. In this regard, Hardin (1968) describes a method used by shepherds to boost their income, which consists of continually increasing the size of their livestock herd. Consequently, the growing number of the livestock leads to the over-exploitation of pastures. To quote Hardin:

> The tragedy of the commons develops in this way. Picture a pasture open to all. It is to be expected that each herdsman will try to keep as many cattle as possible on the commons. Such an arrangement may work reasonably satisfactorily for centuries because tribal wars, poaching, and disease keep the numbers of both man and beast well below the carrying capacity of the land. Finally, however, comes the day of reckoning, that is, the day when the long-desired goal of social stability becomes a reality. At this point, the inherent logic of the commons remorselessly generates tragedy. As a rational being, each herdsman seeks to maximize his gain. Explicitly or implicitly, more or less consciously, he asks, 'What is the utility to me of adding one more animal to my herd?' This utility has one negative and one positive component (Hardin, 1968:1244).

[6] Historically, the colonizer was in search of a legal process capable of formalizing monopoly and control over the land as well as finding legal solutions allowing both land colonization and 'rights' guarantees on the land. The colonizer had set a mechanism towards the melkization according to which the Cadi, who is the judge of Chra'a, must verify if the land is not collective, even if alienable. However, the priority of the Cadi is not to defend the status of collective lands in terms of customary norms, but resorts to the Muslim theory of vivification (Bouderbala, 1996). This has contributed to transitioning toward the Melkization, commodification of collective lands, and subsequently to capitalism and the formation of a colonizing bourgeoisie on the Moroccan territory.

Table 15.1 Grouping of the status of a right and its holder

		Visitor	User	Manager	Owner (−)	Owner (+)
User rights	Access	x	X	x	x	x
	Sampling		X	x	x	x
Governance rights	Management			x	x	x
	Exclusion				x	x
	Alienation					x

Source: Schlager and Ostrom (1992)

Astonishingly, this idea calls for a renunciation of the freedom to produce in order to preserve the resources. It is important to bear in mind that local populations have always developed adequate practices to adapt to their environment. Their capacity to manage the scarcity and fragility of natural resources is one of the manifestations of this ancestral capability.[7]

This view is challenged by Elinor Ostrom whose thesis (1965) deals with the management of phreatic zones in southern California. Ostrom's empirical research points out to a wide variety of operating systems of common properties, which generally combine self-management, customary rules, market mechanisms, and state regulation. Ostrom demonstrates that user communities are often capable of inventing efficient and strong management systems. The author also provided a classification of user rights[8] and resource management rights held by customary groups (Table 15.1). This conception is economic in essence, but it provides insight on the management and access to common natural resources.

As for the Moroccan context, some studies, specifically in the region of Missour (see Kamil, 1992; Ramdan, 1991), have shown that, contrarily to the above hypotheses, the rangelands of the municipality of Missour are rationally exploited and according to the productivity of the resource. Kamil (1992) and Ramdan (1991) developed several tools in their perception of the environment, including the significant level of technical climate, topography, the quality of vegetation and water points, and the quality of animal resting places. Their studies demonstrate their awareness of the damage induced by overexploitation, implicit customary regulations, and the concepts governing collective use.

[7] Breeders use the concept of 'Touga' to designate the species with herbaceous component (leaves, soft stems, etc.) while species with developed woody parts are designated by the term 'Iffessi'. These species represent for plant breeders a phytomass capital available at all times. They also use the term 'Lakhalf' to express breeding by selecting new plants from seed stocks (Bourhim, 2020).

[8] In the Moroccan context, collective land tenure constitutes a common heritage of the entire community which no one owns.

2.2 Legal System and Forms of Access to Natural Resources

The Islamic law acknowledges two types of land: farmlands and uncultivated lands. Farmlands or inhabited lands (*Allard almaamoura*) are considered an ownership (*Melk*[9]), while uncultivated lands are sterile and not owned. The uncultivated lands can be divided as well into two subcategories: 'shared lands' and 'dead lands' (*ARD Almocharaka et ARD Almawat*). Shared lands are a common heritage of neighbouring populations, lands from which it is possible to profit. It should be noted, however, that these lands are different from collective lands,[10] which are considered a tribal or family condominium. 'Dead lands'[11] can be subdivided into two categories: 'Harim lands' (*ARD alharim*) and 'revived lands' (*Al Ihyae*). Harim lands are included in the area bordering lands that are promoted, inhabited, or cultivated. They are reserved for the needs of the owner, such as grazing. However, the owner's right on Harim lands is limited to the right of use (*Haq al intifaa*) and not the right to appropriation. In the case of revived lands (*Kitab Al Ihyae*),[12] the Islamic doctrine states that reviving a 'dead land' is an act that promotes private ownership, leading therefore to the expansion of the owner's rangeland.

The Muslim law[13] specifies that vacant real estate (dead lands) belongs to the state. At some level, forests are believed to be goods belonging to the Muslim

---

[9] In Morocco, the Algeciras Act (1906) has accentuated the pressure of supremacy on Moroccan lands. It stipulates that the makhzen cannot refuse granting authorizations for purchases without convincing reasons. Accordingly, when the signature of the treaty instituting the French protectorate on the Moroccan territory took place on March 30, 1912, the French colonizer adopted the circular of the grand vizier that organizes the control of real estate operations and establishes a list of inalienable goods, namely: public domain, religious land endowments (habous lands), collective lands, forests, and military land grants (guich lands). France had set up a legislative corpus that aims at depriving the tribe of the resources to which it had free access by expanding forests and reducing access and movement on rangelands as well as transferring the collective into titled land (Melk) (Bouderbala, 1996). The objective of the colonial authority is none other than withdrawing collective lands from the market and supervising them. The guardianship exercised on collective lands allowed the colonial authority to exercise political control over collective land; this aim was cloaked under the pretext of protecting the general interest, expropriating for public utility, and maintaining the necessity of the Makhzen authorization before making any change in the status of collective lands when collectivists are concerned.

[10] Collective lands are not private property (Melk). They are the common property of the tribe.

[11] The notion of dead land refers to a vacant land that does not possess an owner. The Dahir n° 1-11-178 of November 22, 2011 promulgating the law n ° 39–08 that cites the code of real rights (BO num 5998 of November 24, 2011) states in article 222 that "dead lands without an owner belong to the State. Taking hold of these lands is permitted only with the authorization of a competent authority and that in accordance with the law." Article 223 of the same Dahir states that "whoever revives a dead land, with the authorization of the competent authority, has the right to make use of it."

[12] Hadith of the Prophet Mohammed according to which: Whoever revives a dead land, it is his, Sahih Alboukhari, Kitab Almazraa, 2210.

[13] The vivification of dead lands is cited in several surahs of the Quran and in hadiths, for instance: Sura Ar-Rum, verse 50; Sahih Alboukhari, Kitab Almazraa, 2210; Sahih Alboukhari, Kitab Ihyae Almawat, 921; Sonan Atirmidhi, 1378;...

community. Profiting from the forest, as a 'dead land', was limited to resupplying in wood and grazing. The needs of the riverside communities, which lived at the vicinity of the forest, could thus be met.[14] This legal nature of 'dead lands' would sometime contribute in stirring conflicts between the sedentary and the transhumant parties. The latter would think of forests as '*Ard Moulana*' (the land of Allah) (Auclair & Simenel, 2013), while the Moroccan State considers them as public goods placed under estate management (*Amlak makhzania*).[15]

2.3 Nature and Extent of Collective Land Rights, User Rights, and Beneficiaries' Rights

In Morocco, the 1917 Dahir, related to the forestry regime, is the only text that that defines the "beneficiaries". According to this Dahir, the beneficiaries are "exclusively Moroccan populations and bordering forests". This text also defines the user rights in the forest as the "right to pasturing cattle, collecting dead wood…".

The beneficiaries gain particular or collective rights on the collective lands belonging to a shared ancestor. These rights may be granted to those considered outsiders to the community, provided that they become part of this community for a specific period of time while contributing to the expenses induced by the management of such collective lands.

In this context, article 4 of the 1919 Dahir, which organizes the administrative supervision of local communities and regulates the management and disposition of collective goods, states that "collective lands are distributed for temporary use between collectivists, following the practices and directives of the supervision". Notwithstanding, the Dahir does not define the meaning of the terms 'collective lands' or 'collectivists'. It presupposes that "these lands can be subjected to a distribution, attributing to each of the heads of families of the community an everlasting right of use", which "is imprescriptible and would be alienated or seized only for the benefit of the community itself".

In this way, the first article of the vizirial decree of August 14, 1945, related to the management of collective property specifies that "any distribution of an apartment block attributing each of the heads of families that belong to the community a permanent right of use on one or several parcels is subject to the approval of the tutorship council".

In this regard, the first article of the regulation n° 2977 of November 13, 1957, related to the collective sharing of land, defines the notion of 'beneficiary' as "the heads of families, that is to say, men who have been married for at least six months or the collectivists' widows who have at least one child. These are entitled to an equal land share". Similarly, Article 4 of the same regulation states that

[14] The Dahir of July 7, 1914 confirms that forests belong to state property.

[15] '*Amlak Makhzania*' in Arabic or 'Domaine Privé de l'État' in French.

"collectivists who have been absent for more than a year are excluded from the share" as well as "the collectivists who have pleaded against the *Jmaa* to 'Melkize' a part of the collective land for their benefit unless they have no source of livelihood". However, the excluded categories can be considered as beneficiaries by referring to the registration on lists established by local authorities. This obscures the concept of the beneficiary and leads to the outbreak of conflicts of legitimacy linked to the use of collective and forestry resources.

In articles 3 and 4 of the Dahir of July 25, 1969, considering the collective lands located in irrigation zones, it is stated that "the assembly of delegates are required to establish the list of beneficiaries. This list has to be sent to the local authority and to the collectivists who are beneficiaries within the time limit of six months from the publication of this Dahir in the official bulletin." The list "can be appealed only before the tutorship council seized by the concerned parties themselves or by the local authority within three months from the notification". This led the Ministry of Agriculture to request from the Ministry of the Interior the designation of beneficiaries. The said list had raised numerous problems that accentuated the process of opposition to the pastoral development project (Tozy, 1989).

Traditionally, the rules related to the user rights of rangelands are designated by the *Jmaa*. In this context, any transhumant can build or use the enclosures constructed on the site whose property may be collective. These enclosures are used by the first transhumant settlers. In other cases, the campsites that host the enclosures are considered, after a continuous and effective presence, as private property (Rachik, 2016). They can even be inherited (Bonte, 2001). In this context, there are moral rules that limit the freedom of movement and constitute a social base for the peaceful use of collective or forestry rangelands, namely keeping a respectful distance (*louqr*) from the tent and the neighbors' rangelands (Mahdi, 1997).

However, this process is not free from difficulties. Within the framework of the pastoral and rangelands development project[16] of the Ait Arfa tribe of Guigou, initiated in the early 1980s, the major difficulty encountered by administrative authorities consisted in providing a definition of the concept of 'beneficiaries. In other words, there have been conflicts concerning the designation of the members who had the right of access to collective or forestry rangelands of the tribe.

According to local customs, each member of the tribe is automatically a beneficiary (Rachik, 2016). Every member whose ancestors belonged to the tribe or defended the tribal territory during the pre-colonial period has a right to claim the status of the beneficiary. Outsiders are excluded from the management of the rangelands and cannot be chosen as representatives of *Jmaa*. Each village chooses a

[16] "Before introducing the pastoral development project, the distinction between the native and the outsider had no tangible repercussions on the use of rangelands. The resident transhumants could lead their herds on the lands. However, the pastoral project has changed everything. What was at stake was the official recognition of the beneficiaries. The members of the tribe could not accept the legal imposition of an informal practice and the alteration of their "generosity" towards foreigners into a right of access to collective goods" (Rachik, 2016).

representative (*Nayb*) whose task is to take care of the rangelands and to assume the role of mediator between the village and the public administration.

Outsiders have not ceased to claim a title as beneficiary citizens deserving the same rights as all Moroccans regardless of their tribal origin. They have also claimed that, in the end, the rangelands belong to the State (Tozy & Mahdi, 1990; Rachik, 2000, 2003). Outsiders adopt modern notions such as 'citizenship' because the notions of tribal collective identity are not inclusive. In most cases, the members of the tribe reject these modern notions[17] because they are rather concerned with personal interests more than defending traditional assets. Anyway, belonging to the ethnic group is the ultimate criterion for obtaining the status of the beneficiary. This criterion is associated with others, namely: the residence within the ethnic community; being head of a family; the exercise of agricultural activity in croplands; or having livestock in rangelands. However, these criteria differ from an ethnic group to another; a situation that creates user rights-related conflicts.

2.4 *Contribution of Endogenous Law in the Sustainability of Social-Ecological Systems*

From time immemorial, local populations had developed customary forms of adaptation to environmental changes. The management of the scarcity and fragility of natural resources has been one of the manifestations of their advanced engineering.

As a concept within endogenous law, the 'custom' denotes numerous meanings. It can be defined as "an unconfined social phenomenon that resists the effects of erosion. It manifests itself by its regularity, its repetition, and above all by its generality. Therefore, even if it encloses a legal rule, it is a social phenomenon that remains ephemeral if it does not meet these conditions" (Bentahar, 2005). The 'custom' can also be defined as a spontaneous product of the legal necessity, a product which functions independently from the legislative body. For this reason, Michaux-Bellaire defines it as the spontaneous product of the social relations evolving within the premises of a primitive tribe. He believes that customs are the result of different uses brought now and then by people of different races and religions (Michaux-Bellaire, 1913).

Robert Montagne suggests in an article devoted to the study of the legal system of southern Morocco that endogenous law is the prerogative of the Siba tribes. This law fades away with the gradual domination of the *Makhzen*, which overrides customary rules (Montagne, 1924).

Endogenous law could serve as a legal basis for a community's legal rights. It could benefit a community's traditional knowledge, particularly when it comes to

---

[17] "The relation to tradition is no longer uniform. Transhumants challenge the old rules based on traditional discrimination. In this case, tradition is maintained only by the group that derives benefits from the traditional law" (Rachik, 2016).

the management and sharing of natural resources as well as conflicts and disputes resolution. The originality of a endogenous law lies in its capacity to construct a set of customs, beliefs, traditional knowledge, modes of use accepted as compulsory rules of conduct, and lifestyles of traditional communities, as well as an integral part of the socioeconomic systems. Customary norms are the product of consensus reached by an assembly or an organization (*Jmaa*). For a custom to exist, the assembly must be alive and independent. Crozier and Friedberg (1977) consider such organization as "a political and cultural construct, a means with which social actors are equipped to regulate their interactions with the aim to obtain the necessary cooperation for the achievement of collective objectives".

When it comes to the Moroccan legal system, it is apparent that it is the fruit of a long and complex history. The onset of colonization had injected the Western positive law in the flexible and varied Moroccans legal regime – which is mainly based on *Fiqh* (the Malekite rite) and customs – with the sole objective to serve the interests of the colonizer. Loaded with a political agenda, the colonization had attempted to divide the country into zones that adopt Muslim law and custom-based zones. The colonization process had even tried to affect the country's legal system by either demonstrating the inferiority of the Muslim law compared to the European one or by sometimes opposing the Muslim law to customs. By wishing to privilege customs, the colonization ended up discrediting it in the eyes of Moroccans (Bentahar, 2005).

In Morocco, customs have played a significant role as a normative source. There are many local customary norms[18] that have been established by the Muslim law throughout the history. Therefore, customs could inspire positive law by suggesting solutions taken from a normative source that is common to Moroccans. These customs enclose a rich legacy capable of providing adequate legal solutions. Such solutions could be compatible with the Moroccan mentality without necessarily undermining national and religious principles.

According to El Khatir (2007), customs are "the product of society according to the contexts and social and political conditions. They have the power to adapt and to model themselves in accordance with social needs. They assimilate social dynamics and function in practice as a system in motion. This interweaving of spatial arrangements and structures provides the capacity to maintain the conditions for the production of law". Indeed, endogenous law is neither static nor defunct. Its structure demonstrates a great capacity for resilience in the face of the various situations to which rural communities are exposed. These communities often develop, produce, and reinvent their customary laws according to new social contexts.

[18] The Constitution of Kenya recognizes that pastoralists and their customs can contribute positively to the management of lands. Additionally, in Mali, positive and customary laws coexist through the recognition of the Malian land tenure system of customary rights. For more information: www.fao.org/3/i5771fr/I5771FR.pdf

3 Conflict Resolution Through Negotiation: The Key Role of Negotiated Law

As a mechanism of conflict resolution and decision and rule-making, negotiation is increasingly becoming a social phenomenon and a legitimate practice. The resulting rules are the product of a collective action and a common regulation. Indeed, negotiation is the meeting of two regulations: the first consists of a top-down control while the second is autonomous, practiced by the concerned stakeholders. The second type of regulation either defies or adjusts the first one (Catlla, 2004).

In addition, negotiation seems to be a legitimate method for structuring organizational strategies, with the objective to increase the freedom of action of each member of the organization (Kerbrat-Orecchioni, 2012). It is also "an activity which brings together several actors who, faced with both divergences and interdependencies, choose to voluntarily seek a mutually acceptable solution" (Kerbrat-Orecchioni, 2012). Negotiation often adopts an in-between stance: neither a norm imposed by state institutions, nor an ambiguous endogenous norm unfavorable to societal changes (Barrière and Faure, 2012); it is positioned at the intersection between the social acceptability of social groups and the state's positive law. "Instead of opposing these systems or forcing them to ignore each other, the construction of a concerted law is defined as a bridge between such regulatory paradigms" (Barrière and Faure, 2012).

In addition, the existence of disagreements does not preclude negotiation, which often generates a common ground to facilitate the achievement of the stated goals. Accordingly, Reynaud (1991) affirms that "what drives an agreement in a social negotiation is not the identification of a possible solution to the negotiated problem, but the capacity to make a common anticipation".

Indeed, negotiating with local communities, for sustainable management of natural resources, can be advantageous to these communities in the future. Morocco has a rich experience in this area, namely the *Agdals* which constitute a traditional form of concerted management of natural resources. The *Agdals* constitute as well a social mechanism which regulates the access to and management of natural resources through the exercise of control and sanctions and the supervision of user rights. More precisely, in the community *Agdals,*[19] collective rules are negotiated within the *Jmaa* which represents the users, beneficiaries, and transhumants, among others, in a particular territory where the decision-making process requires the

[19] In Egypt, Jordan, and Lebanon a traditional management model called '*Hima'*, considered one of the traditional management models that possess a natural resource management capacity, is being valued as a model of sustainable development, particularly in rangelands. It succeeds where various relevant public policies have failed. Indeed, the existence of this traditional land tenure system in Jordanian rangelands has ensured the management and sustainability of resources under the control and supervision of Bedouin tribal institutions. Jordan has been able to develop mechanisms for dialogue, negotiation, participation, and mediation between the community and the government, and it has succeeded in ensuring the link between local legitimacy and national legality (Davies et al., 2012).

concertation between concerned stakeholders. "Different types of institutional arrangements can be identified depending on the relative degree of autonomy of the communities as well as their capacity to manage internal conflicts. In village *Agdals*, where conflicts are less severe, the community frequently adopts autonomous rules on its common territory without necessarily making use of written acts or depending on legal recognition. However, it is noticed that, in the High Atlas, it is becoming common to deposit by the interested parties of a written act, comprising the collective rules of the *Agdal* and the names of the community representatives (*naïb*, *amghar*...), at the office of the local authority" (Genin & Benchakroun, 2007). These acts (or charters) arise as a regulatory tool making it possible to formalize relations between social groups and the territory.

The negotiated law is inclusive since it incorporates local knowledge, customs, and cultures. According to Genin & Benchakroun (2007), "it is indeed a modern law but from within the social body and derived from the realities of local contexts". It also provides an efficient regulatory framework for the management of future conflicts and the protection of common environmental conditions. "The result would be a co-constructed regulation conceived in a way that can be revisable and evolving, or even adaptable and flexible" (Barrière & Faure, 2012). The negotiated regulation is based on divergent interests but often transcends a mere interventionist regulation (Chevallier, 2001).

The challenge of the negotiation in such a perspective is the social acceptance through the formalization of local law, namely territory charters. For Barrière (2015), "the negotiated law is associated to an 'environmental juridical status' located at the intersection of laws and regulations and the norms derived from customs and representations" (Barrière, 2015). Given the multiplicity of legal norms, the state does not monopolize the production of these norms. On the one hand, legality is born from the legislative, executive and judicial trilogy of state powers. On the other hand, there is a law born from the endogenous culture of social groups. Such a law, which is outside the positive legal system, is based on an ancestral legitimacy and constitutes an evolving body, adapted with the realities of local territories and capable of preserving the identity and survival of social groups. This law has the potential to be "an interesting basis for the development of a positive law which, from the very beginning, takes into account the perspectives of local populations" (Genin & Benchakroun, 2007).

Negotiation is an essential tool of conflict resolution and regulation of collective and forestry user rights (Sect. 3.1). It is also the foundation of a legitimate negotiated law (Sect. 3.2) with the potential to enhance innovation in the process of elaborating territory charters (Sect. 3.3).

3.1 Conflict as a Negotiation Trigger

In this research, 'conflict' is understood as an instrument that aims to establish rules and codify the relationships between individuals in order to define the conditions of their coexistence. The charters or compromises set by individuals reveal how they should live together in the future even if they still do not fully share the same interests or values. Conflicts "are only a form of negotiation. What a difficult moment when the interaction of individuals and social groups happens despite the continuous conflicts between them: confrontation and mutual adaptation, struggle and agreement, opposition of rationality, and convivial compromise. This interaction is ultimately, in all its forms, a negotiation" (Catlla, 2004).

Fisher and Smith (2000) state that "it seems more likely that conflicts are dealt with creatively, and that they become constructive in instances where there is a high level of participation, where channels of agreement exist, and where mechanisms of disagreement and consensus management are available. All community stakeholders, including men, women, the youth, and the old have the right to speak" (Fisher & Ury, 1982). Indeed, the conflicts of legitimacy or use can be an essential means of dialogue, consent, and co-production of legal norms, which may help reconcile various divergent interests.

3.2 Concerted Regulation Through a Negotiated Law

Concerted regulation is a complex process which is not limited to a simple implementation of conventional participatory approaches, the result of which is often to propose the same solutions under the pretext of having been discussed with local populations (Genin & Benchakroun, 2007). Concerted regulation is about developing prescriptive mechanisms through the participation of the *real* local actors on the basis of conventions, compromises, charters, among others, that are legitimate in the eyes of the community and the nation. To this end, "to regulate is to guide through the adoption of exemplary behaviors, which would not be a reference or followed by the community members unless a social acceptance legitimizing them is available. The negotiation as a pathway seems, therefore, more relevant than the enforcement of top-down norms that are exogenous to the thinking, culture, practices, and aspirations of the actors supposed to comply with them. The articulation between a local regulation and a supra-local environmental regulation represents a co-construction challenge based on the state of resources and a social acceptance" (Barrière & Faure, 2012).

According to Barrière (2007), the negotiated law "is, by definition, opposed to the imposed law; it is based on compromise, which is the art and result of negotiation" (Barrière, 2007). This law is connected to the state regulation of endogenous norms (customs, local practices, etc.) to reach a form located between law and regulation "in which negotiation is situated at the heart of the rule of law" (Genin & Benchakroun, 2007). Negotiation is defined "by considering, in a functional manner, the representations and understanding of the rules and procedures belonging to parties with different, or even contradictory, objectives in order to reach an agreement on what is possible to set as rules" (Barrière & Faure, 2012). Therefore, by linking the endogenous to the legal norms, the negotiated law establishes bridges between the state-based regulation and the law stemming from customary norms. The negotiated law guarantees an agreement between the sources of norms and the applied legal provisions.

The structure of territorial charters represents an adequate model of the negotiated regulation of collective user rights through the establishment of a concerted governance system. Such negotiated governance requires the emergence of the negotiated law through the production of an act or charter capable of founding the cohesion of social groups based on a common objective of user rights management. This negotiated charter expresses such a social cohesion by defining the rights, duties, and legal relationships of community members regarding the management, the planning, and sustainability of their natural heritage and territory. These negotiated charters constitute a local law and provide a link between local legitimacies and national legality in addition to an ecological pact between local actors and their territory. In addition, negotiated charters are a form of social contract and an expression of a social consensus. They aim to ensure a link between national laws, which do not usually take into account local environmental needs, and the specific needs of the concerned social groups where each member of tribal or ethnolineage communities is confronted with environmental commitment. A negotiated charter "also includes dealing with various forms of misconducts which are defined in terms of offenses associated with fines and/or other types of sanctions" (Genin & Benchakroun, 2007).

3.3 From the Conception to the Implementation of Negotiated Charters: Empirical Cases

In the case of the Ait Ouzighimt tribe of the High Atlas (Fig. 15.1), Ait Rehou of the Middle Atlas (Fig. 15.2), Aït Moussa Oudaoud (Ait Zekri) on the southern side of the High Atlas (Fig. 15.3), and Zaouia Ahensal in the Middle Atlas during colonization (Box 15.1), restrictions have been enacted in order to limit the degradation of rangelands. Such restrictions consist of: restraining the mobility and concentration of herds on rangelands; guaranteeing pastoral land security; implementing concerted charters between concerned tribes; and settling conflicts of use by recognizing and promoting the institutional arrangements which involve modern institutions, namely ethnolineage cooperatives, associations, local communities, and traditional institutions such as *Jmaa* and *Agdals*.

The charters contain clauses identifying the beneficiaries and the outsiders who live within the common territory. They also contain rules related to the access and use of collective and forestry resources, surveillance and control, the creation of corridors in the *Agdals*, the compensation of guards, and the sanctions (fines) in case of non-compliance. Indeed, these charters illustrate the mutual dynamic between local legitimacy and national legality, which is the foundation of a negotiated governance of resources and conflicts. "In other words, between the imposition of national legality and the 'customary' autonomy, the legal pluralism could either

الحمد لله وحده

عقد اتفاق والتراضي

نحن الموقعون اسفله اعيان قبيلة اوزيغمت اصالة عن انفسنا
ونيابة عن جميع افراد القبيلة نشهد على انفسنا بما تصح به
الشهادة شرعا وقانونا اننا اتفقنا على السادة :
1موح بن عدي ايت عيسى بدوار تغرفت الحامل للبطاقة الوطنية رقم 19486اب
2ايت زماد موح بن ايشو بدوار اغرم ازدرن والحامل للبطاقة الوطنية رقم
3915 اب .
3) اهري احمد بن موح بدوار ووشكي والحامل للبطاقة الوطنية رقم 5118 اب
الحلاء وجميع منتوجاته الا النشاء وعليه ان يتصل بالمختار المدة
المذكورين اعلاه لهذا وضعنا غرامة لكل من ضبطت عليه قطع الاشجار
الاشجار بتيسير نظام على الشكل التالي:
كل من ضبط عليه قطع اوحرق شجرة كبيرة يؤدي غرامة قدرها : 10000 ريال
نصف هذا القدر يعطى للصندوق والنصف للحراس .
كل من قطع الاعمدة (تيجدى) يؤدي 200 ريال للواحدة وان وصل المخزن
يؤدي 400 ريال للواحدة . مع العلم ان لاتمنع في ابوونا ومكان .
كل من قسم ليبيع الاعمدة الكبرى (ءاسوتار) يؤدي 5000 ريال للواحد
كل من قطع مايسمى ب (افسين) ليبيعهم يؤدي 5000 ريال
كل من يخزن اويبيع الاخشاب المستعملة للطبخ والتدفئة يؤدي 6000 ريال
لغياب البطاقة . وكل من يريد الاستفادة بشكل قانوني ومنتظم فان اللجنة
4) ايت بوحساين احمد بن حمو ابدوار تشكي اوزيغمت والحامل للبطاقة الوطنية
رقم 3293 اب .
5) ايت مسعود حمو بن موح المزداد بتاريخ 1978/02/08 بدوار ايمي نركت
والمسجل تحت عددرقم 106السنة 1978 بقلعة امكونة .
6) موح بن احمد ايت القائد بدوار تعزوت اوزيغمت والحامل للبطاقة الوطنية
رقم 70127 ب .
7) ابراهيم بن علي افقيرن بدوار افقيرن اوزيغمت
اننا اتفقنا جميعا على ان يكونوا هؤلاء الاشخاص الصفة
حراس منطقتنا وخاصة الخلاء منها فيوف الصابية على
جميع الاضرار التي تلحق بها وخاصة قطع الاشجار الخضراء وكذلك بيع
الفحم الخشبي بدون نظام يعني كل من يريد ان يتصرف في اشجار
رهن اشارة الجميع دون استثناء و
الى اسدرم شاينا سووول

امضاء الموقعون

1)حدوس ابراهيم اوسوة 2) الحاج لحسن اورماد
3) افقيرن موح بن احساين 4) محمد بن الحسين ايت حدو
5سعيد بن موح ابردا 6 محمد بن حمو ايت يمسو
7) موح بن ابراهيم ارماح ايت بن عيس 8)ابراهيم اورماد

الامضاء: عبد الله العباسي

(The translation of the Act)
Concerted Act

Praise be to Allah

We, the undersigned, chiefs of the Ouzighimt Douar, agree to designate the following gentlemen:

1. Oumouh Ben ADDI born on… in Douar… national ID…
2. Aitzmad Moh Ben Ichou born on… in Douar… national ID…
3. Ahri Ahmed Ben Moh born on ... in Douar ... national ID...
4. Ait Bouhssein ahmed Ben Hamou born on… in Douar… national ID …
5. Ait Messoud Hamu Ben Mouh born on… in Douar… national ID…
6. Moh ben Ahmec Ait lqayd born on… in Douar… national ID…
7. Brahim ben Ali Ifqirn born on… in Douar… national ID…

As guards of our territory (land, Agdal…) with the mandate to protect and control the fields and plants. Whoever cuts or sets on fire a large tree will be fined 10000 Rial. Whoever cuts tree trunks to resell them will pay 200 Rial for each piece, and any person who refuses to amicably settle the problem will be fined (Khtia) 400 Rial for each trunk. Any large beam cut will cost 5000 Rial for each piece and whoever cuts thorny plants (firewood) intending to resell them will be fined 5000 Rial.

And for that we signed this deed for all intents and purposes.

Signature of the eight representatives + stamp of the vice-president of the Municipality.

Fig. 15.1 Negotiated Act of the *Ait Ouzighimt Agdal* (Ouarzazate City)

result from a syncretism in form of a yoyo (sometimes legal, sometimes customary), or proceed by an in-between approach, that is a junction between local legitimacies and the public interests of the nation" (Barrière & Faure, 2012). In fact, such charters are uncommon examples of community management based on concertation, negotiation, and co-production of legitimate norms. Upscaling such practices may help ensure the sustainability, or even the coviability, of socioecological systems in the face of global changes.

The drafting of a charter calls for negotiations of traditional powers (*Jmaa*, *Naib*, *Chikh*, *Amghar*…) involving the delegates of territorial communities and local authority. The aim is to set a right to collective ecological interference in order to

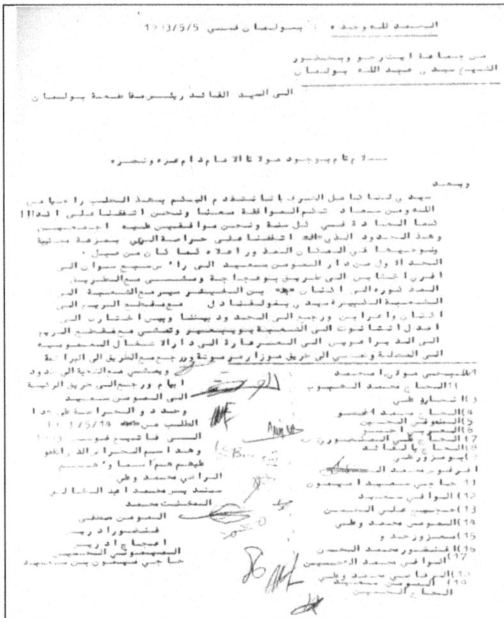
الحمد لله وحده بولمان في 1993/5/5

(The translation of the Act)

Praise be to Allah, Boulemane in 1993

From the chiefs of the Ait Rahho Jmaa and in the presence of Sheikh Sidi Abdullah Boulmane.

To the Caid of the Boulmane district.
We are honored to present this request to you, to solicit your agreement; we have accepted the Agdal as it is the case every year. The Agdal is mapped as follows: from the household of Moumen Said to Ras Sabaa Swan to Avery Akhshabn to the road to Bouajajah… and to Aknan Bin El Khidir…, to the road to Marmousha.

The period of closure: from May 14, 1993, to August 1, 1993.

The names of the 8 Agdal supervisors

Signature of the 20 representatives of the Jmaa ....

Fig. 15.2 Negotiated Act of the Ait Rahou Agdal (Boulemane City)

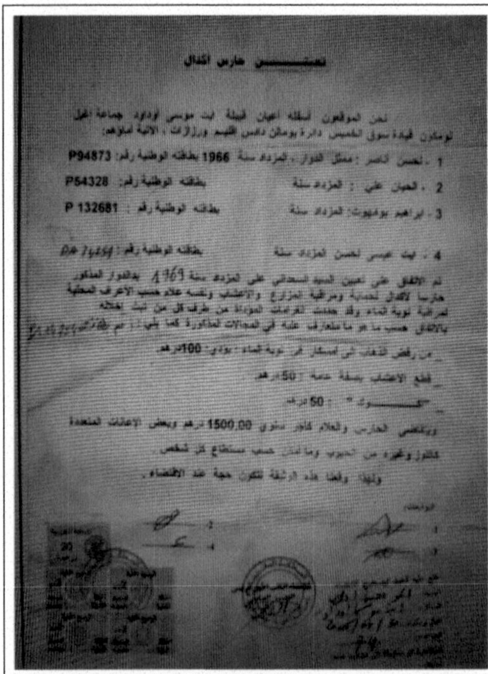
تعيين حارس أكدال

(The translation of the Act)
The appointment of a warden for the Agdal

We, the undersigned, representatives of the douar of Aït Moussa Oudaoud, representing Ighil N'Ougme, Kéada (place of the Caïd) Khemis Dades, Boumalene dades circle, Ouarzazate Province, namely: - Lahcen Ounacer born in 1966 national ID … - Alahiane Ali… - Brahim Boumhhoute… - Aït Aissa Lahcen…
Agree to designate Mr. Essadani Ali, born in 1969 in the same Douar, a warden of the Agdal, to protect and control the fields and the herbage, and also an Allam according to local customs for the control of the water distribution and the fines paid by offenders (those who do not respect this agreement) according to local and customary practices: - those who refuse to go to Amskar during the time of the water distribution pay MAD100 (for deflecting water for the Douar) - wood cutting: MAD50 - small dam ("agoug"): MAD50 (for not having opened or closed it).
The annual salary of the warden and Allam is MAD1500 in addition to a quantity of almonds and other grains given freely by farmers depending on their capacity.

Fig. 15.3 Negotiated Regulatory Act of the Moussa Oudaoud Douar (Ait Zekri)
Source: Barrière et al. (2007)

Box 15.1: Transhumance Charter Drawn Up by the Colonizer in 1941 in the Commune of Zaouiat Ahensal

First rule: Pasture is divided into units allocated to the beneficiaries of different groups, according to a division which respects "the state of the 15 operating forces" prepared by the AI officers in the 1930s. These collectives, however, are not yet registered.

Second rule: These pastoral units or parks are carefully delimited from benchmarks recognized by all and registered on various "procès-verbaux" issued from the arbitrations.

Third rule: The crop line is marked by piles of stones. On the rangeland as such, it is therefore forbidden to cultivate or build sheepfolds; customary law considers that building structures is equal to claiming ownership.

Fourth rule: Wherever water is scarce, livestock watering is meticulously determined: Crossovers, the duration of the stay at the water point ("without even having the time for a cup of tea"), the order of the passage of herds.

Fifth rule: The closure of the best parts of rangelands, especially the high-altitude swards, takes place each spring. This is the practice of Agdal, an essential element of rangeland management. The Agdal knows specific closing and opening dates; the closure period takes place in spring and lasts from one to three months, never more. The Agdal does not welcome foreigners, except through associations. The surveillance of the Agdal is ensured by a warden chosen for his seriousness and paid by the community; offenders are sanctioned by fines paid in cash or in kind.

Source: Bourbouze (1999)

preserve the general interests of the tribe or social group. The implementation of a negotiated charter requires support and verification in order to guarantee its gradual and continuous adaptation. The negotiated charter has thus a juridical and legal value; it could be invoked in the courts and taken into account by the public administration.

4 Conclusion

The state believes in the ownership of the law-making process because of the legitimate violence it is enjoying. The law is supposed to provide solutions to the issues related to the degradation of territories and resources. It must also adapt to environmental change, promote local social negotiations, and enhance traditional institutions and practices whose adaptive capacity, resilience, and sustainability have been demonstrated. The success of traditional institutions and practices – such as the *Agdals* – in preserving collective pastoral and forestry resources stems from the integration of natural, cultural, and economic dimensions while responding to insecurity challenges induced by climate change and biodiversity loss. To tackle these problems adequately, the public administration must intervene immediately to resolve the conflicts of use. These conflicts could be considered so far as a key cause behind the failure of pastoral improvement related policies or projects. To address these issues successfully, the public administration should recognize and promote traditional institutions and concertation while legitimizing the resulting negotiated charters.

The regulation of collective and forestry user rights revolves around the coviability dimension, which transcends the simple rationality of sustainability on behalf of a new negotiated law as a part of a new form of post-modern law, which is not limited to texts drafted by a legislative and executive authority. This new type of law rejects the classical doctrine, which asserts that the state is the ultimate law-maker. The post-modern law promotes the inclusive participation of concerned stakeholders in addition to concertation, dialogue, negotiation, consensus, and the co-construction of legitimate and legal norms.

References

Auclair, L., & Simenel, R. (2013). Quand le forestier endosse les habits du saint... La gestion des frontières communautaires dans le haut Atlas et l'arganeraie (Maroc). *Revue Forestiere Francaise, 65*(4), 341–357. https://doi.org/10.4267/2042/53630

Barrière, O. (2007). Pour l'émergence d'un droit négocié... : comment intégrer la concertation et la gouvernance dans le traitement juridique de la GRN ?. In G. Didier & F. Benchekroun (Eds.) De la parole aux gestes ... : éléments de réflexion sur les dispositifs de gestion concertée des ressources forestières et pastorales au Maroc, *IRD*, pp. 47–50.

Barrière, O. (2015). Repenser le droit de l'environnement dans une conception renouvelée du développement durable: prospective d'un" droit de la coviabilité" des systèmes sociaux et écologiques. In S. Blaise, & C. David, et al. (Eds.), Le développement durable en Océanie : vers une éthique nouvelle ? Aix-en-Provence, E PUP; PUAM, 2015, pp. 215–242.

Barrière, O., & Faure, J. -F. (2012). L'enjeu d'un droit négocié pour le Parc amazonien de Guyane. *Natures Sciences Sociétés*. EDP Sciences, *20*(2), 167–180.

Barrière, O., Aït Hamza, M., Sabir, M., Qarro, M., Najim, L., Magnin-Feysot, F., & Dubois, E. (2007). Maroc: programme de recherche conjoint : gouvernance de la biodiversité pour une gestion patrimoniale de l'environnement (GOVBIO) : rapport d'étape. *IRD*.

Bentahar, M. (2005). Droit et société au Maroc. *IRCAM*, pp. 15–31.

Bonte, P. (2001). Mohamed Mahdi, Pasteurs de l'Atlas. Production pastorale, droit et rituel. Casablanca, Fondation Konrad Adenauer, 1999, *Études rurales*. EHESS, (159–160), pp. 260–263.

Bouderbala, N. (1996). Les terres collectives du Maroc dans la première période du protectorat (1912–1930). *Revue des mondes musulmans et de la Méditerranée*. Edisud, 79(1), pp. 143–156.

Bourbouze, A. (1999). Gestion de la mobilité et résistance des organisations pastorales des éleveurs du Haut Atlas marocain face aux transformations du contexte pastoral maghrébin. In Managing Mobility in African Rangeland: the Legitimization of Transhumant Pastoralism, Niamir Fuller M. (dir.), Londres, Ltd.

Bourhim, R. (2020). La gouvernance négociée : un nouveau mécanisme de résolution des conflits des droits d'usage collectifs dans les espaces arides - Cas du Maroc. *Revue Marocaine d'Administration Locale et de Développement*, N°150–151, Rabat, pp. 171–184.

Catlla, M. (2004) Sociologie de la négociation par Reynald Bourque et Christian Thundered. Paris: La Découverte, Collection Repères, 2002, 124 p., ISBN 2-7071-2953-4.', *Relations Industrielles/Industrial Relations*. Département des relations industrielles de l'Université Laval, *59*(2), 429–431.

Chevallier, J. (2001). La régulation juridique en question. *Droit et société*. Editions juridiques associées, (3), 827–846.

Crozier, M., & Friedberg, E. (1977). *L'acteur et le système*. Seuil Paris.

Davies, J., Poulsen, L., Schulte-Herbrüggen, B., Mackinnon, K., Crawhall, N., Henwood, W. D., & Gudka, M. (2012). Conservation de la biodiversité des zones arides.

El khatir, A. (2007). Droit coutumier amazigh face aux processus d'institution et d'imposition de la législation nationale au Maroc. Etude de l'Organisation Internationale du Travail,2007, disponible sur le site : www.ilo.org, consulté le 27/1/2019.

Fisher, S., & Smith, R. (2000). Cheminer avec le conflit. Compétences et stratégies pour l'action, *IIED*, Londres.

Fisher, R., & Ury, W. Comment réussir une négociation, Paris, Du Seuil, 1982.

Genin, D., & Benchakroun, F. (2007). De la parole aux gestes ...: éléments de réflexion sur les dispositifs de gestion concertée des ressources forestières et pastorales au Maroc, *IRD*.

Hardin, G. (1968). The tragedy of the commons. *Science*. American Association for the Advancement of Science, *162*(3859), 1243–1248.

Kamil, H. (1992). Les pâturages de Missour (Rive gauche) , leur appréciation et leur mode de gestion par les transhumants nomades. Mémoire de fin d'étude IAV, *HassaneII, Rabat*.

Kerbrat-Orecchioni, C. (2012). Négocier la notion de négociation. *Négociations*. De Boeck Supérieur, (1), 87–99.

Mahdi, M. (1997). Le statut collectif des parcours entre le consensus de la collectivite et l'action individualisee. In *Pastoralisme et foncier: impact du régime foncier sur la gestion de l'espace pastoral et la conduite des troupeaux en régions arides et semi-arides* (pp. 31–38). CIHEAM.

Michaux-Bellaire, E. (1913). *Le Gharb*. E. Leroux.

Montagne, R. (1924). *Le régime juridique des tribus du sud marocain*. Paris, Émile Larose.

Naggar, M. (2018). Stratégie sylvopastorale et renouveau des pratiques pastorales en forêt au Maroc. *Revue Forestière française*, N°5, 487–501.

Qarro, M., & Roose, E. (2010). "Zones de parcours", in Gestion durable des eaux et des sols au Maroc : Valorisation des techniques traditionnelles méditerranéennes, *IRD*.

Rachik, H. (2000). Les usages politiques des notions de tribu et de nation au Maroc. *Identity, Culture and Politics: an afro-asian dialogue*, 1.

Rachik, H. (2003). Symboliser *la nation: essai sur l'usage des identités collectives au Maroc*. Le Fennec.

Rachik, H. (2016). Contester le droit: communautés, familles et héritage au Maroc.

Ramdan, A. (1991). Etat des ressources et utilisation des pâturages par les transhumants nomades :Cas de Missour. *mémoire de fin d'étude IAV* , Hassane II , Rabat.

Reynaud, J.- D. (1991). La régulation sociale. *International Review of Community Development/ Revue internationale d'action communautaire*. Lien social et Politiques, (25), pp. 121–126.

Schlager, E., & Ostrom, E. (1992). Property-rights regimes and natural resources: A conceptual analysis. *Land Economics*. JSTOR, pp. 249–262.

Tozy, M. (1989). Quelques cas de production indigène de droit. *Droit et Environnement Social au Maghreb*, pp. 83–103.

Tozy, M., & Mahdi, M. (1990). Aspects du droit communautaire dans l'Atlas marocain. *Droit et Société*. HeinOnline, 15: 219.

Websites: www.fao.org; www.terrescollectives.ma

Chapter 16
Reporting International Conflicts Through the Environmental Discourse: The Moroccan Sahara Conflict as a Case Study

Mohamed Mliless and Mohammed Larouz

Abstract It is usually believed that the role of international NGOs is meant to reinforce environmental justice all over the world. However, it is not the case with the Environmental Justice Organizations Liabilities and Trade's (EJOLT), whose approach seems distorting the real picture in the Moroccan Sahara. Against the expectations and principles of environmental justice, the EJOLT has targeted many achievements made in the region in the area of economic and social development. From its perspective, the exploitation of the region' resources is an environmental justice issue given the fact that the Polisario front – an armed militia that has been fighting Morocco over the territories for more than 45 years – has not been consulted about the extraction and trade of such resources. To argue for this position, the EJOLT preferred taking a pro-Polisario disinvestment stand in the region claiming that environmental justice should prevail, making a separatist movement benefit from the disputed resources. In the same vein, EJOLT uses a political discourse which considers that the exploitation of the region' resources is a 'reinforcement of the occupation by Morocco'. To demystify such a discourse, this chapter examines the EJOLT's homepage and extracts its narratives using the content analysis (CA) method. The first investigation revealed that there are six infrastructures targeted by the EJOLT. The analysis also revealed that the EJOLT uses a biased discourse –

M. Mliless (✉)
Ecolinguistics, Meknès, Morocco
e-mail: m.mliless@edu.umi.ac.ma

M. Larouz
Dean of the School of Arts and Humanities, Moulay Ismail University, Meknes, Morocco
e-mail: m.larouz@flsh.umi.ac.ma

M. Behnassi et al. (eds.), *The Climate-Conflict-Displacement Nexus from a Human Security Perspective*, https://doi.org/10.1007/978-3-030-94144-4_16

mostly distorted and defamatory – that failed to provide the real picture of the population in the region. Moreover, the EJOLT argues that the Polisario militia which lives in the Algerian territory, *NOT* the Moroccan population living in the Moroccan Sahara provinces, must benefit from the existing resources. Similarly, the rhetoric used in the narratives has failed to recognize that one million inhabitants inside the Moroccan Sahara have been directly benefiting from the many projects developed in the region. Equally important, the results show that EJOLT's discourse aims at triggering violence, insecurity, and instability in the region, thus serving potential interested agendas.

Keywords Critical environmental studies · EJOLT narratives · Environmental justice · Moroccan Sahara

1 Introduction

'Environmental justice (EJ)' is a recent form of activism documented in the literature as early as the 1980s (Burch, 1971; Lehtinen, 2009; Mohair et al., 2009). Its opposition to environmental problems has focused on the preservation of the endangered species and the denunciations of hazards linked to the intoxication and the pollution upon the poor, the rural, or the colored communities (Schlosberg, 2007). The promotion of EJ has been the concern of international NGOs to reinforce environmental equity worldwide. However, it should be noted that for some organizations, especially the Environmental Justice Organizations Liabilities and Trade's (henceforth EJOLT), the EJ has acquired another objective. The role of EJOLT is highlighted here especially when it comes to the political orientation of the NGO in opposing industrial and energy projects in the Moroccan Sahara. The discourse used by EJOLT to describe EJ issues in the region under study is characterized by this research as EJ harassments and a disinvestment plan that EJOLT pursues to implement a separatist and anarchist plan in this region. In this context, this research aims to uncover the political agenda that EJOLT monitors regarding the environmental injustice of six projects undertaken in the region. Furthermore, this research attempts to explain how EJOLT's narratives promote insecurity and nourish violence given the fact that the forms of mobilization prescribed by the EJOLT to call for EJ activism are fueled with hatred and defamation. In the virtual platform of the EJOLT (*ejatlas.org*), the research locates 18 mining, fishing, and energy locations in Moroccan among which six have been mapped in the Sahara territory. The six infrastructures, claims the EJOLT, have negative impacts on local communities living in the territory, towns, and tribes. Through its platform, the EJOLT seems to provide an invaluable service to Algeria and the Polisario by launching a viral battle to reinforce boycotts of phosphate, fishery, and renewable energy products from being extracted and exported to foreign markets.

2 Review of the Literature

Since its emergence, EJ has gradually imposed itself in many areas and has acquired different definitions. In this review, the present research attempts to trace the genesis and development of EJ, both as an area of environmental activism and scientific inquiry. It then sheds light on the importance of EJOLT in terms of usefulness and ability to oppose and denounce the negative environmental impacts on communities living near polluted and intoxicated locations. Interestingly, this section shows the extent to which the meaning and aims of EJ have changed. The multifaceted nature of the concept and the way it has been used will be discussed to comment on how the concept moved across the Atlantic and over time.

2.1 The 'Environmental Justice' Concept

In the 1980s, EJ activism developed as a result of the civic movements in the United States to protest against the disproportionate impact of pollution on minorities (Bullard & Johnson, 1997). According to Pellow (2016), the movement originated in the struggles of people of color against toxic waste dumps and waste facility sites in their communities. The event occurred in a rural black community in Warren County, North Carolina, in 1982 to protest against the decision to site a polychlorinated biphenyl (PCB) disposal landfill. The action resulted in the arrest of more than 500 people, including a Congressman from the District of Columbia (Walter E. Fauntroy) and an executive director of the United Church of Christ Commission for Racial Justice (UCC) (Benajamin F. Chavis Jr) (UCC, 1987). In 1991, the 'First National People of Color Environmental Leadership Summit' enlarged the concerns of the movement to include the risks of pollution for public health, occupational safety, transport, resource allocation, and participation community (Bullard & Johnson, 2000). At the end of the Summit, a series of EJ provisions and principles were adopted and various organizations had been developed to structure and set up networks to operate locally and internationally (Faber & McCarthy, 2001).

Given this chronology, the movement emerged as "a response to industry and government practices, policies, and conditions that many people judged to be unjust, unfair, and illegal" (Bullard, 1996: 493). It is mentioned in Beretta (2012) that EJ is now an antitoxic movement composed of a loose network of local groups opposed to the building of hazardous facilities and human exposure to toxic substances. Broadly speaking, EJ has acquired a multi-racial, multi-national, and multi-issue status to foster equal protection of human beings from environmental harms, regardless of their race, ethnicity, origin, and socioeconomic status (Lee, 1992; Pellow, 2016). To implement EJ related forms of opposition, radical ways of thinking about human-environmental relations and the responsibility of activists to ensure a healthy and safe environment for future generations have been developed (Taylor, 2000). In addition to the increasing burden in hazardous environmental conditions, recent concerns of EJ include access to water, energy, and green spaces (Beretta, 2012; Heynen, 2003; Lucas et al., 2004), natural and technological risks (Adger et al., 2003; Pelling, 2005), and the exposure of disadvantaged populations to environmental hazards and health resources (Collins, 2008; Gee & Payne-Sturges, 2004; Morello-Frosch & Jesdale, 2006; Lopez, 2002).

The broader range of issues encompassed by EJ depicts the movement as an ideological protest and a reactive voice that militates against the significant environmental effects on endangered communities (Beretta, 2012). Oftentimes, the EJ discourse describes issues related to disparities, ambient hazards, and the harmful environmental risks that disadvantaged populations are exposed to. The EJ debate has been reinforced by the rise of many fora discussing the impacts of environmental problems and risks on human beings (Callon et al., 2001), the efficiency of such concept for both public policy and action (Faure et al., 1995), and the adequate frameworks of analysis to question the EJ concept within academia (Harvey, 1996). This shift has permitted many environmental NGOs to promote EJ and to ensure that everyone, devoid of race, ethnicity, or social status is protected against the disproportionate impacts of environmental hazards (Beretta, 2012).

As a consequence, EJ networks started putting pressure on governments to mitigate environmental risks and to consider related people's concerns (Faber & McCarthy, 2001). For Foreman (1998), the difficulty to limit to the scope of EJ makes its rhetoric very effective for political use but less operational as an area of public action. In this sense, many assessment processes were provided to systematically analyze the health and social impacts and the benefits of industrial and energy plans and projects (Beretta, 2012). The aim then has been to enable inclusive stakeholder participation and, thereby, contribute to procedural justice measures (Buchan, 2003). Being an EJ impact assessment tool, the EJOLT is considered to be one of the most prominent assessment tools. Its virtual atlas has mapped thousands of hazardous sites in the world, among which 18 are located in Morocco. For this aim, the EJOLT has gathered the necessary data to describe the projects and related impacts on the population. In relation to the six projects located in the Moroccan Sahara region, the present chapter claims that the EJOLT relied on biased documents that were furnished by Algeria and the international NGOs that have been supporting a separatist ideology in the region. The rationale here is that the EJOLT's discourse has been developed based on non-reliable documents. The EJ situation in the Moroccan Sahara region is described by the EJOLT as a matter of 'independence and self-determination of the Polisario', and not as an EJ issue. This is clear in the narratives which fail to provide an adjacent and endangered population that complains from the environmental impacts and risks – such as intoxication and pollution – resulting from the assessed projects.

2.2 Description of EJOLT

Movements for environmental justice have identified two types of justice that are very close and complementary. According to Taylor (2000), the first type is called *distributive justice,* which aims to identify the beneficiaries of the equipment and public services that allow greater environmental efficiency and fair distribution of nuisances or environmental effects according to social or ethnic categories. The second type, called *corrective justice*, is related to environmental hazards inherited from the past, where the commitment of citizens or politicians must be focused on the correction of policies' effects (Taylor, 2000). The EJOLT project is grounded in the second type of EJ to bring together activist and academic organizations to promote mutual learning among stakeholders who use or develop sustainability sciences, particularly on aspects of ecological distribution (EJOLT, 2015). The concern of EJOLT is to establish a wide movement for EJ, to improve activism in this area, and to support researchers in exploring the EJ facets that resist environmental damages caused by multinational corporations. Put under the coordination of the Institute of Environmental Science and Technology at the Universitat Autònoma de Barcelona, the EJOLT collects data on resource extraction and waste disposal in many countries in an inventory of more than 3361 cases that have been reported so far (Fig. 16.1).

The documentation of these cases has been assigned to the EJOLT's Atlas (*ejatlas.org*), an online interactive map that catalogs and localizes stories of resistance against damaging projects related to mining, oil extraction, and plantation forestry among others. The database, which allows searching and filtering across

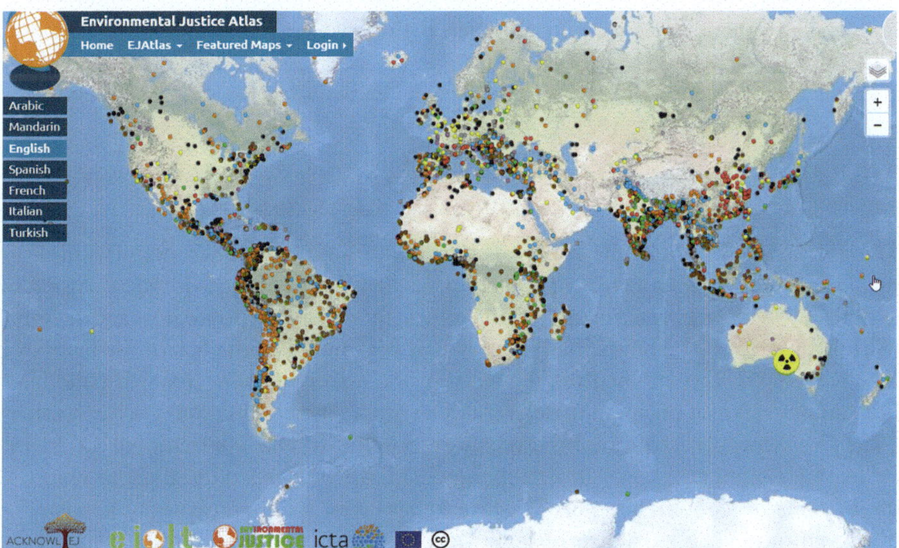

Fig. 16.1 Global identification of cases by the Atlas of EJOLT. (Source: ejatlas.org)

100 fields, has been crafted by a team of over 100 scientists and activists all over the world. The virtual site is a source of knowledge sharing among researchers, journalists, and activists to uncover the impacts of industrial activities on vulnerable populations. In addition to the virtual platform, EJOLT has produced many reports and scientific papers on nuclear energy, oil, gas and climate justice, biomass and land conflicts, mining, shipbreaking and other themes related to industrial waste.

In Morocco, the EJOLT has identified 18 environmental conflicts and EJ projects, among which six infrastructures are located in the Moroccan Sahara. What is abnormal about this is that the linguistic qualifications of the six projects are biased narratives that fall beyond the EJ principles. The argument we advance in this research is that the EJOLT's discourse is not in line with the EJ objectives for many reasons. *First*, the EJOLT's narrative does not provide a locally affected community. *Second*, the EJOLT's discourse is politically oriented since the allegations are built on inputs provided by the Polisario front and pro-separatist international NGOs like the Western Sahara Resource Watch (WSRW) located in Belgium. *Third*, the EJOLT's narratives are supporting separatism, thus nourishing insecurity and instability in the Moroccan Sahara by prescribing radical and destructive forms of protestation. Through the analysis of the content of the narratives, the present research attempts to prove that the EJOLT's discourse towards the Moroccan Sahara is against the development of the region and the wellbeing of the communities living there.

The EJOLT pretends to be leading a struggle and a fight for a just share of the revenues that flow from the region's phosphate and fishery reserves. In the issue at hand, the EJOLT's narratives about the impacts of the industrial infrastructure on populations in the Moroccan Sahara are accounted for within such a perspective. That is to say, the EJOLT's discourse falls beyond the limits of EJ as it embraces a radical and violent stand the aim of which is to serve the political agendas of both Algeria and the Polisario Front.

2.3 *The Moroccan Sahara*

The Moroccan Sahara is a territory of 266,000 km^2 that includes a population estimated at 500,000 inhabitants. The territorial conflict between Morocco, Algeria, and the Polisario Front is one of the oldest conflicts in contemporary history (Mliless, 2020). This conflict had its very first beginnings in the 60s of the last century when Morocco claimed the recovery of the territory under Spanish colonization, and the refusal of Spain to deliver the territory to the Kingdom. On September 18, 1974, Morocco requested the advisory opinion of the International Court of Justice (ICJ) about the case. On October 6, 1975, the ICJ issued its advisory opinion on the Sahara, recognizing that "Western Sahara was not a land without a master", thus confirming the existence of legal ties and allegiance between the Moroccan Sultans and the Sahrawi tribes. Today, the sovereignty over the region has been recognized by the US Presidential decision of Donald J. Trump (Trump, 2020a, b, c).

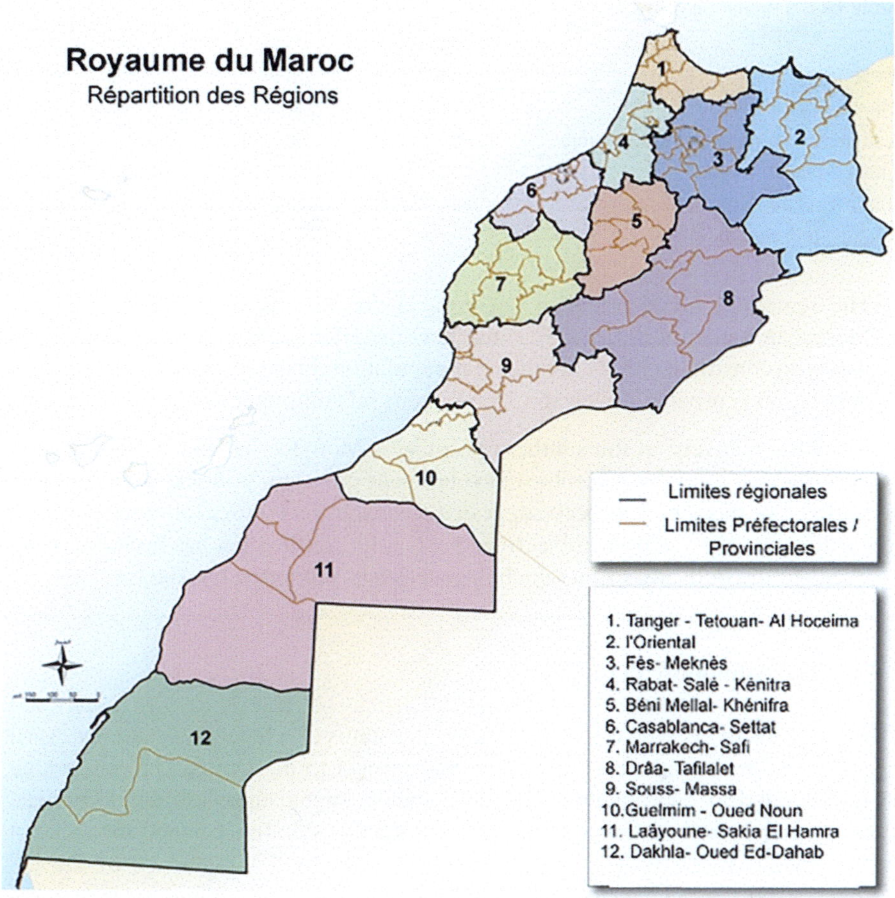

Fig. 16.2 Administrative Map including the Moroccan Sahara. (Source: pncl.gov.ma)

According to the Moroccan administrative division, three regions (10, 11, and 12 in the map below, Fig. 16.2) are what make the territory of the Moroccan Sahara: (1) The region of Oued Eddahab-Lagouira; (2) The region of Laâyoune-Boujdour-Sakia Al Hamra, and (3) The region of Guelmim-Oued Noun.

Concerning the distribution of populations in Moroccan Sahara provinces, the statistics provided by the High Commission for Planning (HCP, 2014), as illustrated in Table 16.1, show the following:

- The region of Guelmim Oued Noun is the most populated era (133.757 or 45.9%). Almost half of the region's population is urban with 280.090 (64.5%), while 153,663 (35.4) live in rural areas.
- The region of Laâyoune-Boujdour-Sakia Al Hamra is ranked second with a population that reaches 367,758 (38.9%), the majority of which are urban (343,362 or 93.3%) and 24,396 (6.6%) identified as rural.

Table 16.1 Population dispersion in the Moroccan Sahara

	Population		
Region	Total	Urban	Rural
Guelmim-Oued Noun	433,757	280,094	153,663
Laâyoune-Boujdour-Sakia Al Hamra	367.758	343.362	24.396
Oued Eddahab-Lagouira	142.955	106.277	36.678
Total	944.470	729.733	214.707

Source: HCP (2014)

- The region of Oued Eddahab-Lagouira is the least populated region in the Moroccan Sahara territories. For instance, the HCP identified 142,955 inhabitants who constitute (15.1%) of the total population. In detail, most of the region's population is urban (74.3%) and 25.6% are rural communities.

Since the recovery of the southern provinces, Morocco has made this region a key priority. To implement a robust development program, political stability must be settled. For this aim, Morocco suggested a consensual political alternative called 'the autonomy plan' with the objective to solve the conflict that has been opposing Morocco to the Polisario Front which is generously supported by Algeria.

2.3.1 The Autonomy Plan

The Plan aims to give the disputed territory the status of a federated state. On April 11, 2007, Morocco submitted the autonomy project to the United Nations, which was elaborated in coordination with civil society organizations, political parties, elected officials of the Moroccan Sahara, and the Consultative Council for Saharan Affairs (Mliless, 2020). The project was supported by 170 members of the US Congress who sent a letter to President George W. Bush asking him to support the autonomy project presented by Morocco, calling it 'promising', 'historic', and 'innovative' (Kasraoui, 2017). The letter adds that the Plan provides a realistic framework for a negotiated political solution. After the American proclamation about the Moroccan character of the Sahara, the USA considers that the Autonomy Plan is the framework for future discussions and the search for a lasting and "mutually acceptable" political solution (Mliless, 2020). Interestingly, the Plan is based on executive, legislative, and judicial prerogatives:

- Exclusive powers of the State (constitutional and religious powers of the King, currency, the defense, and external relations);
- Local administration, police and justice, social, and cultural infrastructure;
- Local taxation collected by the region, the resources allocated by the State, and revenues from natural resources in the region;
- Parliament made up of elected representatives of the Saharawi people, an elected head of government appointed by the King; and
- Courts to execute acts issued by the region institutions.

The Moroccan Sahara territory has practically changed as a result of the political dynamic and the development plan that Morocco has been implementing for decades. On November 6th, 2015, King Mohammed VI delivered a speech from Laayoune, the biggest city in the Moroccan Sahara region, on the occasion of the 40th anniversary of the Green March. The speech laid the grounds for a new development plan for the region. Both the political autonomy plan and the economic achievements will be finalized by a socio-economic start that Morocco has settled around the implementation of large-scale infrastructures.

2.3.2 The Development Plan

In November 2015, the King announced that the Moroccan Sahara would be endowed with a specific development program, called 'the Southern Provinces Development Plan, 2015–2021'. The plan reflects a holistic vision that takes into consideration the economy, society, and the environment but specifically aims at making the region an economic hub between Morocco and the African continent.

On the economic level, the restructuring of the phosphate sector entails the promotion of agricultural and fishing sectors, the development of eco-tourism, and the connection of the region with the rest of the country through road infrastructures. To implement such a plan, the budget set in 2019 was about Euros 7.5 billion dictated to fund more than 600 projects distributed as follows: The phosphate mining industry (Euros 1.6 billion); port infrastructure (Euros 670 million), especially the construction of the Port of Dakhla; Renewable energy (Euros 565 million); road infrastructure (Euros 525 million) Tiznit-Dakhla expressway (Senat Francais, 2021).

On the social level, the plan includes a university hospital center in Laâyoune, a technopole in Foum El Oued, and a center to promote the Hassani culture. Eventually, the environmental dimension involves the protection of water and fishery resources, the development of renewable energy, and the preservation of natural systems and biodiversity.

Today, the development of the Moroccan Saharan provinces is a tangible fact. The economy of the region contributes by 3% to the national GDP. In the short run, many invaluable projects are incessantly planned, namely: the project to connect the city of Dakhla to the national electricity grid; early crop projects; seawater desalination to irrigate 5000 hectares; the promotion of fishery products, and the creation of six seafood development units that might provide 4300 jobs among the local population (Senat Francais, 2021).

Determined to counter the development process of the Sahara territories, the Polisario Front has been offensively attacking Morocco on legal and judicial levels. Added to these battlegrounds, the present research attempts to show that the EJOLT discourse is inscribed within this process to jeopardize the territorial integrity of Morocco, to propagate a disinvestment plan in the region, and to create an atmosphere of violence and instability by nourishing arson and chaos.

2.4 Theoretical Framework

Building on the work of Pellow (2018), the present research is made in the context of the Critical Environmental Justice Studies (CEJ) framework which recognizes that some devalued groups in human societies – including women, immigrants, indigenous peoples, disabled persons, the elderly, children, low-income people, nonhuman species, and people of color – are subject to social inequality and oppression due to environmental risks. In this context, Pellow (2018) points out the role of the EJ movement to make states and legal systems deliver justice and regulate the industry.

The CEJ draws on numerous fields of scholarship to provide rich explanations of how environmental inequalities occur and persist (Pellow, 2018). The discipline is inspired by many theories, including Critical Race Theory and Ethnic Studies (Bell, 1993; Crenshaw, 1991; Delgado & Stefancic, 2012; Goldberg, 2002; Lipsitz, 2006; Márquez, 2014), Critical Race Feminism and Gender and Sexuality Studies (Anzaldúa, 2012; Collins 2008; Connell & Pearse, 2014; Mohanty, 2004; Wing, 2003), and Anti-Statist/Anarchist Theory (Amster et al., 2009; Scott, 2010; Smith, 2011; Torres 2007). These theories altogether have produced "rigorous conceptual and grounded understandings of how social inequality, oppression, privilege, hierarchy, and authoritarian institutions and practices shape the lives of human beings" (Pellow, 2018:36). According to Pellow (2016), these fields are invaluable to strengthening CEJ, which is an area of inquiry concerned with inequality, domination, and liberation.

Furthermore, CEJ draws on other academic fields such as Environmental Humanities (Adamson, 2001, 2011; Adamson et al., 2004; Nixon, 2011), Political Ecology (Bennett, 2009; Blaser & Escobar, 2016; Heynen et al., 2006; Robbins, 2007; Swyngedouw & Heynen, 2003), and Ecofeminism (Gaard, 1993, 2004; MacGregor, 2006; Sandilands, 2016; Sturgeon, 1997). For Pellow (2018:37), these fields of study have explored how "humans and the more-than-human world are entangled, inseparable, and bound up in collaboration, cooperation, conflict, and violence, producing socio-ecological crises as well as responses to them in practices of co-existence, sustainability, solidarity, and justice".

In addition to the above theories and disciplines, this research suggests that critical discourse analysis (Fairclough, 2003, 2010; van Dijk, 2008a, b, 2009; Wodak, 1988, 2002, 2005) could add a lot to the strengthening of CEJ to focus on the links between environmental inequalities and ecological politics. Specifically, critical discourse analysis offers the possibilities for deepening the scope of CEJ to examine how/what environmental inequalities are presented in and through the discourse of actors, victims, and the natural environment.

3 Methodology

The present research attempts to explore the narratives of the EJOLT that describe six industrial and energy projects in the Moroccan Sahara territories (Fig. 16.3). The purpose of this critical analysis is to uncover the non-neutrality of the organization,

Fig. 16.3 EJOLT's identified infrastructures in Morocco as EJ targets. (Source: ejatlas.org)

its biased political stand, and the radical forms of mobilization it supports. The process undergone by the organization may harm the region's security and stability. To analyze EJOLT's narratives, the content analysis (CA) is adopted to examine the data and to confirm the objective of the study. The latter stipulates that the EJOLT's discourse is politically biased since the pragmatic orientation of the narratives aims at seriously perturbing and affecting the Moroccan model of development in the Sahara territories. Moreover, CA, when applied to the narratives, will help prove that the discourse of insecurity and hatred adopted by the EJOLT triggers a climate of tension in the region, thus making it appealing for violence, arson, vandalism, radicalism, and terrorism.

3.1 The Content Analysis (CA)

The CA is a technique adopted by this research to provide a sound assessment of the biased narratives used by the EJOLT to implement EJ principles that might help communities and residents of the Moroccan Sahara territories to curve risks committed by industrial and energy infrastructures. Here, the CA aims to provide a comprehensive analysis of the dynamics within EJ that the EJOLT pleads for to counter and denounce environmental risks and impacts on local communities. For this purpose, six 'descriptive texts' were extracted from the EJOLT's homepage (*ejatlas.org*) in the following locations: Bir-Anzarane offshore oil exploitation;

Cape Boujdour Offshore oil and gas exploration; Phosboucraa and phosphate production; Moroccan and European fishing vessels; Tarfaya Wind-farm infrastructure; and Foum El-Oued wind power.

The CA of the EJOLT's narratives to describe the six projects will help conclude the implications of a defamatory discourse that downgrades the economic and social development achievements made in the region. As such, a qualitative assessment of the EJOLT's narratives on the EJ issue in the Moroccan Sahara would offer valuable insights on how such narratives are communicated to foster one point of view, that of the separatist community living in Algeria. Based on theme categorization (Ryan & Bernard, 2003), CA is also used to segment and classify data or texts into different categories. Furthermore, CA will permit in the present research the extraction of codes (Miles & Huberman, 2007), labels (Dey, 1993), incidents (Glaser & Strauss, 1967), thematic units (Krippendorf, 1980), data-bits (Dey, 1993), or concepts (Strauss & Corbin, 1990). Accordingly, the research draws on Ryan and Bernard (2003) to categorize the narratives into two types: The first includes the narratives named as 'description of Projects'; and the second is put in the Atlas of the EJOLT under the title of 'Groups mobilizing' and 'Forms of mobilization'. In this direction, the EJOLT's narratives will be dealt with in two categories: (1) the linguistics patterns that show the non-neutrality and biased political stands of EJOLT; and (2) the linguistics concepts that portray EJOLT radical orientation that inspires and triggers violence and instability in the region. The adoption of these two dimensions is important to show that the EJOLT's discourse serves an extremist ideology that goes against the development and the wellbeing of the local community in the region under study.

The critical analysis of the linguistic data will permit this study to answer why the EJOLT uses concepts and linguistic structures to take a politically biased stand and to construe a discourse of hatred and insecurity when reporting about EJ issues in the Moroccan Sahara context.

3.2 Research Questions

To examine the issue under investigation, this analysis adopts the technique of 'Key-words in-context', which is a deconstruction model that Ryan and Bernard (2003) employed to observe and identify keywords to systematically search across corpora and texts. Based on this technique, we analyze EJOLT' narratives to answer the following questions:

- Does EJOLT provide a locally afflicted community that suffers from environmental injustice in the Moroccan Sahara?
- Does EJOLT use neutral documentation to promote EJ in the region?
- What pushed EJOLT to implement radical forms of mobilization to stop and shut down the six projects in the region?

3.3 Hypothesis

Based on the above research questions, this research adopts the following hypotheses:

- EJOLT fails to provide a locally afflicted community that suffers from environmental injustice in the Moroccan Sahara.
- EJOLT makes use of separatist-oriented documents to denounce EJ-related practices in the region.
- EJOLT provides radical forms of mobilization to help separatists activists perturb, stop, and shut down the six projects in the region.

4 Interpretation of Results

EJOLT's narratives towards the many Moroccan infrastructures in the Sahara provinces target mineral, energy, and fishing projects. The non-neutrality of EJOLT in addressing EJ in the region affects its image and reputation since it has partially gathered data from the Polisario Front, which is among the key actors involved in the problem. This way, EJOLT becomes the mouthpiece of such an actor who claims EJ for a remote population living in the Algerian territory. In other words, the EJOLT turns, unfortunately, into a tool to execute a political agenda that Algeria and the Polisario Front have settled to counter the development of the region and to undermine the equal sharing of wealth among the local population in the Moroccan Sahara territories. Is it about environmental justice or disinvestment?

For this aim, the CA technique has revealed important results regarding EJOLT's political orientations, which can be categorized into two levels:

- The politically oriented linguistic expressions and concepts show that EJOLT's narrative towards the Moroccan Sahara is an EJ drift that the present analysis qualifies as a pro-separatist and secessionist discourse.
- The procedures and forms of mobilization embraced by EJOLT to improve EJ in the Moroccan Sahara may strongly trigger violence and insecurity in the region.

4.1 EJOLT's Narratives on Oil, Phosphate, Fishery, and Wind Energy

EJ helps change visions and practices of citizenship around the world. In Morocco, people and communities are oftentimes mobilized against social, ecological, and economic inequalities. Some of these injustices are linked to the consequences of environmental degradation and pollution, generally concentrated in regions where the disfavored populations live. It is highly important to investigate the rhetoric of the national and the international organizations on environmental problems and the adequate means they provide to restore EJ. The present research lies at the center of critical environmental studies practices that examine EJOLT's narratives to describe and qualify EJ in relation to the impact of six industrial and energy infrastructures in the Moroccan Sahara. The organization claims that the projects have severely impacted the population on many levels. However, EJOLT, which founded its allegations on separatist and anti-Moroccan references (*Western Sahara Action Forum, Western Sahara Resource Watch, Polisario Front, Algerian Environment Activists, and CODESA among others*), fails to provide a locally infected population. Instead, EJOLT claims that the Polisario Front, inhabiting the Algerian city of Tindouf, has never been '*consulted on the exploitation of the mineral and the energy resources*' and does not '*benefit from the wealth of the natural resources*'. The segmentation of such narratives unfolded important results related to the first element of analysis which indicates that the politically oriented expressions reflect a biased discourse through which EJOLT qualifies Morocco as '*an occupying country*' that '*illegally exploits*' the wealth of the region '*without any prior consent*' of local populations. The implemented projects have caused many environmental hazards the impacts of which are affecting the '*indigenous community*', '*aggravated their health risks*', and prevented them from '*benefiting from the wealth generated by the projects*'.

4.1.1 Offshore Oil and Gas Plants in Bir-Anzarane and Cape Boujdour

The projection of oil and gas offshore prospection and exploitation in the Moroccan Sahara region are described by EJOLT as projects that promote '*environmental inequalities*. Generally speaking, the strategy to improve development in the Moroccan Sahara has never been appreciated by the Algerian government and its proxy the Polisario Front. The everlasting ideology of these actors is felt in EJOLT's discourse. The choice of an autonomy plan and the definitive rejection of the referendum option by Morocco in the early 2000s have marginalized both Algeria and the Polisario Front. Some of their abrupt retaliation has been to denounce the appropriation of the Sahara by Morocco as well as the exploitation of its resources. Many battles were waged to legally pursue and track products originating from the territories, including oil and gas. In this sense, this analysis argues that the biased rhetoric of EJOLT stirs many doubts and questions the neutrality of the organization. Being aligned with Algeria and the Polisario Front, and totally sided with one version of the story, and failing to provide an affected population, this research strongly believes that EJOLT's discourse has been oriented to counter Morocco's legitimate will to fulfill its territorial integrity. In this respect, the descriptions of the six infrastructures (Fig. 16.3) revolve around the principal idea that '*Morocco illegally annexes the territories*' since 1975. Not only that, but EJOLT also voices that Morocco is violating the '*international law*' by allowing '*multinational corporations to exploit the petroleum resource*'.

The political stand is far from being an EJ claim in that EJOLT's allegations are to help the Polisario Front make a '*full referendum*' and gain '*total control over the territory*'. The EJOLT's narrative considers that the '*exploration or exploitation of mineral resources is considered to be illegal*' since the Kingdom '*has never consulted the Sahrawi people before any exploration within the territory'*. In this direction, the analysis assumes that all EJOLT's narratives fail to provide an immediately affected population and keeps alluding to an '*indigenous Sahrawi'* population that lives in Tindouf's refugee camps inside Algeria. Regrettably, EJOLT never mentions that approximately one million (944.470) citizens are located in the Moroccan Sahara (HCP, 2014).

This failure shows that EJOLT's arguments are based on the information provided by the Polisario Front when it claims that '*the polisario is not involved in the decisions being made about the exploitation of their resources*'. Specifically, EJOLT shows a strong alignment with the Polisario Front when it rejects the '*exploration, development, and exploitation of petrol and gas*' in the territory. Unfortunately, the biased and partial discourse lead by EJOLT underscores the key element of EJ which is based on a locally affected population that EJOLT fails to find. This failure to provide a menaced and affected local population is depicted in EJOLT's narratives. Often, EJOLT kept repeating that the affected communities by environmentally hazardous activities are '*the Polisario Front and the indigenous Sahrawi population which suffers health and economic inequalities because of the drilling*'. Meanwhile, such a population is actually located in the refugee camps of Tindouf inside Algeria. It would be unimaginable to consult a non-identified population that lives in another country on whether Morocco must/must not invest and develop the Moroccan Sahara region.

In addition to this, it would be illogical from the part of any country to distribute its wealth among a population that bears arms, holds a separatist ideology, and is not an active collaborator in the development of the country. The unparalleled argumentation provided by EJOLT is built on one side of the story and the fallacious allegations provided in the documentation of the Sahrawi Center for Media and Communication (SCMC) and the Western Sahara Resource Watch (WSRW) to argue that a phantom population '*had not been consulted about the operation on the land and coastal waters of the region'* and that they had 'never given their opinion" on the '*extraction, exploitation, and trade of the resources*'. In order to give much impetus to its allegations, EJOLT builds its arguments on the Peace and Security Council (PSC) of the African Union that the Algerian diplomat Smail Chergui presides and influences from 2013 to 2021 (Alaoui, 2020a, b). Actually, there is a flagrant alignment between the political inclinations of the PCS and EJOLT's biased discourse. Being a primary source for EJOLT, the PSC, which is backed by the South African president Cyril Ramaphosa (Hekking, 2020), was aiming at fostering a pro-Polisario position among the African countries to counter the development plan in the Moroccan Sahara territory (Filali, 2017; Hekking, 2020; Mliless, 2020) (Figs. 16.4 and 16.5).

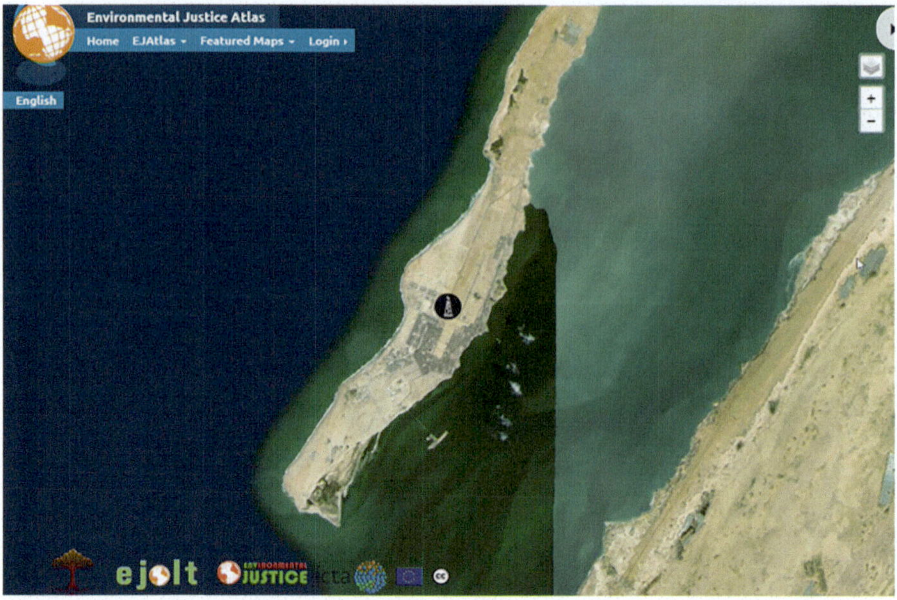

Fig. 16.4 Bir-Anzarane offshore. (Source: ejatlas.org)

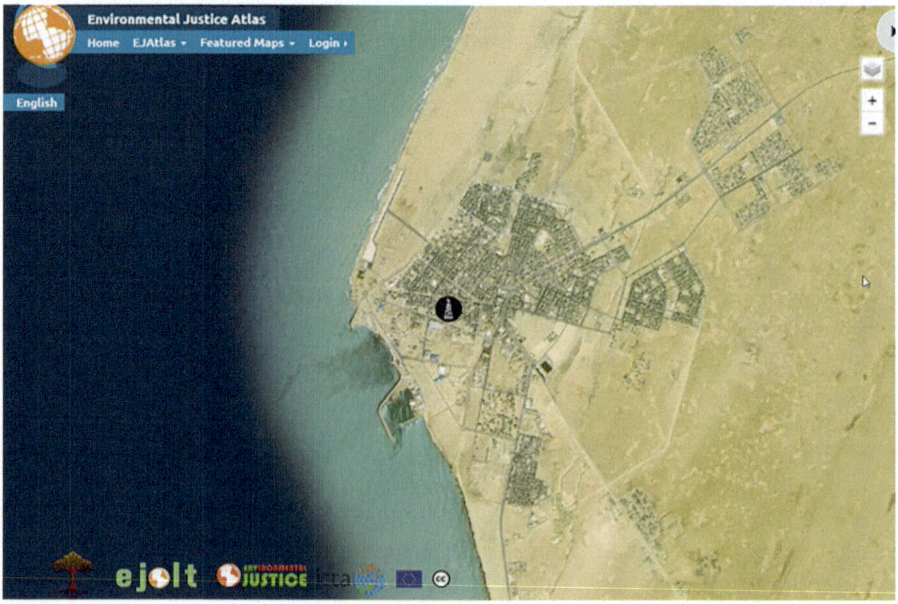

Fig. 16.5 Cape Boujdour offshore. (Source: ejatlas.org)

4.1.2 Phos-Boucraa Phosphate Plant

The example of phosphate infrastructure in the Moroccan Sahara is emblematic. The Phosboucraa Phosphate Plant (Fig. 16.6) is identified and mapped by EJOLT as a highly EJ zone. The place is described by the Atlas based on the information EJOLT collects from the documents provided by the Polisario Front and the WSRW, a pro-NGO established in Belgium.

In the narrative, EJOLT brings together the whole network of companies involved in the exploitation, processing, transport, and trade of the Phosboucraa mine. The description is politically charged in that it focuses on the idea that '*the mine should be stopped*' until a solution is found to the conflict in the region. EJOLT claims that '*the Moroccan Sahara is a former Spanish colony*' and '*a United Nations Non-Self-Governing Territories*' that contains a profitable source of phosphorous situated at Boucra, the exploitation of which continues '*without consulting the Polisario Front on exploiting the mine and selling the extracted resources to many countries around the world*'. In this sense, EJOLT does not provide an explicit account of the so-called '*legitimate population of the region*', does not mention that the refugees in the camps have never been identified, and does not denounce the sequestration of the population by the Polisario Front in Algerian camps. The UN Security Council requests each year that the census must be made in Tindouf camps. Cameron (as cited in UN, 2016) commented on the living conditions in the camps saying that Algerian authorities and the Polisario Front are unable to organize a credible census

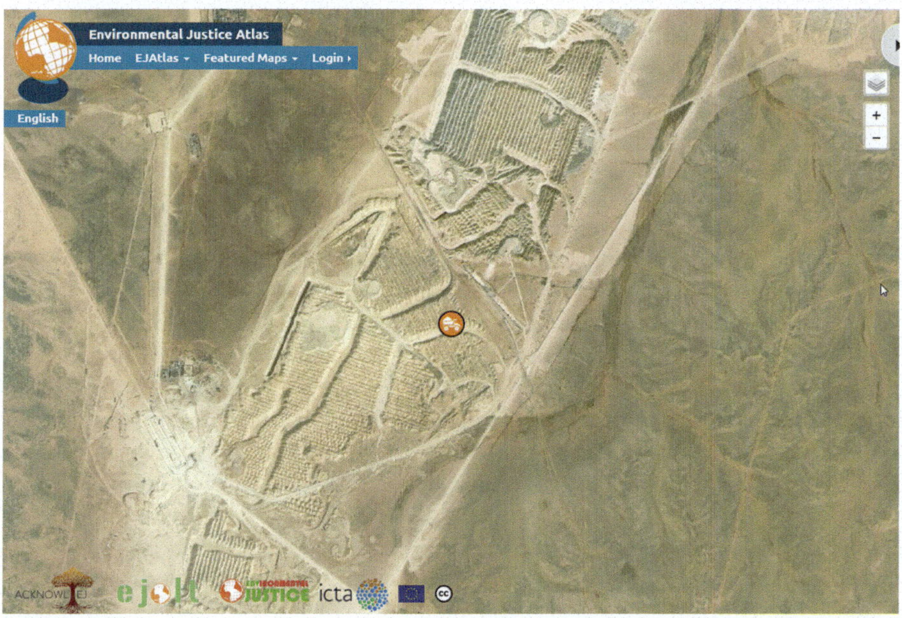

Fig. 16.6 Phosboucraa mine site. (Source: ejatlas.org)

to determine the number of inhabitants of Tindouf camps. In this regard, EJOLT does not observe that the Sahraoui refugees are the only ones in the world who have never been counted by the High Commissioner for Refugees, and the only ones to be placed in a military zone deprived of any right to move within the Algerian territory or abroad. For Dethomas (as cited in UN, 2016), the census is the major condition to avoid the diversion of humanitarian aid to the populations of Tindouf camps. Such aid was diverted and repackaged to make it disappear and went through secret warehouses to be sold in Niger, Mali, and Mauritania – an operation which could not be done without the complicity of the Polisario Front (OLAF, 2015). However, the qualification of the EJOLT in this regard is misleading and fallacious in that the narratives give the impression that the impacted population is living next to the phosphate infrastructures (Fig. 16.7), while the reality shows that the Polisario Front is entrenched inside Algeria.

Such EJ activism discredits the fraudulent propaganda lead by EJOLT as an Algerian EJ lobbying NGO whose activities were behind the blocking in the ports of Panama and Cape Town in 2017 of shipments of phosphate. Again, this analysis assumes that the narratives accompanying this issue are politically oriented in that they rely on one and only side of the story provided by the Polisario Front, WSRW, and Algeria.

4.1.3 Fishing Agreements Between Morocco and Europe

Over the past 3 years, fishing on the Moroccan Sahara coast has experienced tremendous development. It must be said that the Atlantic coast covering the region is characterized by the diversity of fish species, the composition, and abundance of which are largely conditioned by the hydro-climatic factors that prevail on the West African coast. To develop this sector, Morocco has granted to the coastal territories large public and private investments. This dynamic has been made possible due to the strategy and projects, notably the implementation of the 'Halieutis Program' that has concretely contributed to the creation of sea fishing units (artisanal fishing, coastal fishing, or deep-sea fishing) as well as an increase in the number of fishermen working in the domain. The fisheries agreements with the European Union (EU) and Russia are important in that they bring tens of millions of euros per year. The Polisario Front, strongly supported by Algeria and international NGOs (such as WSRW), opposes these agreements. In the narratives, EJOLT mentions the protestations made in 2012 upon which the Polisario Front issued a request for the conclusion of the fishery agreement between the EU and Morocco, considering that its application in the region went against international law and EU commitments.

The description of the fishery sector in EJOLT's narratives is reminiscent of a similar debate on the subject of phosphates. It has been widely demonstrated in its discourse that the exploitation should fully benefit the '*indigenous population*' added to the '*colossal investments made in this area by the phosphate exploitation*'. The rhetoric considers that coastal fishing zones in the Moroccan Sahara under Fisheries Partnership Agreement with the EU '*are not legal due to the contested*

Fig. 16.7 Distance from plant to the nearest population. (Source: ejatlas.org)

nature of Moroccan sovereignty over the territory'. The narrative keeps repeating that fishing deals signed with the EU and now Russia are done '*without the consent of the Sahrawi people*'. The narratives mention that catches made by the European fleets (Fig. 16.8) have '*decreased in the period following the Green March of 1975*'. Since then, EJOLT mentions that the '*Sahraoui population in Tindouf camps does not benefit from the money related to this economic activity that is estimated to be more than Euros 30 million annually*'.

The EJOLT's narrative, based on the version provided by the Polisario Front and WSRW, is blind to facts. The desire to make the fishery sector a strategic axis of economic and social development of the region has been materialized through the realization of many investments to improve the fishing fleet and to raise job opportunities for the local community living in Moroccan Sahara provinces. But that's not all. The local population has been at the center of concerns owing to the implementation of several projects. Due to committed actors, such as the National Fisheries Office (ONP), the sector has achieved several economic, social, and environmental objectives. The South Atlantic area (Laâyoune-Lagouira) covering the Sahara region is characterized by the diversity, composition, and abundance of fish species. The fishery sector has helped the cities of Laâyoune, Boujdour, and Dakhla materialize the following achievements[1]:

- Development of the fishing fleet (190 vessels and 5866 artisanal fishing boats);
- Establishment of 130 units (employs 30% of the total workforce at the national level);

[1] sahara-developpement.com

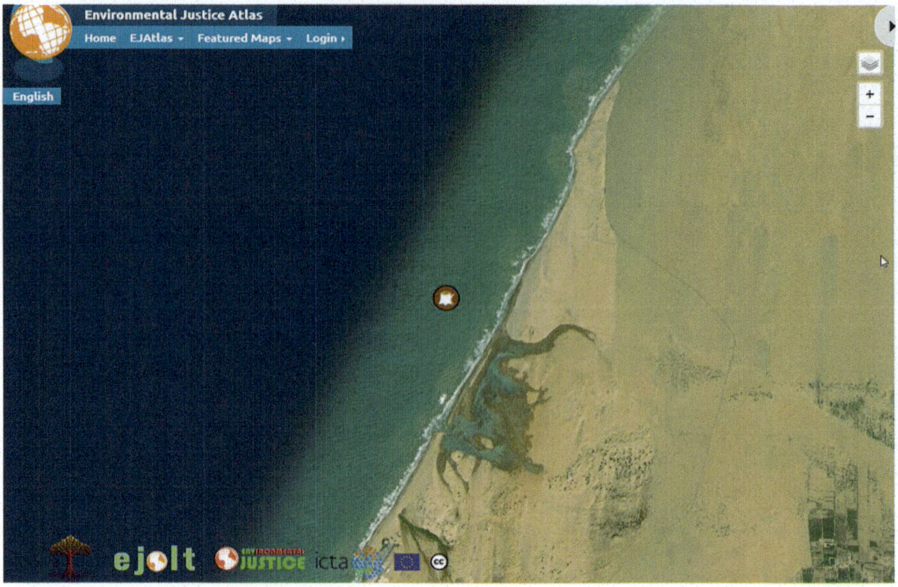

Fig. 16.8 European Fishing Vessels in the Moroccan Sahara. (Source: ejatlas.org)

Fig. 16.9 Fishing infrastructures in the Sahara provinces. (Source: Sahara Developpement, 2021)

- Construction of landing points for the supervision of artisanal fishermen;
- Construction of two training centers in Dakhla and Laâyoune;
- Improvement of fishermen's living and working conditions; and
- Provision of safety and medical material for the benefit of fishers (Fig. 16.9).

4.1.4 Wind Power Plants

The construction of wind power infrastructures in the Moroccan Sahara has raised many controversies among the Polisario Front and EJOLT, which claim that wind energy will deepen the '*occupation of the Polisario's land*'. Concerning the Tarfaya

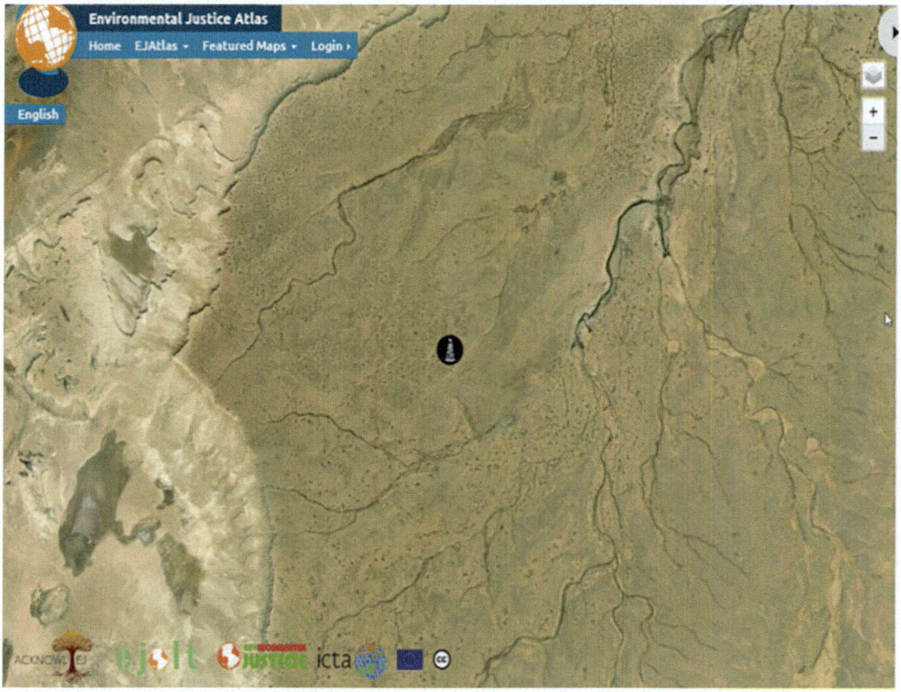

Fig. 16.10 Tarfaya wind-farm. (Source: ejatlas.org)

wind energy plant (Fig. 16.10), EJOLT adopts the same narratives it uses to describe the phosphate and the fishery infrastructures. Again, the CA of linguistic patterns reveals that EJOLT holds a political and separatist ideology that Algeria and the Polisario Front have been embracing for decades. In this sense, it is mentioned in the narratives that Tarfaya is a region that former UN secretary Ban Ki-Moon described to be '*under a situation of occupation*'. EJOLT's discourse qualifies the green energy project as a means which fosters '*colonization and injustice that cements the occupation*'.

As has been the case with the phosphate and fishery, EJOLT says that the Tarfaya wind-farm plant reinforces '*the lack of jobs among the indigenous population*', complaining that the project 'd*oes not provide enough jobs for the local community*'. The description provided by EJOLT for this infrastructure claims that '*among the plant's 60 workers, only three office staff members and 15 security guards are residents of the region*'. To show the EJOLT's biased narratives, the discourse spreads that the relationship between Morocco and the region is that of '*occupation that prevents the Saharawi people who live in refugee camps in Algeria from benefiting from the wealth of the region*'. A flagrant observation in this sense shows that EJOLT fails to map an affected local population and mistakenly considers green wind energy in Tarfaya as an EJ problem that has been negatively affecting a local community. However, the narratives remind us that the Polisario Front, which is entrenched in the Algerian refugee camp of Tindouf, has '*repeatedly demonstrated its limited access to electrical power, with resulting problems in terms of safety, food hygiene, education, and limited social activities*'.

The reality about Tarfaya wind power is that the project was launched in 2015 and stretches over a location of 100 square km. Importantly, the plant involves 131 wind turbines that grind out enough electricity capable of supplying an agglomeration the size of Marrakech city. It should be made clear that wind energy projects in the Moroccan Sahara emit no greenhouse gases and its raw material, the wind, is available all over the area. The Moroccan Sahara's coastal strip is one of the windiest areas in the region – even windier than the Netherlands or Belgium – with a wind load factor of 46%, (Nelson, 2016). According to Nelson (2016), the project has been saving 900 tons of CO2 emissions every year – and around $200 m of oil imports. Additionally, clean power contributes to 42% of the country's needs and provides more than 400,000 Dirhams (£32,000) to the local town of Tarfaya each year. The money is used to sponsor local projects and youth vacations, as well as to fund a street lighting program in Tarfaya. Interestingly, the data released by the Global Wind Energy Council (GWEC) (2020) shows that Africa and the Middle East have installed 894 MW capacity of wind power in 2019, a decrease of 7% on the previous year. According to the same report, the top three markets in Africa and the Middle East in 2019 were: (1) Egypt, with a wind capacity of 262 MW; (2) Morocco, with a wind capacity of 216 MW; (3) Jordan, with a wind capacity of 190 MW; and (4) Ethiopia, with a wind capacity of 120 MW (More details are provided in Fig. 16.11). To refute EJOLT's argument, this analysis draws on Nelson's (2016) which mentions that the project has brought new transmission lines to guarantee power supplies to the local communities. This testimony refutes the allegations made by EJOLT, Algeria, and the Polisario Front to claim that '*indigenous population*' is excluded from wind and solar energy. The reality is that the Algerian ideology, mirrored in EJOLT's narratives, never supported the creation of wind energy in the region that could provide more than a quarter of Morocco's clean electricity.

Considering other benefits of green energy, the significant potential for wind energy remains a source of income and social improvement for the local community. In the light of the recognition of the US administration, under Trump's presidency (Trump, 2020a, b, c), Morocco is reconsidering its energy plan to develop the region and to improve living conditions of the local population that is near one million inhabitants according to the last census (HCP, 2014). Now that the sovereignty of Morocco over its legitimate Sahara territories is increasingly consolidated, the country is engaged in a long battle to participate in international efforts to cut carbon emissions and improve the life quality of the local population. The present analysis assumes that EJOLT has provided fallacious and misleading arguments saying that the industrial and energy infrastructures disqualify the local community from job opportunities offered by the projects. It is mentioned in Nelson (2016), however, that many local inhabitants are integrated into the projects. In this sense,

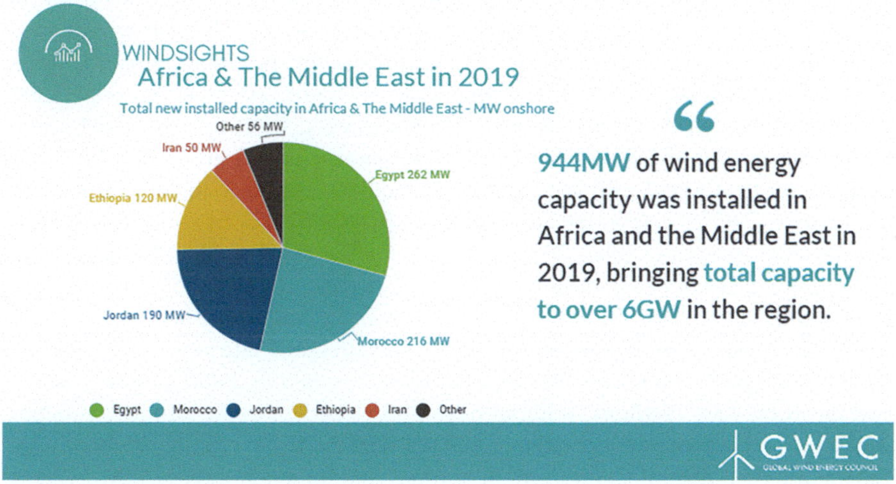

Fig. 16.11 Top markets in Africa and the Middle East. (Source: GWEC, 2020)

Nelson mentions the testimony of Hafsa Tarfaoui, a 25-year-old civil engineer at the plant: '*It is not that Sahrawis don't find jobs in the wind-farm industry. When we want to recruit a person, the priority is finding a person from this area. If you have no diploma, how can they offer you a job*?'

4.2 *EJOLT's Narratives to Promote Violence and Insecurity*

EJOLT's virtual platform is a rich international EJ database that provides invaluable information on the industrial and energy infrastructures as well as their hazardous immediate impacts on the life of local communities. It is claimed in its narratives that the assessed projects not only impact the natural environment due to chemicals and industrial waste, but such projects are also sources of intoxication and pollution of tremendous consequences on public health and the quality of air, water, and soil. In Table 16.2, the present analysis categorized the impacts of six projects in the Moroccan Sahara into three categories: (1) environmental risks; (2) health hazards; and (3) socioeconomic impacts. The results reveal 11 impacts on the environment, 12 impacts related to the socio-economic situation, and 3 impacts related to public health. The feeble number of health risks strongly supports our argument stipulating that the infrastructures are located in remote and unpopulated areas where local populations and even nomad communities are inexistent. This, in turn, supports the hypothesis of this study which assumes that EJOLT (1) fails to identify an immediate population that requires the support of EJ, and (2) fails to provide peaceful measures to establish EJ and TO eradicate health hazards emanating from the projects.

Differently, the distorting and defamatory discourse of EJOLT dwells a lot on environmental and socio-economic impacts. For the latter, EJOLT once again serves as a pro-Polisario Front mouthpiece. The narratives reiterate that the six projects weigh a lot on the socio-economic situation of a volatile and phantom population that lives in another country (Algeria). Inconsiderately, EJOLT constructed a discourse that defends the Polisario Front as a separatist armed group that has been receiving monetary and military support to claim secession from Morocco. Most dangerously, EJOLT celebrates an event that opposes the sovereignty of Morocco over the territories. In October 2010, reports EJOLT, more than 20,000 people were amassed in the Gudim Ezik protest camp near Laayoune. The peaceful dismantling of the camp resulted in the killing of more than 12 Moroccan law enforcement agents who were keeping order in the place. This dramatic fate was not mentioned in EJOLT's report which says that the operation '*resulted in 2,400 people arrested from the part of the civilians*' without paying any concern to the human casualties among the law enforcement personnel. EJOLT's narratives explicitly celebrate separatism, a global phenomenon of modern times and the main threat for Morocco's internal security 'which can cause irreparable consequences for state independence' (Berkii, 2020:109). In this framework, José María Gil Garre (as cited in UN, 2016) warns about the threat posed by the Polisario Front. The distorting and defamatory discourse of EJOLT, added to the climate of tension prevalent in the region, surely will increase a situation of violence and insecurity in the region. Faced with this, the UN does not react, said Garre, calling on the General Assembly to be effective in finding a lasting solution to this dispute. For him, the Autonomy Plan put forward by Morocco guarantees sustainability and secures development in the whole region.

The radical forms of mobilization to block and ban products originating from the Moroccan Sahara are perceived in this research as explicit instigations to violence and insecurity in the region added to the recurrent separatist maneuvers orchestrated by Algeria and the Polisario Front. The Moroccan sovereignty over the territory, as

Table 16.2 Impacts of Infrastructures

Environmental impacts	Health impacts	Socio-economic impacts
1. Biodiversity loss, 2. Global warming, 3. Soil contamination, 4. Waste overflow, 5. Oil spills, 6. Water pollution, 7. Soil erosion, 8. Deforestation, 9. Loss of vegetation cover, 10. Depletion of fish stocks, 11. Noise pollution,	1. Accidents, 2. Exposure to radiation, 3. Occupational disease and accidents,	1. Increase in corruption, 2. Loss of livelihood, 3. Militarization and increased police, 4. Violations of human rights, 5. Displacement, 6. Land dispossession, 7. Increase in violence and crime, 8. Marginalization of Sahrawis, 9. Lack of work security, 10. Labor absenteeism, 11. Firings, 12. Loss of traditional knowledge,

Source: ejatlas.org

recognized lately by the American administration, has been Algeria's prime concern because of the region's wealth in terms of natural assets. The natural resources – available in the soil, wind, and water of the territory constitute a strategic wealth that Algeria and the Polisario Front want to benefit from. Despite its desert aspect, the Moroccan Sahara ensures access to a rich maritime zone that contains a considerable amount of fishery resources. To counter Morocco's investments in the key sectors, Algeria and the Polisario Front have been lobbying and hiring international NGOs to execute what the Algerian Press Service (2021) calls '*the legal battle waged by the Polisario Front against the illegal plundering of natural resources in the territories*. To promote this plan, Algeria and the Polisario Front have been instrumentalizing EJOLT through a radical discourse that calls to violence and instability in the Moroccan Sahara.

In relation to this, the results revealed two main categories of expressions that EJOLT uses to push the pro-Polisario activists, inside the territories and around the world, to undertake radical forms of mobilization to counter introspection, exploitation, and trade of products originating from the Moroccan Sahara. The first radical and violent expressions are put under a title that EJOLT calls '*Group Mobilization*' to name collectives and associations that need to be mobilized to denounce '*the plundering of the Sahrawi people's natural resources*'. Similarly, the second category of expressions called '*Forms of Mobilization*' contains a series of strategies and legal forms of contestation that the Polisario Front and its proxies should undertake '*to stop the exploitation of the region's natural resources*'. For the first category, EJOLT suggests that the following groups should be implicated in this issue, namely the Collective of Saharawi Human Rights Defenders (CODESA), WSRW, indigenous groups or traditional communities, international ejos, local government/political parties, local and international human rights groups, women, journalists, and fisher people. Regarding the second category, EJOLT recommends a series of destructive expressions that dictate the forms of mobilization that the above groups must embrace to lead deliberate, damaging, and destroying EJ actions against industrial properties in the Moroccan Sahara. For this aim, EJOLT recommends the following *modi operandi* in order to stop '*unequal exploitation of the natural resources in the region*':

- Arson
- Sabotage
- Property damage
- Self-immolation
- Hunger strikes
- Occupation of buildings/public spaces
- Land occupation
- Public campaigns
- Public strikes/campaigns/ marches
- Development of a network/collective action
- Media-based activism/alternative media
- Official complaint letters and petitions

- Shareholder/financial activism
- Boycotts of companies-products
- Lawsuits/ Court cases/ judicial activism

From the above violent linguistic patterns (*sabotage, arson, emulation, and property damage among others*), this research asserts that EJOLT's mission promotes violence and insecurity. The alignment between EJ activism and violence is flagrant in EJOLT's discourse. The NGO's platform has become a propaganda tool of war in the region. The contextualization of EJOLT's narratives falls within the framework of agenda-setting aiming to implement the ideology settled by Algeria and the Polisario Front with the objective to disseminate chaos in Morocco. The discourse of harassment implemented by EJOLT perpetuates a constant and sophisticated concern of Algeria that counters Morocco's territorial integrity and undermines its development achievements. Added to the declared propaganda, this analysis identified a new angle of attack launched by pro-environmental justice organizations namely EJOLT. It is crystal clear that EJOLT directs its verbal canons beyond the borders traced by environmental justice principles to execute separatist plans launched four decades ago by Algeria and the Polisario Front. The forms of mobilization enshrined in its rhetoric are meant to nourish separatism in the Moroccan Sahara, to counter the development of the region, and consequently to erase the legitimacy of the Kingdom over the territories recently emphasized by the American recognition. Today, the failures of Algeria on the diplomatic level and the burden that the Polisario Front weighs on its economy are pushing the country to destroy Morocco's ambition to n make the Sahara region a hub for the African economy.

Based on the above, this analysis proposes that EJOLT should reconsider its discourse, at least towards Moroccan Sahara projects. EJOLT must take into account that the fallacious allegations it espouses no longer stand in a region that is witnessing substantial changes. Peace and stability should prevail in the Moroccan Sahara for the benefit of all because there is a vital interest in this territorial continuity being now 'a reality taking into account that 34 of the 54 countries of the African Union have decided to establish an African Continental Free Trade Area" (Mliless, 2020:246). Finally, this research believes that the more neutral and moderate the EJ advocated by EJOLT is, the more successful and sustainable the current Moroccan development can be.

5 Conclusion

Since its emergence, EJ activism has been embraced by international NGOs the mission of which is to identify and fight environmental inequalities in different parts of the world. It should be mentioned that EJ has entered the vocabulary of environmental activism that was highlighted and developed by international environmental movement. The rhetoric of EJ has become omnipresent in the justification of hazards and risks procured by industrial and energy projects on the indigenous population. In the area of analysis, the narrative produced by EJOLT to describe EJ in the Moroccan Sahara was not totally inscribed within the framework of EJ. The discourse used by EJOLT is qualified by the present research as being politically oriented and incites violence and insecurity in the Moroccan Sahara provinces. As evidenced in the findings of the study, the nature of the language used by EJOLT mixes between EJ activism and politics. It has been demonstrated that the kind of EJ embraced by EJOLT serves an ideological agenda settled by the Polisario Front and Algeria to justify radical and extremist actions carried out by pro-separatist activists. EJOLT's discourse on energy and mineral resources in the area under study shows unprecedented convergences between EJ, separatism, and radical forms of mobilization. EJOLT recommends that EJ activists, locally and internationally, must be mobilized to implement its actions carrying out the objectives of EJ.

Having that said, the drift in the notion of EJ supported by EJOLT towards the projects in the Moroccan Sahara has weakened the credibility of EJOLT to fight against environmental inequalities in the world. The same drift was observed in EJOLT's narratives when it failed to provide an immediate locally impacted population, a principle that is supposed to be linked to the goal of EJ. Interestingly, the deficiency is strongly related to EJOLT's inability to build on neutral and diversified resources. Instead, its narratives sound like a Polisario Front discourse that can hardly bring operational solutions to the issue of environmental quality and equity. The analysis finds that the instrumentalization of EJ, in the example of the Moroccan Sahara, brings to the front the credibility and the ability of EJ organizations to handle complex issues. Eventually, EJOLT's narratives assessed in this research unfold contradictory results due to the poor monitoring and management of the goals that EJOLT sets for the implementation of EJ in a region highly threatened by security hazards rather than environmental impacts.

References

Adamson, J. (2001). *American Indian literature environmental justice and ecocriticism the middle place*. University of Arizona Press.

Adamson, J. (2011). Medicine food critical environmental justice studies native North American literature and the movement for food Sovereignty. *Environmental Justice, 4*(4), 213–219.

Adamson, J., Mei, M. E., & Stein, R. (2004). *The environmental justice reader politics poetics and pedagogy*. University of Arizona Press.

Adger, N., Huq, S., Brown, K., Conway, D., & Hulme, M. (2003). Adaptation to climate change in the developing world. *Progress in Development Studies, 3*(3), 179–195.

Alaoui, M. C. (2020a, November 11). *Union africaine nouveau camouflet pour l'algérien Smaïl Chergui* [African union new camouflage of the Algerian Smaïl Chergui]. Retrieved from https://fr.le360.ma/politique/union-africaine-nouveau-camouflet-pour-lalgerien-smail-chergui-228106

Alaoui, M. C. (2020b, November 2). *Union africaine Smail Chergui donne un sens algérien au mot «traduction»* [African Union Smail Chergui gives an algerian sense to the word Translation]. Retrieved from https://fr.le360.ma/politique/union-africaine-smail-chergui-donne-un-sens-algerien-au-mot-traduction-228293

Algerie Press Service [APS]. (2021, February 13). *DW le pillage des ressources halieutiques sahraouies par le Maroc* [DW the plunder of Sahrawi fishery resources by Morocco]. Retrieved from https://www.aps.dz/monde/117525-la-chaine-allemande-dw-evoque-le-pillage-des-ressources-halieutiques-sahraouies-par-le-maroc

Amster, R., Abraham, D., Fernandez, L. A., Nocella, A. J., & Shannon, D. (2009). *Contemporary anarchist studies*. Routledge.

Anzaldúa, G. (2012). *Borderlands/La Frontera*. Aunt Lute Books.

Bell, D. (1993). *Faces at the bottom of the well the permanence of racism*. Basic Books.

Bennett, J. (2009). *Vibrant matter a political ecology of things*. Duke University Press.

Beretta, I. (2012). Some Highlights on the Concept of Environmental Justice and its Use. *e-cadernos ces, 17*, 136–162.

Berkii, T. (2020). Separatism the Main threat for internal security. *Internal Security, 11*(2), 109–117.

Blaser, M., & Escobar, A. (2016). Political ecology. In J. Adamson, W. Gleason, & D. N. Pellow (Eds.), *Keywords for environmental studies* (pp. 164–167). New York University Press.

Buchan, D. (2003). Buy-in and social capital: By-products of social impact assessment. *Impact Assessment and Project Appraisal, 21*(3), 168–172.

Bullard, R. D. (1996). Environmental justice it's more than waste facility siting. *Social Science Quarterly, 77*(3), 493–499.

Bullard, R., & Johnson, G. (1997). *Just transportation dismantling race and class barriers to mobility*. New Society Publishers.

Bullard, R., & Johnson, G. (2000). Environmental justice grassroots activism and its impacts on public policy decision making. *Journal of Social Issues, 56*(3), 555–578.

Burch, W. (1971). *The eregrine Falcon and the urban poor some sociological interrelations*. Ohio University Press.

Callon, M., Lascoumes, P., & Barthe, Y. (2001). *Agir dans un monde incertain essai sur la démocratie technique* [Acting in an uncertain world essay on technical democracy]. Seuil.

Collins, P. H. (2008). *Black feminist thought knowledge consciousness and the politics of empowerment*. Routledge.

Connell, R., & Pearse, R. (2014). *Gender in world perspective*. Polity Press.

Crenshaw, K. (1991). Mapping the margins intersectionality identity politics and violence against women of color. *Stanford Law Review, 43*(6), 1241–1299.

Delgado, R., & Stefancic, J. (2012). *Critical race theory an introduction*. New York University Press.

Dey, I. (1993). *Qualitative data analysis a user friendly guide for social science*. Routledge.

Environmental Justice Organizations, Liabilities and Trade [EJOLT]. (2015). *Final report summary – EJOLT* (Environmental Justice Organizations, Liabilities and Trade). Retrieved from https://cordis.europa.eu/project/id/266642/reporting/de

European Union Anti-Fraud Office [OLAF]. (2015). *Rapport final d'enquete* [Final investigation report]. Retrieved from: FR_OLAF_30 01 15.pdf

Faber, D., & Mccarthy, D. (2001). The evolving structure of the environmental justice movement in the United States new models of democratic decision-making. *Social Justice Research, 14*(4), 405–421.

Fairclough, N. (2003). *Analyzing discourse textual analysis for social research*. Routledge.

Fairclough, N. (2010). *Critical discourse analysis the critical study of language*. Routledge.

Faure, A., Pollet, G., & Warin, P. (1995). *La construction du sens dans les politiques publiques débats autour de la notion de référentiel* [The construction of meaning in public policies debates around the notion of repository]. Harmattan.

Filali, K. (2017, March 23). *L'algérien Smail Chergui mène la charge contre le Maroc à l'union africaine* [Algerian Smail Chergui leads the charge against Morocco at the African Union]. Retrieved from https://ledesk.ma/2017/03/23/lalgerien-smail-chergui-mene-la-charge-contre-le-maroc-lunion-africaine/

Foreman, C. (1998). *The promise and peril of environmental justice*. Brookings Institution Press.

Gaard, G. (1993). *Ecofeminism women animals and nature*. Temple University Press.

Gaard, G. (2004). Toward a queer ecofeminism. In R. Stein (Ed.), *New perspectives on environmental justice gender sexuality and activism* (pp. 21–44). Rutgers University Press.
Gee, G., & Payne-Sturges, D. (2004). Environmental health disparities a framework integrating psychosocial and environmental concepts. *Environmental Health Perspectives, 112*, 1645–1653.
Glaser, B. G., & Strauss, A. (1967). *The discovery of grounded theory strategies for qualitative research*. Aldine.
Global Wind Energy Council [GWEC]. (2020, February 12). *Africa and Middle East add 894MW of wind energy capacity in 2019, market expected to grow by over 10GW by 2024*. Retrieved from https://gwec.net/africa-and-middle-east-add-894mw-of-wind-energy-capacity-in-2019-market-expected-to-grow-by-over-10gw-by-2024/
Goldberg, D. T. (2002). *The racial state*. Blackwell Publishers.
Haut-Commissariat au Plan [HCP]. (2014). *Aspects Démographiques et socio-économiques de la population à travers le RGPH 2014* [Demographic and socio-economic aspects of the population through the RGPH 2014]. Retrieved from https://www.hcp.ma
Harvey, D. (1996). *Justice nature and the geography of difference*. Blackwell.
Hekking, M. (2020, December 7). *South African president defends AU consensus on Western Sahara*. Retrieved from https://www.moroccoworldnews.com/2020/12/328117/south-african-president-defends-au-consensus-on-western-sahara/
Heynen, N. (2003). The scalar production of injustice within the urban forest. *Antipode, 35*(5), 980–998.
Heynen, N., Kaika, M., & Swyngedouw, E. (2006). *In the nature of cities urban political ecology and the politics of urban metabolism*. Routledge.
Kasraoui, S. (2017, November 9). *Timeline Western Sahara dispute from 1859 to 2018*. Retrieved from moroccoworldnews.com/2017/11/233515/morocco-western-sahara-polisario/
Krippendorff, K. (1980). Validity in content analysis. In E. Mochmann (Ed.), *Computerstrategien fur die kommunikationsanalyse* (pp. 69–112). Campus.
Lee, C. (1992). *Proceedings of the first national people of color environmental leadership summit*. United Church of Christ Commission for Racial Justice.
Lehtinen, A. A. (2009). Environmental justice. In R. Kitchin & N. Thrift (Eds.), *International encyclopedia of human geography* (pp. 535–539). Elsevier.
Lipsitz, G. (2006). *The possessive investment in whiteness how white people profit from identity politics*. Temple University Press.
Lopez, R. (2002). Segregation and black/white differences in exposure to air toxic in 1990. *Environmental Health Perspectives, 110*, 289–295.
Lucas, K., Walker, G., Eames, M., Fay, H., & Poustie, M. (2004). *Environment and social justice: Rapid research and evidence review sustainable development research network*. Policy Studies Institute.
MacGregor, S. (2006). *Beyond mothering earth ecological citizenship and the politics of care*. UBC Press.
Márquez, J. (2014). *Black-brown solidarity racial politics in the New Gulf South*. University of Texas Press.
Miles, M. B., & Huberman, A. M. (2007). *Analyse des données qualitatives* (Qualitative data analysis). De Boeck.
Mliless, M. (2020). Agenda-Setting a cross analysis of the Algerian Press Service double discourse towards Morocco Libya and Mali. *Journal of the Geopolitics and Geostrategic Intelligence, 3*(2), 221–253.
Mohair, P., Pellow, D. N., & Roberts, J. T. (2009). Environmental justice. *The Annual Review of Environment and Resources, 34*, 405–430.
Mohanty, C. T. (2004). *Feminism without borders decolonizing theory practicing solidarity*. Duke University Press.
Morello-Frosch, R., & Jesdale, B. (2006). Separate and unequal residential segregation and estimated cancer risk associated with ambient and toxics in U.S. metropolitan areas. *Environmental Health Perspectives, 114*, 386–393.

Nelson, A. (2016, November 22). *Africa's biggest wind-farm sparks controversy in the desert*. Retrieved from https://www.theguardian.com/environment/2016/nov/22/morocco-saharan-windfarms-tarfaya-fan-accusations-of-green-plunder

Nixon, R. (2011). *Slow violence and the environmentalism of the poor*. Harvard University Press.

Pelling, M. (2005). *The vulnerability of cities natural disasters and social resilience*. Earthscan.

Pellow, D. N. (2016). Toward a critical environmental justice studies black lives matter as an environmental justice challenge. *Du Bois Review, 13*(2), 221–236.

Pellow, D. N. (2018). *What is critical environmental justice?* Polity Press.

Robbins, P. (2007). *Lawn people how grasses weeds and chemicals make us who we are*. Temple University Press.

Ryan, G. W., & Bernard, H. R. (2003). Techniques to identify themes. *Field Methods, 15*(1), 85–109. https://doi.org/10.1177/1525822x02239569

Sahara Developpement. (2021, March 5). *Pêche* [Fishery]. Retrieved from http://www.sahara-developpement.com/

Sandilands, C. (2016). Queer ecology. In J. Adamson, A. William, A. Gleason, & D. N. Pellow (Eds.), *Keywords for environmental studies* (pp. 169–171). New York University Press.

Schlosberg, D. (2007). Reconceiving environmental justice global movements and political theories. *Environmental Politics, 13*(3), 517–540.

Scott, J. (2010). *The art of not being governed: An anarchist history of upland Southeast Asia*. Yale University Press.

Senat Francais. (2021, February 11). *Maroc 1999–2019 vingt ans de modernization* [Morocco 1999–2019 twenty years of modernization]. Retrieved from http://www.senat.fr/ga/ga155/ga15512.html

Smith, M. (2011). *Against ecological sovereignty ethics biopolitics and saving the natural world*. University of Minnesota Press.

Strauss, A., & Corbin, J. M. (1990). *Basics of qualitative research grounded theory procedures and techniques*. Sage Publications.

Sturgeon, N. (1997). The nature of race discourses of racial difference in ecofeminism. In K. J. Warren (Ed.), *Ecofeminism women culture nature* (pp. 1–36). Indiana University Press.

Swyngedouw, E., & Heynen, N. (2003). Urban political ecology justice and the politics of scale. *Antipode, 35*(5), 898–918.

Taylor, D. (2000). The rise of environmental justice paradigm. *American Behavioral Scientist, 43*(4), 508–580.

Torres, B. (2007). *Making a killing the political economy of animal rights*. AK Press.

Trump, D. J. [@realDonaldTrump]. (2020a, December 10). *Proclamation recognizing Moroccan sovereignty over the Western Sahara* [Tweet]. Twitter. https://twitter.com/realDonaldTrump/status/1337067019385057290

Trump, D. J. [@realDonaldTrump]. (2020b, December 10). *HISTORIC breakthrough* [Tweet]. Twitter. https://twitter.com/realDonaldTrump/status/1337067073051238400

Trump, D. J. [@realDonaldTrump]. (2020c, December 10). *Morocco recognized the United States of America* [Tweet]. Twitter. https://twitter.com/realDonaldTrump/status/1337067127455539201

United Church of Christ Commission for Racial Justice [UCC]. (1987). *Toxic wastes and race in the United States a national report on the racial and socio-economic characteristics of communities with hazardous waste sites*. United Church of Christ.

United Nations [UN]. (2016, October 5). *Le détournement de l'aide humanitaire destinée aux camps de Tindouf et la légitimité du Polisario pour représenter les Sahraouis au cœur des débats de la Quatrième Commission* [The diversion of humanitarian aid intended for the Tindouf camps and the legitimacy of the Polisario to represent the Sahrawis at the heart of the debates of the Fourth Commission]. Retrieved from https://www.un.org/press/fr/2016/cpsd609.doc.htm

Van Dijk, T. A. (2008a). *Discourse and context a scociocognitive approach*. Cambridge University Press.

Van Dijk, T. A. (2008b). *Discourse and power*. Macmillan.

Van Dijk, T. A. (2009). *Society and discourse how social contexts influence text and talk*. Cambridge University Press.
Wing, A. K. (2003). *Critical race feminism a reader*. New York University Press.
Wodak, R. (1988). *Language power and ideology studies in political discourse*. John Benjamins.
Wodak, R. (2002). *Methods of critical discourse analysis*. Sage Publications.
Wodak, R. (2005). *A new agenda in critical discourse analysis theory methodology and interdisciplinarity*. John Benjamins.

Postface

Climate change is already evident from the rise in the frequency and intensity of extreme weather events around the world. The continued emissions of greenhouse gases are exacerbating climate variability and change which have been bound to adversely affect multiple sectors, especially water, agriculture, and biodiversity, further restricting developing countries' capacities to meet essential sustainable development goals. Such impacts are gradually affecting socio-economic and political dynamics leading to conflicts and displacement. The IPCC's 5th Assessment Report states that climate change will impact forms of migration, compromising human security.

It is in this context that this volume was contemplated to cover issues related to climate-conflict-displacement nexus in areas where such effects are already seen. The chapters received for this volume brought to the fore issues that we were either unaware of or did not know the exact relevance of the implications it could have. As also mentioned in the Preface, this volume is a continuation of the recently published '*Social-Ecological Systems: From Risks and Vulnerability to Viability and Resilience*'. While working on that volume, it was realized that some chapters linked social-ecological imbalances to human migration and conflict. Until now, there are few pieces of evidence of direct linkages of social issues with environmental disturbances. However, the debate is evolving, and the nature-human nexus is being increasingly explored both theoretically and practically. This volume takes a step in that direction, adding to the existing literature on climate-related migration and displacement and showing pathways and options to deal with it.

There are various takeaways from this volume. *First*, it is clear that social-ecological systems are increasingly threatened by many risk dynamics that result in resource conflicts, social inequity, and human displacements and, to some extent, political instability. *Second*, there are many determinants of the harmony of socio-ecological systems and whether it could lead to a conflict. These include political

M. Behnassi et al. (eds.), *The Climate-Conflict-Displacement Nexus from a Human Security Perspective*, https://doi.org/10.1007/978-3-030-94144-4

setup, people's mindset and needs, and developmental phase. On the contrary, politically driven conflict could also affect the environment, further undermining the capacities of the local people to deal with climatic changes. *Third*, crucial social-ecological risks and their implications show that there are inherent linkages between ecosystem changes and social dynamics. Therefore, it is meaningful to engage local communities in the response process by taking into account their specific vulnerabilities and expectations. *Fourth*, there is still a meagre recognition of climate change as a driver of human displacement, and thus, recognition of environmental refugees. This has led to a gap in terms of preparedness and policy action to curb or prevent the damage that could occur as a result of large-scale instability. Meeting sustainable development goals will become a challenge for affected countries, besides further increase in climate vulnerability. *Fifth*, risks to the socio-ecological systems need to be further explored, especially in politically disturbed, climatically vulnerable, and resource deficient regions of the world as they may become the source of unmanageable human displacements in the future, further challenging human security and the stability of countries. There is a need to bring focus of decision-making processes at all scales on the links between climate risks, environmental stress, resource scarcity, and conflicts.

We hope that this volume will play an important role in initiating a timely discussion and recognition of such emerging dynamics. Case studies in this volume already provide relevant insights into the real challenges and opportunities to tackle them at the right time. In light of changing political, economic, and social realities, such challenges may present us with various dilemmas that need to be better understood – such as the current Covid-19 crisis – and tackled for a sustainable and secure future.

Biographical Notes of Authors

Alves, Fátima
Dr. Alves has a Ph.D in Sociology and is an Assistant Professor at Universidade Aberta, Portugal and Coordinator of the Master in Intercultural Relations. She is also the Vice-Coordinator of the Ph.D in Intercultural Relations of the Department of Social Sciences and Management at Universidade Aberta. Dr. Alves is also a Senior Researcher at the Centre for Functional Ecology of the University of Coimbra and coordinator of the Research Group on Societies and Environmental Sustainability.

Andrasko, Becca
Andrasko is a recent graduate of MSc Conflict Studies from the London School of Economics, United Kingdom, where much of the original research included in her chapter was conducted. She has a previous master's degree in MSc Ecological Economics from the University of Edinburgh, United Kingdom. The author is currently working for the Foreign Policy Magazine. Her interests lie in exploring environmental conflict and its consequences, including migration.

Bourhim, Rabiaa
Bourhim is a Ph.D Researcher in Public Law and Political Sciences at the College of Law, Economics, and Social Sciences of Agadir, Ibn Zohr University, Morocco. In 2014, he obtained a Master's degree in Public Law and Administrative Sciences from the College of Law, Economics, and Social Sciences, Hassan II University of Mohammedia, Morocco. In 2012, he graduated from the National School of Administration in Rabat. He is currently an active member of the International Center of strategic research on territorial governance in oases and mountainous areas in Ouarzazate, Morocco. In addition, Bourhim is a consultant to the parallel

M. Behnassi et al. (eds.), *The Climate-Conflict-Displacement Nexus from a Human Security Perspective*, https://doi.org/10.1007/978-3-030-94144-4

government of youth in Morocco. He is an expert in the management and audit of government procurement. His main areas of research include, among others, environmental change, conflict management, regulation, sustainability, territorial governance, and public policy. Bourhim has participated in some national and international conferences and published research papers covering the above areas.

Chougrani, El Houcine
Dr. Chougrani is an Associate Professor of International Law and International Relations at the College of Law, Economics, and Social Sciences, Cadi Ayyad University of Marrakech. He obtained a Doctorate in 2003 with a thesis entitled "The European Investment Bank and the protection of Environment in the Mediterranean". He is the Director of the Moroccan Observatory for Future Generations (MOROFUGE), a Member of the International Research Center for Crisis Management, and the General Director of the Arab Journal of International Law. His main research areas cover international environmental law, water, climate justice, and the rights of future generations. His most recent publications in Arabic are: "The Rights of Future Generations in the Light of Current Developments in the Arab World" (Arab Center for Research and Policy Studies, 2018); and "Contradictions in International Law - A Preliminary Analysis" (Centre for Arab Unity Studies, 2019). Dr. Chougrani has also an extensive professional experience as an Accountant General at the Ministry of Economy and Finance (2002–2014).

Csaplovics, Elmar
Dr. Csaplovics is a professor of remote sensing at the Institute of Photogrammetry and Remote Sensing, Department of Geosciences, TU Dresden. He holds a Doctorate and a Habilitation in remote sensing from TU Vienna (1982, 1992), was a post-doc research fellow at the National Institute of Agronomic Research (INRA), Montpellier, and at the Department of Geology, Geophysics, and Geoinformatics, Free University Berlin from 1988–1992, and was appointed Professor of remote sensing at the TU Dresden in 1993. Dr. Csaplovics was a Visiting Professor at the University College London (UCL) in 2007 and focuses in his research on remote sensing and applied geoinformation analysis for Spatio-temporal monitoring and assessment of land use/land cover with emphasis on wetlands and semi-arid lands (desertification and degradation), as well as on landscape history and world conservation monitoring in transnational project cooperation.

Deafalla, Taisser H. H.
Dr. Deafalla is a Crisis Manager and Expert on climate and environmental change adaptation at many national and international NGOs. Currently, she is a Senior Researcher at the Institute of Photogrammetry and Remote Sensing, Faculty of Environmental Sciences, University of Dresden, Germany. Dr. Deafalla has professional experience in fields of the forestry, environmental change, and rural development where she worked with many local and international organizations. Moreover, she has participated in many conferences and international workshops and has many scientific publications related to environmental change.

Deifalla, Mohamad H. H.
Deifalla is a Lecturer at the College of Engineering and Technology of Industries, Sudan University of Science and Technology, Sudan, where he obtained his Chemical Engineering B.Sc. degree (first class honors) in 2014. Moreover, he has got an M.Sc. degree from the University of Science and Technology, Sudan in 2020. He has professional experience in the fields of the modeling, simulation, and optimization of industrial processes where he worked as an instructor with many local and international organizations, in addition to many scientific research centers. Furthermore, he has participated in many conferences and international workshops.

El-Abbas, Mustafa M.
Dr. El-Abbas is an assistance professor at the Department of Forest Management, Faculty of Forestry, University of Khartoum, Sudan, where he obtained his B.Sc. (Hons) and M.Sc. degrees (2001, 2006). Moreover, he has got a diploma in forest ecology and forest resource management from the University of Helsinki, Finland in 2006. Dr. El-Abbas awards his Ph.D in remote sensing and natural resource management from TU Dresden. He has professional experience in the fields of remote sensing and GIS in general, and object-based approaches in particular with an emphasis on natural resources management. Dr. El-Abbas participated in several international workshops and scientific forums. Additionally, he has many scientific publications.

Ferreira Fernandes, Carla Sofia
Ferreira Fernandes, a Ph.D researcher in Social Sustainability, is interested in the nexus between environmental changes and migration in Morocco. She is affiliated to the Doctoral Programme of the Universidade Aberta on Sustainability and Social Development, Portugal. She is a member of the Research Group on Societies and Environmental Sustainability of the Centre for Functional Ecology of the University of Coimbra, Portugal.

Gill, Gitanjali Nain
Dr. Gill is a tenured Full Professor of environmental law at the Faculty of Law, Northumbria University, Newcastle UK. Her research interests and published work in leading peer-reviewed journals form a coherent body reflecting thematic issues such as access to environmental justice, climate change, biodiversity, SDGs, sustainability, and governance in Asia with a focus on India. Dr. Gill is a recipient of two prestigious British Academy/Leverhulme Research Grants. The first, 2013–16, examined the casework and environmental jurisprudence of the National Green Tribunal of India and the findings were published in her book *Environmental Justice in India: The National Green Tribunal* (Routledge, Earthscan). The second, 2020–23, is titled "Land Acquisition, Rehabilitation and Resettlement of Vulnerable Poor Communities: Criticalities and Scrutiny of Law in Gujarat, India". In 2019, she was granted a GCRF Networking and Collaboration Fund award to undertake original empirical research in India on scientific uncertainty and the precautionary

principle. She acts as an external referee for numerous leading publishers and journals. She has held Visiting Professorships at several universities including the Vermont Law School, USA; Department of Comparative Law, University of Insubria, Como, Italy; Law School, Shanghai University of Finance and Economics, China; Charles University Prague; Symbiosis Law School, Noida India; and Cardiff Law School, Cardiff University UK.

Gopichandran, Ramachandran

Dr. Gopichandran is a Professor at the NTPC School of Business in India. He teaches various aspects of mitigation, adaptation, preventive environmental management, and business communication. These are important elements of public policy, pertaining to which, Gopi has an overall work profile that spans thirty-two years. He has recently contributed to the review of the first and second order drafts of the 6th assessment report of the IPCC; with a special emphasis on chemical ecology for improved adaptation strategies. Gopi's contributions through the Compliance Assistance Programme, OzonAction of the UNEP have been quite significant and consistently so for more than two decades at the regional and global levels. This created the opportunity for him to serve as a Member of the Inter-Ministerial Empowered Steering Committee constituted by the Ministry of Environment, Forest and Climate Change, Government of India, on aspects of the Montreal Protocol. He is a well-known specialist in the areas of science and technology management communication, with a large number of theme-specific editorials and other publications to his credit. He holds two doctoral degrees in the fields of microbial and chemical ecology respectively, with a law degree and is an Alumnus of the International Visitors Leadership Program of the Department of State, United States of America.

Kumar Singh, Neeraj

Dr. Kumar Singh is a Deputy Librarian in A. C. Joshi Library, Panjab University (PU) Chandigarh since 2007. He did PhD from PU, Chandigarh on the topic "Radio Frequency Identification technology application in libraries and information centres of Northern India: A study". He has been awarded the Prestigious Commonwealth Professional Fellowship in 2017 and 2012 at the University of East London, London, UK. He has published twenty-five research papers in international/national peer-reviewed journals.

Larouz, Mohammed

Dr. Larouz is currently the Dean of the School of Arts and Humanities of Meknes. He is also the Vice-President of Academic Affairs issued from the University Council of Moulay Ismail University of Meknes. He earned a Ph.D in Applied Linguistics from Fez University in 2004. He served for two terms as the Head of the Department of English Studies, where he has as Professor worked since 2005. In his department, he has also directed the PhD Program in Applied Linguistics and chairs the Research Group on Applied Linguistics and Language Development (ALLD). Dr. Larouz also holds an MA in Applied Linguistics from Essex University (1996) and has been a Fulbright Scholar to the USA twice (2002 and 2005), in addition to having worked as a secondary school teacher in Morocco for more than fifteen years. He was President of the Moroccan Association of Teachers of English, local Branch (2016–2020) and is an expert and evaluator for the National Agency for Evaluation and Quality Assurance (ANEAQ) and the National Center for Scientific and Technical Research (CNRST) in the Ministry of Higher Education of Morocco. He investigates applied linguistics questions in Morocco and has also research interests in the fields of TEFL, sociolinguistics, research methodology, and communication.

Magdy, Ramy
Magdy is an Assistant Lecturer of Political Science, Cairo University. He had a M.Sc in Political Theory and is currently a PhD Scholar on the same topic. His research interests focus on exploring the questions of power, gender, and displacement in the context of Middle-East and North Africa (MENA) and the Arab region.

Mai, Yasser
Mai is a UN-refugee agency senior resettlement assistant in Cairo. Her research interests focus on exploring the questions of power, gender, and displacement in the context of Middle-East and North Africa (MENA) and the Arab region.

Mliless, Mohamed
Dr. Mliless holds a Ph.D in Applied Linguistics. He published three books on legal and environmental discourse. The many scientific articles he authored and coauthored focus on social and environmental issues. Dr. Mliless is a member of the International Ecolinguistics Association (IEA), the London Journals Press, and the Acta Scientific Open International Library, India. Last but not least, he serves as a review board member of many journals including the Journal of Education Teaching and Social Studies (JETSS), the Journal of Sustainability in Environment (JSE), the International Journal of Sociology and Anthropology (IJSA), the International Journal of English and Literature (IJEL), the African Journal of Political Science and International Relations (AJPSIR), and the Journal of Language and Literature (JOLL).

The manufacturer's authorised representative in the EU is Springer Nature Customer Service Centre GmbH, Europaplatz 3, 69115 Heidelberg, Germany. If you have any concerns regarding our products, please contact ProductSafety@springernature.com

Printed and bound by CPI Group (UK) Ltd, Croydon, CR0 4YY
30/09/2026
02225233-0012